Springer Series in Advanced Manufacturing

The **Springer Series in Advanced Manufacturing** includes advanced textbooks, research monographs, edited works and conference proceedings covering all major subjects in the field of advanced manufacturing.
The following is a non-exclusive list of subjects relevant to the series:

1. Manufacturing processes and operations (material processing; assembly; test and inspection; packaging and shipping).
2. Manufacturing product and process design (product design; product data management; product development; manufacturing system planning).
3. Enterprise management (product life cycle management; production planning and control; quality management).

Emphasis will be placed on novel material of topical interest (for example, books on nanomanufacturing) as well as new treatments of more traditional areas.

As advanced manufacturing usually involves extensive use of information and communication technology (ICT), books dealing with advanced ICT tools for advanced manufacturing are also of interest to the Series.

Springer and Professor Pham welcome book ideas from authors. Potential authors who wish to submit a book proposal should contact Anthony Doyle, Executive Editor, Springer, e-mail: anthony.doyle@springer.com.

INDEXED IN SCOPUS

Wasim Ahmed Khan · Volkan Esat ·
Muhammad Hammad · Hassan Ali ·
Muhammad Qasim Zafar · Rashid Ali

Computer Aided Engineering Design and Manufacturing

A Fourth Industrial Revolution Perspective

Springer

Wasim Ahmed Khan
Faculty of Mechanical Engineering
GIK Institute of Engineering Sciences
and Technology
Topi, Pakistan

Muhammad Hammad
Faculty of Mechanical Engineering
GIK Institute of Engineering Sciences
and Technology
Topi, Pakistan

Muhammad Qasim Zafar
Faculty of Mechanical Engineering
GIK Institute of Engineering Sciences
and Technology
Topi, Pakistan

Volkan Esat
School of Physics, Engineering
and Computer Science, Automotive
Engineering
University of Hertfordshire
Hatfield, UK

Hassan Ali
Faculty of Mechanical Engineering
GIK Institute of Engineering Sciences
and Technology
Topi, Pakistan

Rashid Ali
Faculty of Materials and Chemical
Engineering
GIK Institute of Engineering Sciences
and Technology
Topi, Pakistan

ISSN 1860-5168 ISSN 2196-1735 (electronic)
Springer Series in Advanced Manufacturing
ISBN 978-3-031-77105-7 ISBN 978-3-031-77106-4 (eBook)
https://doi.org/10.1007/978-3-031-77106-4

This Springer imprint is published by the registered company Springer Nature Switzerland AG
The registered company address is: Gewerbestrasse 11, 6330 Cham, Switzerland

To Arqam, Sarah, Muhammad, Sadaf, Merlin, Misu, Pixie, Evay, and Fluff

Wasim Ahmed Khan

To my wife Berna, and my daughter Birce

Volkan Esat

I am dedicating this work to my beloved mother. I recognize her unwavering support and encouragement, which have been the guiding forces in shaping my academic and professional journey.

Muhammad Hammad

To my father Muhammad Ali Khan, and my family

Hassan Ali

To my late father, Muhammad Zafar Choudhary

Muhammad Qasim Zafar

To my students

Rashid Ali

Foreword

The authors, all having decades' rich experience of teaching and research in Advanced Engineering Design and Manufacturing to undergraduate and graduate students, have realized the need for a new title that can give a holistic view of Industry 4.0, Industry 5.0, Quantum Computing as applicable to *Computer Aided Engineering Design and Manufacturing*. The nexus is unique, and discussion opens quite a few new arenas.

There is threadbare synthesis of all subject areas as mentioned above and then the authors provide commentary on each topic in a hierarchical fashion with case studies for all topics. Both conventional and modern practices are examined.

Diverse aspects of today's discrete products such as innovation, customization, use of multi discipline and sustainability are considered. System engineering aspects of discrete products, such as amalgamation of mechanical, electrical, electronics, control engineering, computer software and hardware having reliability, availability, maintainability, and safety, are identified. These discrete product systems may be basic, embedded, cascaded, and/or complex systems with open or feedback configuration.

The methods to analyze and build today's discrete products are graphically depicted using Process Flow Diagram (PFD). PFD allows identification and interrelation between variables of different fields of engineering that is the hallmark of today's discrete products.

The manufacturing and production with intermediate steps of Production Planning and Control (PPC) after Engineering Design in the Fourth Industrial Revolution era are lamented with examples.

Mathematics, Standards, and Databases required to operate the convergence of Industry 4.0, Industry 5.0, and Quantum Computing for performing engineering design and manufacturing with current Technology Readiness Level (TRL) are presented.

The book is structured to have chapters in different parts that are identified to dispense the knowledge on the subject area. Each part and chapter have a prelude to connect with preceding and succeeding parts thus giving a logical progression though any chapter can also be read independently. Each chapter has objectives and Learning outcomes with review questions and bibliography.

The book is a suitable addition to tertiary undergraduate and graduate programs in mechanical, electrical, electronics, control engineering, computer software and hardware degree programs.

Professor Jianguo Lin
FREng, FIMMM, FIMechE, FISME
Professor in Mechanics of Materials Division
Department of Mechanical Engineering
Faculty of Engineering
Imperial College
London, UK

Preface

The advent of Computer Aided Design and Computer Aided Manufacturing with the invention of Numerical Control (NC) machine at MIT, USA led to taking advantage of the use of part programming methods, ASCII and Unicode data representation standards, Computer Numerical Control (CNC), legacy and modern communication standards; and advances in mathematical technology that gave birth to today's Cyber-Physical Systems for engineering design and manufacturing. The invention of the NC machine was itself a landmark for human generations that preceded the development of technologies in the 1st, 2nd, and 3rd industrial revolutions that had their own significance when it came to fighting the evil forces of nature and giving subtle advantage to mankind.

CAD/CAM has widespread use in industry, and universities take pride in excelling in the teaching of the subject. With the advent of the 4th Industrial Revolution, the pillars of Industry 4.0 and Industry 5.0 (Smart Manufacturing in the USA) appear as islands of automation, and the same confusion can be seen at the universities in the teaching of the subject.

The same situation was observed in the 1980s as the commercial hubs were competing to market their products as the best product and commenting that no other product would survive. There was some remarkable work done in Computer Integrated Manufacturing (CIM) by ESPRIT, UK, and NIST, USA. Other countries in Western Europe also gave importance to CIM. ISO took a leading role in standardizing the technology allowing researchers and practitioners to make optimal use of technological advances. Other National and Professional Institutional standard development bodies also created standards and are still doing work in capping and making the Technology Readiness Level at each stage of progression stable.

No single volume is currently available in the market on the subject that can provide a logical and coherent concept to undergraduate students that what is there currently and what to expect in the future in Computer Aided Design and Computer Aided Manufacturing (CAD/CAM) subject domain.

This book aims at bridging the gaps created by commercial and research entities and tries to propose a solution for engineering undergraduate students all over the world. It consists of seven parts and seventeen chapters first structuring the subject areas and later detailing the main topics under consideration. Each part of the book and each chapter has a prelude guiding the reader in a systematic way to the next part or topic.

First conventional methods in design and manufacturing are referred to and then computer aids are introduced. Basic definitions such as function and process and tangible products and intangible services are differentiated. The book gets into the details of Computer Aided Design (CAD) emphasizing model needs analysis to become designed. Contemporary CAD techniques and Analysis Methods are presented.

Manufacturing Processes, both discrete and continuous, and their arrangement in manufacturing systems is given.

Use of types, sizes, accessories, manufacturing parameters, and type of control of manufacturing processes, manufacturing cells, and manufacturing systems is provided. It spans Mechanical, Electrical, Electronics, Control, Computer hardware and computer software aspects of system engineering defined basic systems, embedded systems, cascaded systems, and complex systems. Any other engineering or technology parameter such as ergonomics may also be implanted in the system definition. The basic system comprises Input, Process, Output, and waste from the system. It is important to note that the process in the basic system is different from the process mentioned above with function. Here the process may be any process such as a thermodynamic process, fluidic process, chemical process, biological process, etc. The details are provided in the appropriate section of the book, please refer to the table of contents.

Computer Numerical Control and Programmable Logic Controllers are described from historic and technical perspectives. The point of view related to communication from these devices to integrate heterogeneous machinery using legacy and Industry 4.0 standards is given leading to e-Commerce advantage.

Product life cycle, i.e., the Cradle Grave phenomenon with Value Chain and Supply Chain and their management layers including Manufacturing Engineering Services (MES) and Product Life Management (PLM) with a lot and a single product, is explained in a wholesome and rarified manner. Other production philosophies are also addressed. Design, Life Cycle Assessment, Sourcing, and Manufacturing are other ways of defining the same thing.

The amalgamation of Industry 4.0 pillars and human factor consideration in Industry 5.0 to give socio-economic benefits in a knowledge economy based on startups, small and medium enterprises (SMEs), and large-scale industries (LSEs) are explored.

The evolution of digital computing has progressed from silicon-based technology to the nascent field of Quantum Computing. While still in its early stages, Quantum Computing holds immense promise, as exemplified by the comparison between Intel's 4-bit microprocessor and its 11th Gen Intel® Core™, and Xanadu's 216 qubit quantum chip and IBM's Osprey 433. These advancements evoke memories of working with microprocessors like the MOSTECH 6502, Motorola 6800, and Intel 8086.

Furthermore, the East is witnessing a growing interest in FPGA-based design and verification of silicon chips, as well as photolithography for chip manufacturing. A notable example is the RISC-V open standard instruction set architecture and its chip fabrication by Ali Baba for the ROMA laptop. Additionally, the advent of Graphene is poised to revolutionize the computing landscape. For further details, please refer to the relevant sections in the book.

The last three chapters in Part VII, detail Industry 4.0, Industry 5.0, and Quantum Computing. The last chapter explores the state of the art for the nexus of Industry 4.0, Industry 5.0, and Quantum Computing with Technology Readiness Level for Computer Aided Design and manufacturing teaching and practices.

The reasons for proposing a new book in this area are that as all authors are deeply involved in teaching, assessment, and Continuous Quality Improvement (CQI) of Engineering Design and Manufacturing therefore they can give details of subjects for undergraduate students all over the world. Secondly, no single textbook is currently available that addresses the topic covering all aspects with logical connectivity in an amicable manner. Major features and strengths of the book, which make it unique, are:

1. Description of Conventional Engineering Design and Manufacturing Cycle with Case Study
2. Description of System Engineering Design and Manufacturing Cycle with Case Study
3. Use of technical semantic databases and bots for data usage and data acquisition
4. Use of legacy and Industry 4.0 standards in CAD and CAM and allied interactions
5. Relevance of pillars of Industry 4.0 Industry 5.0 with practices in CAD and CAM
6. Coherent display of the use of Mathematical Technology in CAD and CAM
7. Commercial versus Scientific View of the Technology Readiness Level for CAD and CAM
8. Features of CAD and CAM APIs used in Engineering Design and Manufacturing

The key benefit for potential buyers includes futuristic content explained using state-of-the-art teaching methods with objectives, learning outcomes, and review questions. MS PowerPoint Slides and Solution Manual for instructors are available.

Topi, Pakistan — Wasim Ahmed Khan
Hatfield, UK — Volkan Esat
Topi, Pakistan — Muhammad Hammad
Mardan, Pakistan — Hassan Ali
Topi, Pakistan — Muhammad Qasim Zafar
Topi, Pakistan — Rashid Ali
January 2025

Acknowledgments

Authors are grateful to Prof. D. T. Pham for assessing the basic table of contents, objectives, and learning outcomes of this book. Authors are also grateful to following individuals for helping to construct the argument related to this book:

Dr. Hashim Ali Raja, GIK Institute of Engineering Sciences and Technology
Dr. Taj Muhammad Khan, GIK Institute of Engineering Sciences and Technology
Dr. Taqi Ahmed Cheema, GIK Institute of Engineering Sciences and Technology
Dr. Javaid Rabbani Khan, GIK Institute of Engineering Sciences and Technology
Dr. Ali Atashbar Orang, Middle East Technical University, Türkiye
Ms. Ramesha Tariq, GIK Institute of Engineering Sciences and Technology
Mr. Sarmad Ishfaq, GIK Institute of Engineering Sciences and Technology
Mr. Abdur Razzaq Gauhar, Infinity Engineering (Pvt.) Limited, Pakistan
Mr. Ali Akbar, Infinity Engineering (Pvt.) Limited, Pakistan
Mr. Muhammad Shahzad Malik, EduTech (Pvt.) Limited, Pakistan
Mr. Muhammad Nauman, EduTech (Pvt.) Limited, Pakistan

Score of GIK Institute undergraduate and graduate students some of these are:

Abdul Sami Urooj, Muhammad Taha Khan, Muhammad Ahmed Shakeel, Sheikh Abdul Basit, Ammad ur Rehman Shahid, Ahmed Maqsood, M. Ibad Mughal, Sheharyar, Umer Javed, Ziaullah Jan, and Hafiz Muhammad Rizwan

Middle East Technical University undergraduate students listed below:

Zuhayr Amin, Talal Amin, Walid K. Badwi Albitar, Melisa Altuntas

Disclaimer

Reasonable efforts have been made to publish reliable data and information, but the author and publisher cannot assume responsibility for the validity of all materials or the consequences of their use. The authors and publisher have attempted to trace the copyright holders of all materials reproduced in this publication and apologize to copyright holders if permission to publish in this form has not been obtained. If any copyright materials have not been acknowledged, please write to wasim_khans@hotmail.com, and let us know so we may rectify in any future reprint.

Contents

Part V Future Manufacturing Techniques

Part VI Communication

About the Authors

Professor Wasim Ahmed Khan holds a first degree in Mechanical Engineering from NED University of Engineering and Technology, Karachi, Pakistan. He later obtained a Ph.D. degree in Operations Research from the Department of Mechanical Engineering, University of Sheffield, England, UK. He is a life member of the Pakistan Engineering Council. He is also a chartered engineer (C.Eng.) of the Engineering Council, UK, and a fellow of the Institution of Mechanical Engineers, UK. Professor Khan was recently elected as a senior member of the Institution of Electrical and Electronics Engineers, USA. He is also a member of the IEEE Computer Society. He has diverse work experience including working with the manufacturing industry, software development for local and overseas clients and teaching production engineering, business, and computer science students. He is currently working as a professor, in the faculty of mechanical engineering, at GIK Institute of Engineering Sciences and Technology. He has two books as the principal author by Springer, UK and CRC Press, Taylor and Francis Group, USA. He is also the CRC Press, Taylor and Francis Group, USA Book Series Editor. So far, he has published three books under his series as the principal editor. The book series theme is 'Computers in Engineering Design and Manufacturing'. Professor Khan has more than 30 journal and conference publications. He also has one US patent application under process. The Pakistan patent for the same invention has been granted. He is a reviewer of several internationally reputed journals including *Virtual Reality, Manufacturing Systems*, and the *International Journal of Computer Integrated Manufacturing*. Professor Khan is the founder of the Advance Manufacturing Processes Laboratory (AMPL), Faculty of Mechanical Engineering, GIK Institute. Professor Khan is the principal investigator for two research grants totaling PKR 228 million from the Government of Khyber Pakhtunkhwa, Pakistan. His current research and development interests include the development of indigenous CNC subtractive and additive manufacturing machine tools, 3D scanners, universal testing machines, spinning machines and flexible manufacturing cells (FMCs), indigenous development of revolute robots and collaborative robots, Digital Twining of manufacturing cells and systems, industrial management of SMEs and LSE for the implementation of Industry 4.0 and Industry 5.0

methodologies using customized manufacturing frameworks; and industrial decision making.

Volkan Esat earned his B.Sc. and M.Sc. degrees in Türkiye. He achieved his Ph.D. at the School of Mechanical and Manufacturing Engineering of Loughborough University, UK, in 2006. He worked at Loughborough University until 2010. Between 2010 and 2024, he has been a member of faculty at the Middle East Technical University—Northern Cyprus Campus, Mechanical Engineering Program. He currently works at the University of Hertfordshire, UK, as a member of the faculty in Automotive Engineering. His research interests are metal forming, multi-scale modeling and mechanics of engineering materials, biomechanics, crashworthiness, injury prevention, and computational modeling. So far, he has published around 70 international journal articles, book chapters, and international conference articles altogether; and filed/granted 2 patents.

Dr. Muhammad Hammad completed a bachelor's and master's degree in the Faculty of Mechanical Engineering from the University of Engineering and Technology, Lahore. Following this education, he achieved a Ph.D. degree in the Faculty of Mechanical Engineering from the Ghulam Ishaq Khan Institute of Engineering Sciences and Technology, Topi, Swabi, Khyber Pakhtunkhwa, Pakistan. His doctoral thesis lay in the emerging fields of smart manufacturing and security, a testament to his forward-thinking approach to engineering challenges. As a researcher deeply immersed in the nuances of Industry 4.0, Smart Manufacturing, and the Security of the Industrial Internet of Things (IIoT), he has carved a niche in these rapidly evolving disciplines. His graduate and doctoral studies period was marked by a significant role as a Graduate Research Assistant, contributing extensively to both institutions' academic and research environments. His expertise is not just confined to academic research but extends to practical applications and knowledge sharing. As a reviewer of numerous manuscripts for the IEEE, he has demonstrated a keen eye for detail and a commitment to advancing the engineering field. His dedication to disseminating knowledge was further exemplified through his participation and presentation at the 5th International Workshop on Functional Reverse Engineering of Machine Tools. An engineer, researcher, and scholar, he embodies the spirit of inquiry and innovation that drives the modern world of engineering and technology.

Hassan Ali is currently serving as a graduate research assistant in the Faculty of Mechanical Engineering, Ghulam Ishaq Khan Institute of Engineering Sciences and Technology Topi, Pakistan. He obtained his B.Sc. Industrial Engineering degree from the University of Engineering and Technology Peshawar, Pakistan, in 2021. He later achieved his MS degree in Mechanical Design and Manufacturing Engineering from the Ghulam Ishaq Khan Institute of Engineering Sciences and Technology, in 2023. He has been awarded a university Gold Medal for securing the top position in his B.Sc. Industrial Engineering degree in 2021. He has strong hands-on experience in numerical simulations and additive manufacturing. He has published more than 5 research articles in reputed international journals and conferences. He is currently working on the optimization of support structures in metal additive manufacturing. His research interests are in computational methods in metal additive manufacturing.

Muhammad Qasim Zafar is currently serving as an assistant professor in the Faculty of Mechanical Engineering, Ghulam Ishaq Khan Institute of Engineering Sciences and Technology Topi. He obtained his B.Sc. Mechanical Engineering degree from Bahauddin Zakariya University Multan and graduated with an MS Mechanical Design Engineering degree from the University of Engineering and Technology Lahore, in 2011 and 2016 respectively. He obtained a Ph.D. in Mechanical Engineering specialized in additive manufacturing from Tsinghua University China. He has strong hands-on experience in numerical simulations, advanced material processing technologies, and tribology. He has published more than 15 research articles in reputed International Journals. He is awarded with the prestigious title of PE (Professional Engineer) by Pakistan Engineering Council. He is currently working on several projects including residual stress and crack prediction in metal additive manufacturing.

Rashid Ali is currently serving as an associate professor in the faculty of Materials and Chemical Engineering, at Ghulam Ishaq Khan Institute of Engineering Sciences and Technology Topi. He obtained his B.Sc. in Metallurgy and Materials Engineering from Mehran University of Engineering and Technology, Jamshoro, MS in Mechanical and Automobile Engineering from Myongji University, South Korea, and Ph.D. in Mechanical and Industrial Engineering from the University of Roma Tre, Rome, Italy. He has more than 15 journal publications. He has hands-on experience in numerical residual stress analysis and optimization in materials during material additive manufacturing. His current interests are in computational methods in materials science and engineering.

Part I
Engineering Design and Manufacturing

Prelude

Engineering Design and Manufacturing is the art of producing sustainable, multidisciplinary, customary, and innovative products or a range of products with appropriate user interface (UI), intermediate process/es, and User eXperience (UX) for wealth creation. The product is considered here to be a system that may have multiple energy forms as inputs, can perform defined engineering processes, and can give one or multiple outputs as energy or work forms achieving single or multiple objective functions normally with some type of waste. The concept is elaborated in the following definition of a System.

The system is operating in a structured manner and may be a cyber-physical-biological system in the knowledge economy working as a start-up or SME with E-Commerce facility using contemporary computing tools under a defined legal framework as elaborated in the following paragraphs.

Systems Engineering

To analyze and consequently, control a phenomenon, systems engineering provides a basic understanding of the phenomenon through charts and eventually complex equations to control the phenomenon to the best possible extent through classical and modern control theory.

The system is defined as a basic system, embedded system, cascaded system, and complex system. There is always an input, at least one process, and an output from the system. Normally single or multiple tangible or intangible objective functions are achieved by the operation of a tangible system. There is always some kind of waste associated with the system out of systems developed by humans. All types of systems are explained pictorially in Fig. 1.

All types of systems are representable by a mathematical equation called Control Function. Please refer to the appropriate literature for a deeper understanding of Control Function.

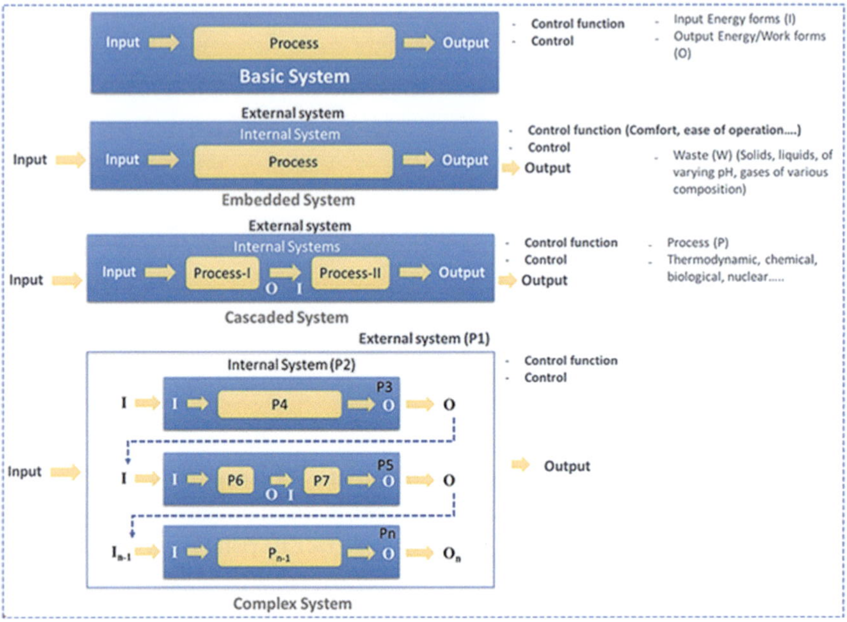

Fig. 1 The Definition of a System

The processes in the systems can be graphically represented through its graphical depiction using a process flow diagram (PFD) that provides ease in identifying process variables and relating it to other variables in system engineering design of the object for:

1. Mechanical Engineering Aspects
2. Electrical Engineering Aspects
3. Electronic Engineering Aspects
4. Computer Hardware Aspects
5. Computer Software Aspects
6. Any other Engineering Aspects depend on the nature of the product
7. Reliability, Availability, Maintainability, and Safety (RAMS)
8. Ergonomics including UI and UX.

Figure 2 provides a detailed process flow diagram of a 5 kg capacity Collaborative Robot (Cobot). The cobot is presented in Fig. 3.

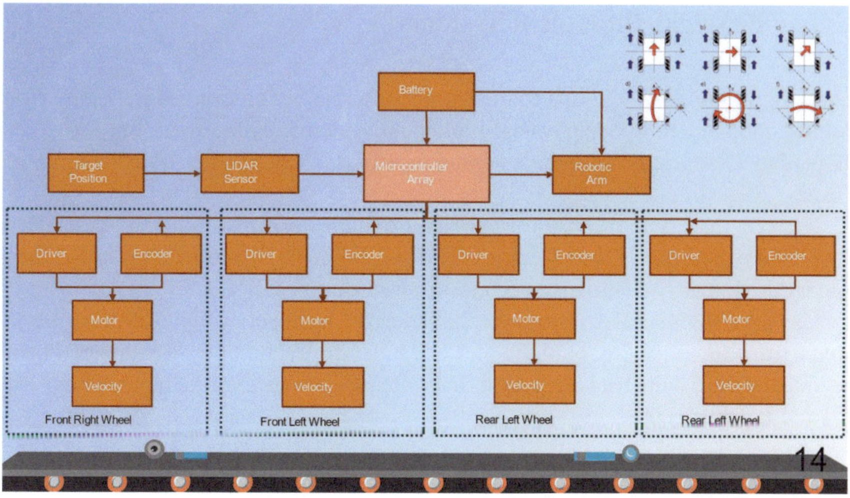

Fig. 2 Process Flow Diagram of a Cobot. Courtesy Muhammad Taha Khan, Muhammad Ahmed Shakeel, Sheikh Abdul Basit; Faculty of Mechanical Engineering, GIK Institute of Engineering Sciences and Technology, Topi—23640, Khyber Pakhtunkhwa, Pakistan. Reproduced with Permission

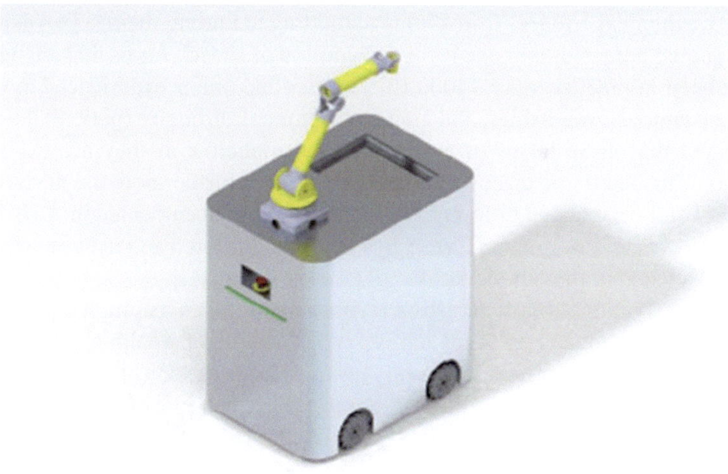

Fig. 3 Cobot

Cyber-Physical-Biological Systems

The progression from the first to the fourth industrial revolution led humanity from cottage industry to cyber-physical-biological systems. The terminology was best defined by Claus Schwab, Founder and Executive Chairman of the World Economic Forum (2024) in 2016.

The practical demonstration of the fourth industrial revolution is well established today and one of the examples is known as Smart Manufacturing in Northern Americas (Industry 4.0/Industry 5.0 in Western Europe).

The fourth industrial revolution is characterized by the consolidation of numerous fields from previous Revolutions and advances in Technology Readiness Levels beyond the third industrial revolution. This includes advances in the pillars of Industry 4.0 and Industry 5.0.

Chapters 14 and 15 in Part VII provide details on these concepts.

Knowledge-Economy

Acquisition of knowledge is as old as that of human beings. Primitive humans used it to improve their experiences from dining to fighting to finding shelter. Two examples can be quoted here. Number one is the recognition of speech forms that are believed to have been in practice since 14000 BC. The second one is experienced in dining. From raw animals, vegetables, and grains, the humans through experiential learning determined that these forms of food taste more delicious if they are exposed to lightning. This led to the discovery of how to fire and furthermore the development of pottery and weapons in different ages as thoroughly documented in *A History of Knowledge—Past, Present and Future* by Charles Van Doren and in numerous other volumes as listed in the bibliography.

This journey in the fourth industrial revolution has taken a completely new pace where new knowledge is patented and is a major source of wealth creation. People who are not scientists or practitioners rely on knowledge of some modern technology (tangible or intangible) and work from home, which is a gift of COVID-19 and is being heavily relied upon after its aftermath. Other groups of people rely on the gig economy which is short-term benefits from the knowledge of some chosen fields and then moving on to new technology.

Start-ups and SMEs

A new form of knowledge gathering and using it for incubation, germination, and acceleration leading to a start-up that is the formal business resulting from a raw idea in germination, experimenting with an idea for a positive outcome in incubation, and giving the idea a tangible shape in acceleration normally using a source of funding. If the tangible idea is attractive to a venture capitalist, it leads to the formal launching of a start-up that is a recognized business earning profit and paying taxes.

Small and medium enterprises (SMEs) are as old as the first industrial revolution when the cottage industry was formalized into factories but the idea of a small setup of few human resources, few machines, and a little capital is still there with different countries having their own definition of SME.

SMEs are the backbone of the country's economy as they build the industrial echo system that either becomes a vendor for larger factories or is a direct supplier to the public for services or products with a difference. Schematics for Advanced Manufacturing Processes Laboratory mapping and technology are provided in Fig. 4. A non-unique process flow diagram of an SME's operation is given in Fig. 5.

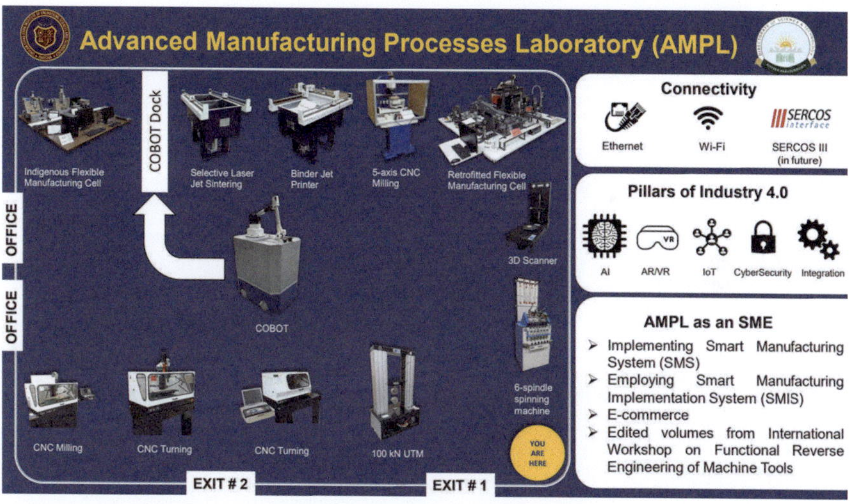

Fig. 4 Advanced Manufacturing Processes Laboratory (AMPL) Mapping and Technology

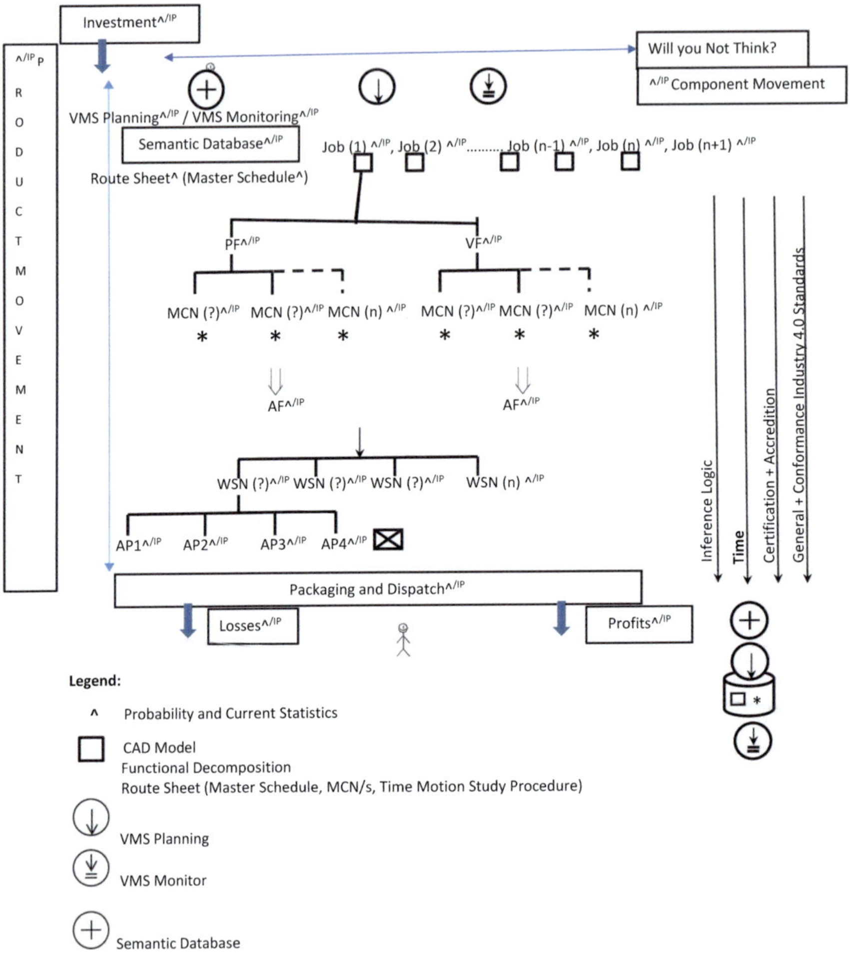

Fig. 5 Non-Unique Process Flow Diagram of an SME's Operation

E-Commerce

The third and fourth industrial revolutions have brought convenience to consumers and suppliers to communicate efficiently in B-C E-Commerce or B-B E-Commerce. This communication for business may be supplemented with graphics, point-to-point signals through Industrial Internet of Things (IIoT), and Communication through networks that may be physical or cellular and financial transactions through electronic transfer if everything is done in a legal framework.

* Procedure Sheet, Process Sheet, Part Program, QC Procedure, Tangible & Intangible Stores, LTAVNVHR^, Mains/Generator

 Semantic Database

 Assembly Procedure (AP), QA Procedure (QAP)

VMS	Virtual Manufacturing System
PF	Production Facility
VF	Vendor's Facility
AF	Assembly Facility

Route/Procedure/Process Sheets	Process Plan
Master Schedule	Master Schedule of the Production Setup
Product Schedule	Product Schedule for Individual Product
MCN	Machine Tool Number
WSN	Workstation Number
IP	Intuition Process

^LTAVNVHR—Lead Time by Average /Novice/Master Human Resource

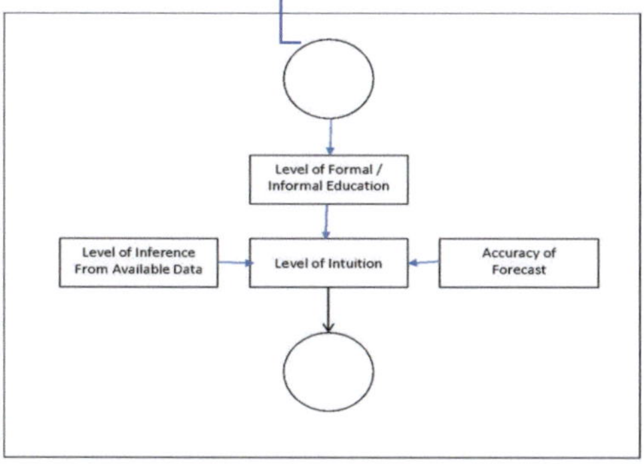

Fig. 5 (continued)

Chapter 1
Engineering Design

Prelude
Design, Life Cycle Assessment, Sourcing, and Manufacturing by a team working on multiple projects in a constrained timeline is the hallmark of the 4th Industrial Revolution. With clients spread all over the world and strict adherence to Quality Standards for manufacturing facilities located in diverse nations to cut the cost but not to compromise on human factors and reliability, availability, maintainability, and safety is the challenge. A very high level of customization is demanded by the client, and to compete in the market the design team must come up with innovative ideas. The cost of the product is the key factor in the demand–supply model. Low production volumes are preferred to not to tie up capital in different inventories. Intricate technologies are involved to achieve all of it as detailed below. Sustainability is a popular demand.

Three types of Design are considered i.e., original design, adaptive design, and variant design. Example of original design is Mars Rover, adaptive design is different applications of siphon mechanism and examples of variant design are e.g., products with variable size and arrangement like in a set of wrenches, hand tools, and garden tools. Figures 1.1, 1.2 and 1.3 provide schematics related to original design, adaptive design, and variant design.

Supplementary Information The online version contains supplementary material available at https://doi.org/10.1007/978-3-031-77106-4_1.

W. A. Khan et al., *Computer Aided Engineering Design and Manufacturing*, Springer Series in Advanced Manufacturing, https://doi.org/10.1007/978-3-031-77106-4_1

Fig. 1.1 Mars Exploration Rover (MER)—an example of original design

1.1 Objectives

The Engineering Design process today is a well-structured activity. However, the lessons from the past are the basis of current development especially due to the Technological Readiness Level (TRL) in computing and mathematics. The objective of this chapter is to introduce the readers to the elements of Engineering Design as applicable to innovative, customized, multidisciplinary, and sustainable discrete products with appropriate ergonomic considerations depending on the type of product. The concept of component, structure, assembly, and mechanism forming a machine (product) is introduced here.

The purpose of including content on analytical and experimental methods of design analysis is to remind the young readers that while currently, human beings are living in the era of the 4th Industrial Revolution, there are design analysis tools that were being used not in too distant past while at least analytical analysis based on the balance of forces, etc. is still used for basic design analysis and forms the basis of numerical methods of design analysis while strain gauges are still extensively used to verify the design, especially in life-critical applications. Knowledge of these methodologies will render these audiences to resort to analytical and experimental design analysis tools giving them an extra edge.

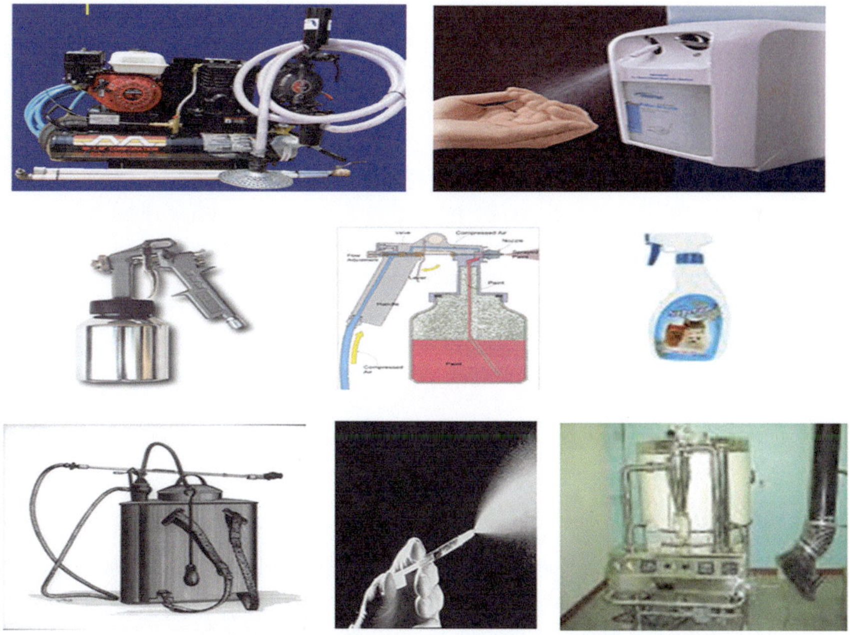

Fig. 1.2 Examples of adaptive design (browsed from WWW at various instances)

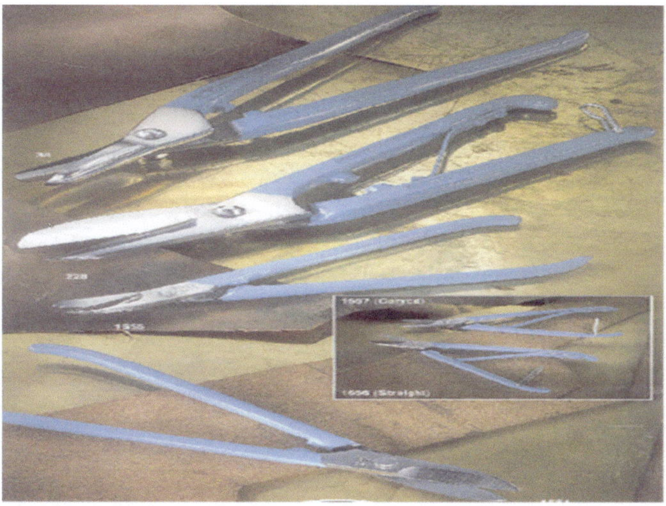

Fig. 1.3 Examples of variant design (courtesy—footprint tools—defunct)

An introduction to Design for Manufacturing (DFM) and Design for Assembly (DFA) is needed. The readers must also understand the concepts related to Reliability, Availability, Maintainability, and Safety (RAMS). There is also a need for understanding concepts like Axiomatic Design and Concurrent Design. The product life cycle is introduced here to reflect on the sustainable living of humans on Earth as defined by United Nations Development (UNDP) Sustainable Development Goals (SDGs). These different concepts are needed at this stage to reflect later how these concepts still have roots in modern design implemented in the era of the fourth industrial revolution with cyber-physical systems playing an important role.

1.2 Function (Individual's Activity) and Process (Team's Activity)

In engineering product design, designers as individuals have various roles such as identifying the problem and need, working on various phases of engineering design, and developing and testing prototypes. This means designers make *the function* (Verb) of the product possible and pave the way towards a successful product. Any person performing any of the above functions (noun) individually is another perspective. However, a team of managers, designers, operators, and other personnel should act together during various stages of the design and development efforts so that the remaining steps of the design *process* can be completed with all views amicably incorporated. In other words, the engineering design process can only be satisfactorily accomplished when individuals and the whole team perform their functions in collaboration. The individual team member function is different from the product's function. It is the vocational and intellectual ability of an engineer to contribute to the engineering design process. It is a positive contribution when it is performed for the team (process) and harms the objectives of the team performing engineering design if no collaboration or confused collaboration is offered to the team.

In the current 4th Industrial revolution era, consumers present a challenge to designers in a way that they not only require functional products, but also pay attention to aesthetics, demanding them to be much more durable, trendy, and relatively cheap. Therefore, designers as individuals and the designers as teams need to be much more systematic, applying the latest scientific developments and implementing advanced engineering design procedures.

1.3 Innovation, Customization, Sustainability and Use of Multi Discipline in Engineering Design

Industry 4.0 (I4.0) age demands a more innovative, customized, sustainable, and multidisciplinary engineering design. To address the challenges of the I4.0, in which problems are more multidisciplinary in essence, innovative solutions are needed for the smart factory to be more efficient and generate highly customized products. The product needs to have sustainability characteristics in its use and during its production.

Industry 5.0 (I5.0) brings in the human factor to supplement Industry 4.0 (I4.0). Innovation in engineering design means generating novel, exceptional design ideas that can potentially result in unprecedented, quality products for consumers. Companies adapting innovation in design aim to address a demand for a customized product or commodity. Smart factories of Industry 4.0 and Industry 5.0 have the capability to conform to innovative products flexibly and in relatively shorter lead times.

Customization in engineering design becomes a more popular topic every day, concentrating on delivering products in line with the requested specifications of the customer. Customization via CAD tools bring numerous advantages to the designer, beginning with efficient product modeling and improvement, resulting in better productivity. The challenge with product customization is mainly about the company's competence to adjust and stabilize product range, responsiveness, and cost-effectiveness concurrently. These issues can potentially be addressed via machine learning methods which can handle big data through sophisticated computational facilities so that state-of-the-art opportunities for customization are possible.

The nature of engineering problems in the age of I4.0 and I5.0 are highly multidisciplinary, solutions to which require expertise and involvement from various engineering disciplines such as mechanical, electrical, electronics, computer software, computer hardware, control engineering, any other specific subject area related to the product, RAMS, UI and UX etc. Multidisciplinary approach suggests integrated engineering systems, and therefore, systems engineering. The 4th Industrial Revolution era is a period in which systems engineering becomes more and more important. Multidisciplinary engineering design can be realized through teams of expert designers from the disciplines. However, expertise in a certain field for a certain designer may not be enough for a successful multidisciplinary effort; establishment of a common, shared vocabulary, and a certain amount of knowledge regarding other team members' areas are also essential factors. Please refer to Figs. 1.1, 1.2 and 1.3.

Life Cycle Assessment (LCA) of the product under design considerations and proposed methods to produce it add up to efforts towards sustainable living on planet Earth.

1.4 Ergonomics

Ergonomics (E) along with the discipline of Human Factors (HF) are design related schemes through which ease of use, user (human) health and comfort, safety, quality of life, rapport, and efficacy are achieved and advanced. It has been reported that current challenges of health and safety as well as productivity within the manufacturing sector have not been satisfactorily resolved by the prevailing advancement of technology. Even in the 4th Industrial Revolution age, it is apparent that humans participated vigorously in design and manufacturing activities. Therefore, ergonomics in these fields, emerging from physical ergonomics embodying anatomical, biomechanical, anthropometric, mental, and physiological notions, plays an important role in explaining and finding solutions to issues involving use of products and physical activity.

Ergonomists taking part in so-called *participatory ergonomics* (PE) design processes should satisfy the requirement that all participants are able to contribute to the design process. Objects can assist significantly with the integration of ergonomic knowledge into the design process. During a design process, an object can help in direct communication between involved parties as well as in transferring ergonomic knowledge to downstream participants later in the design stages.

Recent research claims that using two classes of objects can contribute significantly to the integration of ergonomic knowledge into design. For example, in a processing plant layout design problem, layout sketches can be the *boundary object*, which aid in establishing dialog between parties, whereas an ergonomics guideline document introduces necessary ergonomics knowledge into the design process as the so-called *intermediary object*. *Boundary objects* in ergonomics can be abstract or tangible; using a common language for designers to utilize their knowledge and expertise, helping them to identify their discrepancies and similarities over a certain boundary, and expediting a means for knowledge transfer amongst them. Characteristics of *boundary objects* are defined as follows:

- They are not ready-made and must be generated by designers.
- They should possess built-in affordances: possibilities for action supplied by the environment.
- Some can be flexible and malleable so that new processes/changes can be introduced to the older version of design.
- They require rules and instructions.
- A supervisor in the form of a facilitator must dictate rules and instructions.
- A temporary learning medium is established.
- The location of all these design practices is important: a familiar place for all participants.
- The boundary objects turn into outputs forming pieces of the design.

Efforts in ergonomics may range from individual level (e.g., an operator using a device) to system level such as operating advanced machinery or a complex product, or even managing the organization of a company. In Industry 4.0 era, some forthcoming challenges regarding ergonomics have been reported by researchers, which warrant further research and improvement:

- analytics of processes connected with ergonomics.
- businesses' capabilities and sophistication regarding ergonomics analytics.
- capability and sophistication of integrated knowledge management and its harmonization.

1.5 Reliability, Availability, Maintainability and Safety (RAMS)

The engineering concept of RAMS; namely, *reliability*, *availability*, *maintainability*, and *safety*, is an analysis tool that helps estimate the condition of an engineering system in the future, figure out any requirements that would improve it, and carry out necessary decision-making. In other words, RAMS assesses and estimates the integrity of an engineering system. Adequate application of RAMS helps refrain from possible failures even from the earlier stages of the design efforts. Via RAMS, practical proof can be collected for any necessary decision-making.

Reliability can be defined as the capability of carrying out a specific task without failure within a prescribed duration, whereas *availability* refers to the system continuing to be operational under the expected working conditions. The engineering systems should be maintained regularly and smoothly before and during service, which is ensured by a well-structured *maintainability*. Maintainability can be conducted during downtime of the system, which are two types; *preventive maintenance* (PM), and *immediate corrective maintenance* (ICM); and can be carried out during uptime of the system, which is referred to as *deferred corrective maintenance* (DCM). Even without any sign of failure or problem, systems should also be subjected to periodic maintenance during their life cycle. Finally, *safety* considerations are essential so that the system does not injure people, damage other machinery, or impair any other property in the course of its service.

1.6 Modeling and Design

Modeling can be defined as creating mathematical expressions and statements that explain natural or artificial phenomena. From the engineering perspective, various forms of modeling serve as the basis of our design and analysis efforts; namely, *analytical*, *computational*, and *experimental*. These contemporary modeling techniques are covered in detail in the oncoming sections.

Design plays a central role in engineering disciplines. Through scientific engineering design, useful products are created for the benefit of the people and societies across the globe. It can be said that all engineers are essentially designers. Researchers claim that design is concurrently *art* and *science*, and it is *synthesis*.

1.6.1 Need Analysis

There is a basis for everything on which certain things are to be done to fulfill specific needs. Such that, in Scientific Research, need is the basis to fabricate an idea and then put it in action. Establishing need is the base of research which covers different aspects such as why a study is important, relevant, necessary, and what it can contribute within the parameters of existing knowledge.

There are three types of need. These are Natural Need, Derived Need and Generated Need. Natural need is like hunger and thirst for human beings individually and it extends to the environment on the earth for humans collectively. The derived need is simply an engineering artifact that can either be operated manually or through a source of power. Human energy or energy derived from fossil fuel, non-fossil fuel, or alternate energy sources are the derived need. Generated need is the boosting the market of a product through modern marketing tactics. All three types of need remain in transition i.e., what was required yesterday is not required today. These needs are also interdependent.

Determining customers' and stakeholders' needs is a key stage in designing and manufacturing a product. To analyze the need for a certain product, which can be a single simple part, or a complicated system composed of several parts, information must be collected and interpreted. In the Industry 4.0 era, knowledge management systems are utilized to effectively compile and assess related data. It is a common practice and a logical path to follow to apply previous knowledge regarding former designs of the product to come up with new, improved versions of it. A challenge appears here in the way that the available data may be scarce, and possibly only in the possession of some companies and/or governments. In this regard, various researchers concentrate on building and benefiting from open data resources, mainly through official governmental domains, so that the required information can be easily obtained, and used in engineering design. Being one of the pillars of Industry 4.0, the concept of *big data* has increasingly drawn attention. Utilization of *big data* increased the speed and effectiveness of engineering design and associated product development endeavors. Still, techniques to explore and mine *big data* warrant further research and development for them to reach their full potential.

Mentioned below are the parameters, with the help of which the need can be established, and work could be started upon reasonable research.[1]

---

[1] Originally Prepared by Ammad ur Rehman Shahid (2023) under the guidance of Hassan Ali and Wasim Ahmed Khan, Faculty of Mechanical Engineering, GIK Institute of Engineering Sciences and Technology, Topi—23640, Khyber Pakhtunkhwa, Pakistan. Reproduced with Permission.

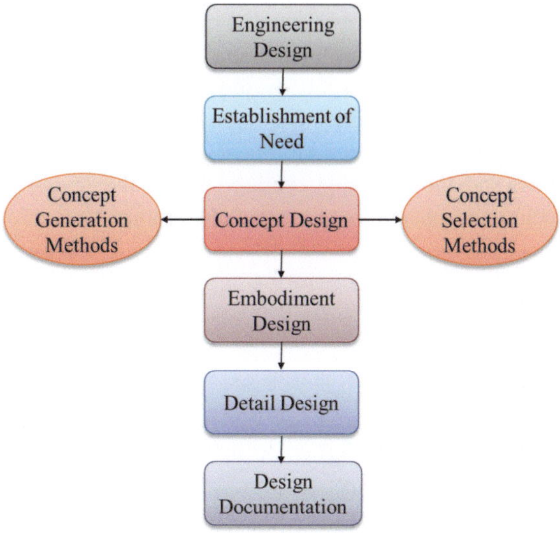

Fig. 1.4 Pillars of mechanical engineering design

- Gap in knowledge
- Research questions or objectives
- Significance and relevance
- Practical applications
- Literature review
- Justification for methodology
- Contribution to existing knowledge.

A flow chart exhibiting a typical mechanical engineering design algorithm commencing with establishment of need is provided in Fig. 1.4.

1.6.2 Concept Design

Following the establishment of the need stage, the concept (or conceptual) design phase comes about. Today, it is a team effort.

This step in design generally manifests itself in two consecutive methodologies: concept generation, and concept selection.

1.6.2.1 Concept Generation Methods

After the first phase is cleared, the next phase is the most important and the initial phase of the design process. In Concept Design, the design team must estimate the cost, specification, marketing, and concept generation and selection considering multidisciplinary features and libraries in the form of relational databases e.g., geometry, material, machine elements, control elements, etc. The concept design is an iterative process specifically required for original design. The specification of the product is raw to firmer after concept design and it is a systematic iterative process.

Concept Generation Methods/Techniques are methods with which the help of new ideas and concepts are generated during the design phase. Mentioned below are the techniques used in the Concept Generation Method.

- Brainstorming
- Mind Mapping
- Scamper (Substitute, Combine, Adapt, Modify, Put it into use, Eliminate, Reverse)
- Analogies and Metaphors
- Forced Connections
- Sketching and Doodling
- Reverse Engineering
- Prototyping and Hands on Exploration
- Design Games and Challenges
- Expert Interviews and User Involvement.

1.6.2.2 Concept Selection Methods

Then, the next phase is Concept Selection Methods/Techniques, in which we use to choose the most promising design concept during the design phase. In this phase we choose the most practical design possible after deducting many theories. These techniques help the designer to design under defined criteria and users need. Mentioned below are the techniques used in Concept Selection Method:

- Design Matrix
- Pugh Matrix
- Morphological Matrix
- Concept Scoring
- Concept Testing and Feedback
- Cost Benefit Analysis
- Risk Assessment
- Multi-Criteria Decision Analysis (MCDA)
- Analytical Hierarchical Process.

1.6.2.3 Analytical Hierarchical Process for Concept Selection—5 Axis CNC Milling Machine[2]

a. **Material Selection (Analytical Hierarchy Process)**

The procedure follows an initial comparison for material usage after which the available alternatives, in this case, the three developed concepts for a prototype 5 Axes CNC Milling Machine, are compared. This will help in selecting the final concept design. The engineering criteria chosen are Stiffness, Damping, Thermal Stability, Material Cost, Density, Strength, and Durability.

- The **Density** of the machine is a priority as it would allow the machine to be easy to carry and lightweight.
- **Material Cost** was determined and needed to be as modest as possible.
- **Stiffness** is not the priority because machines may get damaged under heavy vibrations while knowing that today's machines are lighter and stiffer. It was planned that the Vibration problem will be catered through performing natural frequency analysis at a later stage.
- The material should be **Durable** with considerable **Strength**.
- **Damping** materials reduce or eliminate the damaging forces caused by mechanical or electrical energy and it was considered. Damping characteristics also help in minimizing vibration.
- **Thermal Stability** is the measure of workability under high temperatures.

Final Decision on Material Selection is as shown in Table 1.1.

The key features used for material selection are as per following order:

Stiffness, Damping Thermal Stability, Material Cost, Density, Strength, and Durability

The variable values were taken from ASM Handbook.

The final matrix obtained from AHP ranks **Cast Iron 30** as the highest and most suitable, according to the set engineering criteria. This method allows initiation of concept design.

b. **Three different concepts were planned**

The problem with this concept is that the maintenance of such concept design is very difficult to perform, and the material used is also large. This concept design is difficult to manipulate as its mechanism introduced several complexities in the overall system. Higher manufacturing cost and less safety are the two factors that contribute to this concept design and serve as anomalies from our main objective. On the contrary this design has the highest durability, but complications do occur when it comes to modification of the design. The z axis movement does not involve

[2] Work part of Capstone Design Project (2020) by Ahmed Maqsood; M. Ibad Mughal; Sheharyar; Umer Javed under the supervision of Wasim Ahmed Khan and Khalid Rahman, Faculty of Mechanical Engineering, GIK Institute of Engineering Sciences and Technology, Topi—23640, Khyber Pakhtunkhwa, Pakistan. Updated by Wasim Ahmed Khan. Reproduced with permission.

Table 1.1 Final decision on material selection

Material	Stiffness	Damping	Thermal stability	Material cost	Density	Strength	Durability	Total score	Rank
Stainless steel 303	0.0091312	0.020805	0.0201613	0.0224039	0.0423149	0.006938	0.016202	0.1475174	3
Aluminum 6061	0.0238297	0.041654	0.049018	0.0519363	0.0182541	0.002993	0.033508	0.25786444	2
Cast iron 30	0.0052628	0.010158	0.008216	0.1945532	0.1585036	0.025988	0.108825	0.59490543	1
				WINNER = cast iron 30!!					

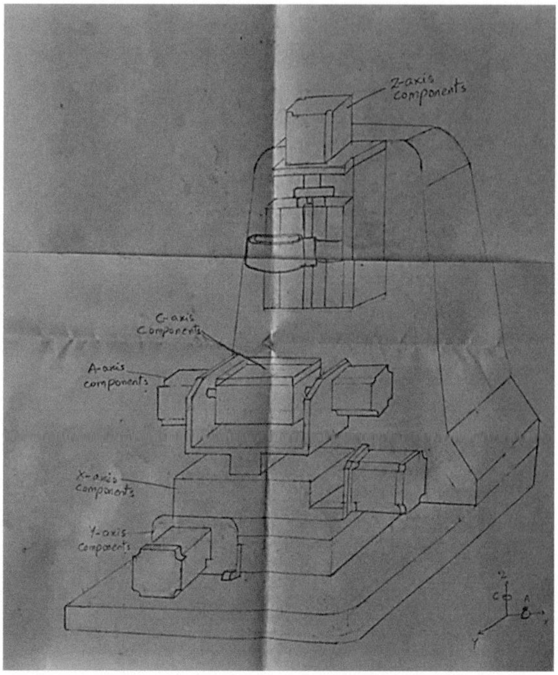

Fig. 1.5 Isometric view of concept design 1

Table 1.2 Axis orientation (ISO 841-2001)

Axis on worktable	Axis mounted to tool holder
C	Z
A	
X	
Y	

a lead screw which also contributes to the further complexities. The lubrication and oil film requirements for this design are difficult to execute as this would require disassembling the assembly to uncover the parts (Fig. 1.5; Table 1.2).

1.6.2.4 Second Concept Design

A second concept design was formulated using the same concepts but with consequent changes and alterations. The second concept design has changed a few complications by introducing lubrication easily possible. The second concept design consisted of the 3-axis mounted upon each other in the following descending order A, X, Y axis and the remaining two above these. This design decreases the complexity of parts assembly and makes the machine more maintainable and reliable for operations. The mass of the base top has also been decreased and a base top with links is used, the manufacturing and material cost decreases as a benefit.

The isometric view of the proposed concept design model, together with the axis of rotations is detailed below along with summarizing the relevant data (Fig. 1.6; Table 1.3).

The mentioned concept design presented a very well-constructed configuration but there were some major problems faced with the durability of the base top. The maintainability and clearance between the axis are still a problem. The use of links to form the Z-axis did not create a perfect concept design for the Z-axis. The geometry and Axis construction is still unreliable since any damage to the base would affect the axis mounted upon it. The number of vibrations in the system are still present

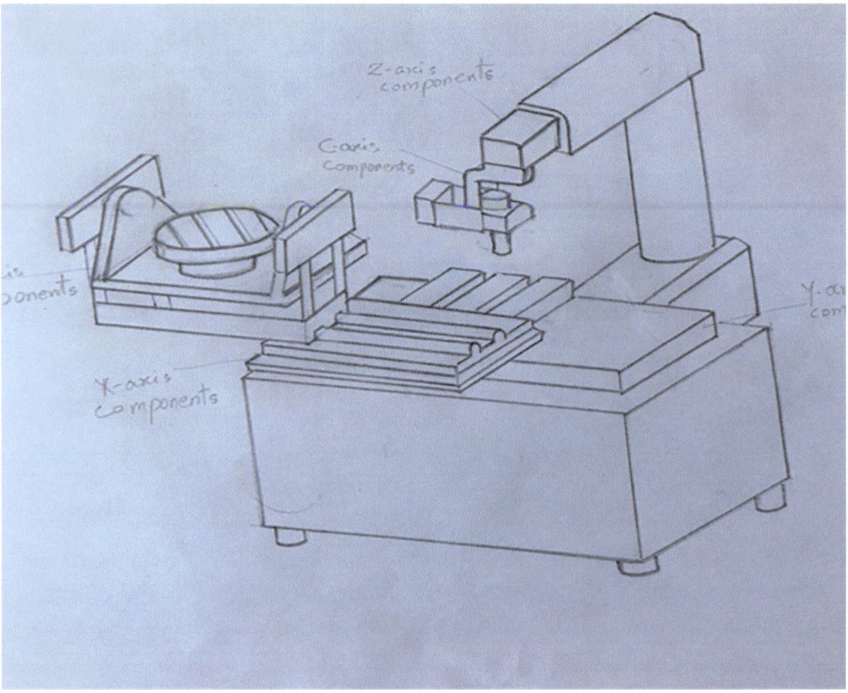

Fig. 1.6 Isometric view of concept design 2

Table 1.3 Axis orientation (ISO 841-2001)

Axis on worktable	Axis mounted to tool holder
	Z
A	C
X	
Y	

due to the movement of A-axis and the Z-axis. Efficiency and reliability are also compromised due to limited use of lead screws and could result in completely uneven and damaged work parts (Fig. 1.7; Table 1.4).

The first two concepts provided a greater insight into some common and some exceptional problems to be observed in a 5-axis CNC machine. The final concept design was generated keeping in mind the core complications and obstacles from the first two concept designs. The concept designing process followed the same pattern as the above configurations which kept in check all the aims and objectives that were taken from the first two concept designs. The major changes included increasing the maintainability and clearance, while decreasing the manufacturing cost, material cost and the complexities. The new concept design is certainly more reliable. The

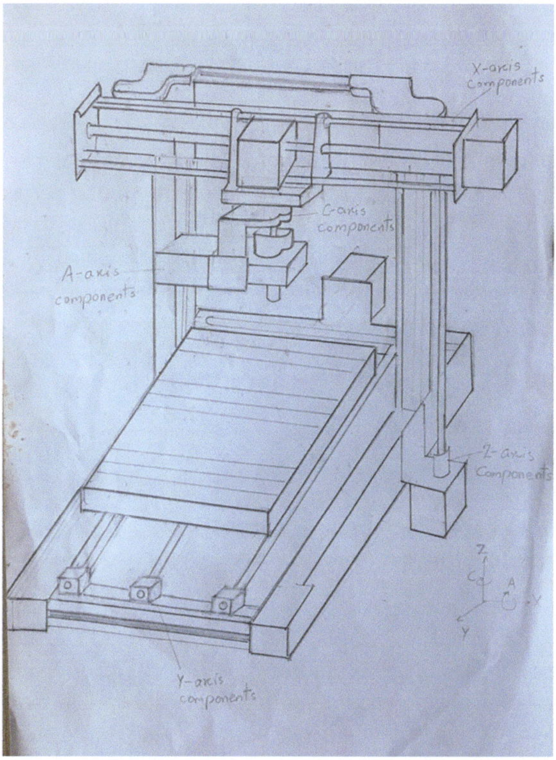

Fig. 1.7 Isometric view of concept design 3

Table 1.4 Axis orientation (ISO 841-2001)

Axis on worktable	Axis mounted to tool holder
	Z
	C
	A
Y	X

new design will ensure easier removal of chips from the system and a better supply of lubrication applied to the system. This design model was based on the modern units being used in industry.

1.6.2.5 Concept Design Selection Using Analytical Hierarchy Process (AHP)

Using AHP as a supporting tool for decision making will help to gain a better insight into complex decision problems. As the problem is structured in a hierarchy, it forces decision makers to think through the problem, considering possible decision criteria and selecting the most significant criterion with respect to the decision objective. Using pairwise comparisons helps to discover and correct logical inconsistencies. The method also allows to "translate" subjective opinions, such as preferences or feelings, into measurable numeric relations. AHP helps to make decisions in a more rational way and to make them more transparent and better understandable. The table shows the final decision matrix constructed after a series of matrix and vector calculations. The procedure follows an initial comparison of criteria, after which the three alternatives, which in this case are the three developed concept designs for a 5 Axes CNC Machine, are compared against each criterion to give a final decision for most versatile design fulfilling each engineering criteria. The key features upon which the selection is based are as follows:

- Material cost
- Manufacturing cost
- Maintenance
- Reliability
- Safety
- Vibrations
- Clearance between parts
- Durability.

The Final AHP table is as shown in Table 1.5.

The final matrix obtained from AHP ranks Concept **DESIGN-3** as the highest and most suitable, according to the set engineering criteria. This method allowed the concept development team to initiate the embodiment design while getting concurrence from team members from marketing, sales, production planning and shop floor. The data collected for the matrix was from expert opinion and mainly from literature reviews of already developed 5-axis CNC machines (Fig. 1.8).

Table 1.5 Final AHP table

Machine	Material cost	Manufacturing cost	Maintenance	Safety	Reliability	Vibrations	Clearance	Durability	Total score	Rank
5 axis CNC-1	0.0091312	0.0208054	0.020613	0.0224039	0.0423149	0.0095605	0.006937886	0.016202313	0.1475174	3
5 axis CNC-2	0.0238297	0.0416544	0.049018	0.0519363	0.0182541	0.0366707	0.002932909	0.033508326	0.25786444	2
5 axis CNC-3	0.0052628	0.0101583	0.008216	0.1945532	0.1585036	0.0833987	0.025938023	0.108824802	0.59490543	1
				WINNER = 5 axis CNC-3!!						

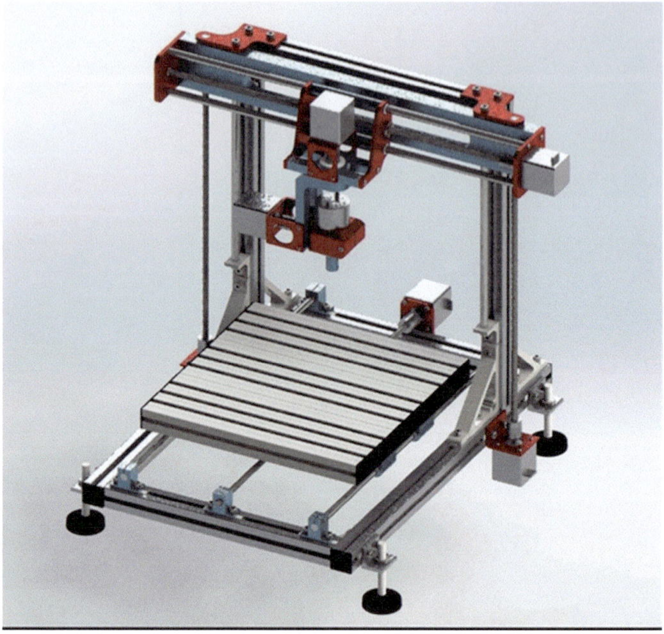

Fig. 1.8 CAD model of selected concept design

1.7 Marketing and Sales

Marketing and sales are two highly important activities of a business that work side by side, eventually aiming for the same objective of finding and keeping customers. Sales concentrates on arranging business deals, and therefore, creating funds and revenue, whereas marketing deals with informing the public (potential consumers) regarding products, raising awareness, and constructing relationships at all levels. Methods applied by these two departments vary. Sales generally target a smaller audience, composed of likely consumers, trying to influence them into buying the products. Marketing, on the other hand, is preoccupied with and takes on a bigger market with activities such as branding, various sorts of advertising market research, content creation, and campaigning. On tactics level, sales prefer individual contact, whereas marketing adopts all around comprehensive operations such as social media advertisements. Both departments are very valuable to a company that determines much of its performance over time.

The product is available for marketing and presales once the concept design is complete as the Computer Aided Design (CAD) software application packages do provide image those are an excellent resource for marketing combined with modern sourcing methods that gives the cost of any product through sourcing it for machine elements, control elements or other parts premade or made to order (as discussed above in the case of relational databases for concept design). These software packages include but are not limited to AutoCAD, aPriori, PTC Creo, and SolidWorks.

1.8 Design Analysis Technique

Engineering design, materials selection and substitution, and manufacturing/production are heavily interconnected activities. Contemporary mechanical engineering design rests on three pillars; sizing/dimensioning, choosing appropriate material(s), and addressing specific attributes as can be seen in Fig. 1.9. As mentioned earlier, a mechanics of materials-based design philosophy, embodying and utilizing calculations of force intensities (stresses) and normalized deformations (strains), dominates the sizing/dimensioning stage. The second pillar of selection of appropriate material(s) requires at least a basic understanding of how engineering materials behave. Considering a careful selection of materials, dimensions of the machine elements that constitute our designs can be determined. With these meticulously decided dimensions, along with the introduction of necessary design factors (also known as *factor of safeties*) that can potentially change the sizes even further, the designed parts are trusted to withstand all sorts of mechanical and/or thermal loads (such as static, quasi-static, dynamic, etc.) that they will come across during the lifespan of the product. The third step of addressing specific attributes helps achieve a customized product that focuses on delivering customers' needs and performs desired functions.

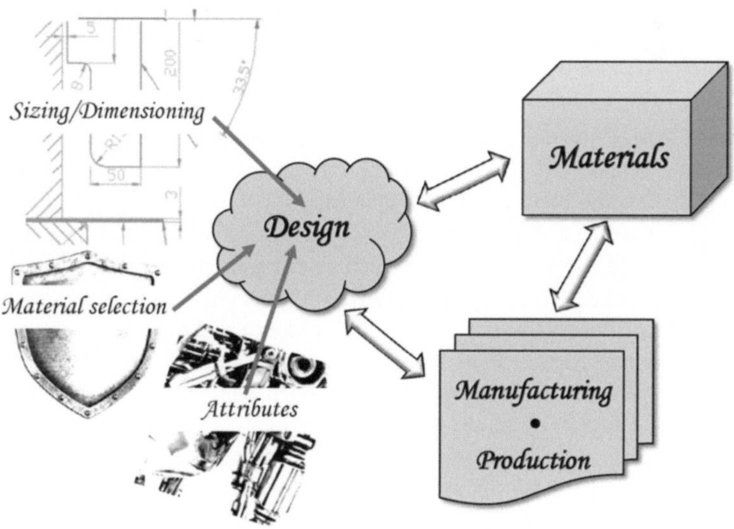

Fig. 1.9 Pillars of mechanical engineering design in relation to materials and manufacturing

Design Analysis Technique is considered an umbrella concept that brings together *design*, which is merely *synthesis* that aims to generate a part or parts of a system (i.e., a product), and analysis, which is the assessment of an already designed part or a system under given specific conditions, aiming to appraise its structural integrity and functioning. The following sections delve into the details of how these efforts are carried out traditionally and currently in the 4th Industrial Revolution era.

1.8.1 Contemporary Analytical Techniques

Mathematical models, also called metamodels, based on analytical mathematical expressions, are an effective, yet simplified technique to define the engineering problem at hand. Although they can be easy to implement and use, they may lack essential details of the designed products. If the complexities of the problem are to be reflected and embedded in the analytical model, it can be extremely difficult and perplexing to come up with a realistic *metamodel*. On the other hand, contemporary numerical modeling techniques such as Finite Element Analysis (FEA) and Computational Fluid Dynamics (CFD) can offer high-fidelity, detailed models and solutions; however, depending on the problem and its details, they may be highly computationally expensive. In the 4th Industrial Revolution era, various researchers engaged in improving the performance of *metamodels* while decreasing the associated cost of constructing them via novel artificial neural network techniques.

1.8.2 Contemporary Numerical Techniques

Numerical techniques are based on the use of a mathematical numerical methodology utilized via computer to solve the mathematical model of a phenomenon, in our context, the design process of a machine element, product, or a mechanical system. Numerical techniques are highly cost-effective, saving lead times of projects. They are by far the leading method employed in Industry 4.0 era as design and analysis of almost any product or system is possible, for most of which analytical solutions are not possible due to various factors such as complex geometries, nonlinearities (in geometry, loading, material properties), and boundary conditions (i.e., loads, joints, etc. over the domain of problem). Most common numerical techniques can be grouped into two classes: firstly CAD, namely computer aided design, tools, and secondly, the other computational analysis methods. The main theme of this book is to convey detailed information and discuss various aspects of CAD. This section aims to deliver information on the second group of other computational/numerical analysis methods. Some of these widely used computational analysis methods, particularly through advanced commercial software, can be summarized as follows:

- Finite element method/analysis (FEM/FEA)
- Computational fluid dynamics (CFD)
- Finite volume method (FVM)
- Boundary element method (BEM)
- Finite difference methods.

The finite element method (FEM), formally established around the 1960s, is arguably the most powerful computer-based numerical method to be utilized in solving practical engineering problems. The main idea that lies at the core of FEM is *discretization*, i.e., representing the certain geometry of a product (or a system) as a collection of discrete parts, namely, *finite elements*. The whole problem is first divided into those finite number of elements forming a so-called *mesh*, followed by expressing the particulars of the problem (geometry, material properties, loadings, etc.) over each element, assembling the equations of the elements constructing a global matrix equation, and solving it to find primary unknowns, from which all desired solutions for the problem can be estimated. Computational analysis software using FEM thereby follows an analysis cycle with three distinct steps:

- Preprocessing (CAD model/geometry, mesh, material properties, boundary conditions, other attributes)
- Solver/Analysis (solving the constructed numerical model)
- Post Processing (obtaining all desired solutions from the calculated primary unknowns).

4th Industrial Revolution era requires efficient use of the numerical techniques as part of the smart factory and the real-time processing requirements. Addressing this matter, current commercial software is becoming more and more advanced, capable of handling *Multiphysics* simulations, which represent the multidisciplinary nature of today's complex engineering problems. In Fig. 1.10, an FEA computational model of a crash test barrier design during crash simulation, and a hybrid computational model of a typical automobile interior with a 50th percentile male driver composed of FEA and CFD module components during an impact can be seen.

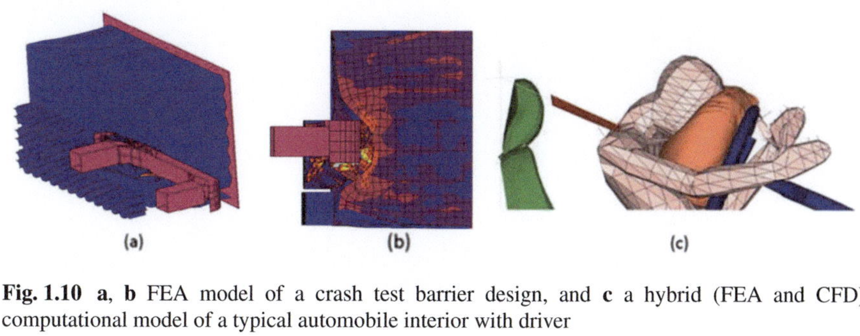

Fig. 1.10 **a**, **b** FEA model of a crash test barrier design, and **c** a hybrid (FEA and CFD) computational model of a typical automobile interior with driver

1.8.3 Contemporary Experimental Techniques

Contemporary experimental techniques can be grouped into two groups: *material characterization and testing*, and *prototype generation and testing*. Materials selection and substitution exist both within mechanical engineering design methodologies, and as an external independent scientific activity on its own, with which design frequently interacts. To facilitate design of a product, and then improve, optimize, and maintain it, mechanical properties of candidate materials must be known, which define their behaviors under mechanical load and other necessary conditions. Various experiments have been researched and constructed over centuries, and eventually established across the globe as standard tests to characterize materials for these purposes. We can list some of the most common macro-scale mechanical material characterization tests as follows:

- Tensile testing
- Compression testing
- Impact tests (Izod, Charpy)
- Hardness tests (Brinell, Rockwell, Vickers, Knoop)
- 3-, 4-point bending/flexure tests
- Torsion testing
- Creep testing
- Bulge test
- The Erichsen deep-drawing (cupping) test
- Cook and Larke compression test.

Figure 1.11 depicts several test machines specialized for material characterization: a typical computerized tensile/compression testing machine, an Izod/Charpy impact testing machine, and a digital/computerized hardness tester capable of performing various tests such as Brinell, Rockwell, Vickers, and Knoop tests.

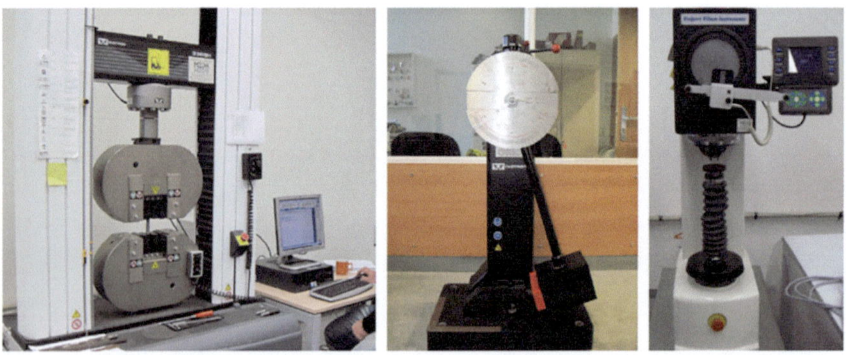

(a) (b) (c)

Fig. 1.11 **a** A typical computerized tensile/compression/bending testing machine, **b** an Izod/ Charpy impact testing machine, and **c** and a digital/computerized hardness tester

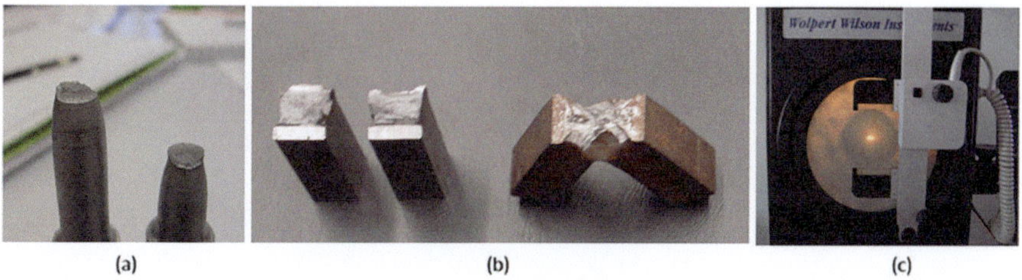

Fig. 1.12 **a** A fractured cylindrical steel tensile testing specimen showing partial cup-and-cone fracture surfaces on both pieces, **b** two Charpy impact test specimens; one broken into two due to low energy absorption capability of the material; other deformed but not broken due to high energy absorption capability, and **c** the indentation on the metal measured during hardness testing

In the 4th Industrial Revolution era, specimens for these tests were highly standardized, and can be found in specific standards published around the world. These standards also describe the testing conditions and protocols so that reproducibility, required quality and necessary legal requirements can be achieved among institutions and relevant industrial sectors. Figure 1.12 shows various test specimens used in some of the characterization methods.

The other group in contemporary experimental approaches is *prototype generation and testing*. Once the conceptual design is carried out based on the selected material, physical prototypes are manufactured and tested to validate and verify the functioning and performance of the product. If no numerical technique is used, many prototypes must be manufactured and tested as part of a physical trial-and-error procedure, which is highly time-consuming and expensive. In the 4th Industrial Revolution times, carbon footprint (i.e., total greenhouse gas emissions) of the production stages for an experimentally developed and manufactured product may be unnecessarily high. Therefore, in addition to reasons above, using experimental techniques entirely for design is not a viable option due to sustainability considerations. Employing experimentations only after a design is analytically and/or computationally complete is highly necessary; however, just to come up with a small, limited number of prototypes to test as a final step for the design efforts. Figure 1.13 demonstrates the prototype for a specialized tool holding fixture design.

1.9 Embodiment Design

Embodiment design is universally considered to be an intermediate stage in between conceptual design and detail design. Embodiment design consists of a substantial number of iterations through which incremental analysis and design steps are carried out continuously. Although the iterations are inherently part of the embodiment design process, the extensive number of these improvement steps may have adverse effects on the whole design process in terms of time, cost, and labor. Embodiment design works in such a way that a certain conceptually designed preliminary product

Fig. 1.13 Prototype for a special fixture design

is chosen, and its variations are generated from it considering not only function, cost, and attributes, but also manufacturing and assembly methods associated with it. In embodiment design, designers should pay attention to the following factors:

- Linking type of material, size of the part, and manufacturing processes
- Cost considerations in between processes
- Cost appraisals considering ambiguities.

Some researchers suggest that the negative effects that may be experienced during embodiment design can be mitigated through *collaborative CAD networks*, through which designers can easily exchange ideas, and perform much more efficiently when compared to the typical methods currently used.

The embodiment design is the first step of converting a model (Concept Design) into a viable design of components, structures and assemblies after due analysis using contemporary analysis techniques as described above. The specification of the product is firmer at this stage and both time and cost involved so far in engineering design requires no turning back. Use of conformance and technical standards is recommended for both the product and the operation of the product. This stage is recommended for all products that fall in the adaptive and variant category of design.

1.10 Detail Design

In the detail design stage, the ideas formed, and decisions taken in the prior conceptual and embodiment design stages are refined and realized into the design of the final product with all intricate details so that CAD and CAE steps of manufacturing can be completed, and manufacturing can commence. Many engineering design problems manifest themselves as optimization problems. Optimization problems can be of single- or multi-objective types. Depending on the number of objective functions that are evaluated and assessed, optimization becomes one of the types mentioned above. As an example, various real-life problems demand the optimization of several clashing objectives. Therefore, these problems are handled as multi-objective optimization problems, which compel a set of equally optimal solutions called Pareto fronts. However, it is not an easy task to build the model and reach a better or best solution. Another way of approaching these problems can be through constrained optimization problems, which can provide good solutions in a relatively short period of time. For this reason, this approach can be used as a decision-making tool by professionals for tackling the engineering design problems at hand. Another valuable optimization approach which became increasingly popular in the 4th Industrial Revolution era is the metaheuristic algorithm. They possess the potential to reach reasonable solutions in a shorter time. Different from the optimization algorithms, they may not produce the best solution to a design problem; however, they can come up with a close enough, good enough solution.

Optimization systems according to the problem type can be grouped as follows:

- **Topology optimization**, also known as layout optimization or topological optimum design, is in search of an optimal layout of material within an engineering system.
- **Shape optimization** assumes fixed topology, while pursuing optimal contour, and/or shape.
- **Sizing/dimensioning optimization** assumes fixed topology and shape, while investigating for optimal sizes, dimensions, or cross-sections within parts of a structural system.

Detail design stage is done for mechanisms and machines (collection of components, structures, and assemblies). Mechanisms and machines have movement in them. The specification at this stage is frozen. And the product must be analyzed for e.g., modal analysis and foundation requirements (if required). This stage is also important for products falling in the category of adaptive and variant design. Conformance and technical standards are pertinent to both the product and its operation.

1.11 Design for Manufacturing (DFM)

Design for Manufacturing (also referred to as Design for Manufacturability, or in short, DFM) can be defined as a novel methodology for designing parts for a machine, system, or in general for a product focusing on making manufacturing easier and more efficient, aiming for superior products with reduced cost. This process is usually carried out early at the design stage through elucidating, optimizing, and

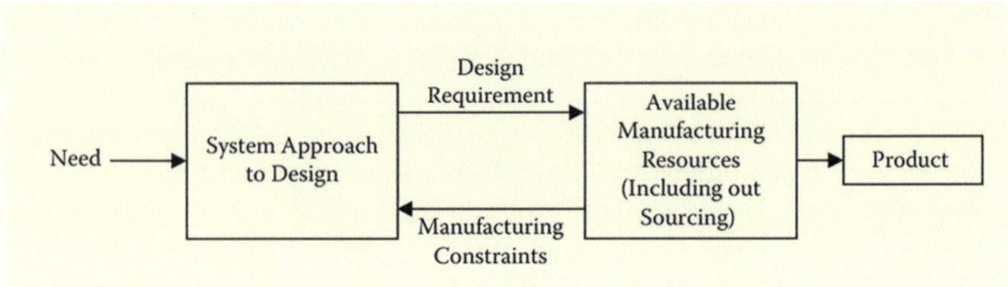

Fig. 1.14 Design for Manufacturing[3]

cultivating the design of the product before the commencement of actual manufacturing. One of the principles of DFM is to incorporate inputs from all involved parties and stakeholders (i.e., designer, in-house manufacturer, contract manufacturer, material supplier, customers, etc.) so that the product can be further improved via group discussions. These discussions involve investigating the design at various levels such as part, sub-system, system, and/or holistic perspectives. Utilizing DFM, the final improved product should ideally be manufactured more easily, faster, and cheaper. Five DFM principles are universally established as follows:

- Manufacturing process
- Engineering product design
- Material(s)
- Working environment
- Standards compliance and experimentation/testing.

Advantages of DFM methodology can be summarized as:

- Cost can be predicted realistically keeping in mind product goals.
- Possible manufacturing problems can be mitigated so that manufacturing can be carried out more rapidly and economically.
- Product design can be done efficiently so that potential future changes in design can be made easier and cheaper.
- Design can be refined eliminating unnecessary attributes and costs.
- DFM can bring advantages in prior cost items such as supplier prices.

Design for Manufacturing concept can sometimes be mistakenly used as Design for Assembly (DFA) methodology, which are in fact two different methods. Having said that, a combined methodology of Design for Manufacturing and Assembly (DFMA) is also established and used in industry and academia.

Outsourcing for DFM is normally allowed as it allows addressing orders from a wider market target. Vendor development is common in such cases. A typical DFM algorithm can be seen in Fig. 1.14.

[3] Reproduced with permission from CRC Press, Taylor & Francis Group, USA.

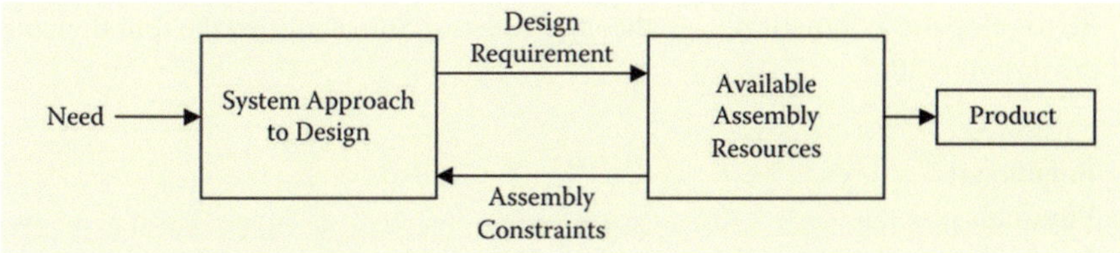

Fig. 1.15 Design for Assembly[4]

1.12 Design for Assembly (DFA)

Design for Assembly (DFA), using a similar analogy to DFM, can be defined as a process for designing an engineering system with the main goal of making assembly easier and more efficient. At the heart of the process, using a smaller number of components that constitute the system is a main objective, ultimately decreasing assembly time and cost. Another remedy DFA offers is to embody mechanical fasteners into the components themselves such as using fit instead of bolts and nuts. Some other guidelines can be to create such features in the assembled parts so that no other way of positioning the parts is possible, or if possible, to use symmetric features in parts so that easy assembly is possible in one way or another. Tolerance plays a big role in DFA. They need to be selected carefully providing ease of assembly, meanwhile reducing the cost of achieving tight high-quality tolerances. Employing off-the-shelf commercial parts and utilizing a smaller number of tooling and machinery for assembly are other critical recommendations for successful implementation of DFA.

Assembly is normally done at the manufacturer's site as this allows a single point for packaging and dispatch to the distributor/retailer/client. Figure 1.15 depicts DFA.

1.13 Axiomatic Design

Axiomatic Design is a systems engineering approach developed by Professor N. P. Suh at MIT, USA in the 1990s. The methodology applies matrix equations that relate customer needs and expectations with design parameters and process features and requirements. Concepts of axiomatic design can be summarized as follows:

- Analysis of customer needs and expectations
- Specification of the problem considering needs analysis
- Constructing a solution
- Analysis and optimization of the solution
- Assessment of the design regarding initial needs and expectations.

---

[4] Reproduced with permission from CRC Press, Taylor & Francis Group, USA.

To develop the axiomatic design theory, Professor Suh identified the four domains of the design realm:

- Customer
- Functional
- Physical
- Process.

These domains are linked to one another in order from top to bottom. Then, functional requirements (FRs) are associated with design parameters (DPs) in the form of equations, and necessary axioms are formed. These constructed equations dictated by the design problem at hand form the global matrix equation for the entire design problem.

Through axiomatic design, many initial designs may be brought down to one optimal design in a systematic way. Then, detailed design stages can take place, possibly ending with a better, higher quality product.

1.14 Concurrent Design

Concurrent design (or also called concurrent engineering) is a design approach where the stages of design and/or manufacturing can overlap and be carried out simultaneously, with which disadvantages of sequential design and manufacturing can be avoided. Via integration of the product development stages, time-to-market for the product can be significantly reduced. Concurrent design is generally a multidisciplinary effort, conducted by cross-functional teams of experts from different areas. To make these efforts efficient, a suitable means of knowledge sharing must be established.

1.15 UNDP Sustainable Development Goals (SDGs)

The United Nations describes the Sustainable Development Goals (SDGs) (also known as the Global Goals) as a universal action call to end poverty, protect the planet, and ensure that by 2030 all people enjoy peace and prosperity. The SDGs were introduced by the United Nations in 2015. There are 17 goals that are integrated affecting one another, ultimately aiming to stabilize and foster social, economic, and environmental sustainability. The SDGs are claimed to be constructed to end global challenges such as poverty, hunger, AIDS, discrimination against women and girls, etc. The 17 SDGs can be seen in Fig. 1.16.

As can be seen from the goals, innovative engineering design and manufacturing plays a central role for the direct and indirect realization of most of the SDGs.

Fig. 1.16 UNDP Sustainable Development Goals (SDGs)

1.16 Life Cycle Assessment

Life cycle assessment (or analysis) (LCA) is a method developed to evaluate the environment-related effects of a product, system, or process on its surroundings throughout the course of its economic life, namely, in its cradle-to-grave life cycle. Universally, an LCA is considered to have four phases:

- Phase 1. Goal and scope of analysis.
- Phase 2. Inventory analysis: material and energy flows, interaction with surroundings, raw materials used, carbon footprint.
- Phase 3. Impact assessment.
- Phase 4. Critical review, data sensitivity, and results.

1.17 Sourcing for Engineering Design

Establishing effective supply chains is a highly critical issue in the 4th Industrial Revolution era. In the contemporary design and manufacturing sectors, procurement and engineering efforts must exist together. There are various reasons for this close collaboration (rather than cooperation) starting from cost savings, efficient business operations, and quality product design. With Industry 4.0, technology dominates almost every area of these production efforts. Sourcing is no different. User-friendly software, cloud-based knowledge sharing facilities, etc. are some examples. Examples of sourcing using manual, and software packages-based rules embedded operation are provided in Part I, Chap. 2.

1.18 Design Documentation

Generated designs, mainly in the form of CAD models, need to be documented clearly and efficiently for manufacturing purposes. CAD models contain modeling data that can be visualized in 3D, along with various annotations including dimensions, tolerance symbols, other necessary symbols, and explanatory text. By use of the CAD model with annotations, technical and/or working drawings are produced to explicitly convey the design information to the engineers, operators, and workers so that manufacturing of the part can be accomplished effectively and accurately.

An interesting approach called Sketch-Based Modeling (SBM) has become increasingly popular through which 3D models can be automatically generated from hand drawings and sketches. Two versions are reported to exist side-by-side in the form of input generation for 3D subjects and transferring engineering designs into computer aided design and manufacturing.

1.19 Prototyping

As mentioned, and discussed earlier, prototyping is an experimental approach where physical models are manufactured from conceptual designs to test various aspects of the product physically. In older versions of engineering design, a full-on experimental design approach can be adapted; however, contemporary engineering design is mostly based on computational techniques (along with analytical techniques, where necessary), requiring prototyping only when the design is believed to reach its final form. Therefore, it can be concluded that prototyping evolved from being an almost standalone design and testing activity to an essential but a final step for contemporary engineering design. Through prototyping, crucial data and feedback are collected, verifying, and validating the design, establishing communication between shareholders, and amending the design, if necessary.

The AR/VR pillar of industry 4.0 is the current and future method of prototyping at a fraction of cost.

1.20 Product Life Cycle and Engineering Design

The product life cycle concept has been touched upon in the LCA section. LCA is a prediction while product life cycle covers the transition of a product from cradle to grave whereby it goes through many phases explained later.

In a manufacturing framework, also explained in different sections in this book with different perspectives, following definitions hold:

1. **Product Lifecycle (PLC)** covers the complete Cradle to Grave Cycle starting from extraction of raw material from ore to end of life where humans opt for reuse (including remanufacturing), recycle and waste disposal of components, structures, assemblies, mechanism, and machines.
2. **In the 4th Industrial Revolution era**, businesses are required to respond to their customers' needs and requirements in a prompt and efficient way, creating value and satisfaction. This reveals the fact that product life cycles (PLC) continuously and increasingly evolve. With this evolution, innovation in design is inevitable. A classification summarizes the six phases of product life management (PLM) as follows:

    i. Strategy/innovation
    ii. Requirement analysis and product design
    iii. Product development
    iv. Product testing/quality assurance (QA)
    v. Product deployment
    vi. Product support and maintenance

3. **Product Lifecycle Management** also encapsulates product development, product testing/quality assurance and product deployment. This part of PLM is called Manufacturing Engineering Services (MES) and is an identifiable organ of today's Engineering Design and Manufacturing of Discrete products especially as it also applies to continuous manufacturing. This is the value chain which is embedded in PLM and is part of PLC supply chain—a production philosophy and most important part of 4th Industrial Revolution.

1.21 Case Study[5]—1: Engineering Design Process of Single Plate Clutch Assembly

In the following case study, the authors will be providing the example of the Engineering Design Process with a practical example of "Single Plate Clutch Assembly" for a Toyota Corolla car step by step.

---

[5] Originally Prepared by Ammad ur Rehman Shahid (2023) under the guidance of Hassan Ali and Wasim Ahmed Khan, Faculty of Mechanical Engineering, GIK Institute of Engineering Sciences and Technology, Topi—23640, Khyber Pakhtunkhwa, Pakistan. Reproduced with Permission.

1.21.1 Establishment of Need

(i) **Gap in Knowledge**: The historical background of clutch mechanisms in the automotive sector reveals the limitations of earlier designs, such as multiple plate clutches and cone clutches. These designs faced challenges in terms of efficiency, durability, and complexity. Recognizing these limitations highlighted the necessity for a more streamlined and effective clutch mechanism.

(ii) **Research Questions or Objectives**

- What improvements, innovation, and benefits does it give over previous clutches?
- What are the deficiencies of previous clutch mechanisms, and how do they fall short in terms of efficiency, durability, and complexity?
- In what ways does the single plate clutch address these identified limitations and provide a superior solution?
- What are the fundamental design principles and materials that make the single plate clutch an efficient power transmission component?

(iii) **Significance and Relevance**: The importance of the single-plate clutch lies in its role in optimizing power transmission within vehicles. This optimization, in turn, significantly influences fuel efficiency, driving comfort, and overall performance. Its relevance is shown by its widespread use in modern automobiles, making it critical to understand its functionality and benefits.

(iv) **Practical Applications**: Single-plate clutches find extensive use in the manual transmissions of a wide array of vehicles, including cars, trucks, and motorcycles. Their widespread use and practical importance in daily transportation are demonstrated. Additionally, their influence on how the automotive industry operates highlights their relevance and significance.

(v) **Literature Review**: A comprehensive review of the literature includes digging into the historical evolution of clutch designs, analyzing the progression that resulted in the development of the single plate clutch, comparing its advantages to earlier designs, and comprehending the materials and engineering principles that comprise its construction. This review provides useful insights into the evolution and advantages of single-plate clutches over their predecessors.

(vi) **Justification for Methodology**: The chosen methodology encompasses a multi-faceted approach, combining literature review, analytical studies, experimental testing, and potentially computational simulations. This approach is well-suited to gaining a comprehensive understanding of the design, functionality, and advantages of single plate clutches.

(vii) **Contribution to Existing Knowledge**: This research seeks to contribute by consolidating historical knowledge regarding clutch mechanisms, conducting a detailed analysis of single-plate clutches, and emphasizing their efficiency, durability, and versatility. Additionally, the study aims to deepen the understanding of materials and engineering principles, thereby influencing the design and performance of single-plate clutches.

1.21.2 Concept Design of a Single Plate Clutch Assembly[6]

(a) **Concept Generation Method**

(i) **Brainstorming**: The first step in imagining a single plate clutch assembly involves brainstorming. Encourage the team to think in innovative ways. Alternative clutch plate materials, creative engagement mechanisms, or new ways to dissipate heat created during operation are all possibilities.

(ii) **Mind Mapping**: Create a mind map to visualize the various components and functions of a single plate clutch assembly. This will help identify potential areas for improvement or innovation. For instance, map out the interactions between the clutch plate, pressure plate, and flywheel, and consider how these components can be optimized for performance and durability.

(iii) **SCAMPER**: Apply the SCAMPER technique to each element of the clutch assembly:

- **Substitute**: Explore the possibility of substituting traditional friction materials with advanced composites for improved wear resistance.
- **Combine**: Consider combining the clutch and flywheel into a single integrated unit to reduce weight and increase efficiency.
- **Adapt**: Adapt technology from other industries, such as aerospace or automotive, to enhance clutch performance.
- **Modify**: Modify the engagement mechanism to allow for adjustable clutch engagement points, adapting to different driving styles.
- **Put It into Use**: Investigate how the clutch assembly could be utilized in unconventional applications, such as industrial machinery or electric vehicles.
- **Eliminate**: Identify redundant or unnecessary components and eliminate them to simplify the design and reduce manufacturing costs i.e., removing redundant plates, minimizing springs and dampers.
- **Reverse**: Reverse engineer existing clutch assemblies to gain insights into their weaknesses and opportunities for improvement.

(iv) **Analogies and Metaphors**: Make comparisons between the clutch assembly and standard equipment or mechanisms. Take the clutch's

[6] Originally Prepared by Ammad ur Rehman Shahid (2023) under the guidance of Hassan Ali and Wasim Ahmed Khan, Faculty of Mechanical Engineering, GIK Institute of Engineering Sciences and Technology, Topi—23640, Khyber Pakhtunkhwa, Pakistan. Reproduced with Permission.

engagement mechanism, for instance, and compare it to a door latch to get ideas for a more effective and dependable design.

(v) **Forced Connections**: Motivate team members to form links between seemingly unrelated ideas or things. This might involve linking the clutch to concepts like magnetic levitation or hydraulic systems to generate innovative ideas that may not be immediately obvious.

(vi) **Sketching and Doodling**: Allow designers to express their ideas through sketches and doodles. Visualizing concepts can lead to breakthroughs in the design process. Sketching can help convey ideas for the shape, size, and arrangement of clutch components.

(vii) **Functional Reverse Engineering**: Study existing clutch assemblies to understand their design principles and limitations. Identify areas where improvements can be made, such as the materials used, the production process, or the tolerances.

(viii) **Prototyping and Hands-on Exploration**: Make physical models of various clutch assembly ideas. Hands-on exploration enables designers to put their concepts to the test practically, finding possible problems and continually improving the design.

(ix) **Design Games and Challenges**: To promote creativity and competitiveness among the team, organize design challenges and games. Encourage engineers to think outside the box by finding creative solutions to specific clutch-related challenges.

(x) **Expert Interviews and User Involvement**: Consult with professionals in the field of clutch design for advice and obtain opinions from possible customers. Real-world requirements and difficulties can be understood to direct the idea design process toward workable and user-friendly solutions (Fig. 1.17).

(b) **Concept Selection Method**

(i) **Design Matrix**: A design matrix is a tool used to list and organize design requirements and specifications for the single plate clutch assembly. It outlines the parameters such as torque capacity, durability, size, weight, and material properties needed for the clutch assembly.

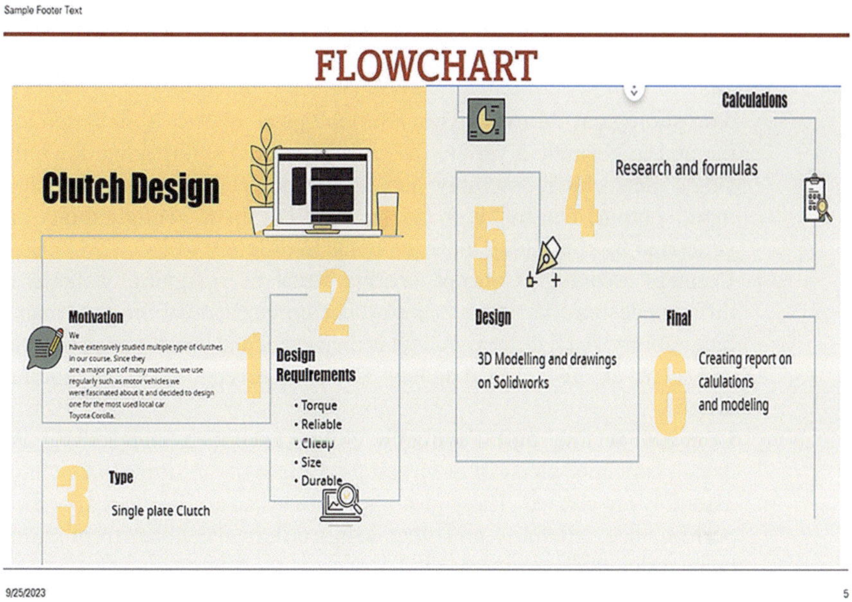

Fig. 1.17 Flowchart for clutch design

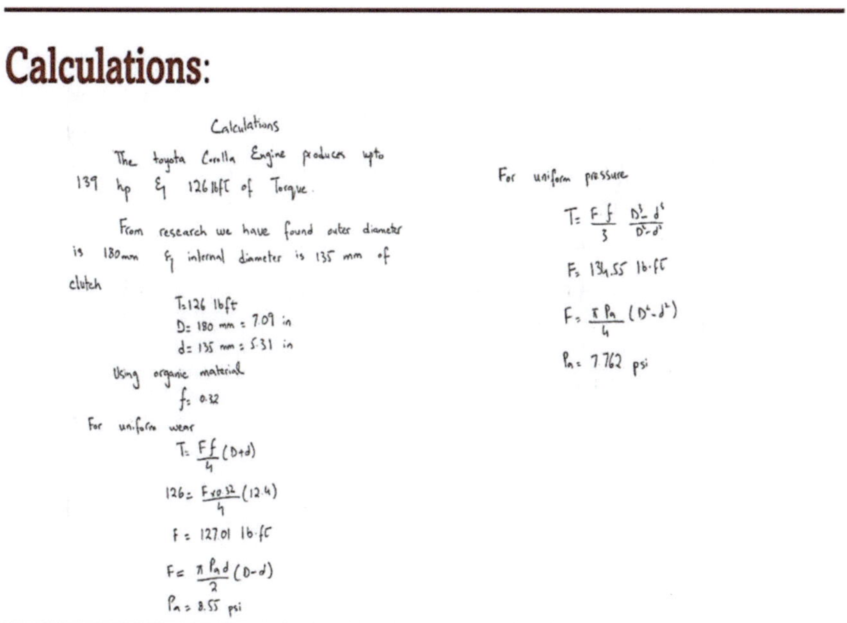

(ii) **Pugh Matrix**: The Pugh Matrix is utilized to evaluate and compare multiple design concepts against a baseline or reference design. The criteria, such as ease of manufacturing, cost, performance, and durability, are compared for each design concept to select the most suitable one.

(iii) **Morphological Matrix**: The morphological matrix is a systematic method to generate a variety of design concepts by breaking down the single plate clutch assembly into its fundamental elements, including input, output, control, and mechanism. Different configurations are generated by combining these elements in different ways.

(iv) **Concept Scoring**: Concept scoring involves assigning weights to different design criteria, such as cost, performance, reliability, and manufacturability. Each design concept is then evaluated against these criteria, and scores are given based on how well each concept meets the required criteria.

(v) **Concept Testing and Feedback**: After generating and scoring the design concepts, physical or virtual prototypes of the single plate clutch assembly are created. These prototypes are tested and evaluated for their performance, reliability, and other design criteria. Feedback from testing helps refine the concepts and identify areas for improvement.

(vi) **Cost Benefit Analysis**: Cost–benefit analysis involves assessing the costs associated with each design concept, including operational costs, manufacturing costs, material costs, and manufacturing costs. To determine the overall cost-effectiveness of each proposal, this analysis is matched with the advantages, such as enhanced performance, efficacy, or durability.

(vii) **Risk Assessment**: Risk assessment involves identifying potential risks associated with each design concept, such as manufacturing complexity, component failure, or market acceptance. To create risk mitigation methods for each concept the seriousness and likelihood of these risks are analyzed.

(viii) **Multi-Criteria Decision Analysis (MCDA)**: MCDA is a method that helps in evaluating and comparing design concepts based on multiple criteria simultaneously. The criteria include technical, economic, environmental, and social aspects. Weightings are assigned to each criterion, and the scores for each concept are aggregated to determine the best overall design concept. The best-given design for the Single Plate Clutch Assembly is finalized here.

So, after keeping in consideration, the new design is shown in Fig. 1.18.

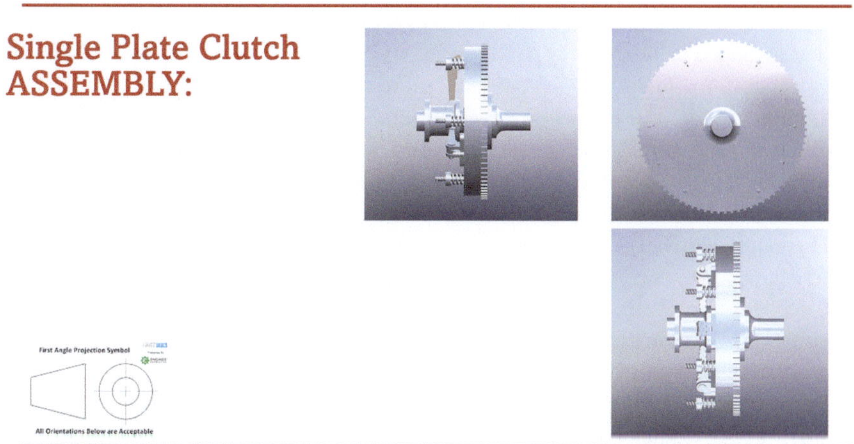

Fig. 1.18 Single plate clutch assembly

1.22 Case Study[7]—2: Design and Assembly of a Radial Engine Using SolidWorks

This case study encompasses the comprehensive process of modeling and designing a radial engine assembly. The step-by-step procedure for modeling and designing each constituent using SOLIDWORKS is elaborated upon with precision and thoroughness.

- **Radial Engine**: A radial engine has the assembly of five gas chambers numbered from 1 to 5 on a single plane. To effectively regulate the airflow, a pair of one-way valves are interconnected with each gas chamber. In modern applications, radial engines are not widely used, but they had historical and specialized uses in various industries. Some of the key industries in which radial engines are used are aircraft, marine, automotive, industrial, and agriculture (early tractors and agricultural machinery) etc. Radial engines hold a prominent position within the aerospace sector. The complete removal of the water-cooling system from radial engines yields a significant 20% decrease in engine malfunctions, a development that contributes positively to aviation safety.

  Radial engines come in various sizes and configurations, so their specifications can vary widely depending on the specific engine model and application. However, some general specifications and parameters that are commonly associated with radial engines can be provided as follows:

[7] Originally Prepared by Ziaullah Jan, Faculty of Mechanical Engineering, GIK Institute of Engineering sciences and Technology, Topi—23640, Khyber Pakhtunkhwa, Pakistan. Reproduced with Permission.

- **Cylinder Configuration**: Radial engines typically have an odd number of cylinders arranged in a circular or star-like pattern around a central crankshaft. Common configurations include 5, 7, 9, or even more cylinders.
- **Displacement**: The displacement of a radial engine is the total volume swept by all its pistons during one complete engine cycle. It is usually measured in cubic inches (CID) or liters (L) and can range from a few liters to several liters or more, depending on the engine size.
- **Horsepower**: Radial engines can produce a wide range of horsepower, from a few hundred to several thousand horsepower, depending on their size and design.
- **Compression Ratio**: The compression ratio is the ratio of the volume of the cylinder when the piston is at bottom dead center (BDC) to the volume when it's at top dead center (TDC). Compression ratios for radial engines typically range from 5:1 to 7:1 or higher.
- **Fuel Type**: Radial engines can run on various fuels, including aviation gasoline (avgas) for aircraft engines and diesel fuel for some industrial and marine applications.
- **Cooling**: Radial engines can be either air-cooled or liquid cooled. Many historical radial aircraft engines are air-cooled, while some modern radial engines may use liquid cooling for better temperature control.
- **Weight**: The weight of a radial engine depends on its size and materials used in its construction. Larger radial engines can weigh several hundred kilograms, while smaller ones may weigh less than a hundred kilograms.
- **RPM (Revolutions Per Minute)**: Radial engines typically operate in a relatively low angular speed RPM range compared to some other engine types. The RPM can vary depending on the engine's design and application but is often in the range of 1000–2500 RPM.
- **Fuel Efficiency**: Radial engines are known for their reliability and torque but may not be as fuel-efficient as some modern engine designs. Fuel efficiency can vary based on factors such as the engine's design, load, and operating conditions.

These specifications provide a general overview of radial engines, but it is important to note that there can be significant variations and customization based on the intended use and engineering decisions made by manufacturers. Different radial engine models may have unique features and performance characteristics.

(a) **Applications**

Radial engines are used in various applications, including aircraft, marine vessels, industrial machinery, and historical vehicles. The specific design and specifications can be tailored to meet the requirements of each application.

(b) **Working Principle and Governing Equations**

The basic working principle of a radial engine involves the conversion of chemical energy from fuel into mechanical energy to drive an aircraft's propeller. Here's a simplified overview of its operation:

- **Air and Fuel Mixture**: The engine draws in a mixture of air and fuel into the cylinders in appropriate application.
- **Compression**: As the piston moves upward in the cylinder, it compresses the air–fuel mixture, increasing its pressure and temperature.
- **Ignition**: A spark plug ignites the compressed air–fuel mixture, causing an explosion.
- **Power Stroke**: The explosion drives the piston downward with considerable force. This is the power stroke, and it generates mechanical energy.
- **Exhaust**: After the power stroke, the exhaust valve opens, and the piston moves upward again, pushing out the spent gasses from the combustion process.
- **Intake**: Finally, the intake valve opens again, and the cycle repeats as the piston moves downward to draw in a fresh mixture of air and fuel.
- The governing equations for the operation of a radial engine involve principles of thermodynamics, fluid dynamics, and mechanical dynamics. Some of the key equations and principles include:
- **Ideal Gas Law**

Describes the relationship between pressure (P), volume (V), and temperature (T) of a gas:

$$PV = nRT$$

where

P Pressure
V Volume
n Number of moles of gas
R Universal gas constant
T Temperature in Kelvin.

- **Thermodynamic Cycle**: The engine operates on a thermodynamic cycle, typically an Otto cycle for spark-ignition engines or a Diesel cycle for compression-ignition engines. The equations governing these cycles describe the processes of compression, combustion, expansion, and exhaust.
- **Newton's Laws of Motion**: These laws govern the motion of the pistons, connecting rods, and crankshaft, which convert the reciprocating motion of the pistons into rotational motion for driving the propeller.
- **Fluid Dynamics**: The behavior of air and fuel mixture inside the cylinders is governed by fluid dynamics principles, including the equations of motion and conservation of mass.
- **Heat Transfer**: Heat transfer equations describe the exchange of thermal energy between the combustion chamber, cylinder walls, and the surrounding environment.
- **Combustion Chemistry**: The combustion process involves chemical reactions, and equations related to stoichiometry and combustion kinetics determine the combustion efficiency and emissions.

These equations and principles collectively define the performance, efficiency, and behavior of a radial engine as it converts fuel into mechanical energy to propel an aircraft. The specifics of these equations can vary depending on the type of radial engine, such as whether it operates on a spark-ignition or compression-ignition cycle.

The following procedure entails the complete detail of each step used during the modeling, design and assembly of the radial engine using SOLIDWORKS.

(c) **2d Drawings**

See Figs. 1.19, 1.20, 1.21 and 1.22.

(d) **Modeling of Bushes, Piston Ring, Piston Pin Plug, Link Pin, and Master Rod Bearing**

All these components are modeled in SOLIDWORKS using one command i.e. Extrude (Figs. 1.23, 1.24, 1.25, 1.26, 1.27 and 1.28).

(e) **Modeling of Articulated Rod**

Basic sketch of articulated rod is first sketched which is then extruded as shown in Figs. 1.29 and 1.30.

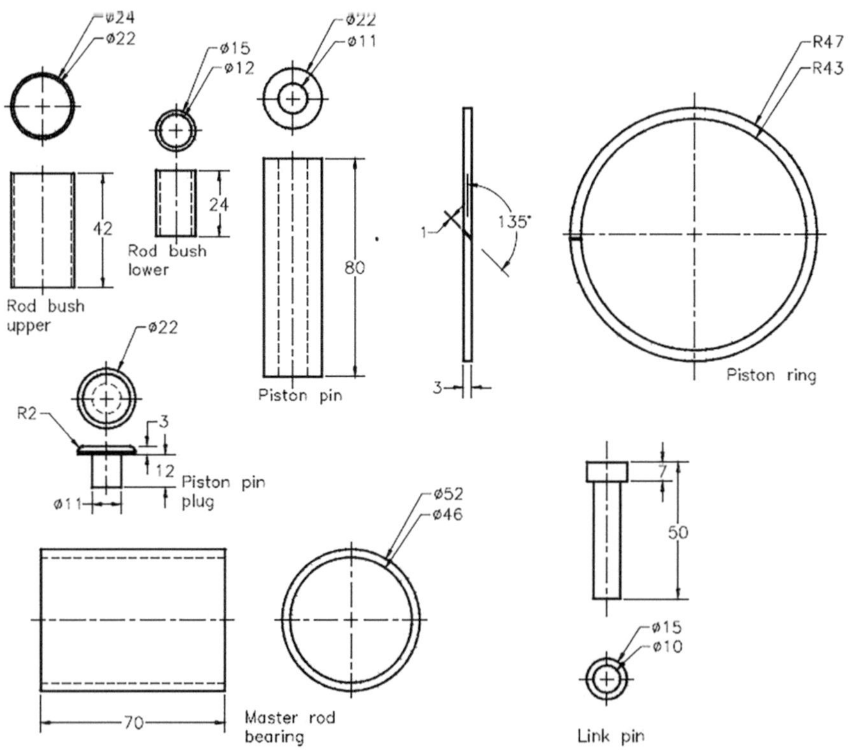

Fig. 1.19 Top views of bushes, piston ring, piston pin plug, link pin, and master rod bearing

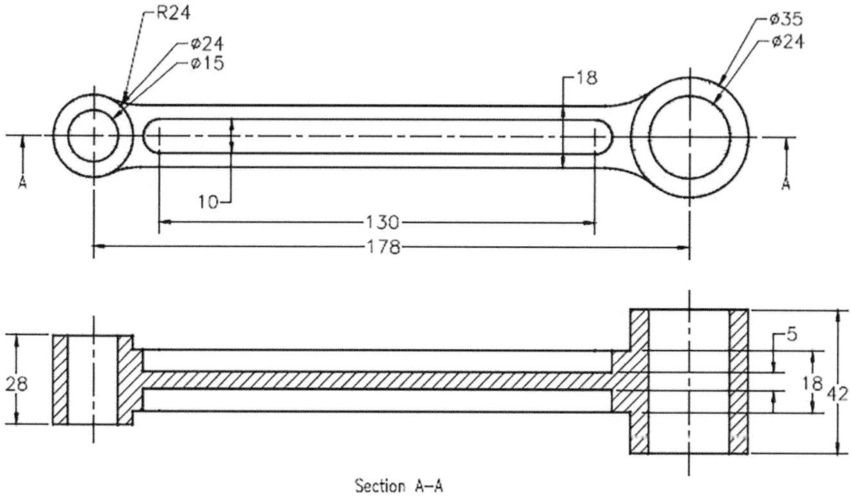

Fig. 1.20 Top and cross-sectional view of articulated ring

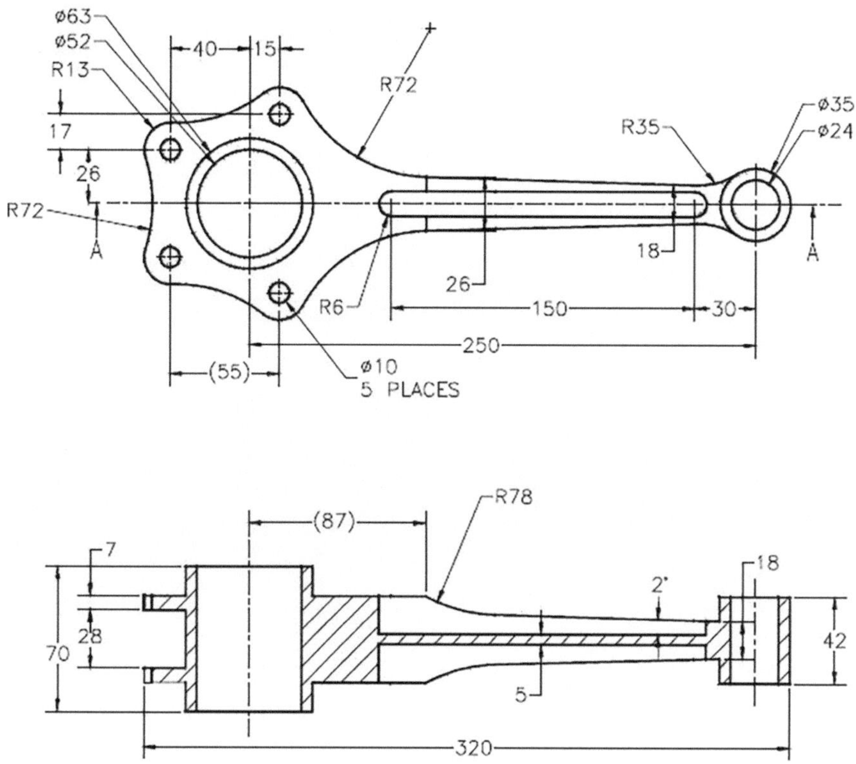

Fig. 1.21 Top and cross-sectional view of master rod

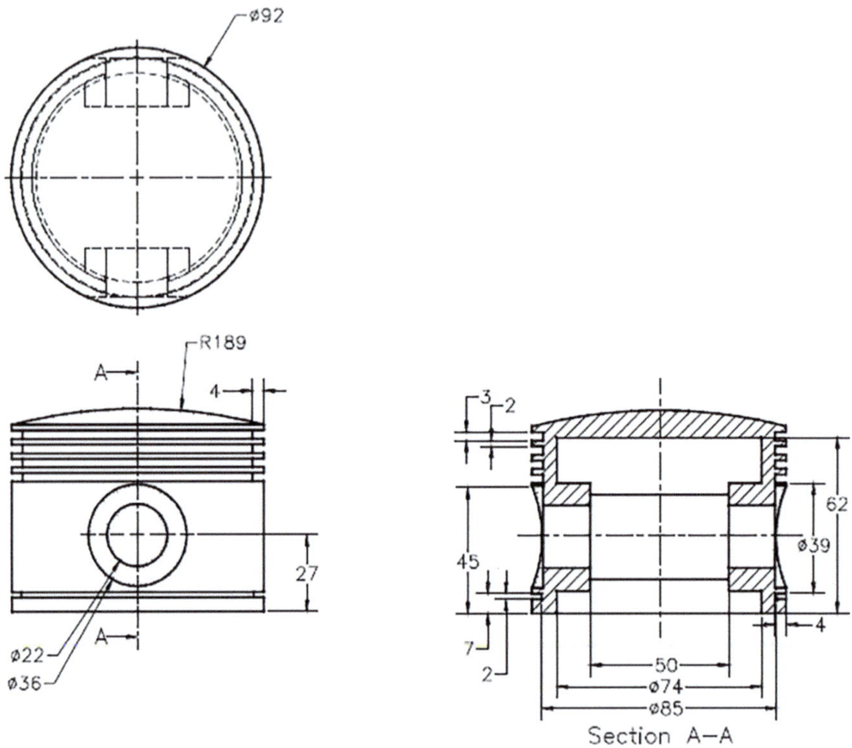

Fig. 1.22 Top, front, and cross-sectional view of piston

Fig. 1.23 Upper rod bush

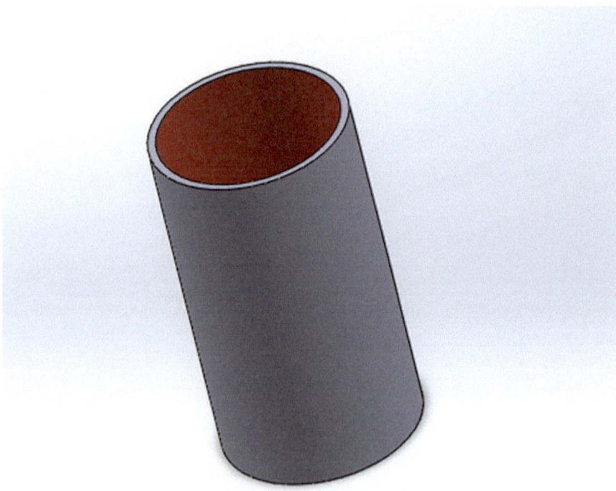

Fig. 1.24 Lower rod bush

Another sketch is drawn over the top surface and the extrude cut is performed as shown in Fig. 1.31.

(f) Modeling of Master Rod

The modeling starts with a sketch and then extruded using extrude command. For the cut between two faces of the master rod a sketch parallel to the first one is drawn and then using extrude cut command the side surfaces are cut as shown in Figs. 1.32 and 1.33.

(g) Assembly in SOLIDWORKS

After modeling all the components of the radial engine, the next step is to assemble them. In SOLIDWORKS, an assembly is a digital representation of a physical product or mechanism composed of multiple parts and components. It is a fundamental concept in computer-aided design (CAD) software like SolidWorks and is used to create complex designs by assembling individual parts and defining their relationships and interactions. Here a more detailed explanation of what an assembly is in SolidWorks:

Components: In an assembly, the individual parts or components of a design are represented. These components can be parts created within the same SolidWorks project or imported from external files. Parts can include anything from simple geometric shapes to highly complex 3D models.

Relationships: Assemblies allow one to define how components relate to each other in terms of position, orientation, and motion. One can establish relationships such as

Fig. 1.25 Link pin

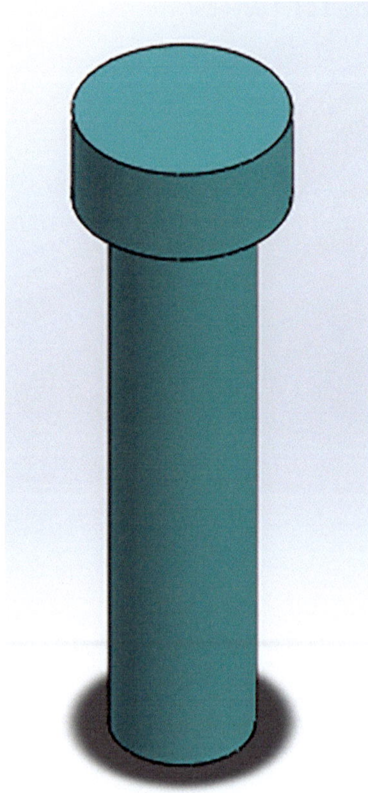

Fig. 1.26 Piston ring

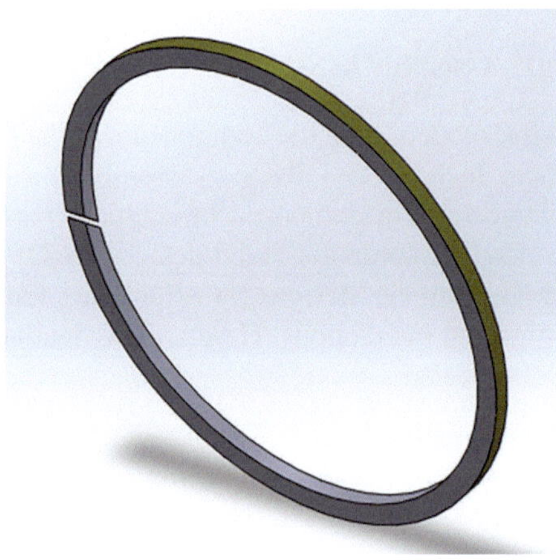

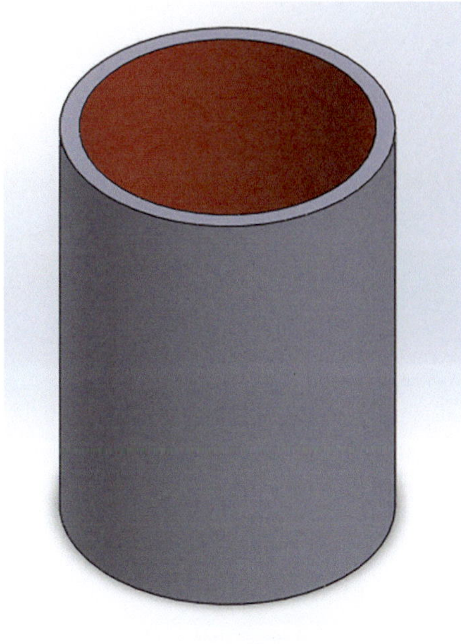

Fig. 1.27 Master rod bearing

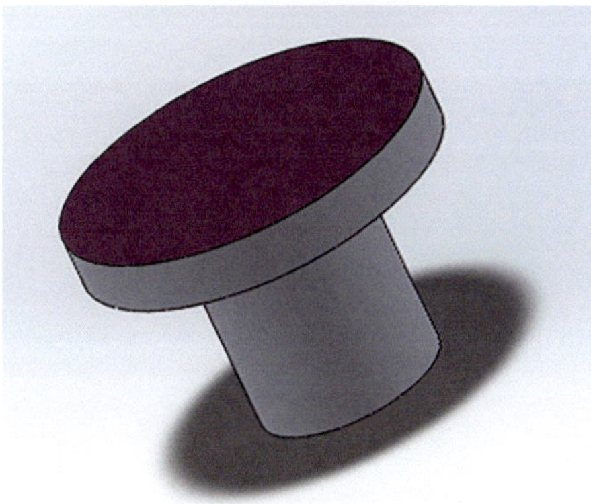

Fig. 1.28 Piston plug

mates, coincidences, alignments, and more to ensure that components are assembled correctly and can move as intended.

Motion and Animation: SolidWorks assemblies enable one to simulate the movement and interactions between components. This is useful for testing the functionality of a design, checking for interference or collisions, and creating animations or motion studies.

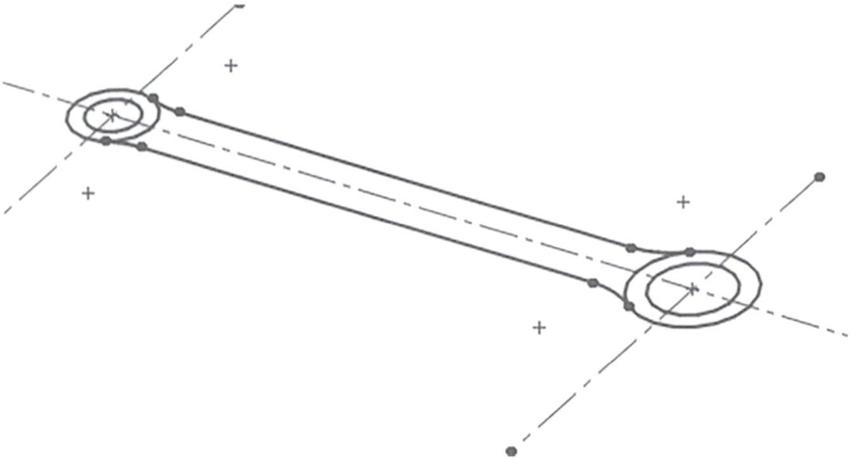

Fig. 1.29 Basic sketch of articulated rod

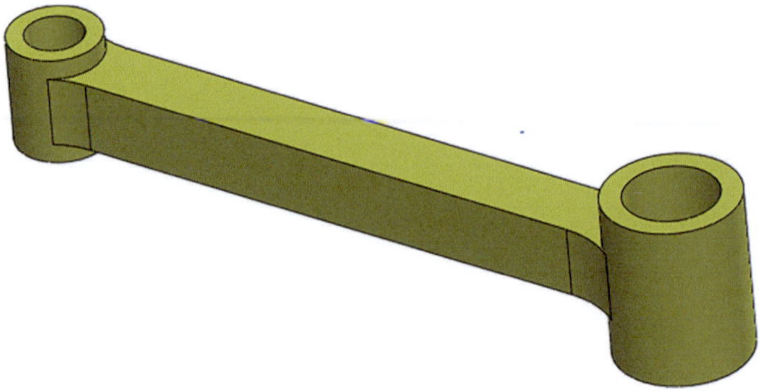

Fig. 1.30 Extruded rod

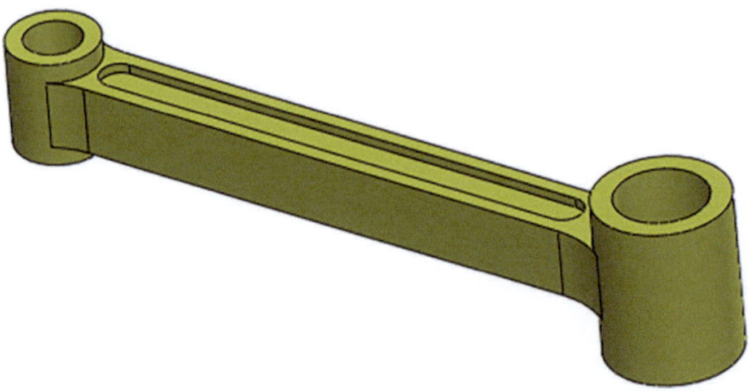

Fig. 1.31 Final design of articulated rod

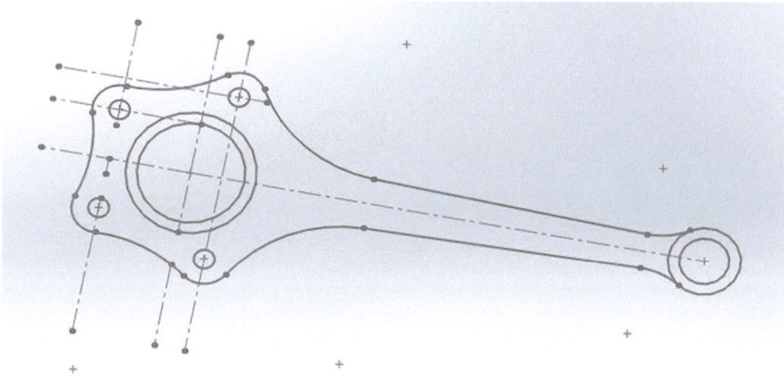

Fig. 1.32 Basic sketch of the master rod

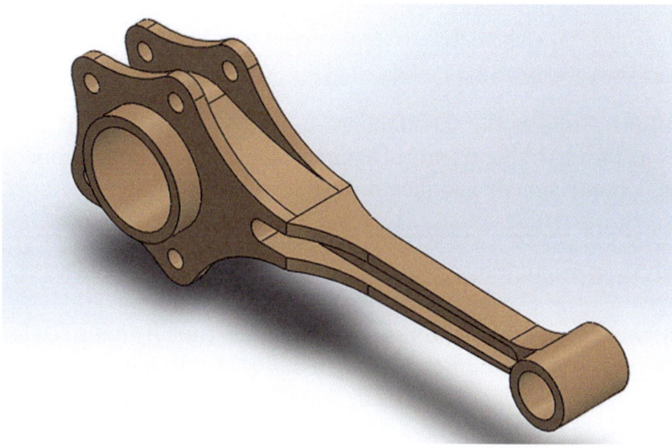

Fig. 1.33 Final design of the master rod

Bill of Materials (BOM): One can generate a Bill of Materials (BOM) from an assembly, which is a list that details all the components used in the assembly along with quantities, part numbers, and other information. BOMs are essential for manufacturing and documentation purposes. Souring can also be performed.

Visualization: Assemblies allow you to visualize how a product or mechanism will look when fully assembled. You can use various viewing tools to examine the assembly from different angles, sections, and perspectives. SolidWorks 3D Experience can also help in visualization.

Design Changes: When one makes changes to individual parts within an assembly, those changes can propagate through the assembly. This means that if one modifies one component, the assembly can update automatically to reflect the changes, which is crucial for maintaining design integrity.

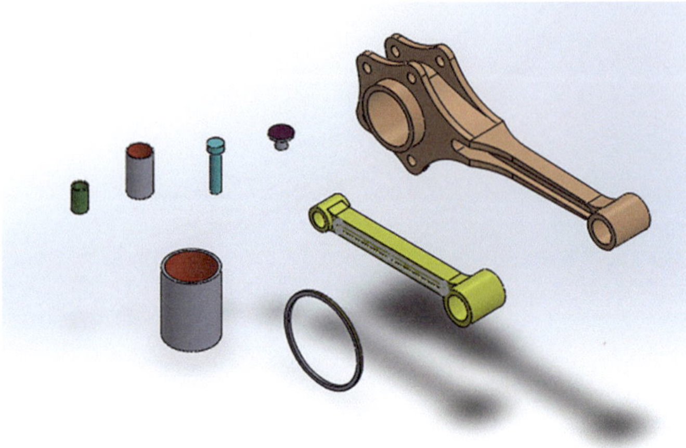

Fig. 1.34 Importing the components to the assembly module

Collaboration: SolidWorks assemblies facilitate collaboration among team members working on different parts of a project. Multiple users can work on different components simultaneously and then bring them together in the assembly.

Analysis: One can perform various analyses on assemblies, such as stress analysis, motion analysis, and interference detection, to evaluate the performance and behavior of the design.

CAD software such as SolidWorks provides a user-friendly interface for creating, editing, and managing assemblies, making it a powerful tool for designing complex mechanical systems, machines, products, and more. Assemblies are a core aspect of parametric modeling in SolidWorks, allowing engineers and designers to build and refine intricate designs efficiently.

Assembly of Radial Engine

For assembly, the first step is to import all the components to the assembly module as shown in Fig. 1.34.

For assembly these parts are mated, and different relationships are developed for each part with each other as shown in Fig. 1.35. Figure 1.36 illustrates the assembled radial engine while Fig. 1.37 exhibits the rendered image of the engine.

1.23 Learning Outcomes

The reader is expected to understand the difference between function (Individual Work) and Process (Team's Work). The reader should also have the capability to differentiate between modeling using mathematical expressions to a sketch with an

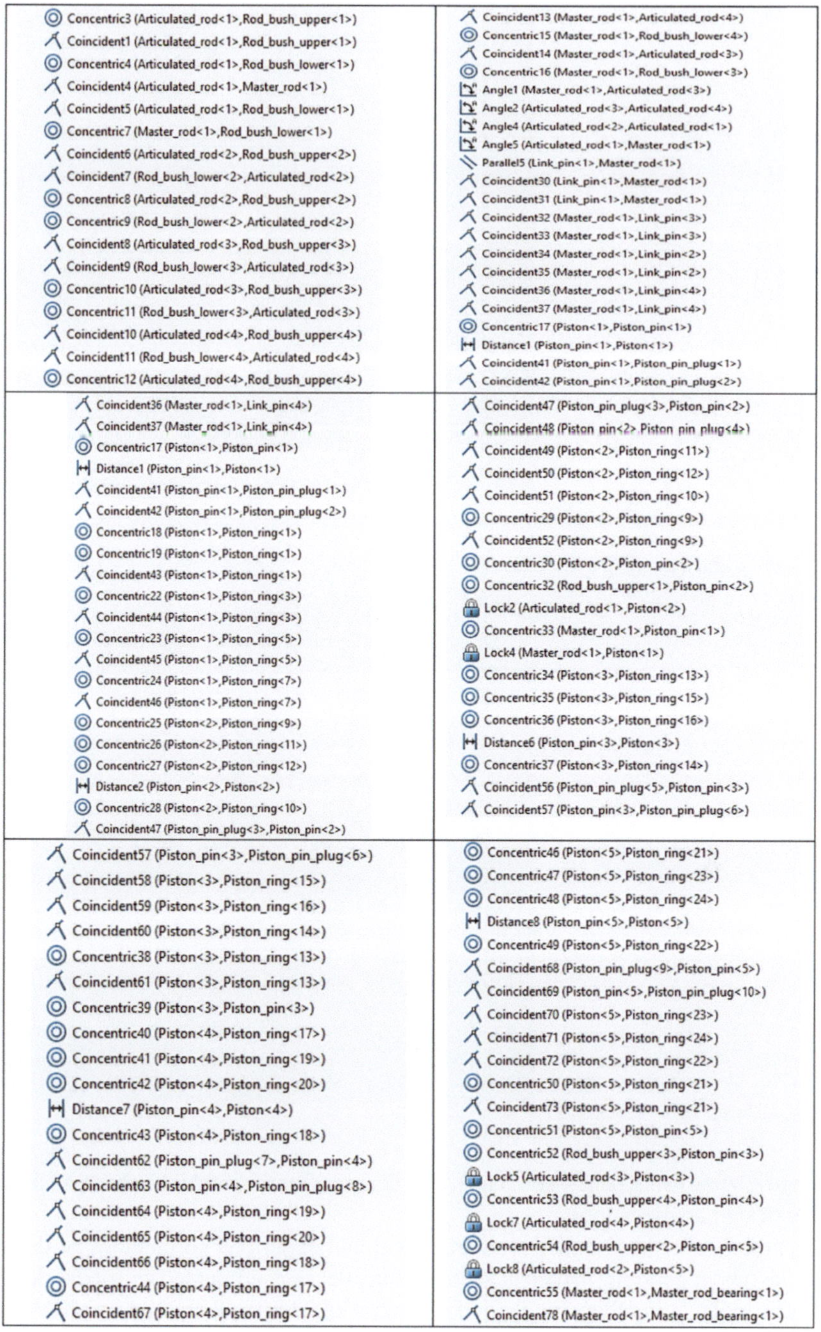

Fig. 1.35 Mates of different components

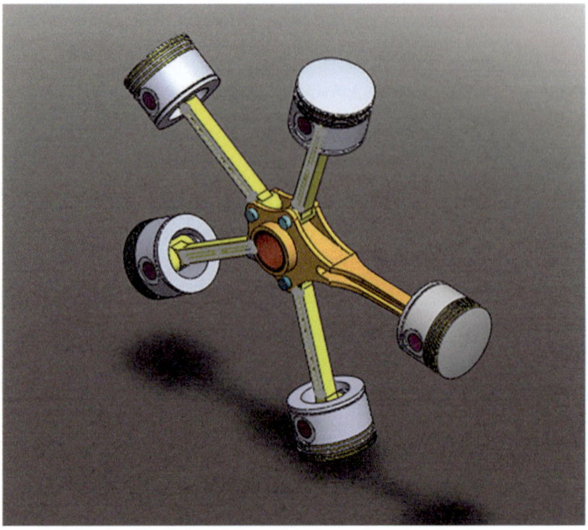

Fig. 1.36 Assembled radial engine

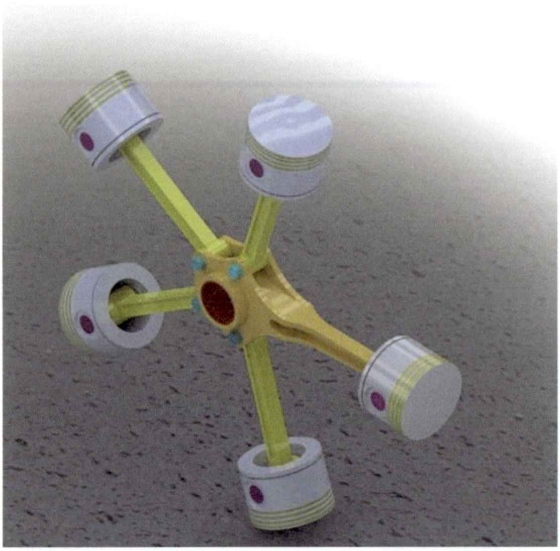

Fig. 1.37 Rendered radial engine

isometric view using pencil and paper, and design after the model has been analyzed using design analyses tools.

Learners must have understood analytical and experimental methods of design analysis, and if possible, in their academic setup, have experimented with these design analysis tools giving them hands-on knowledge of additional methods available for design analysis which are:

- Analytical Methods of Analysis
- Photoelasticity

- Brittle Coating
- Strain Gauges.

The reader is introduced to the complete design process using the Conventional Approach to Design, and it is the concept that the reader must remember to compare with System Approach to Computer-Aided Design described in Part II, Chap. 5 with having ability to differentiate Conventional CAD/CAM cycle to Modern CAD/CA Cycle. Life Cycle Assessment, Sourcing in design practices in the less developed part of the world and the modern world, and design documentation for manufacturing are introduced.

The case studies highlighted 'Concept Generation and Selection Methods' and 'Modeling and design of a Radial Engine using SolidWorks'. Major concepts presented in this chapter are the Conventional Approach to Design, DFM, DFA, Axiomatic Design, Concurrent Design, Ergonomics, RAMS, LCA and Product Life Cycle.

1.24 Review Questions

Q1. What is the importance of objects that can contribute significantly to the integration of ergonomic knowledge into design?
Q2. Define modeling and list three main modeling techniques.
Q3. Name at least five parameters required to fulfill *need analysis* for design.
Q4. Describe at least five of the concept generation methods.
Q5. Describe at least five of the concept selection methods.
Q6. How is RAMS practically assured in discrete products?
Q7. Identify the pillars of mechanical engineering design?
Q8. Describe the most common contemporary numerical analysis techniques.
Q9. What are the three main steps of the finite difference method (FDM)? Give at least three applications.
Q10. What are the three main steps of finite element analysis (FEA)? Give at least three applications.
Q11. What are the three main steps of boundary element analysis (BEA)? Give at least three applications.
Q12. What are the three main steps of finite volume analysis (FVA)? Give at least three applications.
Q13. Name and describe at least five of the macroscopic mechanical material characterization tests.
Q14. Name and describe at least two of the microscopic mechanical material characterization tests.
Q15. In the Industry 4.0 era, should we pursue an entirely experimental design technique? Why/why not?
Q16. List the three factors in embodiment design that designers should pay attention to.

Q17. Are today's engineering problems mostly optimization problems? If so, are they generally single- or multi-objective problems?
Q18. Explain the five principles of DFM with one application in mind.
Q19. Name at least three principles for DFA with one application in mind.
Q20. What is the concept of axiomatic design?
Q21. How can the disadvantages of sequential design and manufacturing be avoided?
Q22. Name at least 10 of the UNDP Sustainable Development Goals (SDGs) each with its impact on cost.
Q23. List the four phases of life cycle assessment (LCA) with one application in mind.
Q24. How should the relation between sourcing and engineering efforts be in Industry 4.0?
Q25. Why is design documentation needed?
Q26. Is it possible to generate computer 3D models from hand drawings automatically? If possible, what is the name of the technique?
Q27. What are the advantages associated with prototyping? What will be the carbon footprint if prototyping is done for every discrete product?
Q28. Name the six phases of the product engineering life cycle. Take a discrete product as an example.

Bibliography

1. Li X, Chen C-H, Zheng P, Jiang Z, Wang L (2021) A context-aware diversity-oriented knowledge recommendation approach for smart engineering solution design. Knowl-Based Syst 215:106739
2. Wang X, Liu A, Kara S (2022) Machine learning for engineering design toward smart customization: a systematic review. J Manuf Syst 65:391–405
3. Giordano V, Coli E, Martini A (2022) An open data repository for engineering design: using text mining with open government data. Comput Ind 142:103738
4. Zhu S, Wu Q, Jiang Y, Xing W (2021) A novel multi-objective group teaching optimization algorithm and its application to engineering design. Comput Ind Eng 155:107198
5. Band SS, Ardabili S, Danesh AS, Mansor Z, AlShourbaji I, Mosavi A (2022) Colonial competitive evolutionary Rao algorithm for optimal engineering design. Alex Eng J 61(12):11537–11563
6. Gupta SK, Chen Y, Feng S, Sriram R (2003) A system for generating process and material selection advice during embodiment design of mechanical components. J Manuf Syst 22(1):28–45
7. Broberg O, Andersen V, Seim R (2011) Participatory ergonomics in design processes: the role of boundary objects. Appl Ergon 42(3):464–472
8. Hall-Andersen LB, Broberg O (2014) Integrating ergonomics into engineering design: the role of objects. Appl Ergon 45(3):647–654
9. Pietrykowski K, Tulwin T (2014) Aircraft radial engine CFD cooling model. SAE Int J Engines 8(1):82–88
10. Han K et al (2020) Wind-driven radial-engine-shaped triboelectric nanogenerators for self-powered absorption and degradation of NO_X. ACS Nano 14(3):2751–2759

11. Camba JD, Company P, Naya F (2022) Sketch-based modeling in mechanical engineering design: current status and opportunities. Comput-Aided Des 150
12. Xiao J, Lou X, Shi B, Xiao J, Niu Y (2023) An improved hybrid membrane algorithm based on hormone modulation mechanism for engineering design problems. Expert Syst Appl 227
13. Putnik GD, Putnik Z, Pinheiro P, Alves C (2022) Engineering is design and only design—part I: the value of making a distinctive sign. Procedia CIRP 109:678–683
14. Hasby FM, Roller D (2016) Sharing of ideas in a collaborative CAD for conceptual embodiment design stage. Procedia CIRP 50:44–51
15. Wu D, Wang GG (2021) Causal artificial neural network and its applications in engineering design. Eng Appl Artif Intell 97
16. Wiese P, John P (2003) Engineering design in the multi-discipline era: a systems approach. Wiley
17. Kicinger R, Arciszewski T, de Jong K (2005) Evolutionary computation and structural design: a survey of the state-of-the-art. Comput Struct 83(23–24):1943–1978
18. Schmid SR, Kalpakjian S (2006) Manufacturing engineering and technology. Pearson Prentice Hall
19. Öztürk Y, Erdoğan HC, Esat V (2012) Finite element modelling and analysis of a frontal offset deformable crash test barrier design. In: Proceedings of UMTIK2012—15th international conference on machine design and production, Pamukkale, Denizli, Türkiye, 16 pp
20. Suh NP, Suh NP (2001) Axiomatic design: advances and applications, vol 4. Oxford University Press, New York
21. UNDP. The sustainable development goals (SDGs)

Chapter 2
Manufacturing and Production

Prelude

Discrete and Continuous Manufacturing

With reference to the definition of the system, manufacturing is done for products using discrete or continuous processes. The manufacturing performed with the discrete manufacturing processes produces products that are countable. There is a wide variety and number of such products that are produced all over the world. Some waste is associated with discrete manufacturing. The sustainability factor is an important element here and is explained as a case study: Product Life Cycle Code—PLCC in Part II, Chap. 5.

The products produced at continuous processes are powder, grain, or liquid and their byproducts. Such processes also produce products that are not recommended to be sensed by haptic senses; rather these give output that directly or indirectly provides comfort, leisure, and convenience like the high flux magnetism or thrust of a jet engine. These processes also fit well in the definition of a System, and it is just that the process may be thermodynamic, biological, nuclear processes, etc. Sustainability again plays an important role here, too.

In both cases, databases of knowledge acquired by human beings over centuries and standards defined by regional, national, international, and professional bodies are commonly used in the total engineering design and manufacturing of the product.

Tangible and Intangible

The tangible gain is the basic gain of ultimate historical possessions which are capital, potable water, and beautiful women. In the context of this book, the process of value added by manufacturing is something tangible.

Supplementary Information The online version contains supplementary material available at https://doi.org/10.1007/978-3-031-77106-4_2.

W. A. Khan et al., *Computer Aided Engineering Design and Manufacturing*, Springer Series in Advanced Manufacturing, https://doi.org/10.1007/978-3-031-77106-4_2

The services required to maintain ultimate historical possessions are all but intangible. In the context of this book manufacturing supplemented by all the services achieves the ultimate gain called production.

In the era of industrial revolutions, a character is added to tangible and intangible parts by the solemn force [if you believe in it], or by the sovereign force backed by armed muscle, or by the owner of the business who has a controlling share in the business that is the corporate authority. The corporate authority executes tangible functions and intangible services by powers vested in it and effectively uses different types of management hierarchies to get the job done with multiple trained human resources.

Demand, supply, and population economic models have a direct effect on the production of products that today are innovative, customized, multidisciplinary, and sustainable.

Manufacturing and Production

Tangible Manufacturing supplemented by Intangible Services is Production.

2.1 Objectives

For readers, this chapter relates to theoretical and hands-on knowledge of Manufacturing and Production. An attempt is made to differentiate between Manufacturing and Production, normally the Value Chain, in the production of Discrete Products and for products from continuous manufacturing.

An introduction to Manufacturing Processes and Manufacturing Systems will lay down the foundation of Manufacturing and Production including discrete and continuous production. The interface between design and production, the production planning and control is introduced and exemplified in Part IV, Chaps. 10 and 11. Production philosophies, a process that human beings reflect by comparing different scenarios once every function of an exercise is running smoothly, will also be introduced. Comments on sourcing in the less developed world compared to the modern world are highlighted.

2.2 Manufacturing Processes Classification

Manufacturing processes are physical mechanisms used to alter the shape and form of engineering materials (Metal, Ceramic, Polymer, Plastic, Rubber, and Polymer Matrix Composites). Manufacturing Processes. A general classification is as follows:

1. Forming Processes
2. Bulk Deformation Processes
3. Sheet Metal Forming Processes
4. Metal Removal Processes
5. Joining and Fastening Processes
6. Heat Treatment Processes
7. Surface Treatment Processes
8. Fabrication of Micro-electronics and Micro-mechanical Devices
9. Non-Conventional Machining
10. Additive Manufacturing Processes.

Components of general classification of manufacturing processes are listed/ illustrated in figures. Details of operation of these manufacturing processes can be found in any source of literature (Like the Library or e-Library). Please refer to the bibliography as well (Figs. 2.1, 2.2, 2.3, 2.4, 2.5, 2.6, 2.7, 2.8, 2.9, 2.10, 2.11, 2.12, 2.13 and 2.14).

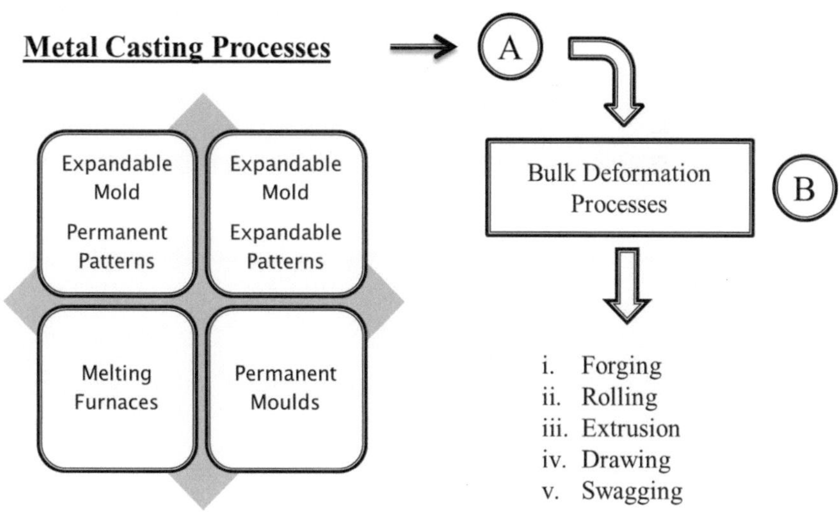

Fig. 2.1 A list of common casting processes[1]

[1] Reproduced with permission from Springer UK.

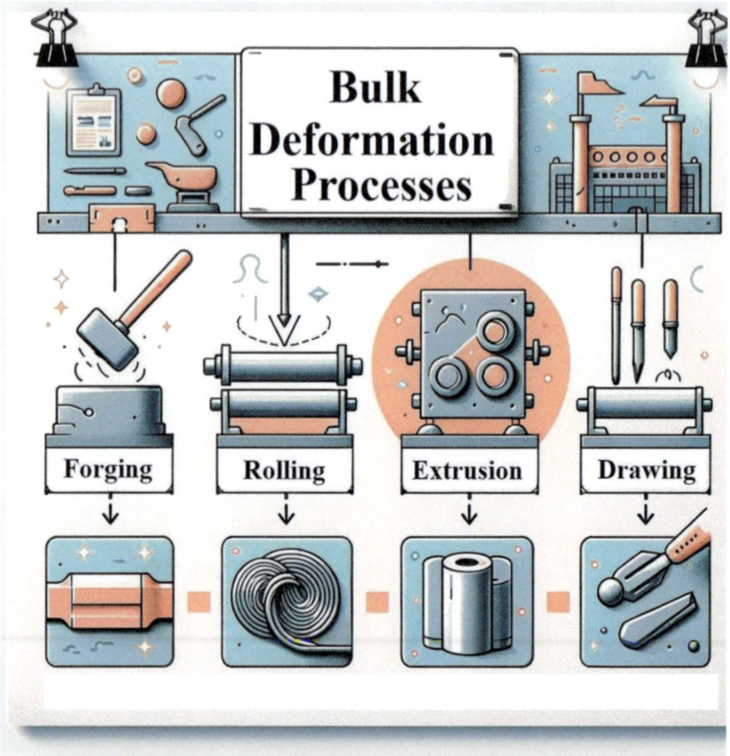

Fig. 2.2 Overview of bulk deformation processes: a visual guide

Humanity, in pursuit of continuous development has indulged and will keep on indulging in conceptualizing new requirements of materials and manufacturing processes, be it executed on earth, in space or at moon or at another planet for the sake of experimenting, perfecting and mass-producing new novel products.

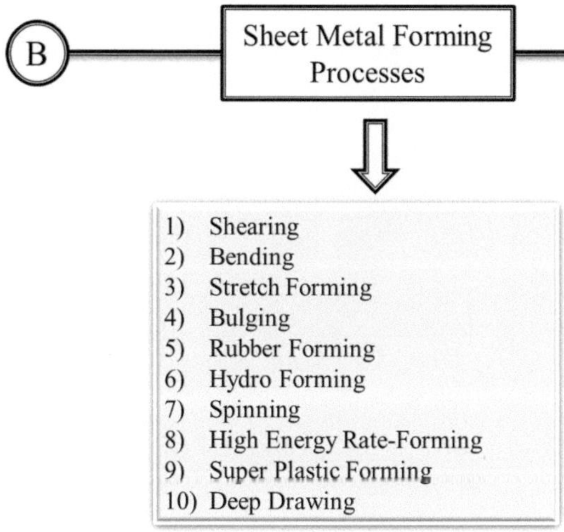

Fig. 2.3 A list of common sheet metal forming processes[2]

[2] Reproduced with permission from Springer UK.

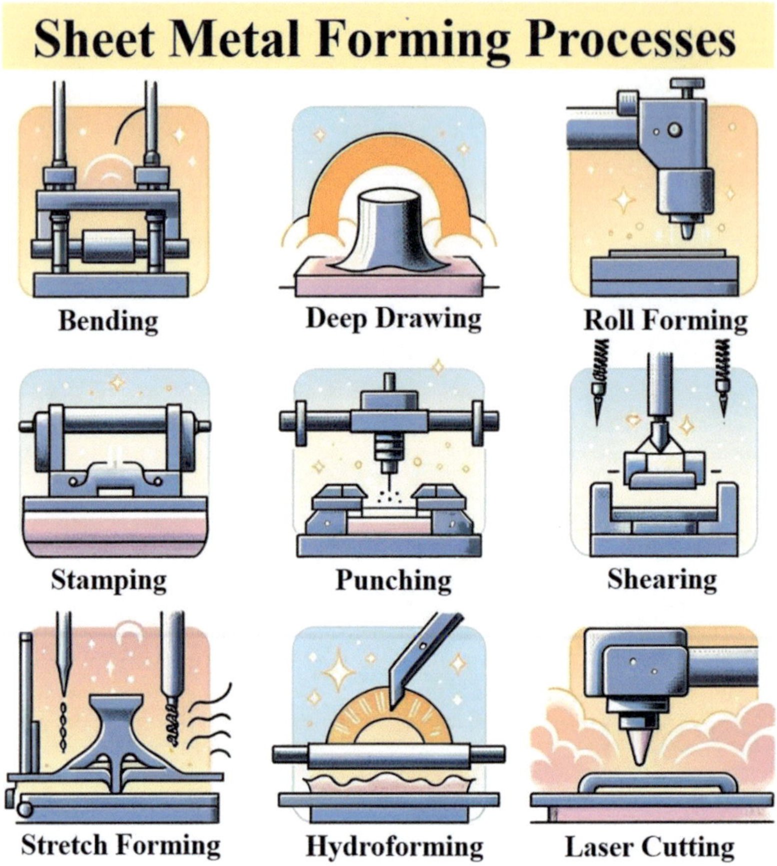

Fig. 2.4 Overview of sheet metal forming processes: a comprehensive figure

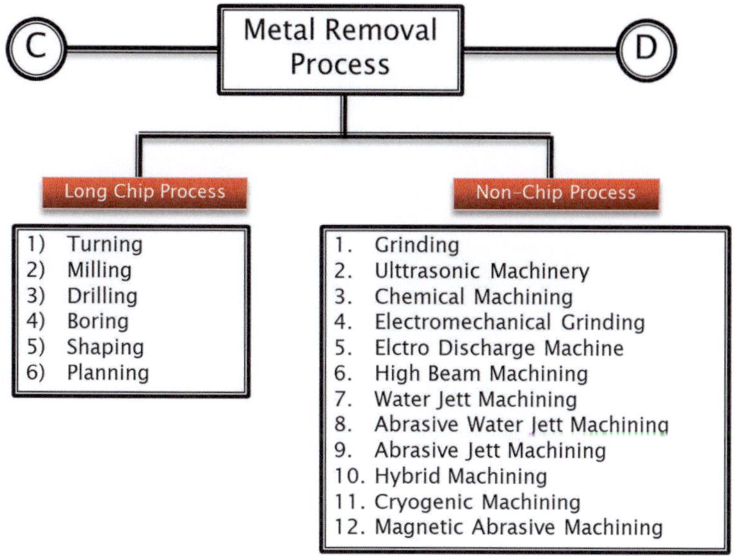

Fig. 2.5 A list of common metal removal processes[3]

[3] Reproduced with permission from Springer UK.

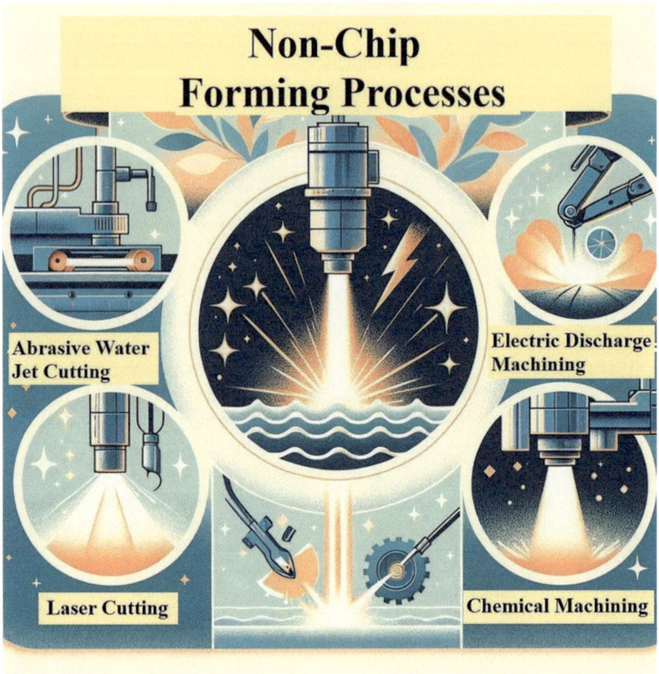

Fig. 2.6 Illustration of non-chip forming processes in manufacturing

Fig. 2.7 Illustration of chip-forming processes in metalworking

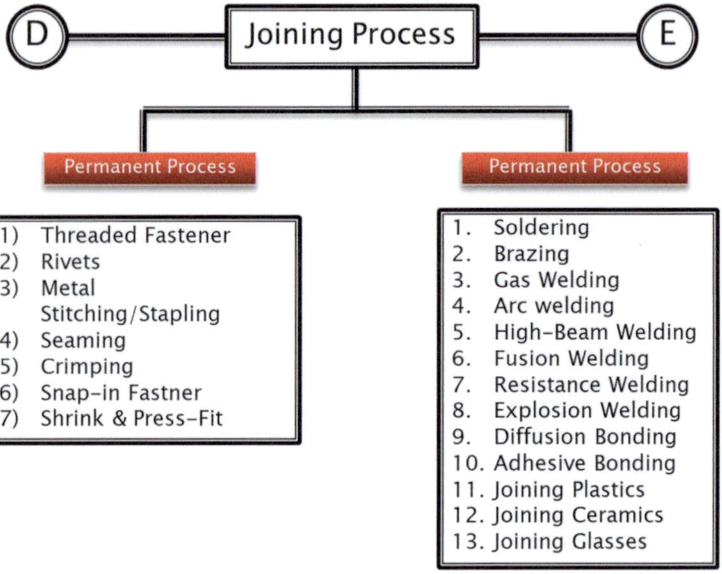

Fig. 2.8 A list of common joining processes[4]

[4] Reproduced with permission from Springer UK.

Fig. 2.9 Overview of key temporary joining methods

Fig. 2.10 Illustration of permanent joining techniques

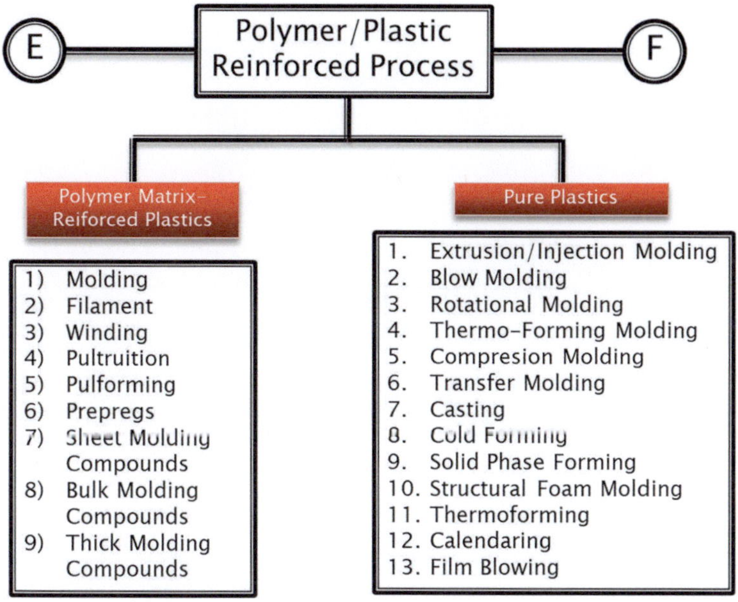

Fig. 2.11 A list of common polymer/reinforced plastics parts manufacturing processes[5]

[5] Reproduced with permission from Springer UK.

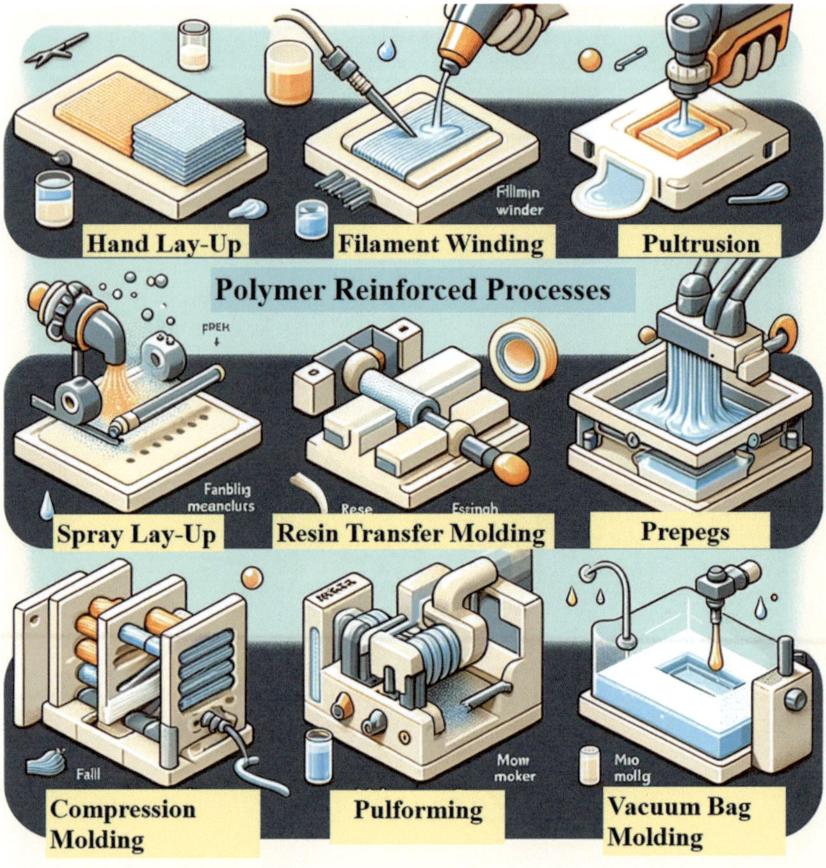

Fig. 2.12 Polymer reinforcement processes: a 3D overview of enhanced material strength and durability

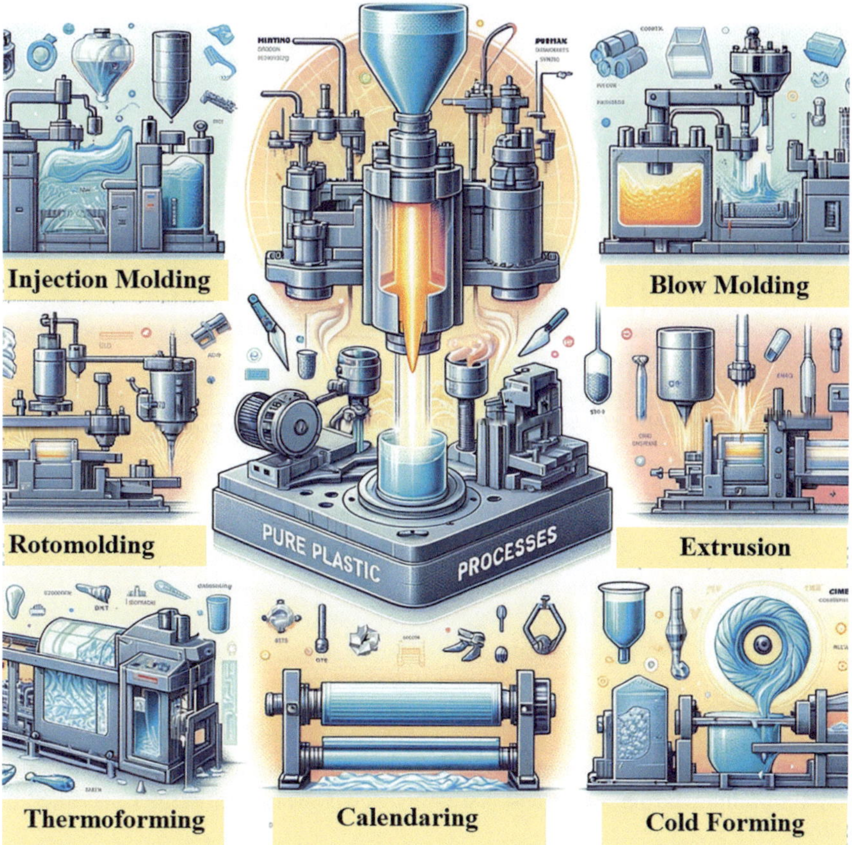

Fig. 2.13 The art of plastic manufacturing: a 3D insight into precision and performance

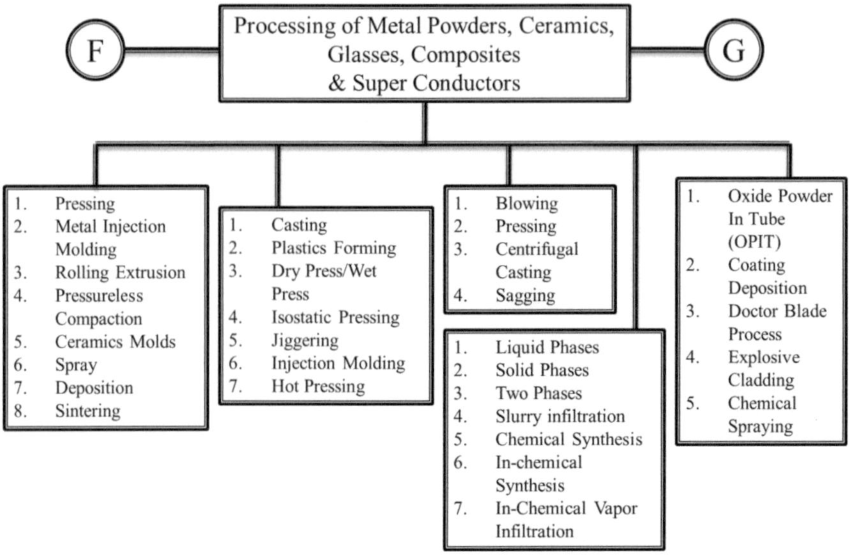

Fig. 2.14 General classification of manufacturing processes applied to processing of metal powders, ceramics, glasses, composites and superconductors[6]

2.3 Manufacturing Systems

Chryssolouris (2006) theoretically defines manufacturing systems as those which contain Manufacturing Areas and Assembly Areas. These are further classified as follows:

1. Job shop
2. Project shop
3. Cellular system
4. Flow line
5. Continuous Manufacturing System
6. Flexible Manufacturing System
7. Reconfigurable Manufacturing System
8. Virtual Manufacturing System
9. Assembly Systems.

All of these theoretical manufacturing areas' basic configuration is as follows while remembering that these manufacturing systems always require flow of information, material and energy as input and assembled/packed product as output therefore operate on system engineering principle where process may be embedded, cascaded or complex and vary in nature giving tangible output that provides intangible comfort with waste that must be controlled for the sake of sustainability (Figs. 2.15 and 2.16).

[6] Reproduced with permission from Springer UK.

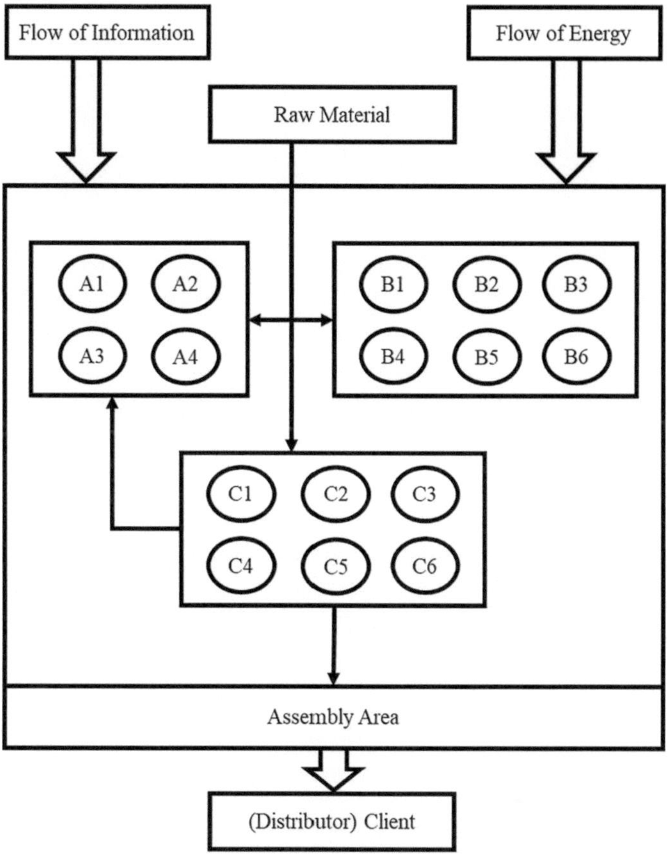

Fig. 2.15 Job shop[7]

Legend for a Job Shop

1. Highly flexible
2. Low Production Volume
3. Normally conventional machines setup (Exceptionally may be supported by CNC Machines)
4. Process plans which are slightly difficult
5. Simple manufacturing Instructions
6. Simple part Program (if any)
7. Varying configurations of parts are produced resulting in the collection of many Jigs, fixtures, and clamping devices at the job shop
8. Large variety of tools, tooltips, and measurement devices are collected over long periods of time

[7] Reproduced with permission from Springer UK.

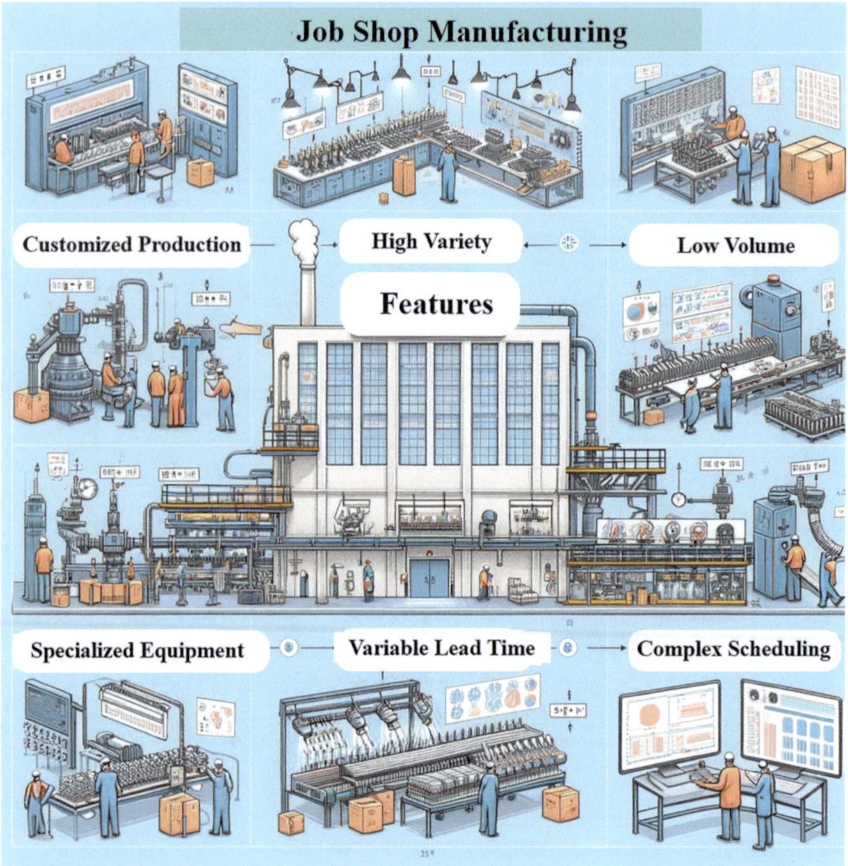

Fig. 2.16 Three-dimensional view of job shop manufacturing

9. Collection of manufacturing machinery at Job Shop is based on the nature of the main industry supported e.g., a Cement Plant, Oil Refinery, Educational Institution etc.
10. Small but versatile human resource
11. Automation is desired but seldom supported
12. Assembly may be complex
13. Operational Safety is primary
14. Quality of Product is paramount
15. Number of manufacturing processes are arranged as per the optimal production sequence of a product or range of products in a regular or irregular perimeter
16. Small product Inventory is maintained
17. A Small to Medium Enterprise (SME) attached with an Large Scale Enterprise (LSE)

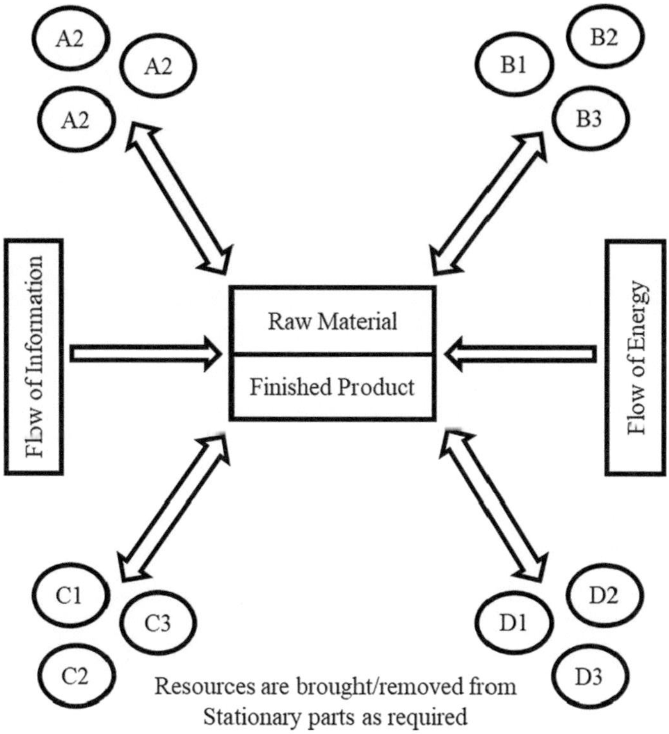

Fig. 2.17 Project shop[8]

18. Technical and compliance standards are applicable
19. High Potential for using Industry 4.0 and Industry 5.0 (Figs. 2.17 and 2.18).

Legend for a Project Shop

1. Large/Huge/normally Life Sensitive products such as airplanes, spacecraft, missiles, ships, submarines, Buildings, and Silos etc.
2. Low Volume Production with longer production times for discrete products.
3. Raw materials and Machines move to the product to build it.
4. Design and manufacturing steps simulated at each stage to maintain accuracy due to very high cost and human life sensitiveness during operation.
5. Empirical methods of testing are performed to ensure safety.
6. Complex Production Planning and Control.
7. Use of multiple production philosophies such as JIT, ERP, and Kaizen.
8. Complex testing of parts, structure, assembly, and mechanisms.
9. Complex testing of the assembled machine without humans and with humans.

[8] Reproduced with permission from Springer UK.

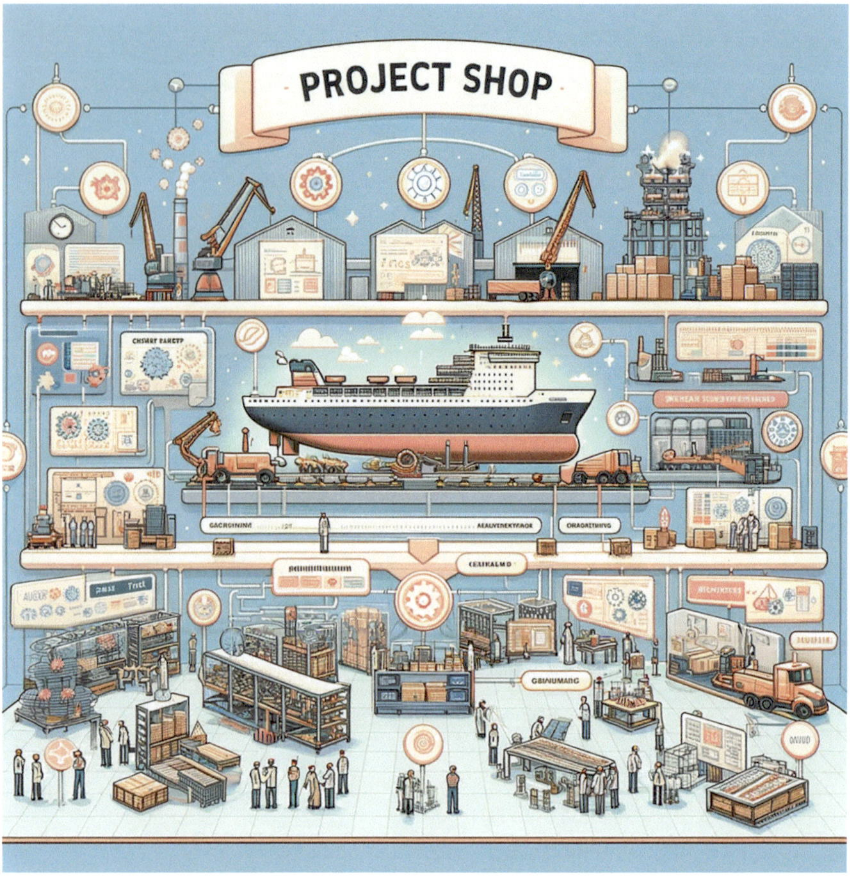

Fig. 2.18 Project shop manufacturing in 3D: crafting solutions for unique industrial challenges

10. The design, Manufacturing, and Operation history of the machine is maintained for its serviceable life.
11. Scheduled Maintenance.
12. Replacement of sensitive parts as recommended by the designer or by the operator whenever observed during testing and operation.
13. All staff involved including labor is highly technically trained.
14. Safety of Product Operation is primary.
15. Quality of Product is paramount.
16. Number of manufacturing processes are arranged as per optimal production sequence of a product or range of products in a regular or irregular perimeter.
17. A Large-Scale Enterprise (LSE) configuration with low production volume.
18. Sustainability requires Green Manufacturing. Remanufacturing towards the end of life of the manufacturing facility should be an option.
19. Automation requires Cloud Manufacturing.

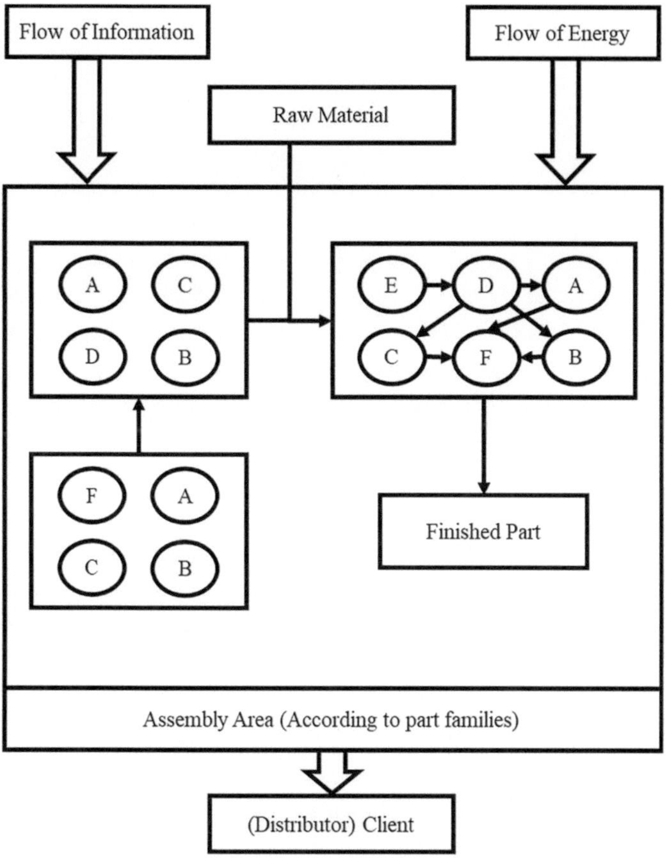

Fig. 2.19 Cellular (batch) manufacturing[9]

20. Multiple production philosophies may be applicable.
21. Huge inventory of products may be maintained in case of products including powder, grains and liquids.
22. Highly Inflexible.
23. Products are innovative, customized, and multidisciplinary.
24. Technical and compliance standards are applicable.
25. High Potential for using Industry 4.0 and Industry 5.0 (Figs. 2.19 and 2.20).

Legend for Cellular (Batch) Manufacturing

1. Manufacturing of Products is based on similarity in shape and size.
2. Coding and classification methods such as OPTIZ and DCLASS may be used.
3. Manufacturing Philosophy, ‘Group Technology’, has its applications.
4. Small to medium batches are produced.

[9] Reproduced with permission from Springer UK.

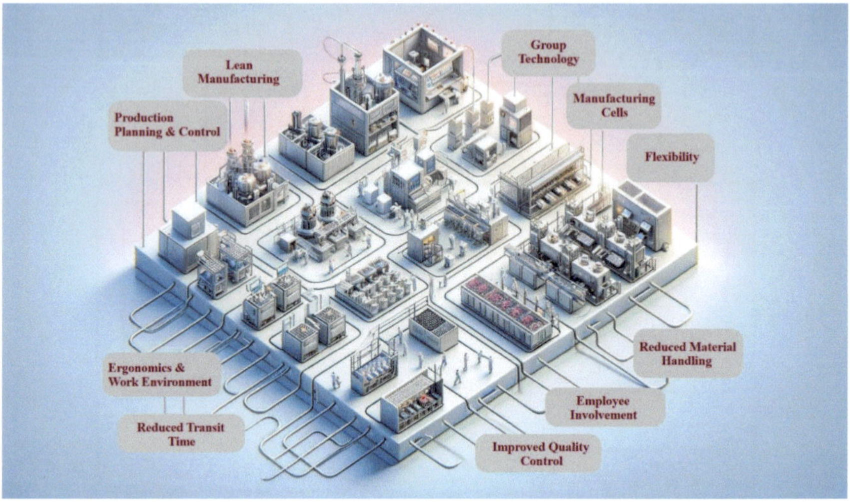

Fig. 2.20 A detailed three-dimensional model of the cellular manufacturing system

5. Total production volume, including seasonal demand, may be high.
6. Automation has ample use.
7. Production Planning and Control for similar batches is the same. Due to the use of Group Technology large databases are maintained for possible future referral. The Production Planning and Control (PPC) department should have vitality.
8. Maintenance schedules can be comparatively easily planned as seasonal demand permits space for maintenance.
9. Inventory buildup with respect to demand needs to be planned.
10. Frequent model changes thus keeping the design office forward-looking.
11. Product Life Cycle Assessment has high importance due to the shorter useful life of the product.
12. Remanufacturing is important to decrease waste. Safety of Operation is primary.
13. Quality of Product is paramount.
14. Number of manufacturing processes are arranged as per the optimal production sequence of a product or range of products in a regular or irregular perimeter.
15. A configuration that is used in both SME and LSE.
16. Comparatively, more flexible than Project Shop.
17. Sustainability requires Green Manufacturing. Remanufacturing leads to minimal waste.
18. Automation through cloud manufacturing is a viable option.
19. Products may be innovative, customized, sustainable, and multidisciplinary.
20. Technical and compliance standards are applicable.
21. High Potential for using Industry 4.0 and Industry 5.0 (Figs. 2.21 and 2.22).

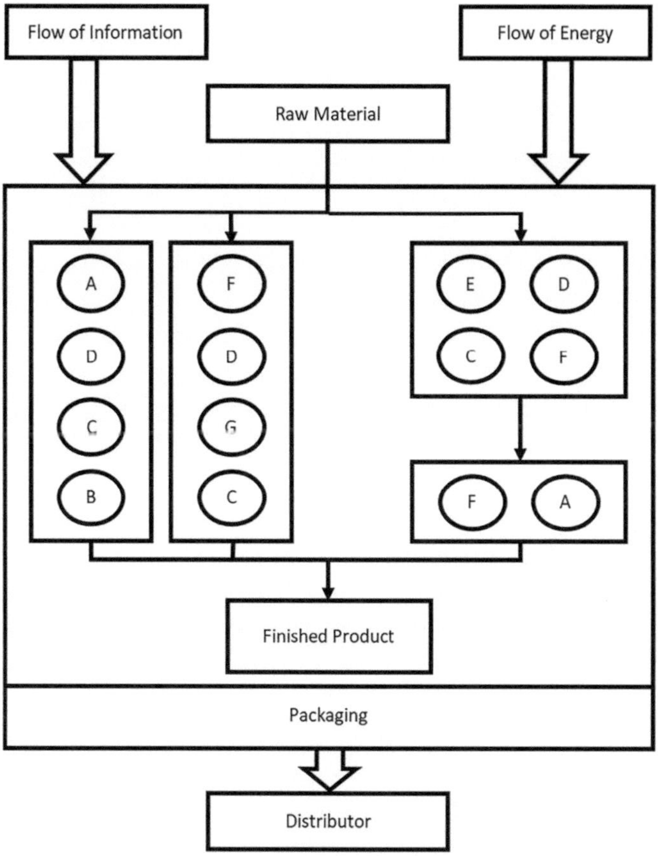

Fig. 2.21 Flow line[10]

Legend for a Flow Line

1. Historical transition from cottage industry to factory production.
2. One of the early design and operation is attributed to Henry Ford in 1904 to produce automobiles in large volumes.
3. High production volume.
4. Inflexible.
5. Today has a very high level of automation as per existing Technology Readiness Level.
6. Multiple production philosophies are used.
7. Innovative, multidisciplinary, and customized products are possible.
8. Cellular monitoring of the vehicle by manufacturer.
9. Cellular monitoring of the vehicle for supply chain purposes.

[10] Reproduced with permission from Springer UK.

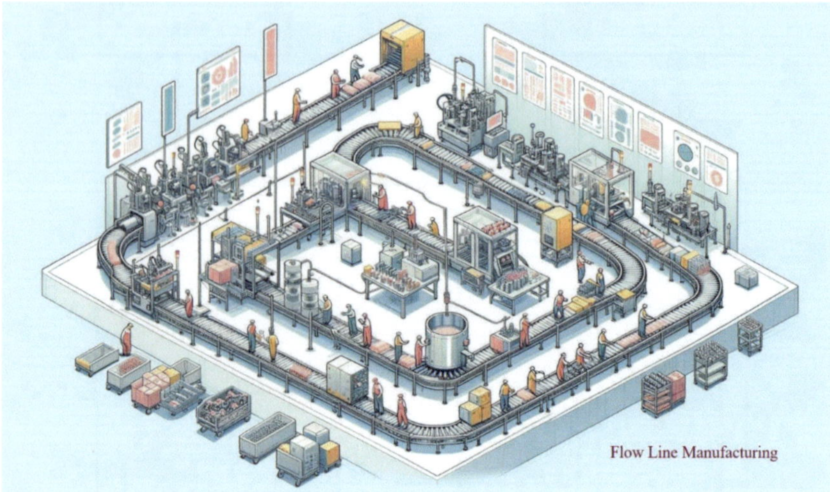

Fig. 2.22 Exploring the flow line manufacturing system: a 3D perspective

10. Electrical batteries are becoming common though the lifecycle problem of the batteries is not resolved.
11. New battery charging points are being introduced widely.
12. Vehicles have a very high second-hand/renovated resale market.
13. Manufacturing Engineering Services and Product Life Management is well defined by most international manufacturers.
14. General slack in the market due to economic conditions prevailing after Russia–Ukraine War and unstable Middle East.
15. Smart manufacturing practices are being transferred from the automobile industry to other small and medium enterprises in a similar way to what happened after the cold war era.
16. Safety of operation is primary. Self-driving vehicles will become common.
17. Quality of product is paramount.
18. Number of manufacturing processes are arranged as per the optimal production sequence of a product or range of products in a regular or irregular perimeter.
19. Products are innovative, customized, sustainable, and multidisciplinary.
20. Technical and compliance standards are applicable.
21. High Potential for using Industry 4.0 and Industry 5.0.
22. Large Scale Enterprise (LSE) configuration (Figs. 2.23 and 2.24).

Legend for a Continuous Manufacturing Setup

1. Used for high volume production of liquids, powders and grains like in oil refinery, Cement plant, and fertilizer factory.
2. Highly inflexible.
3. Requires startup and shutdown procedure spanning over days.
4. Safety precautions are of prime importance.

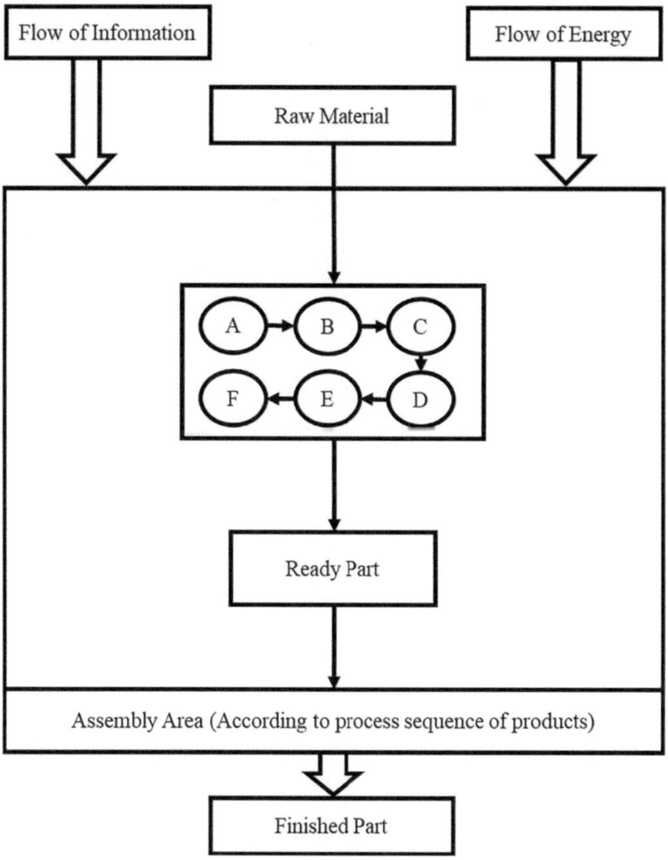

Fig. 2.23 Continuous manufacturing[11]

5. Quality of Product is paramount.
6. Number of manufacturing processes are arranged as per optimal production sequence of a product or range of products in a regular or irregular perimeter.
7. High Level of automation desired.
8. Intricate Production Planning and Control.
9. Maintenance Schedules are intricate
10. Some products are seasonal while others are produced round the year.
11. Production Planning and Control is complex.
12. Huge inventory of finished product.
13. Import and export sensitive raw material/product.
14. Byproducts are produced.
15. High Potential for use of Industry 4.0 and Industry 5.0.
16. Green Manufacturing is essential for sustainability.

[11] Reproduced with permission from Springer UK.

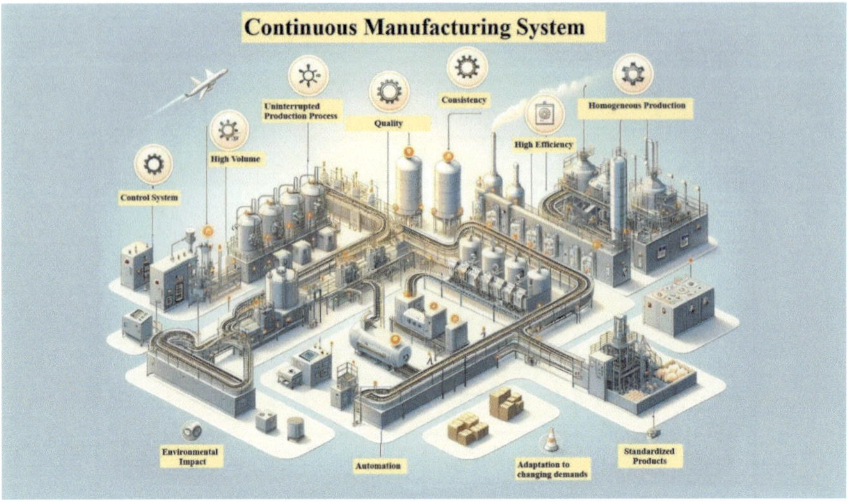

Fig. 2.24 Seamless production: the three-dimensional view of continuous manufacturing system

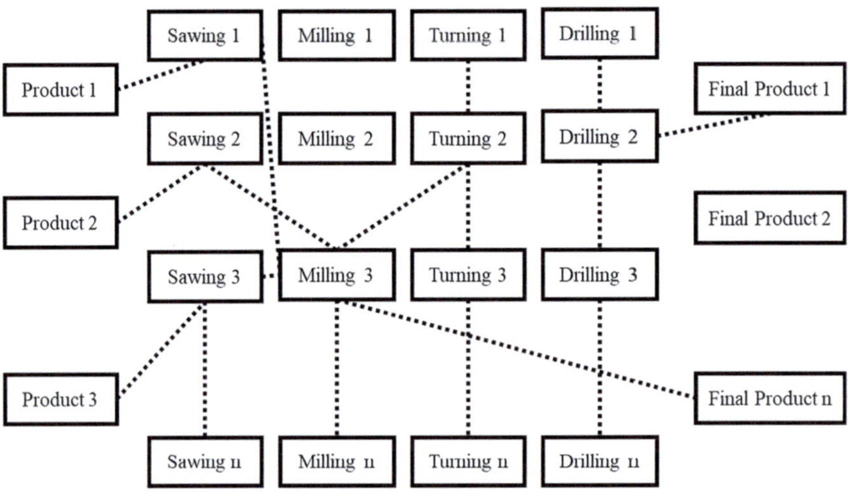

Fig. 2.25 Reconfigurable manufacturing system[12]

17. Technical and compliance standards are applicable.
18. A Large-Scale Enterprise (LSE) configuration (Figs. 2.25 and 2.26).

Legend for a Reconfigurable Manufacturing System

1. Akin to Job Shop
2. Order Intensive

[12] Reproduced with permission from Springer UK.

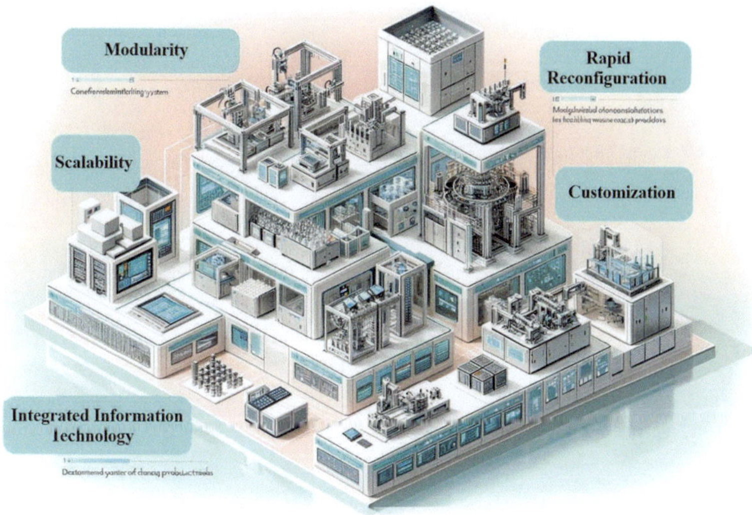

Fig. 2.26 A three-dimensional glimpse into the reconfigurable manufacturing system

3. Smaller Batches
4. Setup Changes may be required
5. Complex Production Planning and Control
6. Complex Maintenance Schedule
7. Mix of Conventional and Computer Controlled Manufacturing Processes
8. Quality intensive
9. Innovative, customized, and multidisciplinary products
10. Shorter Lead Time
11. Sizeable inventory of consumable and non-consumable production support accessories
12. Small in process inventory and finished product inventory
13. Theory of Constraints production philosophy has application while other production philosophies like lean manufacturing and Kaizen are also applicable
14. High Potential for Application of Industry 4.0 and Industry 5.0
15. Sustainability in all respect is essential
16. Technical and compliance standards are applicable
17. A Small to Medium Enterprise (SME) configuration (Figs. 2.27 and 2.28).

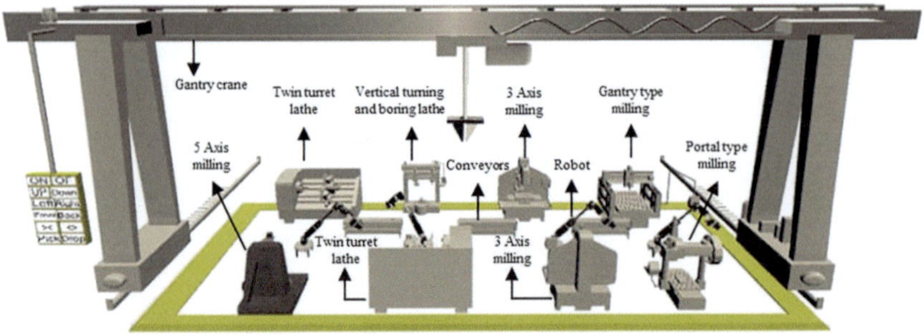

Fig. 2.27 Flexible manufacturing system[13]

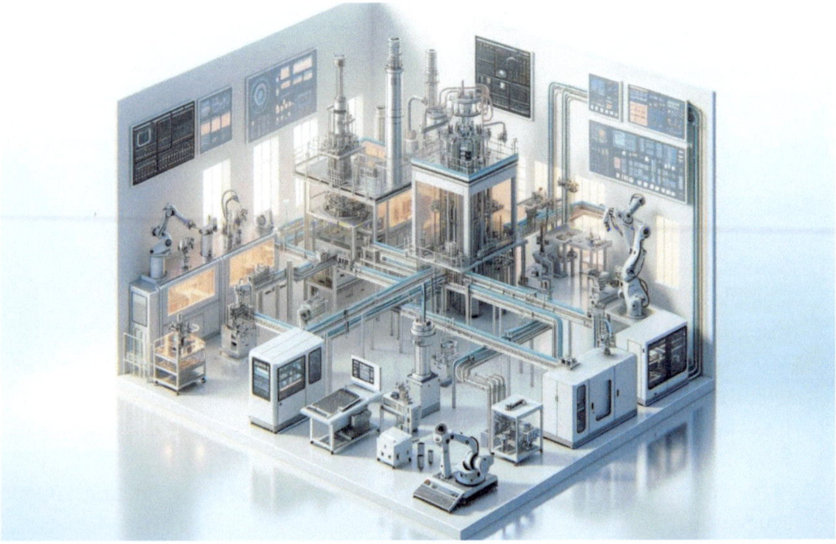

Fig. 2.28 Visualizing a flexible manufacturing system in three dimensions

Legend for a Flexible Manufacturing System

1. Flexible Discrete Manufacturing
2. Very high level of automation
3. Higher production rate
4. Smaller Batches
5. High Quality Products
6. SME Configuration
7. Technical and compliance standards are applicable
8. Industry 4.0 and Industry 5.0 implementations are comparatively easier to apply

[13] Reproduced with permission from Springer UK.

9. Complex Production Planning and Control
10. Small in process and finished product inventory as the lead times are small due to automation
11. Expensive in building and operating
12. Highly skilled manpower required to operate the FMS
13. High level of horizontal and vertical integration within the SME
14. Innovative, customized and multidisciplinary products
15. E-Commerce from shop floor possible.

Virtual Manufacturing System

URL-1: Virtual Manufacturing System[14]:

https://drive.google.com/file/d/1JO3ePKl1qpQcUcDmU0T_284e_MEOe5OV/view?usp=sharing

Legend for Virtual Manufacturing System

1. Virtual replica of physical Manufacturing System
2. Integration of Design Office, Finance, Accounts, Administration, Security and outside microdomains such as local government, Judiciary, Banks and DFIs etc. Possible
3. Offline Planning for production parameters of an order
4. Real time monitoring of manufacturing System
5. Management control of lowest to topmost tier of the manufacturing organization (Manufacturing ERP—mERP)
6. Maintenance of physical gadgets is easy to implement
7. Real time availability of production statistics
8. Training of relevant machine, process, cell, or whole manufacturing system without disturbing actual production or causing any physical damage
9. Technical and Compliance standards easily implementable
10. Applicable to discrete and continuous manufacturing
11. Applicable to Small to Medium Enterprises (SMEs)s and Large-Scale Enterprises (LSEs)
12. Controlling the manufacturing system remotely in case of pandemics and colonies at the Moon and Mars (Figs. 2.29, 2.30, 2.31, 2.32 and 2.33).

The second part of the manufacturing systems comprises the assembly area. Chryssolouris suggests all four possible scenarios as diagrammatically provided above.

---

[14] Protected by Pakistan patent number 37/2022 Acceptance Number 144217 and US patent applications number 1786442 (Pending); Logic Design by Wasim Ahmed Khan. Programmed by Aqeel Ahmad. Reproduced with permission.

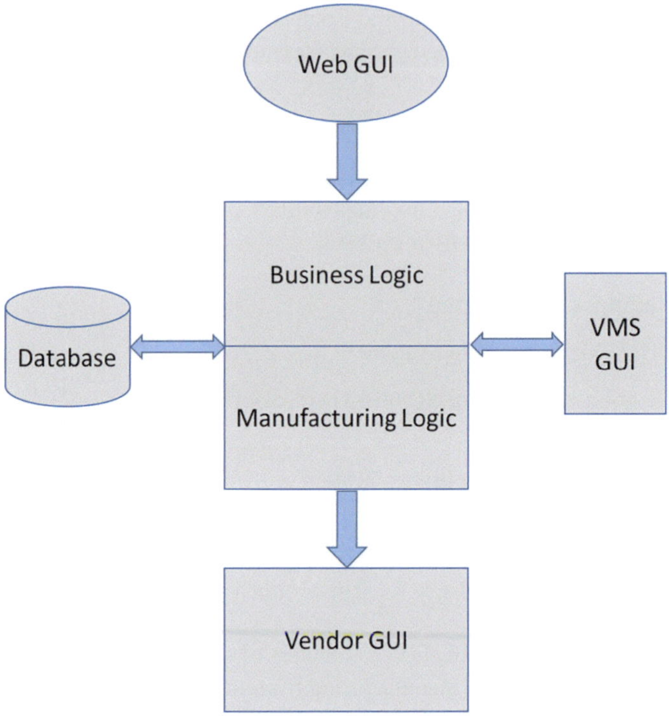

Fig. 2.29 Overall virtual manufacturing system description[15]

2.4 Production Planning and Control

Production Planning and Control are mainly intangible services that augment manufacturing to produce a product or range of products. A detailed description of production planning and control is provided in Part IV, Chaps. 10 and 11.

2.5 Production Documentation

Production documentation is the tangible medium through which the intangible instructions are passed on to the production floor to produce a product or a range of products. It includes production drawings, process plans, part programs, requisition forms for raw material, tools, and measurement gadgets; short-term and master schedules, maintenance schedules; time motion study for each job, quality control and quality assurance procedures, assembly and packaging instructions as well as

[15] Reproduced with permission from Springer UK.

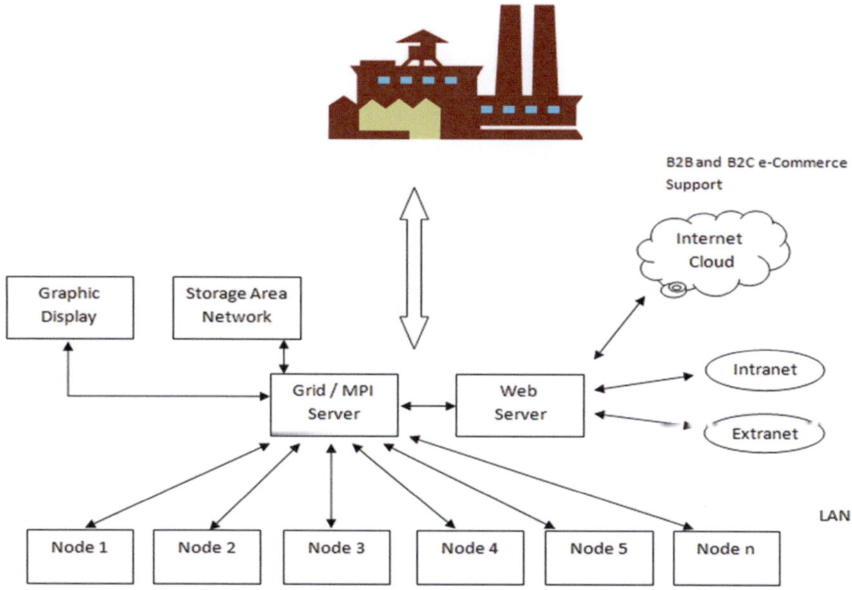

Fig. 2.30 Hardware components of virtual manufacturing system[16]

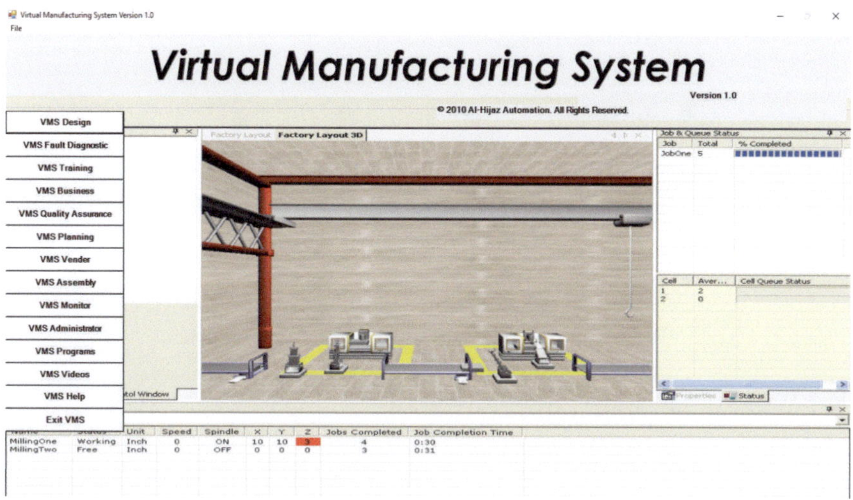

Fig. 2.31 mERP User Interface (version 1.0)[17]

transportation instruction; general and spatial safety instruction for producing, assembling and handling part/s; information related to various forms of inventories, and;

[16] Reproduced with permission from Springer UK.

[17] Reproduced with permission from Springer UK.

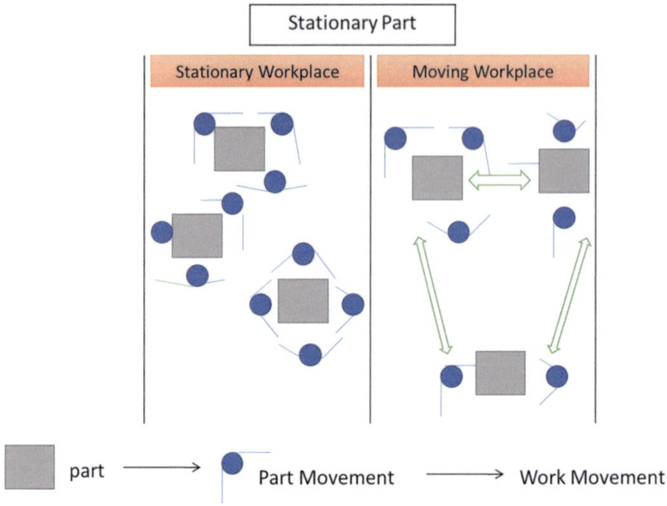

Fig. 2.32 Typical assembly procedures for stationary parts[18]

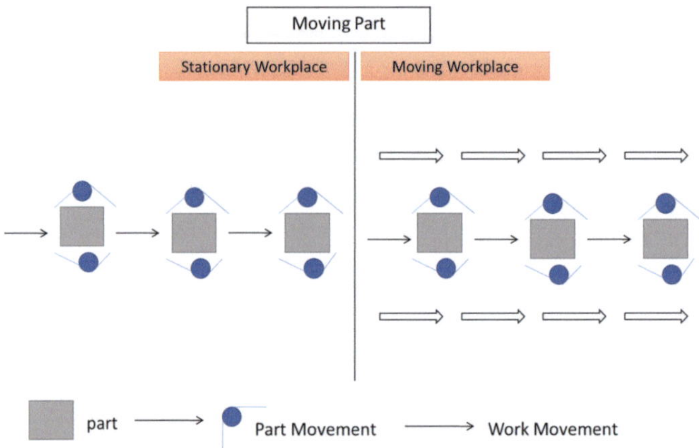

Fig. 2.33 Typical assembly procedures for moving parts[19]

costing information are delivered to a worker electronically or on a piece of paper. The worker provides feedback information related to each action to the appropriate supervisory cadre. Self-explanatory production drawings of some parts of 5-axis milling machine are reproduced (Figs. 2.34 and 2.35).[20]

[18] Reproduced with permission from Springer UK.

[19] Reproduced with permission from Springer UK.

[20] Five Axis Milling Machine Parts. Courtesy of Edutech, Lahore, Pakistan. Reproduced with Permission.

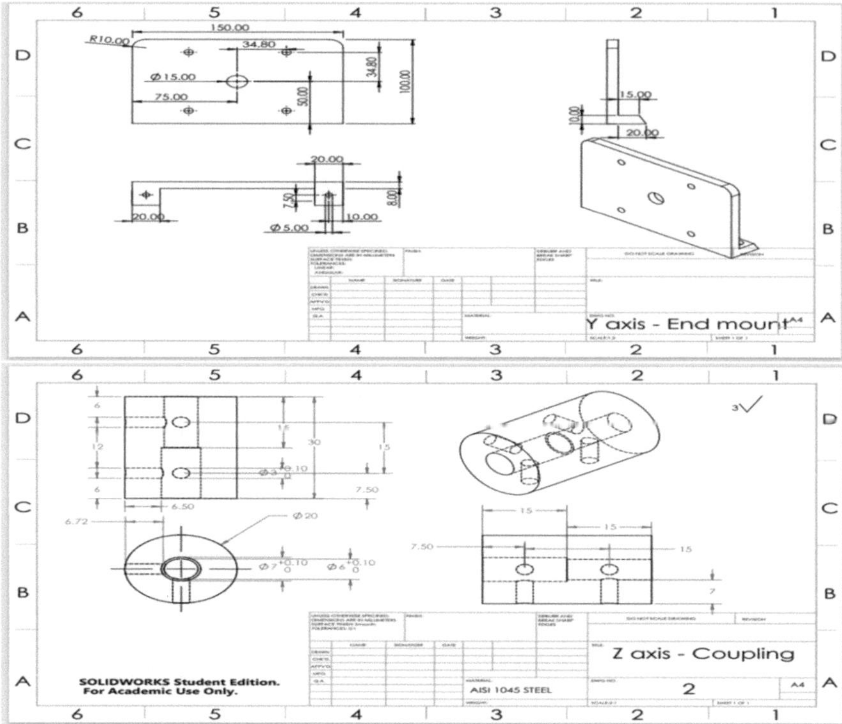

Fig. 2.34 Production drawings for 5-axis CNC machine

The Design to Production Interface encompasses Product Life Cycle Assessment and Manufacturing Execution Services which are cradle to the after sales service of a product and actions for value addition respectively. Product Life Management starts from the need in design function to once the product has been floated in the market. Complete Production documentation is as enlisted below and is provided to the appropriate cadre on paper or electronically:

1. System Engineering Design
2. Process Plan
3. Production Plan

    3.1 Job Schedule
    3.2 Maintenance Schedule

4. Raw Material Inventory
5. Tool Inventory
6. Measurement Gadgets Inventory
7. Jigs and Fixture Inventory
8. Coolant Inventory
9. Lubricants Inventory

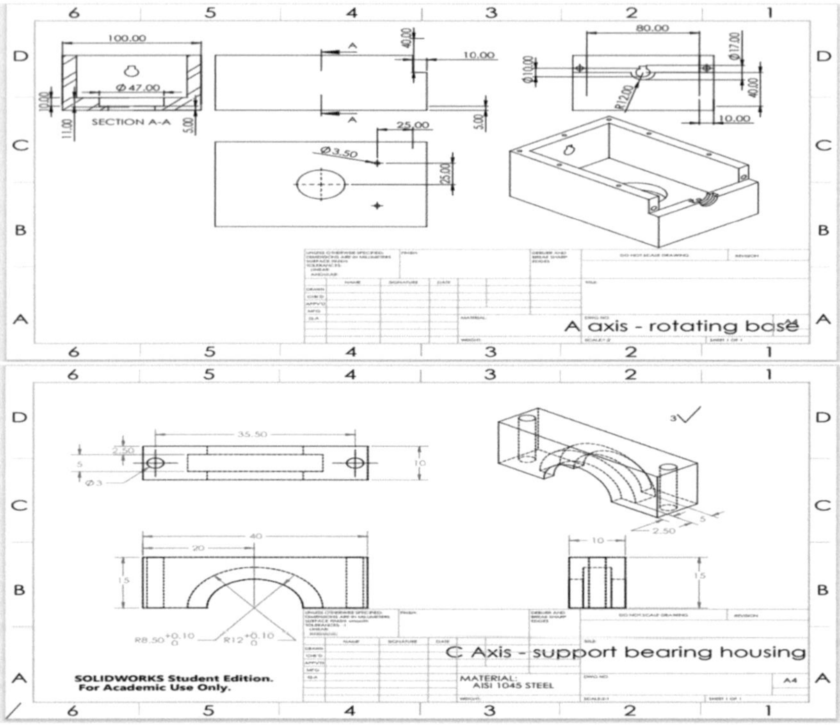

Fig. 2.35 Production drawings of a 5-axis milling machine[21]

10. In Process Inventory
11. Finished Product Inventory
12. Transit Inventory
13. Requisition Forms
14. IIoT for electronic transmission with Computerized Manufacturing/Assembly Process or Manual Data Input (MDI) terminal for Conventional Manufacturing/ Assembly Processes (Known as 'Horizontal and Vertical Integration' in Industry 4.0)
15. Packaging Instructions
16. Delivery Instructions for delivery to Distributor (includes Business to Business e-Commerce) and Delivery Instructions for delivery to Client (Includes Business to Client e-Commerce).

[21] Reproduced with permission from EduTech Lahore, Pakistan.

2.6 Sourcing[22]—A Detailed Description of Sourcing for Engineering Design

Sourcing is an important element of the Engineering Design and Manufacturing cycle. Consumables and non-consumables are required at competitive rates and prescribed time to meet the strict Lead Time requirements by the clients.

Manual and computer aided sourcing schemes are implemented as per the practices in an organization. Both Manual and Computer Aided Sourcing procedures are described below with the help of examples. Manual Sourcing is described first that uses lengthy clerical work and is very intricate even if the prescribed rules are followed. In the following manual scheme implementation of Khyber Pakhtunkhwa Public Procurement Regulatory Authority's (KPPPRA) procedures are followed:

GHULAM ISHAQ KHAN INSTITUTE OF ENG. SCIENCES & TECHNOLOGY

Indent No. ------------------------ Dated: ------------------------

Name of Faculty/Department/Section: --

It is confirmed that specifications, quantities etc. have been checked and found in order. The indent has been completely filled in all respects (including Annex-A, B and C) and recommended for further processing.

Signature of Indenter: ------------------------

Name of Indenter : ------------------------

Designation ------------------------

Countersigned by: Dean / Head of Department

Signature ------------------------

Name ------------------------

Designation ------------------------

[22] An example of manual sourcing for consumable and nonconsumable items as per rules of Khyber Pakhtunkhwa Public Procurement Regulatory Authority, Pakistan; is provided as a case study. The organization taken as an example is GIK Institute of Engineering Sciences and Technology. Only forms are reproduced in logical sequence and no consumable or nonconsumable item or their cost is mentioned thus no financial comparative statement is available. Technical evaluation for development of electromechanical hardware is provided.

Some of the sources of Software Packages Capable of Performing Computer Aided Sourcing are listed afterwards.

Director (Procurement)

Please note if any information is missing inform FE within two working days.

INDENT FORM ANNEXURE "A"

GHULAM ISHAQ KHAN (GIK) INSTITUTE OF ENG. SCIENCES AND TECHNOLOGY

Indent No. -------- **Dated:** ------------------------

Name of Faculty/Department/Section --

1. Description / Specification of required stores (in case of more than one items list may be attached.

2 Quantity Required

----

3. Specific Justifications for purchase of stores (attach a self contained note duly approved by the Dean / H. O. Dept. ---------------------------------------

4. Whether to be Imported / Purchased locally ---------------------------------------

5. Estimated cost in F.E. / Local Currency ---------------------------------------

6. Mode of Purchase i.e. Open/Limited/Single tender/Proprietary basis. ---------------------------------------

   If stores are to be purchased on single quotation basis then name of manufacturer and necessary justifications must be provided as to why competitive tender should not be issued to other suppliers. If stores are to be purchased on proprietary basis justifications should be given (in a self-contained note duly signed by the Head of Dept. for selecting particular Proprietary brand and also certify that items of desired specifications/ quality being procured is/are *Proprietary* and is/are manufactured solely by (Name of the Firms) and is/are not available from any other source.

7. Suggested mode of shipment along with justification if recommended by Air.

8. Last supply order No. / Name of Manufacturer/Supplier and rate at which purchased (if applicable)

INDENTER	9. Certified that Stock position shown in Annex-"B" is correct.
Signature:	Signature
Name of Indenter	Name of Dean/Head of Department
Designation & Date	Designation & Date

10. Report of Procurement Department:

11. Certified that sufficient funds are available under the Head in the Budget Estimates for the Year .... to cover the expenditure in Para-5 above for current proposal.

Budgetary Head	Year	Allocation	Expenditure

Pipeline	Balance	Current Proposal

Signature ------------------------------

Name

Head of Finance Department

Dated:

12 Above stated position is confirmed and approved.

Name & Signature of Competent Authority

Dated:

(Delete which is not applicable)

13

GHULAM ISHAQ KHAN (GIK) INSTITUTE OF ENGINEERING SCIENCES AND TECHNOLOGY

Last purchase and existing Stock Position of Machines, Spare Parts, and Consumable as recommended for purchase.

Date	Indent No.	Faculty / Department/Section:

S. #.	Description Store Items with Specification	QuantityRequired	A/C Unit	Stock Available in Main Store	Date Of Last Purchased	Last Purchased Cost in Rs. (Per Unit)	Consumed During Last 02 Year	Proposed Unit Cost	Percentage Increase Over Last Price	Remarks
1										
2										
1										
2										
3										
4										
5										
6										
7										
8										
9										

MAIN STORE	INDENTER	COUNTERSIGNED BY DEAN / HEAD OF DEPARTMENT
Signature	Signature	Signature
Name	Name:	Name
Designation	Designation	Designation
Dated	Dated	Dated

Annex-C

GHULAM ISHAQ KHAN (GIK) INSTITUTE OF ENGINEERING SCIENCES AND TECHNOLOGY

Name of Faculty / Department: ....

Title of the Store/Equipment demand ____________________________________

____

Whether store / equipment is being demanded as replacement or as addition to the existing one or as a new addition. Please describe the purpose in detail:

i. **If as replacement, give details of the existing equipment; when purchased; at what cost and its uses. Stipulated life? Can it not be repaired? Was it demanded in the Budget Proposal of the Faculty/Department?**

ii. **If it is an addition to the existing equipment, please give justification for additional requirement(s) – whether due to increase in number of students or class size or new section(s)? It is an additional requirement for producing projects and students learn more.**

iii. **If it is a new requirement; then why it was not purchased previously and why it is being demanded now and what are the justifications – new course or program etc. It is an additional facility for students to produce their projects and also for research work.**

Sig. of Indenter: ____________________	Sig. of Dean /HoD Deptt.:________________
Name of Indenter: ________________	Name: ______________________________
Designation:________________________	Designation: _______________________
Date: ____________________________	

TENDER NOTICE FOR MANUFACTURING OF INDUSTRIAL SCALE MACHINES

Tender # GIKI/PD/FME-RE/806/21

Ghulam Ishaq Khan (GIK) Institute of Engineering Sciences and Technology, Topi, District Swabi, Khyber Pakhtunkhwa, under the project 'Promoting Enterprises of Functional Reverse Engineering' sponsored by Directorate General of Science and Technology (DoST), Government of Khyber Pakhtunkhwa, invites sealed bids (Single Stage, Two envelopes) according to KP-PPRA Procurement Rules 2014 from the reputed manufacturers of equipment for manufacturing of following Industrial Scale Machines as per given specifications. The bidders who are on Active Taxpayers List and are registered with FBR are eligible to apply. The machines are:

1. **3D Scanner 2. Binder Jet Printer 3. 5-Axis CNC Milling Machine 4. Electro Hydrodynamic Jet Printer 5. Selective Laser Jet Sintering Machine 6. Coordinate Measuring Machine 7. Pneumatic Jet Printer 8. Spinning Machine 9. Revolute Robot 10. Tensile Testing Machine.**

The vendors having proven capacity can apply for manufacturing all the machines jointly; or can bid for the individual machines also.

Submission Procedure: The terms and conditions are given in the tender documents which can be downloaded from GIK Institute and KPPRA websites (www.giki.edu.pk) / (www.kppra.gov.pk). The individual machine specifications will be provided with the tender documents. Tender fee (non-refundable) of **PKR 2,000/-** and sealed bids along with 2% earnest money of the total bid cost should reach on or before **1500 Hours on March 31, 2021** on the following address. The technical proposals will be opened in the presence of bidders at **1530 Hours** on the same day. For mobilization advance, the bidder will have to provide the bank guarantee. The development will be in close collaboration with the investigators of the project. For further information please contact:

Deputy Director (Procurement)
GIK Institute, Topi-23640, Swabi, Khyber Pakhtunkhwa
Ph: 0938-232323 Ext: 3333 Fax No. 0938-2833333
E-mail: abc@giki.edu.pk

Tender Serial No. GIKI/PD/RE-843/21

GIK INSTITUTE OF ENGINEERING SCIENCES AND TECHNOLOGY, TOPI

(PROCUREMENT DEPARTMENT)

SCHEDULE OF REQUIREMENTS FOR PURCHASE/SUPPLY OF LAB EQUIPMENT / COMPONENTS

The delivery of services and supply will be made in 60 days after issuance of Purchase Order (without penalty) and with the prescribed penalty as per following schedule of requirement: -

MODE OF PENALTY	NO OF DAYS	TOTAL DELIVERY PERIOD
Without Penalty	60 days	60 days
With Penalty @ 0.10 % per day for 10 days, 0.15% per day for 20 days, 0.20% per day for 30 days, 0.25% per day after 30 days.		

PRICE SCHEDULE FOR THE GOODS IN PAK RUPEES:

Name of Bidder: ______________________________

Tender No: GIKI/PD/RE-843/21

PURCHASE OF LAB EQUIPMENT / COMPONENTS- tabulated as under:

Tender No.	Sr. No.	Description/Specification	Qty	Unit Price (RS)	Unit Price with 17% GST (RS)	Total Cost including GST (RS)	Bid Security (2% of the Total Cost)
GIKI /PD/ RE- 843/ 21	1	Digital Torque Meter (0~100Nm)					
	2	Print Head for Binder Jet Printer					
	3						
	4						
	5						

Issued by: Assistant Director Procurement

Signature of the Bidder: ______________________________

Stamp of Bidder: ______________________________

Note: In case of discrepancy between unit price and total, the unit price shall prevail.

Annexure-A

SPECIFICATIONS

TENDER NO. (GIKI/PD/RE-843/21)

PURCHASE OF LAB EQUIPMENT / COMPONENTS

Tender No.	Sr. No.	Description/Specification	Qty
GI KI/ PD /R E- 84 3/2 1	1	**Torque Meter: (With Display):** Application: Servo/Stepper Motor Torque meter Type: Rotary Torque Meter Range: 0 ~ 100 Nm Input Voltage: 220 V AC/24 V DC Working temperature: 0 ~ 70 Degree Celsius Maximum Rotation Speed: 5000 RPM Accuracy: 0.5%	01 Nos
	2	**Print Head for Binder Jet Printer:** Origin: XAAR UK or equivalent Model: 382/35PL Printhead Compatible with: Wit-color Ultra3000 4H, WIT-COLOR ULTRA3000 6H, WIT-COLOR ULTRA3000 8H, MYJET WD3216XR60, MYJET WD3212XR60 Active nozzles: 382 Print swath width: 53.7 mm Nozzle pitch: 141 μm Nozzle density (nozzles per inch): 180 npi Drop velocity *: 5 m/s Printhead weight (dry): 150 g Dimensions (W × D × H): 86 × 21 × 63.5 mm Ink type: Solvent Drop volume*: 35 pl Typical firing frequency: 9.5 kHz	01 Nos
	3		

Assistant Director Procurement
GIK Institute, Topi

TENDER BIDS SUBMISSION & OPENING DATE AND TIME

TENDER NOTICE: GIKI/PD/RE-843/21
PURCHASE OF LAB EQUIPMENT / COMPONENTS

Tender No.	Description/Specification	Date of Submission of Bids	Date Opening of Tender
GIKI/ PD/RE-843/21	Purchase of LAB EQUIPMENT / COMPONE (**As per Annexure-A**)	**1500 pm on November 10, 2021**	**1530 pm on November 10, 2021**

Assistant Director Procurement
GIK Institute, Topi

GIK INSTITUTE OF ENGINEERING SCIENCES AND TECHNOLOGY, TOPI

Technical Evaluation Criteria for Purchase of LAB EQUIPMENT / COMPONENTS (Tender # GIKI/PD/RE-843/21)		
S. No.	**Item Name and Description**	**Maximum Marks**
1	**Relevant Experience of the Bidder**	**10**
1.1	1 - 3 year experience	
1.2	4 - 8 year experience	
1.3	9 - 15 year experience	
1.4	Above 15	
2	**Financial Soundness of bidder**	**10**
2.1	Bank Statement of Last three months	
2.2	Registration with tax authorities (GST/NTN)	
2.3	No Litigation History	
3	**Technical Evaluation of quoted items**	**40**
3.1	Specification as per attached Annex-A	
3.3	Delivery schedule as per requirement	
3.4	List of clients / where such equipment/items delivered	
3.5	Guarantee / Warranty	
TOTAL		**60**

Note: Please submit documents as per above mentioned criteria

- ***Technical qualification status shall be decided on the basis of Pass/Fail basis. The applicant must secure at least 50% accumulative score.***
- ***Order/contract will be awarded on the basis of highest score considering 60% technical and 40% Financial.***

Assistant Director Procurement
GIK Institute, Topi

GIK INSTITUTE OF ENGINEERING SCIENCES AND TECHNOLOGY, TOPI

TENDER NO.

GIKI/PD/RE-843/21

TENDER DOCUMENT

F.O.R. BASIS

TENDER FEE NON-REFUNDABLE
Rs. 2,000/-

PROCUREMENT DEPARTMENT

Ph: 0938-281026 EXT (2444) Fax: 0938-281045

Email: abc@giki.edu.pk

Web: www.giki.edu.pk

GIK INSTITUTE OF ENGINEERING SCIENCES AND TECHNOLOGY, TOPI PROCUREMENT DEPARTMENT

INTRODUCTION

GIK INSTITUTE OF ENGINEERING SCIENCES AND TECHNOLOGY, TOPI, having (hereinafter referred to as the Purchaser) invites sealed bids from eligible bidder(s) under Single Stage Two envelope bidding system (i.e. Technical and Financial Bids in separate envelopes) for supply and delivery of goods, materials, equipment and/or services described in the Schedule of Requirement. All the Annexures from "C" to "L" are an integral part of the Tender Document and its compliance is mandatory and must be signed and stamped by the vendor/supplier and attached with the relevant proposal, otherwise bids will be rejected.

The Technical Bids will be opened first on the date mentioned in the Tender Notice. The financial bids of the only technically qualified bidder(s) will be opened publicly. The Technical Evaluation Report will be published on GIK INSTITUTE, the website before financial bid opening.

1. **INSTRUCTIONS TO BIDDER(S)**

   1.1. The bidder(s) shall manufacture all machines according to the drawing and models and bring also samples and bear all costs associated with the preparation and delivery of their bids, and the Purchaser will in no case be responsible or liable for those costs, if required.

   1.2. In case of any doubt as to the meaning of any portion of the Specifications or other Terms and Conditions, bidder(s) may seek clarification of the same in writing, but not later than 5 days prior to the deadline for submission of bid(s) prescribed by the Purchaser.

   1.3. At any time prior to the deadline for submission of bids, a modification in bidding document in the form of an addendum can be issued in response to a clarification requested by a respective bidder(s) or even whenever the Purchaser considers it beneficial to issue such a clarification and/or amendment to all bidder(s). The clarification/amendment issued (in general) shall be placed on the GIK INSTITUTE web site for the intimation to all the prospective bidder(s).

   Bidder(s) are required to submit their bid(s), which will bear words "CONFIDENTIAL" and for **TENDER NO. GIKI/PD/RE-843/21** as per Tender Notice.

   1.4. The Purchaser will not take any responsibility for collecting the bids from any Agency.

   1.5. The Purchaser reserves the right to increase or decrease the quantities or may cancel any or all items shown in the Schedule of Requirement without assigning any reason thereof.

   1.6. The Purchaser reserves the right to have the items inspected by its own representative, or by 3rd party at its own cost (if required).

   1.7. It must be indicated in the offer that the Bid fully conforms to Technical Specifications and Terms and Conditions of the Tender.

   1.8. The Purchaser reserves the right to accept or reject any bid and to annul the bidding process and reject all the bids at any time prior to award of Purchase Order without thereby incurring any liability to the affected bidder(s) or any obligation thereunder. However, GIK Institute shall upon request communicate to any bidder(s) who submitted a bid, the grounds for its rejection, but it is not required to justify those grounds.

   1.9. Unsolicited advice/clarifications and personal approaches by the bidder(s) at any stage of evaluation are strictly prohibited and shall lead to disqualification.

   1.10. The bidder(s) are expected to examine all instructions, forms, terms and specifications in the Bidding Documents. Failure to furnish all information required by the Bidding Documents or submission of a bid not substantially responsive to Bidding Documents in all respects will result in the rejection of the bid(s).

1.11. All prospective bidder(s) are advised to carefully read all terms and conditions mentioned in the Tender Documents prior to filing/submission of their Bid.

1.12. The blacklisting procedure of the Company is attached as Appendix-L. The bidders, contractors, suppliers, agents, consultants, partnership, or firms shall be dealt accordingly.

2. **BID PRICES**

2.1. Firm bid prices shall be quoted in Annexure-B duly filled in, stamped, and signed by authorized representative of Bidder(s). Prices must be quoted as per format of Schedule of Requirement (SOR).

2.2. Bidding Forms and Bid Summary Sheets from Annexure-"C" to "F" and "H" to "K" must also be duly filled-in, stamped and signed by authorized representative of bidder(s).

2.3. Prices must be quoted in Pakistani currency inclusive of all taxes and duties and in foreign currencies, indicating unit price and total bid prices. GST must be quoted separately along with a copy of GST Certificate only for Pakistani currency. Bids offering prices in foreign currency will be rejected. Conditional bid/bid having any price adjustment formula will be rejected.

2.4. Prices must be quoted including all charges like handling, loading, transportation, and unloading, Octroi and Zila Tax etc. for delivery of material at Purchaser's stores located at **required delivery locations**.

2.5. Quoted prices will be valid for 120 days from the opening date of the **Financial Bid**.

2.6. Technical brochure/literature confirming Size, Brand and Country of Origin of quoted items/material must be attached with the Technical Bid (where deemed necessary).

2.7. Any Bid not confirming Clause 2.1 to 2.6 shall be rejected without any right of appeal.

2.8. Discount (if any) shall only be entertained on Schedule of Requirement of Bidding Document (Financial Proposal). If the discount is mentioned elsewhere in the bid, the same shall not be entertained.

2.9. Tenders submitted without Tender Fee Rs. 2,000/- (in shape of challan form/pay order/Demand Draft) and earnest money CDR should be rejected at the time of opening of Tender. Earnest money and challan form will be submitted against each Tender separately.

3. **BID SECURITY MONEY**

3.1. The sealed bids complete in all respect must reach in the office of Assistant Director Procurement GIK Institute along with Bid Security Money 2% of the total cost in the name of GIK INSTITUTE OF ENGINEERING SCIENCES AND TECHNOLOGY, TOPI in shape of CDR to be attached with financial offer.

3.2. Short Bid Security Money or Crossed Cheque as Bid Security Money is not acceptable.

3.3. The Bid Security Money of unsuccessful bidder(s) will be returned after award of Purchase Order to successful bidder(s). The Bid Security Money of successful bidder(s) will be discharged/returned upon bidder(s) furnishing the executing the order/contract (Subject to the satisfactory report without any objection from the end user).

3.4. The Bid Security Money of bidder(s) whose bid is rejected under Clause 2.7 will be returned forth-with.

3.5. The Bid Security Money shall be forfeited:

3.5.1. If a bidder withdraws its bid during the period of bid validity.

3.5.2. In the case of a successful bidder, who fails (a) to sign the contract; (b) to furnish performance security (if any).

4. **SEALING AND MARKING OF BIDS**

4.1. The bidder(s) shall furnish Technical and Financial Bid(s) in two separate sealed envelopes.

4.2. Technical Bids will be opened first.

4.3. Earnest money as mentioned in the Press Tender Notice must be enclosed with Financial Bid.

4.4. Only Financial Bid(s) of Technically responsive bidder(s) will be opened publicly.

4.5. Financial Bids of technically non responsive bidder(s) will be returned after award of Purchase Order to successful bidder(s).

4.6. The bid shall be submitted in a sealed envelope marked as under:

PROCUREMENT DEPARTMENT

GIK INSTITUTE OF ENGINEERING SCIENCES AND TECHNOLOGY, TOPI

4.7. The envelope shall also bear the word "CONFIDENTIAL" and following identifications:

TENDER NO. GIKI/PD/RE-843/21

DON'T OPEN BEFORE *(AS PER SCHEDULE MENTIONED IN THE TENDER NOTICE/SOR)* TECHNICAL/FINANCIAL BID(s) (INDICATE ONE ON EACH ENVELOPE)

4.8. If the envelope is not marked, as instructed above, the Purchaser will assume no responsibility for the misplacement or premature opening of bid(s).

5. **DEADLINE FOR SUBMISSION OF BIDS**

5.1. The bids must be delivered to the Procurement Department, GIK Institute, Topi on or before as per schedule mentioned in Press Tender Notice and the same will be opened accordingly.

6. **LATE BIDS**

   6.1. Any bid received by the Purchaser after the prescribed deadline for submission of bids mentioned in Press Tender Notice will not be accepted and returned unopened to the bidder(s).

7. **BID OPENING**

   7.1. The purchaser will open the Technical Bid, in the presence of the bidder's representative(s) who may choose to attend on the date, time and location mentioned above.

   7.2. The bidder(s) names, bid prices including bid price modification and bid withdrawals, if any, will be announced at the time of financial bid opening.

   7.3. The Purchaser will examine the bids to determine whether they are complete, no computational errors have been made, the required sureties have been furnished, documents have been properly signed, and the bids are generally in order.

   7.4. Arithmetical errors found will be corrected as follows:

      7.4.1. Where there is a discrepancy between amount in figures and word, the amount in word will prevail.

      7.4.2. Where there is a discrepancy between the unit price and the total amount derive from the multiplication of the unit price and the quantity, the unit price as quoted will govern, unless there is an obvious gross arithmetical error in unit price, in which event, the extended amount quoted would prevail.

      7.4.3. In case the due date of bids opening falls on a holiday, the bids shall be opened on the next working day.

8. **DETERMINATION OF RESPONSIVENESS**

   8.1. After opening of the bids, the purchaser will determine whether each bid is substantially responsive to the requirements of the bidding documents.

   8.2. For this clause, a substantially responsive bid is one which conforms to all the terms and conditions of the bidding documents without material deviation. A material deviation is one which means in-consistent with the bidding document, affects in any substantial way the scope, quality, or prescribed delivery schedule or which limits in any substantial way, the Purchaser's right or the bidder's obligation under the contract.

   8.3. A bid determined to be non-responsive will be rejected by the purchaser and shall not subsequently be made responsive by the bidder(s) by correction of the non-conformity.

   8.4. The purchaser may waive any minor deviation, non-conformity or irregularity in a bid, which does not constitute a material deviation, provided that the waiver does not prejudice or affect the relative standing order of other bidder(s).

   8.5. To assist in determining a bid's responsiveness the bidder(s) may be asked for clarification of his bid. The bidder(s) is not permitted, however, to change the bid price or substance of his bid.

8.6. The bidder(s) may lodge a written complaint for redress of their grievances and disputes to the Dispute Resolution Committee within fifteen days of the placement of the Technical Evaluation report on the web.

9. **ISSUANCE OF PURCHASE ORDER (PO)**

9.1. The Purchaser will issue Purchase Order to the successful bidder(s) whose bid is determined to be the best evaluated bid.

10. **PAYMENT**

10.1. Payment will be made as follows inspection and acceptance of goods, materials, and equipment etc.:
10.1.1. Payment after delivery
10.1.2. Payment Against Delivery
10.1.3. Part Delivery Part Payment
10.1.4. ____% advance payment against Bank Guarantee

11. **DELIVERY SCHEDULE**

11.1. Delivery of the required material against this tender is to be completed as soon as possible from the date of receipt of firm Purchase Order.

12. **EXTENSION IN THE DELIVERY PERIOD**

12.1. Delivery of the goods shall be made by the supplier in accordance with the delivery schedule given in the Purchase Order.
12.2. The supplier may claim extension of the time limits as set forth in the Purchase Order in case of.
12.3. Changes in the specifications of goods, material and equipment by the Purchaser.
12.4. Delay in provision of clarifications regarding material, drawings, and services by the Purchaser.
12.5. Force Majeure pursuant to Clause # 14.
12.6. Justified reasons subject to a written request highlighting the same within a period of seven working days prior to expiry of delivery period mentioned in Purchase Order.
12.7. It should be noted that a request for an extension in delivery period shall be considered only if the supplier agrees in writing to pay any increase in taxes or any other charges levied by the Government during the extended delivery period. The purchaser shall not bear any additional price increase during the extended period.
12.8. In case of extension in the delivery period, the supplier will extend validity of Performance Bond accordingly at his cost.
12.9. If the supplier fails to supply the ordered material for any reason, within the stipulated time, the Performance Bond shall be forfeited, and material shall be purchased from elsewhere at his risk and cost.

13. **LIQUIDATED DAMAGES**

13.1. If you fail to supply the material in the above specified period, GIK Institute reserves the right to blacklist you for future business or forfeit the security money or impose penalty as per above rates of the contract value or may impose any other financial penalty as deemed fit.

13.2. Even after imposition of LDs, if the supplier fails to materialize the delivery (material and or services); the Purchaser reserves the right to cancel Purchase order/contract and to forfeit the earnest money (if applicable) after intimating the supplier for such cancellation/forfeiture.

14. **FORCE MAJEURE**

14.1. The supplier shall not be liable for penalty for delay in delivery of ordered goods, if, and to the extent delay in delivery or other failure to perform his obligation under the Purchase Order, of being the result of occurrence of Force Majeure i.e. causes such as natural calamities, war, civil disturbance, military action, fire as well as other circumstance proved to the satisfaction of the Purchaser to be beyond the reasonable control of the supplier, which may impede the fulfillment of the obligations under the Purchase Order.

14.2. The supplier shall notify the Purchaser promptly of the occurrence of Force Majeure and submit his case in writing within 07 days of such occurrence.

14.3. If any of the parties is prevented from fulfilling its assumed obligations by Force Majeure of constant duration of at least one month, the party shall meet for negotiation. If no satisfactory agreement is reached within a period of two months from the Commencement of the Force Majeure conditions, either party shall have the right to cancel the Purchase Order with immediate effect.

15. **INSPECTION AND TEST**

15.1. The Purchaser's representative(s) shall have the right to inspect and/or test the goods to confirm their conformity with respect to specifications mentioned in the Purchase Order. The representative of the Procurement Cell will witness the inspection (where so required).

15.2. Should any inspected or tested goods fail to conform to the specifications the purchaser shall reject them, and supplier shall replace the rejected goods. All costs incurred on such replacement shall be entirely borne by the supplier.

15.3. Rejected material shall be moved/replaced by the supplier within 07 days from the receipt of letter/fax issued by the Procurement Department. The supplier shall be liable for the **storage** charges @ 0.2% (half percent) of the cost of rejected material on an everyday basis, if the same is not removed within seven days.

15.4. Acceptance/rejection of the material by 3rd party (where required) will be final and binding on both the parties.

16. **WARRANTY**

16.1. The supplier shall warrant that all goods supplied under Purchase Order shall be according to specifications given in Purchase Order and approved drawings/design etc. Any deviation in material, drawing/design (where applicable) will be replaced by the supplier at his cost.
16.2. The Purchaser shall promptly notify the supplier in writing, of any claims arising under this warranty.
16.3. The supplier will invariably provide warranty/guarantee of spare and consumables.

17. **QUALIFICATION OF SELECTED BIDDER(S)S**

17.1. The Purchaser will determine to its satisfaction whether the bidder(s) selected as having submitted the lowest evaluated, technically responsive bid qualifies to satisfactorily perform the order.
17.2. The determination will consider the bidder(s) financial, technical and production capabilities, availability of items ordered for. The bidder(s) shall provide necessary documents as proof along with the bid.
17.3. Any affirmative determination will be a prerequisite for award of the Purchase Order to the bidder(s). A negative determination will result in rejection of the bid.
17.4. The Purchaser reserves the right at the time of award of order to increase or decrease to a reasonable extent the quantity of goods specified in the Tender Document without any change in price or other terms and conditions.

18. You are encouraged to inform Procurement Department on the following addresses/contacts, in case where any GIK Institute employee ask for any type of favor whether monetary or in kind:

DESIGNATION	ADDRESS	TEL #	FAX #	E. MAIL
Assistant Director Procurement	Ghulam Ishaq Khan Institute, Topi	0938-281026 (Ex: 2444)	0938-281045	abc@giki.edu.pk

ANNEXURE - C

BIDDING FORM (TECHNICAL BID)

Assistant Director Procurement
GIK Institute of Engineering Sciences
and Technology, Topi

Gentlemen,

1. Having examined the Bidding Documents including the specifications, the receipt of which is hereby acknowledged, we the undersigned offer to supply & deliver «Description» in conformity with drawings, specifications of goods and conditions of Tender Document.

2. We undertake if our bid is accepted, to commence delivery within 30 days (Please specify days) from the date of receipt of your confirm Purchase Order.

3. If our bid is accepted, we will provide the Performance Bond equal to 10% of the Purchase Order excluding GST, for due performance of the purchase order.

4. We agree to abide by all the terms & conditions of the tender for the period of 120 days from the opening of financial proposal & it shall remain binding upon us and may be accepted at any time before the expiry of that period or any extension thereof agreed by us.

5. Until a formal Purchase Order is placed, this bid, together with your written acceptance thereof, shall constitute a binding contract between us. We understand that you are not bound to accept the lowest price or any bid you may receive.

Dated this ______________________ day of _______________ 20_____

(Signature) (In the capacity of)

Duly authorized to sign Bid for and on behalf of __________________________________

(Signature of Witness)
Name: - __________________________________
Address: - ________________________________
____

ANNEXURE -D

BIDDING FORM (FINANCIAL BID)

Assistant Director Procurement
GIK Institute of Engineering Sciences
and Technology, Topi

Gentlemen,

1. Having examined the Bidding Documents including the specifications, the receipt of which is hereby acknowledged, we the undersigned offer to supply & deliver «Description» in conformity with drawings, specifications of goods and conditions of Tender for the sum of Rs.______________________ (Total bid amount in words) (inclusive of all taxes) or such other sum as may be ascertained in accordance with the said conditions.

2. Until a formal Purchase Order is placed, this bid, together with your written acceptance thereof, shall constitute a binding contract between us. We understand that you are not bound to accept the lowest price or any bid you may receive.

Dated this ____________________ day of ____________ 20_____

(Signature) (In the capacity of)

Duly authorized to sign Bid for and on behalf of ___________________________________

(Signature of Witness)
Name: - ___________________________________
Address: - ________________________________
____

ANNEXURE – E

BID SUMMARY SHEET

TENDER NO. GIKI/PD/RE-843/21

(TO BE ATTACHED WITH TECHNICAL BID)

1. Bidder Name Address, Phone & Fax No E-mail address ________________________________ ________________________________ ________________________________

2. Manufacturer Name & Country of Origin: ________________________________ ________________________________ ________________________________

3. Items Quoted:(give serial no. only): ________________________________ ________________________________

4. Price Validity: ________________________________

5. Offered Delivery Period: ________________________________

6. GST Registration No: ________________________________

7. Bidding form (Annexure-C attached with Technical bid: Yes ( ) No ( )

8. Any Deviation: ________________________________

Signature ______________________

Name & Designation__________________________

ANNEXURE – F

BID SUMMARY SHEET

TENDER NO. GIKI/PD/RE-843/21

(TO BE ATTACHED WITH FINANCIAL BID)

1. Bidder Name; ____________________

Address, Phone & Fax NO.____________________

E-mail address ____________________

2. Manufacturer Name and Country of Origin: ____________________

3. Items Quoted:(give serial no. only): ____________________

4. Price Validity: ____________________

5. Total FOR («Deliver To») Price: (Inclusive of all taxes except GST) Rs ____________

6. Total FOR («Deliver To») Price (With GST): Rs ____________

Total C&F («Deliver To») Price: ____________

7. Offered Delivery Period: ____________________

8. Payment Terms: ____________________

9. GST Registration No. ____________________

10. Bidding Form (Annexure-C-1 attached with financial bid): Yes/ No

11. Any Deviation: ____________________

Signature ____________________

Name & Designation____________________

ANNEXURE - G

DECLARATION OF FEES, COMMISSION AND BROKERAGE etc.
CERTIFICATE

Submitted to GIK Institute with the reference to Purchase Order No. ____________ hereby declares its intention not to obtain or induce the procurement of any contract, right, interest, privileges or other obligation or benefit from Government of Pakistan or any administrative subdivision or agency thereof or any corrupt business practice.

Without limiting the generality of the foregoing, the Seller/ Supplier represents and warrants that it has fully declared the brokerage, commission, fees etc., paid or payable to anyone and not given or agreed to give and shall not be given or agree to give to anyone within or outside Pakistan either directly or indirectly through any national or juridical person, including its affiliate, agent, associate, broker, consultant, briber, finder's fee or kickback, whether described as consultant fee or otherwise, with the object of obtaining or including the procurement of a contract right, interest, privilege or other obligation or benefit in whatsoever form from GOP except that privilege or other obligation or benefit in whatsoever form from GOP except that which has been expressly declared pursuant hereto.

The Seller/ Supplier certificate that it has made and will make full disclosure of all agreements and arrangements with all persons in respect of or related to the transaction with GOP and has not taken any action or will not take any action to circumvent the above declaration, representation, or warranty.

The Seller/ Supplier accepts full responsibility and strict liability for making any false declaration, not making full disclosure, mis-representing facts or taking any action likely to defeat the purpose of this declaration, representation, and warranty. It agrees to any contract, right interest. Privilege or other obligation or benefit obtained or procured as aforesaid shall, without prejudice to any other rights and remedies available to GOP under any law, contract, or other instrument, be voidable at the option of GOP.

Notwithstanding any rights and remedies exercised by GOP in this regard, the Seller/ Supplier agrees to indemnify GOP for any loss or damage incurred by it on account of its corrupt business practices and further pay compensation to GOP in an amount equivalent to ten times the sum of any commission, gratification, bribe, tinder's fee or kickback given by the Seller/ Supplier as aforesaid for the purpose of obtaining or inducing the procurement of any contract, right, interest, privilege or other obligation or benefit in whatsoever form from GOP.

For & On Behalf of
Seller/ Supplier

ANNEXURE - H

(WHERE APPLICABLE, TO BE SUBMITTED ON DEMAND OF GIK INSTITUTE)

FAIR PRICE CERTIFICATE

WE M/S____ HEREBY CERTIFY THAT PRICES QUOTED BY US AGAINST GIK INSTITUTE TENDER NO. **GIKI/PD/RE-843/21** ARE THE LOWEST AND MOST COMPETITIVE AS GENERALLY APPLICABLE TO ALL OTHER BUYERS AND OR SOLD THROUGH OUR AGENTS AS PER PREVAILING INTERNATIONAL MARKET AS ON THE DATE OF QUOTE AND IF IT IS ESTABLISHED AT ANY STAGE THAT THESE WERE HIGHER WE SHALL BE HELD RESPONSIBLE AND AGREE TO PAY IMMEDIATELY THE DIFFERENTIAL TO GIK INSTITUTE, TOPI.

SIGNATURE AND SEAL OF THE
MANUFACTURER / SOLE AGENT/AUTHORIZED DEALER

NOTE: FAIR PRICE CERTIFICATE WILL BE REQUIRED, IN CASE SINGLE BID IS FOUND TECHNICALLY RESPONSIVE IN THE RESPECTIVE TENDER.

ANNEXURE - I

(On official letterhead of the contractor)
To be signed by the Chief Executive of the bidding company or a representative duly authorized by board resolution.

Integrity and Ethics Undertaking

We hereby commit and undertake to observe the following principles during our participation in the contract execution:

a) That we will not directly or through any other person or firm, offer, promise or give to any of the employees of GIK INSTITUTE involved in or relevant to the execution of the contract any gain, pecuniary benefit, or facilitation payment to obtain in exchange any advantage of any kind whatsoever during the execution of contract or at any stage thereafter.

b) That we did not enter with any bidder into any undisclosed agreement or understanding either formal or informal to restrict competitiveness or to cartelize in the bidding process.

c) That we will ensure that the remuneration of agents *(if engaged)* is appropriate and for legitimate services only.

d) That we will not use subcontracts, purchase orders or consulting agreements as means of channeling payments to employees of GIK INSTITUTE.

e) That we will not and have not committed any offense under the Pakistan Penal Code, Prevention of Corruption Act or National Accountability Ordinance to achieve any advantage, gain or benefit during the tender process or the execution of contract.

We further understand and acknowledge that any violation or transgression of the above-mentioned principles will attract disqualification from doing business with GIK INSTITUTE and may also result in permanent exclusion from future contact award processes.

We also accept and undertake to respect and uphold GIK INSTITUTE's absolute right to resort to and impose such disqualification, debarment, or exclusion.

For and on behalf of ___________

Tender No. ____________

Contract No ________________

ANNEXURE - J

(On official letterhead of the bidder)
To be signed by the Chief Executive of the bidding company or a representative duly authorized by board resolution.

Integrity and Ethics Undertaking

We hereby commit and undertake to observe the following principles during our participation in the tender process and during the contract execution:

a) That we will not directly or through any other person or firm, offer promise or give to any of the employees of GIK INSTITUTE involved in the tender process or execution of the contract any gain, pecuniary benefit or facilitation payment in order to obtain in exchange any advantage of any kind whatsoever during the tender process or during the execution of contract.

b) That we have not and will not enter with other bidders into any undisclosed agreement or understanding either formal or informal to restrict competitiveness or to cartelize in the bidding process.

c) That we will ensure that the remuneration of agents *(if engaged)* is appropriate and for legitimate services only.

d) That we will not use subcontracts, purchase orders or consulting agreements as means of channeling payments to employees of GIK Institute.

e) That we will not commit any offense under the Pakistan Penal Code, Prevention of Corruption Act or National Accountability Ordinance to achieve any advantage, gain or benefit during the tender process or the execution of contract.

We further understand and acknowledge that any violation or transgression of the above-mentioned principles will attract disqualification from the tender process and may also result in permanent exclusion from future contract award processes.

We also accept and undertake to respect and uphold the Institute's absolute right to resort to and impose such disqualification, debarment, or exclusion.

For and on behalf of _________

Tender No. GIKI/PD/RE-843/21

ANNEXURE - K

Form 4

Mandatory for participation in Bidding Process

AFFIDAVIT

I,____________________________ S/o ____________________________ aged __________ years ___________ working as Proprietor/Managing Partner/Director of M/s ____ having its registered office at____ do hereby solemnly affirm and declare on oath as under:

1. That I am competent to swear this affidavit being proprietor/one or the partners/ Director of *M/s* ________________
1. That M/s ____ is a proprietorship/partnership firm/company is participating in tender process conducted by Institute.
2. That I hereby confirm and declare that none of my/our group/sister concern/associate company is participating/ submitting this tender.
3. That I hereby confirm and declare that my/our firm/company M/s __________________________ and my/our firm/group/company/ sister concern / associate company have not been black listed/de-listed any Institutional agencies/Govt. Deptt/ Public Sector Undertaking.
4. That there is no change in the Name & Style, Constitution and Status of the firm, after Pre-qualification.
5. That I further undertake that in case any of the facts contained above and in-our application is round other-wise or incorrect or false at any stage, my/our firm/company/ group/sister concerns/ associate companies shall stand debarred from the present and future tenders of the GIK INSTITUTE.

(Signature of the Proprietor/ Managing Partner/Director with Seal)

DEPONENT

Verified at _______________ on ______________________ that the contents of paras 1 to 6 of this affidavit are true and correct to best of my knowledge and no part of this is false and nothing material has been concealed or falsely stated therein.

(Signature of the Proprietor/ Managing Partner/ Director with Seal)

DEPONENT

(Signature & Seal of Notary

Annexure - L

BLACKLISTING PROCEDURE

1. Blacklisting

   Blacklisting means: temporarily or permanently, barring an entity or a person against whom proceedings have been initiated including but not limited to bidder, contractor, supplier, agent, consultant, company, partnership, or firm; hereinafter referred to as, Respondent from participating in any future procurement of goods and services. The Respondent(s) individually or collectively as a consortium may stand blacklisted if found to have been involved in any or all of the following acts:

(a) Undermines or adversely affects the operations of the company through any of the following:

- Withdrawing a bid during the bid validity period.
- Failure or refusal to:

  i. sign the Contract.
  ii. accept Purchase Order/Service Order Terms.
  iii. execute work.
  iv. submit 2% earnest money as per tender terms.
  v. make supplies as per specification agreed.
  vi. fulfill contractual obligations as per contract.
  vii. meet purchase order/service order terms and conditions; and/or,
  viii. to remedy underperformance as per contractual obligations.
  ix. Or any other non-compliance of obligations vital for the execution/compliance of the contract.

(b) Repeated non-performance.
(c) Indulgences in corrupt or fraudulent practices while obtaining or attempting to obtain contracts in the company.
(d) Convicted of fraud and corruption. Tax evasion or criminal misappropriation by a court of competent forum.
(e) Notified blacklisted/debarred/cross debarred by any public sector organization or international agency. In this regards the contractor must submit a certificate that he or his principal is not declared blacklisted/debarred/cross debarred by any public sector organization or international agency.
(f) Furnished information that was false and materially inaccurate or submitted forged or fake documents.

2. Debarment of Natural Terms

The following shall stand disqualified and debarred from participating in GIK INSTITUTE tenders or contracts pursuant to blacklisting proceedings:

i. In case of a company; all directors including its Chairman, Chief Executive

and Chief Financial Officer.
ii. In case of partnership; all partners.

3. Debarment of Associated Companies or Entities

Where a company or an entity has been blacklisted, all its associated or subsidiary undertaking whether by way of common directorship, common management and control, shareholding or direct or indirect control through directors of blacklisted entity etc. shall also stand disqualified from participating in GIK INSTITUTE tenders or contracts.

4. Proceedings for Blacklisting

(a) Director (Procurement) of his own accord or on receipt of information or a complaint shall refer the matter to the Procurement Committee.
(b) The Committee, after examining the material placed before it, shall determine whether it is necessary and appropriate to initiate formal blacklisting proceedings.
(c) In case the Committee decides to initiate blacklisting proceedings, Respondents shall be formally intimated in writing about the nature of complaint/ matter and initiation of blacklisting proceedings.
(d) The Respondent shall be intimated by giving 15 days' notice through courier services, registered post or fax or email and shall be provided an opportunity of furnishing response either through written representation or personal hearing or both.
(e) In case the Respondent fails to furnish his defense or representation the committee may proceed based on information, record, and material available before it provided that two subsequent notices at seven working days intervals have been given.
(f) The Committee shall complete its proceedings within 60 days from the date of first notice given pursuant to paragraph (d) above.
(g) The Respondent against whom blacklisting proceedings have been initiated may be represented through:
    i. In case of an individual or sole proprietorship; in person.
    ii. In the case of a firm or partnership; by the Chief executive or the

Managing Partner, duly authorized.

iii. In the case of a company, by the Chief Executive or a Director or any officer duly authorized by the Company.

5. Findings and Decisions

(a) The Committee after finalizing its proceedings shall record its findings in writing and decide one of the following measures for implementation:

i. Temporarily debarred, specifying the period.

ii. Blacklisted if the Respondent fails to take remedial action within the specified time; or,

iii. Blacklisted.

(b) The order of the Committee shall be communicated in writing as per Form 3.

(c) The documents, deliberations reduced in writing and all records of proceedings are to be maintained for five years by the Procurement Cell.

6. Appeal

(a) The Respondent may appeal within 15 days working days of intimation of decision of the Committee to the Pro-Rector (Admin and Fin) in writing of any irregularity in the decision of the Committee.

(b) If there exists any substance in the appeal the Pro-Rector (Admin and Fin) after recorded reasons may direct that Committee may review the case by providing opportunity of hearing to the Respondent provided that the review shall be completed within 30 days.

(c) The Committee shall record its findings in writing and refer the same to the Pro-Rector (Admin and Fin) for appropriate orders.

Assistant Director Procurement

GIK Institute, Topi

GIK INSTITUTE OF ENGINEERING SCIENCES AND TECHNOLOGY, TOPI					
Technical evaluation criteria of for manufacturing of mechanical hardware of prototypes and industrial scale machine tools suppliers					
Evaluation parameters			Bidders		Remarks/rational for scoring
S. No.	Item name and description	Maximum marks	M/s Hafiz Lathe Machine	M/s Standard Engineering Works PVT LTD	
1	Financial soundness	30	20.5	25	In light of the provided information, the committee has decided that both prequalified vendors M/s Hafiz Lathe Machine and M/s Standard Engineering Works are technically qualified for opening of financial bids for manufacturing of mechanical hardware of prototypes and industrial scale machine tools
2	Experience record/general experience	35	19	28	
3	Equipment capabilities	35	25	28	
Total		100	64.5	81	
Technical qualification status shall be decided on the basis of Pass/Fail basis. The applicant must secure at least 40% score in each category and must secure at least 50% accumulative scores					
1	*Financial soundness*	*30*	M/s Hafiz Lathe Machine	M/s Standard Engineering Works PVT LTD	1. Both vendors have a good credit history as well as bank statements. 2. M/s Standard Engineering Works has greater working capital than M/s Hafiz Lathe Machine in the last 3 years. 3. M/s Standard Engineering Works has paid tax above 3 million in the last 3 years
1.1	Available bank credit line	7.5	7	6	
1.2	Working capital in the last 3 years	7.5	5.5	7	
1.3	Amount of income tax paid in the last 3 years	10	3	7	
1.4	No litigation history	5	5	5	

(continued)

(continued)

GIK INSTITUTE OF ENGINEERING SCIENCES AND TECHNOLOGY, TOPI					
Technical evaluation criteria of for manufacturing of mechanical hardware of prototypes and industrial scale machine tools suppliers					
Evaluation parameters			Bidders		Remarks/rational for scoring
S. No.	Item name and description	Maximum marks	M/s Hafiz Lathe Machine	M/s Standard Engineering Works PVT LTD	
2	*General experience*	*35*			Multiple similar projects have been done by both vendors. Multiple projects in hand. M/s Standard Engineering Works and M/s Hafiz Lathe Machine have done multiple projects as per pre-qualification criteria
2.1	Similar office projects completed in past 5 years	10	8	9	
2.2	Similar office projects in hand	8	4	6	
2.3	Projects of general manufacturing supply of Rs. 15 million and above completed over the last 5 years	10	4	8	
2.4	Projects of general manufacturing supply of Rs. 15 million and above in hand	7	3	5	
3	*Equipment capabilities*	*35*			M/s Hafiz Lathe Machine is a machine tool manufacturer while M/s Standard Engineering Works is leading discrete product manufacturer. They are also experienced with machine tool manufacturing and both vendors are technically qualified
3.1	Casting capability	4	3	4	
3.2	Welding capability	4	3.5	4	
3.3	Metal cutting capability	5	4	4.5	
3.4	Metal working capability	6	3	4	
3.5	Surface treatment	6	4	4	
3.6	Inspection facility	6	4	4	

(continued)

(continued)

GIK INSTITUTE OF ENGINEERING SCIENCES AND TECHNOLOGY, TOPI					
Technical evaluation criteria of for manufacturing of mechanical hardware of prototypes and industrial scale machine tools suppliers					
Evaluation parameters			Bidders		Remarks/rational for scoring
S. No.	Item name and description	Maximum marks	M/s Hafiz Lathe Machine	M/s Standard Engineering Works PVT LTD	
3.7	Assembly facility	4	3.5	3.5	

Vendor Details **M/s EduTech Lahore** 601, Al-Hafeez Business Center, 89 B/III Gulberg III, **Lahore**. Tel: 042-35756043, 0321-8835034	PO #: 206 Indent #: Nil Department: Reverse Engineering (FME) Date: September 18, 2023 STRN #: 3277876152318 NTN #: 3380595-4

Subject: PURCHASE ORDER FOR SUPPLY and INSTALLATION OF DIGITAL CONTROLLERS

With reference to your Proposal No. FQ/ET/2023/43 dated 29 May, 2023 against Tender # GIKI/PD/RE-953/23, we are pleased to place work order to you for supply and installation Digital Controllers for the following two machines as per agreed terms and conditions:

S #	Items description	Qty	Unit	Unit price (PKR)	Total amount (PKR)
1	UTM-100 KN Control and Software	01	Each	Rs. 1,271,000/-	Rs. 1,271,000/-
2	6-Spindel Spinning Machine Control	01	Each	Rs. 1,271,355/-	Rs. 1,271,355/-
The details are mentioned in Annexure-A					
Total amount exclusive GST					Rs. 2,542,355/-
18% GST					Rs. 457,624/-
Total amount inclusive GST					Rs. 2,999,979/-

1. **Place of Delivery**: You will deliver and install the ordered controllers to GIK Institute, Topi along with your delivery challan a copy of which will be provided by you to the Procurement Section.
2. **Date of Delivery**: 60 days (Before 20 November 2023).
3. **Partial Supply**: Allowed.
4. **Documents**: Original Invoices with packing list.
5. **Inspection**: The equipment will be inspected by the GIK team at our site.
6. The order shall not be varied or modified without mutual agreement in writing by both parties.
7. **Replacement**: The vendor will be responsible for arranging free of cost and inclusive of delivery charges, prompt replacement of the equipment or any part if found defective during inspection by the GIK Institute.

8. **Penalty Clause**: If you fail to supply the ordered materials in the above specified period, GIK Institute reserves the right to blacklist you for future business and or may impose a financial penalty as per the following rates. *0.25% per day if the delivery exceeds one month.*
9. **Mode of Payment**: Payment will be made to you on submission of your original bills in duplicate through crossed Cheque within one month after acceptance of the items subject to deduction of taxes as per law. Or advance payment against bank guarantee.
10. **Guarantee/Warrantee**: Warranty of the equipment for a period of one year.
11. **After-Sales Service**: Complete support for after-sales services will be provided by your technical expert.
12. Kindly acknowledge receipt of this purchase order and return a scan copy of PO duly signed and stamped.
13. **Installation/Commissioning**: You will arrange free of cost installation and commissioning of equipment at GIK Institute.
14. In case of any dispute the decision of the Rector will be final and binding on you.
15. A formal Agreement will be signed by both parties.

Yours faithfully,

For and on behalf of GIK Institute, Topi

ABC
Dy. Director (Procurement)

M/s Edutech Lahore

ANNEXURE – A

➢ **UTM-1000 KN Control and Software Specification**

- Structure material = SS 304
- Control Panel of machine along with using a standard keypad
- Motherboard Core i5 with following specifications
    - Processor: Intel – I5
    - RAM: 4 GB
    - Storage: 500 GB
    - Wi-Fi Card, Bluetooth, Keyboard
    - Color LCD Display
    - Windows 10 Operating System
- UTM Data Acquisition (DAQ) System
- UTM Control Software

➢ **6-Spindel Spinning Machine Control**

- STM 32 Microcontroller Based Data Acquisition (DAQ) System
- DAQ Includes: Electric Panel (tower type)
- Touch Screen HMI, LCD display Motherboard Core I5 with following specifications.
    - Processor: Intel – I5
    - Ram: 4 GB
    - Storage: 500 GB
    - Wi-Fi Card, Bluetooth, Keyboard
    - Windows 10 Operating System
- Data Monitoring and saving Module.
- Emergency stop function.
- Component to integrate with the control system.
- Motor for spindle rotation, 1hp, 2800rpm (available),
  Inverter for spindle rotation
- 3 x stepper motors with drives for drafting rollers(available)
  3.5A (1 x stepper motors with drives for Lifting of ring rail – available)
- Vacuum pump-800W, 0-18,000 rpm (available)
- DC Power supply 20A, Volt: 24V (available)

Yours faithfully,
For and on behalf of GIK Institute, Topi

ABC
Dy. Director (Procurement)

M/s Edutech Lahore

DATE:	GIKI - STORE RECEIPT & INSPECTION REPORT	SR&IR #.

Purchased from / by Procurement Indented by: For Faculty/ Department/ Section Ref. Bill/Inv. No. Ref. IV#.

S #	Description of Store Items	QUANTITY	A/C UNIT	UNIT PRICE	TOTAL PRICE	CODE NOS.	LEDGER
1							
2							
3							
4							
5							
6							
7							
8							
9							
10							

It is certified that the store received as mentioned above, are in accordance with specification. (All concerned are requested to kindly put the Name, designation and date at the time of singing this Inspection)

Name of Inspector Designation Signature Name of the Head of Dept. Designation & Signature

COUNTERSIGNED:

RECIPIENT NAME, DESIGNATION & SIGNATURE (STORE) **Director (Facilitation)**

(Kindly keep the packing material in original shape/ conditions of all warranted equipment's at the end period of warranty time please.)

As per decision of the Governing Council chaired by the Rector on 20th, November 2008, every Indenter is responsible/ bound to provide an Acceptance / Replacement/ Rejection report within FOUR DAYS on receipt of an item, in QUALITY, QUANTITY, and SPECIFICATION of the MATERIALS received in/ delivered to Indenter.

Suite of software facilitating computer aided sourcing that can help design exercises with best possible Sourcing are listed below. This suite of software can configure rules of procurement or may also suggest rules to adopt, thus simplifying the intricate clerical procedures. A pool of suppliers/vendors can be developed for all classifications of purchases that a company makes.

Their servers can access prices of miscellaneous products with technical detail such as DXF, IGES design data of the product that can be amalgamated with total design and many possibilities can be explored e.g., design concepts, cost comparison, and manufacturability.

This suite of software lacks the digital twining concept given as an example under CSG modeling in Part III, Chap. 7.

The links to different software sites are only given for information. These are public sites of the software packages, and their terms of use are clearly defined at each website.

SAP Ariba (Module of General-Purpose ERP Software Package)

https://www.sap.com/products/spend-management.html

Oracle (Module of General-Purpose ERP Software Package)

Applications/Oracle E-Business Suite/Oracle Sourcing

Oracle Sourcing | Oracle Products

Autodesk (Discrete Engineering Design and Manufacturing)

Autodesk | Autodesk responsible sourcing

SolidWorks (Discrete Engineering Design and Manufacturing)

https://www.solidworks.com/3dexperience-marketplace

aPriori Solutions for Sourcing

https://www.apriori.com/solutions/roles/sourcing-professionals/#:~:text=aPriori%20Solutions%20for%20Sourcing%20Professionals&text=Provide%20detailed%20cost%20data%20for,avoid%20capacity%20and%20quality%20risks.&text=Generate%20detailed%20cost%20estimates%20including%20secondary%20processes

2.7 Production Philosophies

Hitomi (1975) defines manufacturing as comprising essentially all the tangible operations and production as encompassing both tangible and intangible operations. These operations for the sake of maximizing productivity and profitability are the essence for manufacturing philosophies. A contemporary list of these operations are as follows:

1. Group technology (GT)
2. Just in time (JIT)
3. Material resource planning (MRP)
4. MRPII
5. Enterprise resource planning (ERP)
6. Supply Chain Management
7. The theory of constraints (TOC)
8. Lean manufacturing
9. Six Sigma
10. Computer-integrated manufacturing (CIM)
11. KAIZEN
12. Agile manufacturing.

A manufacturing system may operate under a single production philosophy or under multiple production philosophies. The purpose of using single or multiple production philosophies is to maximize productivity to increase profitability and thus maximize wealth creation. Big names in manufacturing while working with their production system under some guidelines for the quality and safety of the product have developed variants of the above-named philosophies and incorporated them at their main production line, satellite production line, and vendor's production line to have consistency in operability and to reduce lead time and maximize the profits.

The production philosophies named above sometimes have a distinct feature for their application e.g. Group Technology is applicable to the manufacturing of parts that can be classified in some forms of groups based on geometry, assembly, or operation. The theory of constraints is applicable to any production line where a momentary or permanent bottleneck is present. Six Sigma mobilizes all elements of production including the workers in a quality-based environment and generates profits out of it.

2.8 Production

Types of Discrete Production Machinery are:

1. Strategic Machine Tools
2. Non-Strategic Machine Tools.

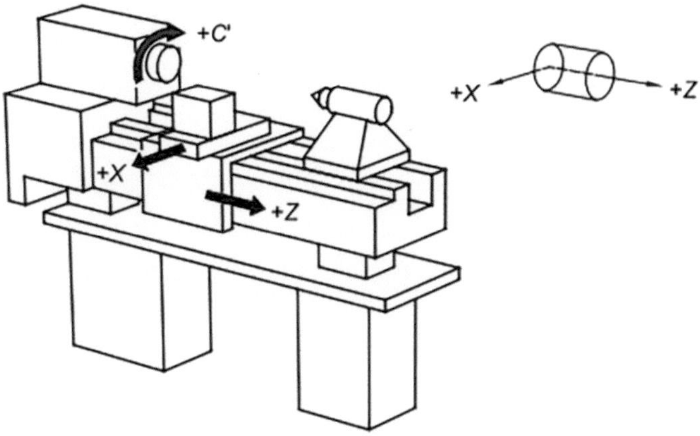

Fig. 2.36 Engine Lathe with axes orientation and machine tool configuration[23]

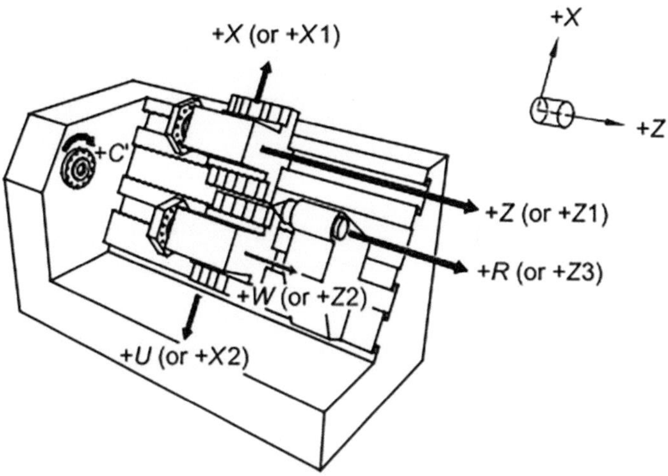

Fig. 2.37 Twin Turret Lathe with axes orientation and machine tool configuration[24]

Strategic Machine Tools

A Machine Tool that can produce any discrete product including Non-Strategic Machines is called a strategic machine tool. It includes both subtractive and additive discrete manufacturing processes. It also includes continuous manufacturing processes (Figs. 2.36 and 2.37).

[23] Extracted from ISO 841: 2001: Axis orientation and machine tool configuration of Engine Lathe. Standard licensed to Wasim Ahmed Khan.

[24] Extracted from ISO 841: 2001: Axis orientation and machine tool configuration of Twin Turret Lathe. Standard Licensed to Wasim Ahmed Khan.

Fig. 2.38 Milk packing machine[25]

Non-strategic Machines

A Machine that produces food, shelter, apparel, and other items of general use is called a non-strategic machine tool. It may be a discrete or continuous manufacturing machinery (Figs. 2.38 and 2.39).

URL-2:

https://nhmpak.com/equipment/by-package/pouch/milkpack-6000

URL-3:

https://mfipk.com/product/wheat-thresher/

---

[25] Browsed on 17th February 2024.

Fig. 2.39 Wheat thresher[26]

The systems engineering design of discrete products is composed of:

1. Mechanical Engineering Aspects of the Machine
2. Electrical Engineering Aspects of the Machine
3. Electronics Engineering Aspects of the Machine
4. Computer Hardware Engineering Aspects of the Machine
5. Computer Software Engineering Aspects of the Machine
6. Control Engineering Aspects of the Machine
7. Any other Engineering Aspect related to a specific product
8. Any Enforcement Aspect (Such as Carbon Footprint or emission of gasses affecting Ozone layer)
9. Reliability, Availability, Maintainability and Safety (RAMS) of the Machine
10. Ergonomics of the Machine (MMI, HCI, and other Parameters).

The systems engineering design to production life cycle has the following functions:

1. System Engineering Design
2. Design Documentation
3. Simulation
4. Physical Model
5. Prototyping
6. Pilot Production
7. Production Documentation
8. Production (Fig. 2.40).

[26] Browsed on 17th February 2024.

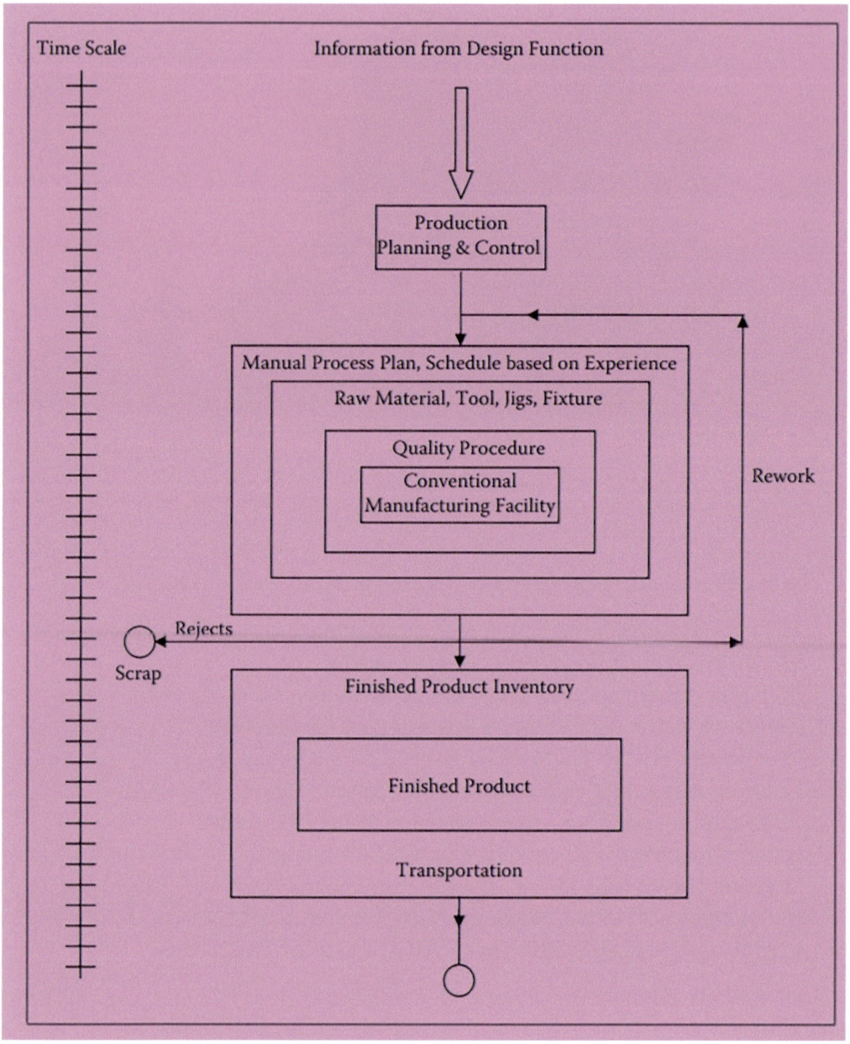

Fig. 2.40 Conventional approach to manufacturing[27]

The conventional approach to manufacturing is an obsolete method of discrete manufacturing. The work is performed as a 'Function (Individual's Work)' rather than a 'Process (Team's Work)'. Also, there is no clear demarcation between different modules of discrete manufacturing.

[27] Figure reproduced with Permission from CRC Press, Taylor and Francis Group, USA.

The system approach to manufacturing is a modern method of discrete manufacturing. The work is performed as a 'Process (Team's Work)'. Also, there is a clear demarcation between different modules of discrete manufacturing. Time spent on engineering design is of essence and conformance, quality control/assurance and environmental standards play an important role by getting public attention though the price of the product is higher than the products produced otherwise. There is a clear demarcation between different modules of system approach to discrete manufacturing and interfaces between different modules portray the efficiency of the approach (Fig. 2.41).

2.9 Case Study: Design and Manufacturing of a 5-Axis CNC Milling Machine[28]

(a) Control specifications.
(b) Design report.
(c) Video link of Proposed Functionality (User eXperience).

5-Axis CNC Milling Machine

APPROVED PART LIST (APL)

Prepared by: (Edutech Lahore with advice from FME, GIK Institute)

Name: XYZ Function: Mechanical Design Engineer

Approved by: (Edutech Lahore)

Name: ABC Function: Project Manager
Name: DEF Function: Consultant

DOCUMENT REVISION RECORD

Pages not listed remain at the released revision level.

Pages affected by the current ECN reflect the present document's revision level.

[28] A collaborative design by FME, GIK Institute and M/s Edutech. Reproduced with permission from M/s Edutech, Lahore, Pakistan.

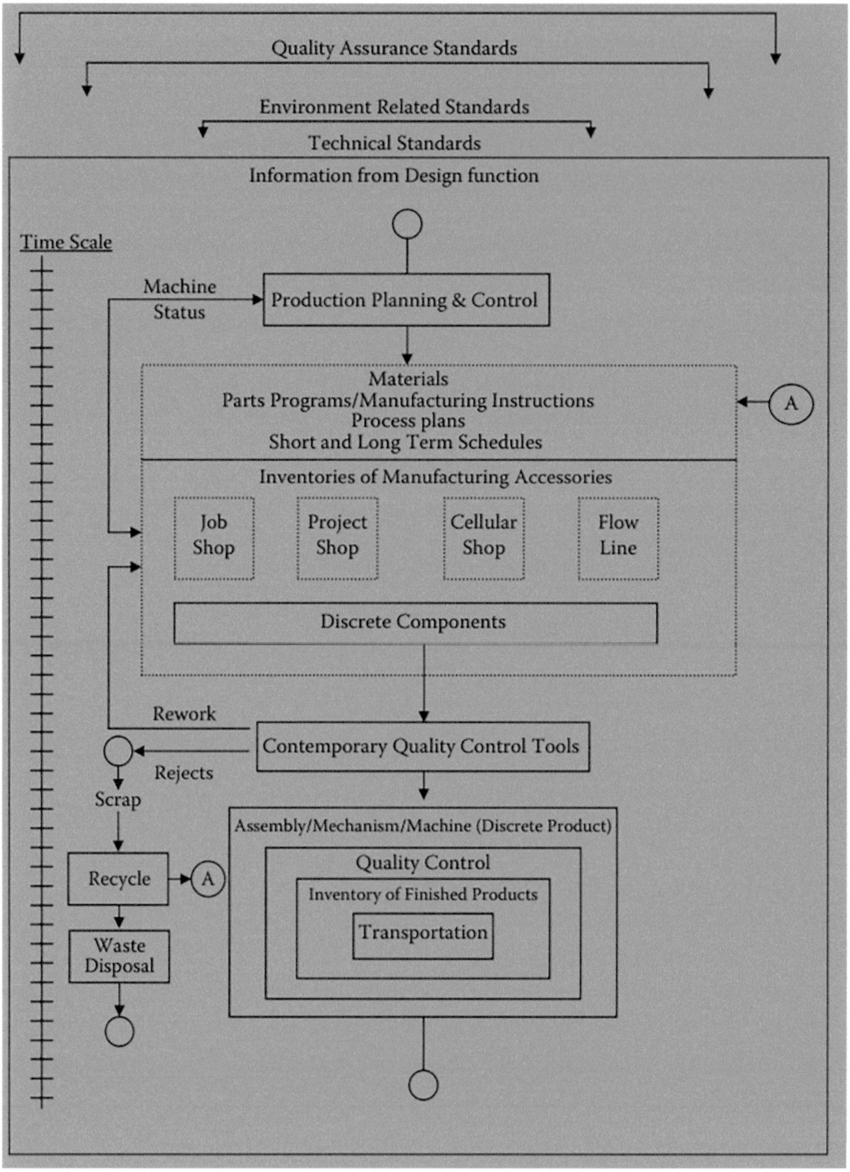

Fig. 2.41 System approach to manufacturing[29]

[29] Figure reproduced with permission from CRC Press, Taylor and Francis Group, USA.

Revision Level	ECN Number	Date	Checked by	CM Release	Pages Affected

Evolution of Concept: 5-Axis CNC Milling Machine

The development of the 5-Axis Milling machine took an evolutionary cycle whereby the system engineering design and manufacturing of the product was considered and implemented. All aspects of system engineering design i.e., mechanical, electrical, electronics, control, computer software and computer hardware had a role to play

As a first step concept design was performed (please refer to Part I, Chap. 1, Sect. 1.17) and concept generation and selection methods were applied. Initially a wooden model was developed to check the concept in terms of following aspects:

I. Mechanical
II. Electrical
III. Electronic
IV. Computer Hardware, and
V. Computer Software.

The picture of the knee and column type concept physical implementation in wood is presented in Fig. 2.40.

After the successful operation of the wooden model as per desired criterion, in the next step special emphasis was given to complete mechanical engineering design and a metal working prototype of 5-Axis milling machine with all its intricate design implementation was developed (please refer to Fig. 2.41). SolidWorks was heavily used for the purpose while ANSYS was used for Numerical Analysis. The required electronic circuitry was also given emphasis and the circuit board was developed using Proteus software. The User Interface (UI) was developed in Python and control of all 5-Axis was demonstrated.

In the third stage of evolution, different specifications of industrial scale 5-Axis Milling Machine are in hand with improved engineering design (please refer to Fig. 2.42a–c) with more intricate mathematics as presented in Chap. 7, Sect. 7.5. Industrial grade controller design has been performed and can be accessed through following URL along with specifications and exploded views as provided below (Figs. 2.43 and 2.44):

URL[30] Arrangement of Components for Industrial Scale 5 Axis CNC Milling Machine Controller:

[30] Developed by DevoMech, Islamabad, Pakistan in consultation with Wasim Ahmed Khan.

Fig. 2.42 Wooden prototype of 5 axis CNC milling machine

https://drive.google.com/file/d/1w6RPev5FtjwRr7NEYTDDru47yw7E-JgF/view?usp=drive_link

a. **Control Specifications**

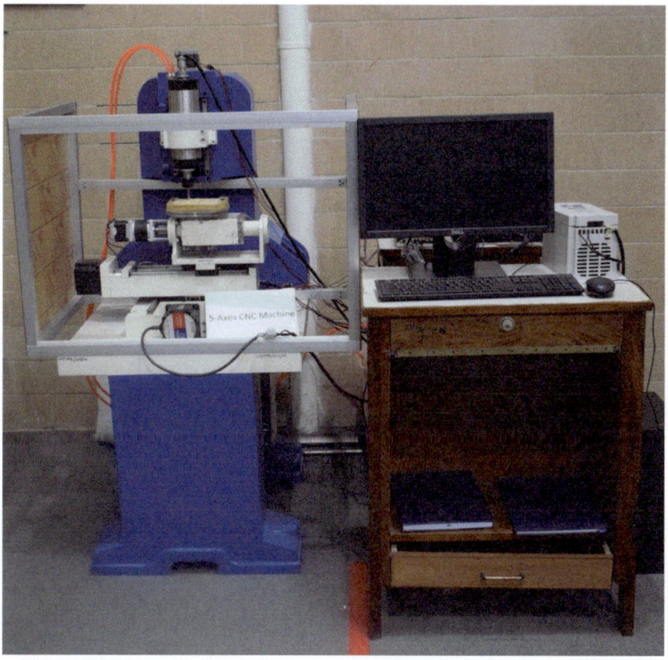

Fig. 2.43 Working metallic prototype of 5-axis milling machine

<table>
<tr><th>Sr#</th><th>Item Name/ Specification</th><th></th></tr>
<tr><td>1.</td><td>Travels X x Y x Z: 15" x 14" x 12"
(381 x 356 x 305 mm)</td><td rowspan="4">B AXIS - TILT; 120 ° to 35- °
C AXIS - ROTATIO N: 360 °</td></tr>
<tr><td>2.</td><td>Travels X x Y x Z: 24" x 16" x 16"
(610 x 406 x 406 mm)</td></tr>
<tr><td>3.</td><td>Travels X x Y x Z: 30" x 20" x 20"
(762 x 508 x 508 mm)</td></tr>
<tr><td>4.</td><td>Travels X x Y x Z: 40" x 25" x 25"
(1016 x 635 x 635 mm</td></tr>
</table>

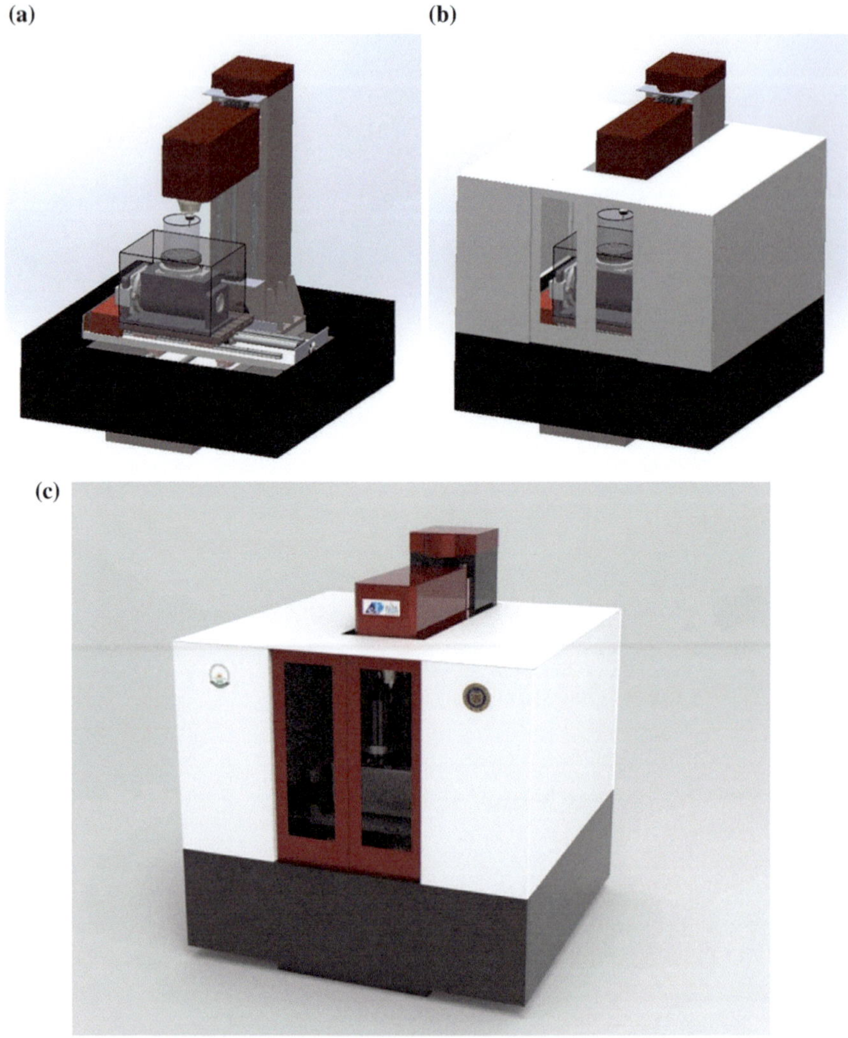

Fig. 2.44 **a** Industrial scale 5-axes CNC milling machine (Stage-1). **b** Industrial scale 5-axes CNC milling machine (Stage-2). **c** Industrial scale 5-axes CNC milling machine (Stage-3)

b. **Design Report**

b. Design Report

Contents

1. Scope

This document serves as the EEE Approved Parts List (APL) for components approved for use on 5-axis CNC Milling Machine by EDUTECH LAHORE.

2. Applicable Documents

The key EDUTECH LAHORE procedures used for this project are as follows:

AD1: Project Management Procedure
AD2: Risk Management Procedure
AD3: Design Procedure
AD4: Purchasing Procedure
AD5: Process Control (Manufacture, Assembly and Test)
AD6: Material Handling, Storage, Packaging, Preservation and Delivery.

All EDUTECH LAHORE procedures and work instructions will be followed as applicable on this project and are subject to audit by the company Quality department.

3. Applicable Standards

In general, complies with the applicable sections of the following standards:

3.1 Standards for Safety of Machine

- EN ISO 12100 Safety of machinery—General principles for design—Risk Assessment and risk reduction
- EN 60204-1 Safety of machinery—Electrical equipment of machines—Part 1: General requirements
- EN ISO 13849-1 Safety of machinery—Safety-related parts of control systems—Part 1: General principles for design
- EN ISO 13850 Safety of machinery—Emergency stop function—Principles for design
- 2002/44/EC: European parliament and of the council
- 2009/104/EC: Safety and health requirements for use of work equipment by workers at work
- 2001/95/EC Directive on General Product Safety
- 2014/35/EC: Low voltage directives.

3.2 Standards Related to 5-Axis CNC Milling Machine

Depending on the type of machinery, other standards may also be applicable on QAP documents.

4. Components

4.1 Main Component

4.2 Spindle

The spindle servo motor is a critical component in a CNC milling machine that drives the spindle, which rotates the cutting tool. It is a type of servo motor that is designed specifically for high-speed and high-precision machining applications.

Spindle servo motors are typically designed with high torque and low inertia to provide fast and accurate responses to changes in cutting conditions. They are also designed to handle the high radial and axial loads generated during cutting, while maintaining high accuracy and stability.

In a CNC milling machine, the spindle servo motor is controlled by the CNC controller, which adjusts the speed and direction of rotation based on the cutting conditions and the program instructions. The spindle speed is typically adjusted to optimize cutting performance and tool life, while minimizing cutting forces and surface roughness (Fig. 2.45).

Model number	Items description	Qty
SZGH 10-2-35 5.5/7.5-4-1500	Spindle servo motor, 5.5 kW, 35 Nm, max speed 8000 rpm, rated speed 1500 rpm, B5 FLANGE type	01
SZGH-S4T5P5	Driver, phase 380 V (matched with 5.5 kW 3 spindle)	01
SZGH 1CX-5M	Cable, 5 m, control cable	01

(continued)

(continued)

Model number	Items description	Qty
SZGH1EX-5M	Cable, 5 m, encoder cable	01
SZGH1FX-5M	Cable, 5 m, feedback cable	01
RJ-1KW	Braking resistors, 1 kW/32 RJ	01

In CNC milling machines, pneumatic cylinders are commonly used for clamping and unclamping the workpiece, as well as for tool changing and other operations that require precise and repeatable linear motion. The use of pneumatic cylinders allows for fast and reliable operation, as well as simplified control and maintenance (Fig. 2.46).

Spindle	Product name	BT40 6000 r/min belt driven spindle
	Diameter	120 mm
	Driven power	5.5 kW
	Driver	Belt drive
	Cooling	Water cooling
	Lubrication	Oil grease
Accessories	380 V three phase servo motor	
	3.5 T pneumatic cylinder	
	Driver, resistor, pulley, belt, bracket, cable	

Belt Driven Spindle

1. High precision: Dynamic balance grade is G0.4.
2. High configuration: Adopt internationally famous brand bearing and high specification standard parts.
3. High stability: The cool clamping and unclamping mechanism design which is unique, stable, and long life (Fig. 2.47).

Servo Motor

1. With Rust proof housing
2. Using a separate fan for cooling
3. The motor shaft with special key
4. Supply suitable pulley and belt (Fig. 2.48).

Driver for Servo Motor

1. 220 V or 380 V for choose
2. With durable use key
3. Screen is waterproof and oil-protection
4. With contact cable (Fig. 2.49).

Pneumatic Cylinder

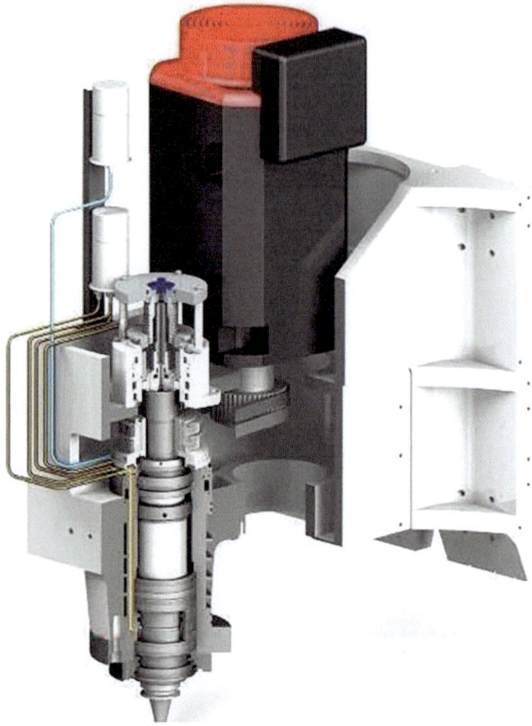

Fig. 2.45 5-axis milling spindle servo motor

1. Blow air hole is equipped with a valve to adjust air volume.
2. Solenoid valve is DC-24 V.
3. 84 × 84-hole site, use 4 pieces M10X1.5 or 3/8 16 thread screws (Fig. 2.50).

4.2.1 Ball Screw

A **ball screw** is a mechanical linear actuator that translates rotational motion to linear motion with little friction. A threaded shaft provides a helical raceway for ball bearings which act as a precision screw. As well as being able to apply or withstand high thrust loads, they can do so with minimum internal friction. They are made to close tolerances and are therefore suitable for use in situations in which high precision is necessary. The ball assembly acts as the nut while the threaded shaft is the screw. In contrast to conventional lead screws, ball screws tend to be rather bulky, due to the need to have a mechanism to recirculate the balls (Figs. 2.51 and 2.52).

Part No.	Name/description	Model/supplier part No.	Make	QTY
PTN-100306	Ball screw, x	DFUR-3210-1185	SHAC Motion Technology Co., Ltd.	01
PTN-100307	Ball screw, y	DFUR-3210-1015		01
PTN-100308	Ball screw, z	DFUR-3210-960		01

(continued)

(continued)

Part No.	Name/description	Model/supplier part No.	Make	QTY
PTN-100309	Ball nut	DFUR-3210-4		03
PTN-100310	Ball nut support	DSG32H		03
PTN-100311	BK, BF support units	BK25, BF25		06

- Static load rating: 6101 kg
- Dynamic load rating: 3092 kg.

The ball screw of nominal dia. 32 mm in the design is selected, two of the ball screws of the same specification are used (Fig. 2.53).

4.3 Linear Guide Rail

A linear guide rail, also known as a linear motion guide or linear slide, is a mechanical component used to provide linear motion to a load. It consists of a rail or track and a carriage or block that moves along the rail, providing smooth and precise linear motion.

Linear guide rails are commonly used in various industrial and engineering applications, such as machine tools, robotics, automation, and material handling systems. They are also used in consumer applications, such as sliding doors and drawers.

The rail of a linear guide rail is typically made of steel or aluminum and can be either cylindrical or square in shape. The carriage or block is usually made of a material such as aluminum or stainless steel and contains rollers or ball bearings that move along the rail, providing low friction and high accuracy.

Linear guide rails can be either preloaded or non-preloaded. Preloaded guide rails have a small amount of clearance between the rail and the carriage, which is taken up by a spring or other mechanical component, providing increased accuracy and stiffness. Non-preloaded guide rails have more clearance between the rail and the carriage, making them less stiff but more tolerant to misalignment and shock loads (Figs. 2.54 and 2.55).

Fig. 2.46 Five axis milling belt driven spindle

Part No.	Name/description	Model/supplier part No.	Make	QTY
PTN-100312	Linear rail guide, x	GHW-30HA-1050	SHAC Motion Technology Co., Ltd.	02
PTN-100313	Linear rail guide, y	GHW-30HA-870		02
PTN-100314	Linear rail guide, z	GHW-30HA-870		02
PTN-100315	Linear rail bearing	GHW-30HA		08
PTN-100316	Linear rail bearing	GHH-30HA		04

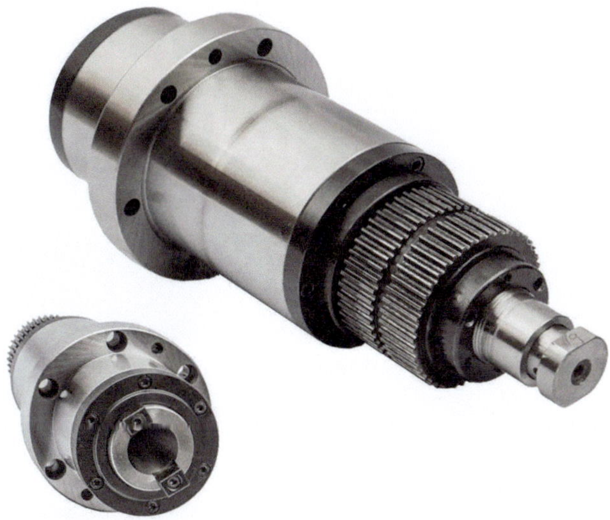

Fig. 2.47 Belt driven spindle servo motor

Fig. 2.48 Servo motor with belt

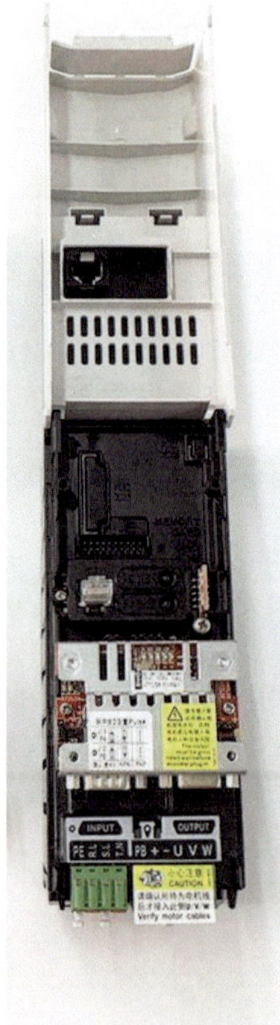

Fig. 2.49 Driver for servo motor

型號	組件尺寸 (mm)			滑塊尺寸 (mm)														滑軌尺寸 (mm)							滑軌的固定螺栓尺寸	基本動額定負荷	基本靜額定負荷	容許靜力矩			重量	
																												M_R	M_P	M_Y	滑塊	滑軌
	H	H_1	N	W	B	B_1	C	L_1	L	K_1	K_2	G	M	T	T_1	H_2	H_3	W_R	H_R	D	h	d	P	E	(mm)	C(kN)	C_0(kN)	kN-m	kN-m	kN-m	kg	kg/m
GHW 15CA	24	4.3	16	47	38	4.5	30	39.5	60.5	8	4.85	5.3	M5	6	8.9	3.95	3.7	15	15	7.5	5.3	4.5	60	20	M4x16	11.38	16.97	0.12	0.10	0.10	0.17	1.45
GHW 20CA	30	4.6	21.5	63	53	5	40	50.5	76.7	10.25	6	12	M6	8	10	6	6	20	17.5	9.5	8.5	6	60	20	M5x16	17.75	27.76	0.27	0.20	0.20	0.40	2.21
GHW 20HA								65.2	91.4	17.6																21.18	35.9	0.35	0.35	0.35	0.52	
GHW 25CA	36	5.5	23.5	70	57	6.5	45	58	84	10.7	6	12	M8	8	14	6	5	23	22	11	9	7	60	20	M6x20	26.48	36.49	0.42	0.33	0.33	0.59	3.21
GHW 25HA								78.6	104.6	21																32.75	49.44	0.56	0.57	0.57	0.80	
GHW 30CA	42	6	31	90	72	9	52	70	98.4	14.25	6	12	M10	8.5	16	6.5	10.8	28	26	14	12	9	80	20	M8x25	38.74	52.19	0.66	0.53	0.53	1.09	4.47
GHW 30HA								93	121.4	25.75																47.27	69.16	0.88	0.92	0.92	1.44	
GHW 35CA	48	7.5	33	100	82	9	62	80	112.4	14.6	7	12	M10	10.1	18	9	12.6	34	29	14	12	9	80	20	M8x25	49.52	69.16	1.16	0.81	0.81	1.56	6.30
GHW 35HA								105.0	138.2	27.5																60.21	91.63	1.54	1.40	1.40	2.06	
GHW 45CA	60	9.5	37.5	120	100	10	80	97	137.4	13	10	12.9	M12	15.1	22	8.5	20.5	45	38	20	17	14	105	22.5	M12x35	77.57	102.71	1.98	1.55	1.55	2.79	10.41
GHW 45HA								128.8	169.2	28.9																94.54	136.46	2.63	2.68	2.68	3.69	
GHW 55CA	70	13	43.5	140	116	12	95	117.7	166.7	17.35	11	12.9	M14	17.5	26.5	12	19	53	44	23	20	16	120	30	M14x45	114.44	148.33	3.69	2.64	2.64	4.52	15.08
GHW 55HA								155.8	204.8	36.4																139.35	196.2	4.88	4.57	4.57	5.96	
GHW 65CA	90	15	53.5	170	142	14	110	144.2	200.2	23.1	14	12.9	M16	25	37.5	15	15	63	53	26	22	18	150	35	M16x50	163.63	215.33	6.65	4.27	4.27	9.17	21.18
GHW 65HA								203.6	259.6	52.8																208.36	303.13	9.38	7.38	7.38	12.89	

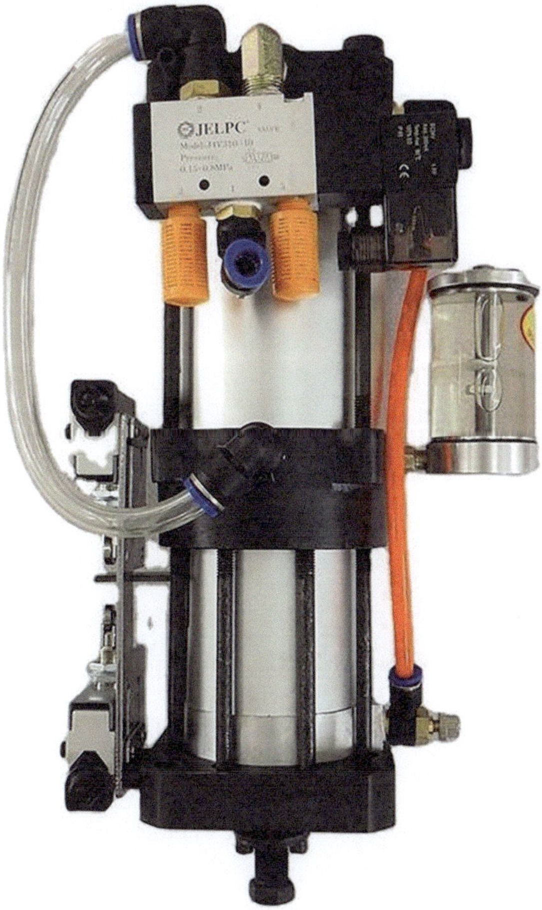

Fig. 2.50 Pneumatic cylinder

4.4 Servo Motor

A servo motor is a type of electric motor that is widely used in industrial and engineering applications to provide precise and accurate control of motion. They use a feedback control mechanism that allows them to rotate to a specific position and maintain that position with high accuracy. Servo motors are available in various sizes and power ratings and can be configured for different types of motion. They are commonly used in robotics, CNC machines, automation systems, and consumer applications, such as RC cars and drones. The servo motors are a critical component in providing precise and accurate motion control in a wide range of applications (Figs. 2.56, 2.57 and 2.58).

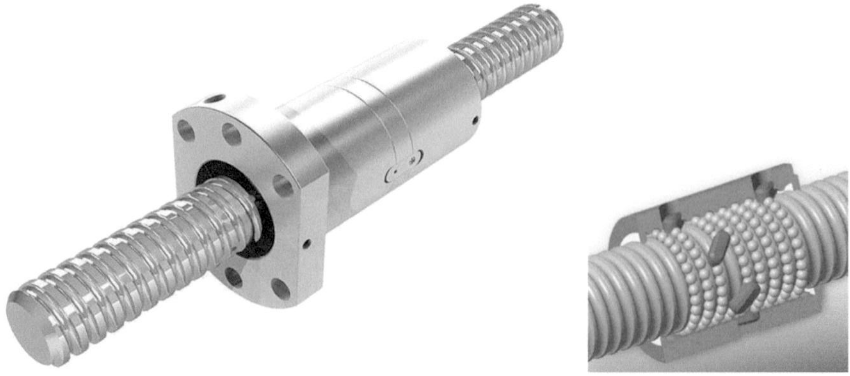

Fig. 2.51 Ball screw

Part No.	Name/description	Model/supplier part No.	Make	QTY
PTN-100317	Servo motor	130DNMA2-0002C	Dorna Technology Co., Ltd.	03

5. Components Selection

The components listed in Appendix A are approved for use on EDUTECH LAHORE 5-axis CNC Milling Machine product.

All components are purchased in accordance with the listed part number and the applicable procurement specification. The parts are approved for use in 5-axis CNC Milling Machine hardware only when screened and qualified as indicated by the part number and the governing specifications.

6. Explanation of Acronyms

Acronyms represent the part.

6.1 Acronyms Under Part Number

Acronyms under "Part Number" required for parts listed for the 5-axis CNC Milling Machine from Edutech Lahore.

6.2 Acronyms Under Name/Description

Acronyms under "Name/Description" represent the name and description of the SCD parts.

6.3 Acronyms Under Model/Supplier Part Number

Acronyms under "Model/Supplier Part Number" describe the Model and/or part number from the Supplier.

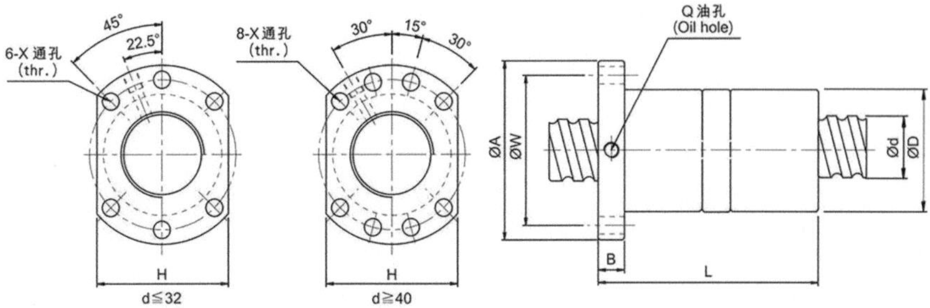

單位:mm

	公稱型號 Model no.	軸徑 d	導程 l	珠徑 Da	螺帽尺寸									動額定負荷 Ca	靜額定負荷 Coa
					D	A	B	L	W	H	X	Q	n		
	DFUR1604-3	16	4	2.381	28	48	10	80	38	40	5	M6×1	1×3	488	940
☆	DFUR1605-4		5	3.175	28	48	10	100	38	40	5	M6×1	1×4	885	1525
☆	DFUR1610-3		10	3.175	28	48	10	118	38	40	5	M6×1	1×3	716	1232
	DFUR2004-3	20	4	2.381	36	58	10	80	47	44	6.6	M6×1	1×3	541	1187
☆	DFUR2005-4		5	3.175	36	58	10	101	47	44	6.6	M6×1	1×4	999	1994
	DFUR2504-3	25	4	2.381	40	62	10	80	51	48	6.6	M6×1	1×3	605	1534
☆	DFUR2505-4		5	3.175	40	62	10	101	51	48	6.6	M6×1	1×4	1119	2581
☆	DFUR2510-4		10	4.762	40	62	12	145	51	48	6.6	M6×1	1×4	1927	2771
☆	DFUR3205-4	32	5	3.175	50	80	12	102	65	62	9	M6×1	1×4	1264	3402
	DFUR3210-4		10	6.35	50	80	12	162	65	62	9	M6×1	1×4	3092	6101
	DFUR4005-4	40	5	3.175	63	93	14	105	78	70	9	M8×1	1×4	1407	4341
	DFUR4010-4		10	6.35	63	93	14	165	78	70	9	M8×1	1×4	3480	7979
	DFUR5010-4	50	10	6.35	75	110	16	171	93	85	11	M8×1	1×4	3898	10325
	DFUR6310-4	63	10	6.35	90	125	18	182	108	95	11	M8×1	1×4	4401	13611
	DFUR6320-4		20	9.525	95	135	20	290	115	100	13.5	M8×1	1×4	7404	19008
	DFUR8010-4	80	10	6.35	105	145	20	182	125	110	13.5	M8×1	1×4	4900	17366
	DFUR8020-4		20	9.525	125	165	25	295	145	130	13.5	M8×1	1×4	8403	25345

備註：有標註☆記號者可製作左螺纹。

Fig. 2.52 Ball screw selection (DFU 3210-4)

6.4 Acronyms Under Make

Acronyms under "Make" describe the authorized Company to be selected for parts listed for the 5-axis CNC Milling Machine.

6.5 Acronyms Under QTY

Acronyms under "QTY" describe the quantity required for parts listed for the 5-axis CNC Milling Machine.

7. Appendix-A Approved Components

Ball Screw Support

BK
Fixed Side

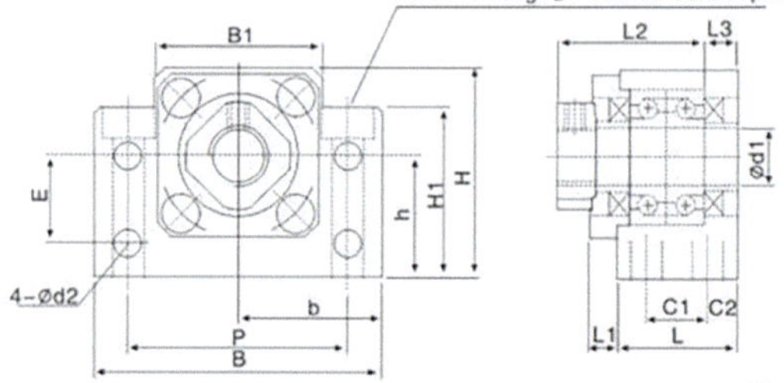

Unit:mm

Model Number	d1	L	L1	L2	L3	C1	C2	B	H	b ±0.02	h ±0.02	B1	H1	E	P	d2	X	Y	Z
BK 10	10	25	5	29	5	13	6	60	39	30	22	34	32.5	15	46	5.5	6.6	10.8	5
BK 12	12	25	5	29	5	13	6	60	43	30	25	34	32.5	18	46	5.5	6.6	10.8	1.5
BK 15	15	27	6	32	6	15	6	70	48	35	28	40	38	18	54	5.5	6.6	11	6.5
BK 17	17	35	9	44	7	19	8	86	64	43	39	50	55	28	68	6.6	9	14	8.5
BK 20	20	35	8	43	8	19	8	88	60	44	34	52	50	22	70	6.6	9	14	8.5
BK 25	25	42	12	54	9	22	10	106	80	53	48	64	70	33	85	9	11	17.5	11
Bk 30	30	45	14	61	9	23	11	128	89	64	51	76	78	33	102	11	14	20	13
Bk 35	35	50	14	67	12	26	12	140	96	70	52	88	79	35	114	11	14	20	13
BK 40	40	61	18	76	15	33	14	160	110	80	60	100	90	37	130	14	18	26	17.5

BF
Floated Side

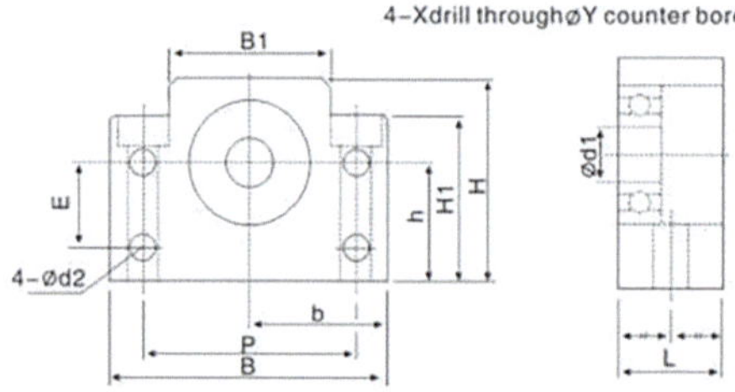

Unit:mm

Model Number	d1	L	B	H	b ±0.02	h ±0.02	B1	H1	E	P	d2	X	Y	Z
BF 10	8	20	60	39	30	22	34	32.5	15	46	5.5	6.6	10.8	5
BF 12	10	20	60	43	30	25	34	32.5	18	46	5.5	6.6	10.8	1.5
BF 15	15	20	70	48	35	28	40	38	18	54	5.5	6.6	11	6.5
BF 17	17	23	86	64	43	39	50	55	28	68	6.6	9	14	8.5
BF 20	20	26	88	60	44	34	52	50	22	70	6.6	9	14	8.5
BF 25	25	30	106	80	53	48	64	70	33	85	9	11	17.5	11
BF 30	30	32	128	89	64	51	76	78	33	102	11	14	20	13
BF 35	35	32	140	96	70	52	88	79	35	114	11	14	20	13
BF 40	40	37	160	110	80	60	100	90	37	130	14	18	26	17.5

SHAC® 124

Fig. 2.53 Ball screw support selection (BF25 and BK25)

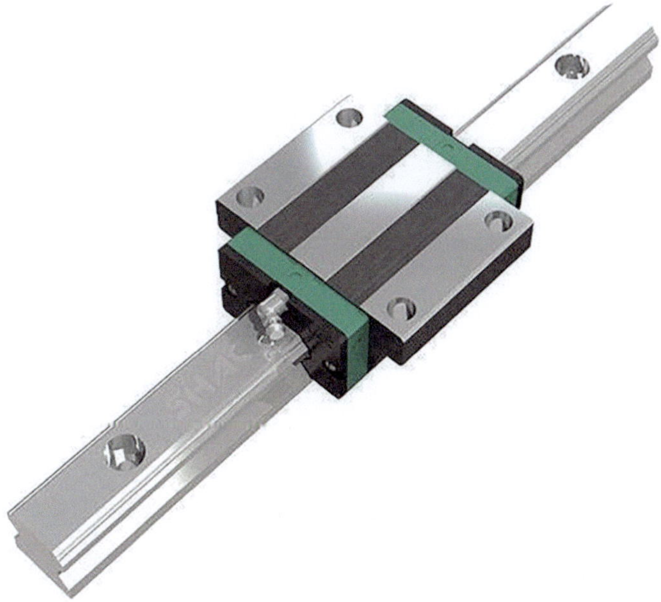

Fig. 2.54 Linear guide rails

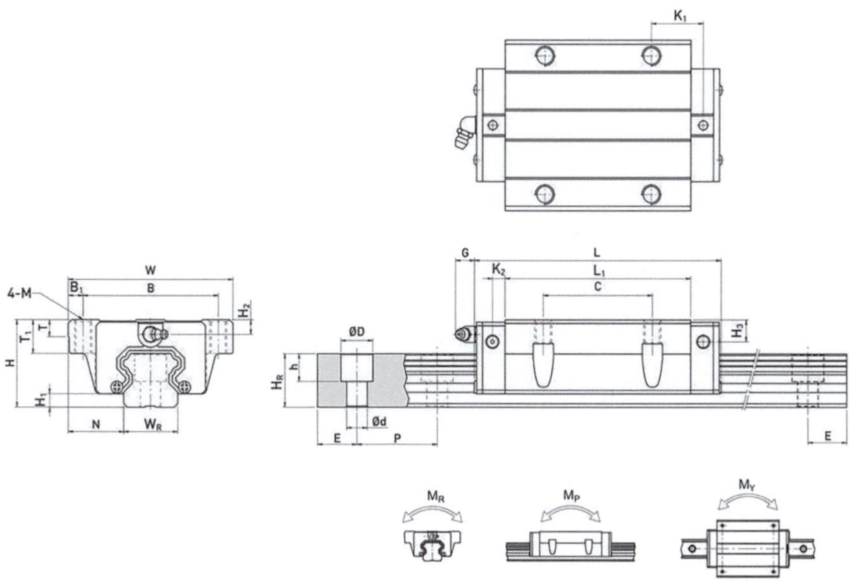

Fig. 2.55 Catalog description of linear guide rail and dovetail support

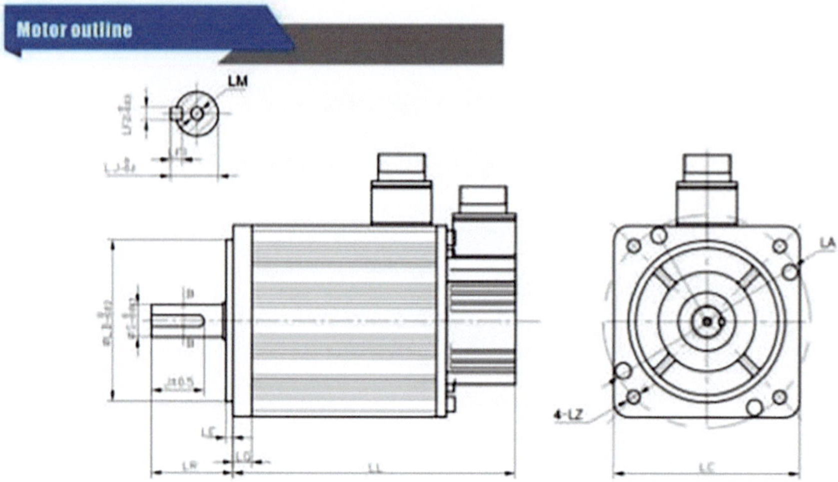

Fig. 2.56 Selection of servo motor

Fig. 2.57 Servo motor

SERVO MOTOR SPECIFICATIONS & DIMENSIONS

130 Series

Servo Motor series	130 Series				
Servo Motor model	130DNMA2-0001C	130DNMA2-01D3B	130DNMA2-01D5C	130DNMA2-0002C	130DNMA2-0003C
Input voltage	220VAC	220VAC	220VAC	220VAC	220VAC
Inertia			Medium		
Rated power (W)	1000	1300	1500	2000	3000
Rated torque (N*m)	4.77	8.27	7.16	9.55	14.33
Rated current (A)	5	9.32	8.4	10.3	13.5
Maximum current (A)	15	28	25.2	30.1	40.5
Rated speed (rpm)	2000	1500	2000	2000	2000
Maximum speed (rpm)	3000	3000	3000	3000	3000
Torque constant (N*m/Amp)	0.95N.m/Arms	0.89N.m/Arms	0.85N.m/Arms	0.93N.m/Arms	1.07N.m/Arms
Back EMF constant(V/Krpm)	66V/Krpm	61.7V/Krpm	59.8V/Krpm	72.6V/Krpm	76V/Krpm
Rotary inertia (with brake)(10^{-4}Kg*m^2)	7.1 (7.5)	12.1 (12.5)	10.6 (11.1)	13.8 (14.3)	20.4 (20.9)
Resistance (line-line) (Ω)	1.08	0.45	0.543	0.52	0.32
Inductance (line-line) (mH)	12.8	5.7	6.3	6.8	4.7
Mass (with brake) (kg)	6.5 (8.8)	7.6 (9.9)	8 (10.5)	9.6 (11.9)	12.6 (14.9)
LL (with brake) (mm)	154 (198)	182 (226)	173 (217)	192 (236)	230 (274)
LR (mm)	58	58	58	58	58
LE (mm)	6	6	6	6	6
LG (mm)	12	12	12	12	12
S (mm)	22	22	22	22	22
LJ1 (mm)	0	0	0	0	0
LJ (mm)	18	18	18	18	18
J (mm)	36	36	36	36	36
LF1 (mm)	7	7	7	7	7
LF2 (mm)	8	8	8	8	8
LM (mm)	M6 deep 15	M6 deep 15	M6 deep 15	M6 deep 15	M6 deep 15
LA (mm)	145	145	145	145	145
LB (mm)	110	110	110	110	110
LC (mm)	130	130	130	130	130
LZ (mm)	9.5	9.5	9.5	9.5	9.5

Fig. 2.58 Servo motor selection based on torque (connecting variable) and other parameters

Part No.	Name/description	Model/supplier part No.	Make	QTY
PTN-100301		SZGH 10-2-35 5.5/ 7.5-4-1500		01
PTN-100302				01
PTN-100303				01
PTN-100304				01
PTN-100305				01
PTN-100306	Ball screw, x	DFUR-3210-1185	SHAC Motion Technology Co., Ltd.	01
PTN-100307	Ball screw, y	DFUR-3210-1015		01
PTN-100308	Ball screw, z	DFUR-3210-960		01
PTN-100309	Ball nut	DFUR-3210-4		03
PTN-100310	Ball nut support	DSG32H		03
PTN-100311	BK, BF support units	BK25, BF25		06
PTN-100312	Linear rail guide, x	GHW-30HA-1050		02
PTN-100313	Linear rail guide, y	GHW-30HA-870		02
PTN-100314	Linear rail guide, z	GHW-30HA-870		02
PTN-100315	Linear rail bearing	GHW-30HA		08
PTN-100316	Linear rail bearing	GHH-30HA		04
PTN-100317	Servo motor 130 DNMT1-0001ABAM	130DNMA2-0002C	Dorna Technology Co., Ltd.	03
PTN-100318	Driver, servo motor	EPS-B1-0001AA-0000		03
PTN-100319	Timing belt pulleys	ID19, OD32, W19mm	Local	06
PTN-100320	Rotary and tilting table	–	Taiwan	01
PTN-100321	Telescoping bellow cover, x	xx	Cangzhou Shenghao Machine Tool Accessories Co., Ltd.	02
PTN-100322	Telescoping bellow cover, y	yy		02
PTN-100323	Armored bellow cover, z	zz		02
PTN-100324	Leveling pad for main body	D40L150	Local	06
PTN-100325	Leveling pad, for covering sheet	D8L75	Local	04
PTN-100326	Belt timing pulley, x	W18Lx	Local	01
PTN-100327	Belt timing pulley, y	W18Ly	Local	01
PTN-100328	Belt timing pulley, z	W18Lz	Local	01

B. 5-Axis CNC Milling Machine Design Report (DR)

Prepared by: (EduTech Lahore in collaboration with FME, GIK Institute)

Name: ABC
Function: Mechanical Design Engineer

Approved by: (Edutech Lahore)

Name: DEF
Function: Technical Approval

Approved by: (Edutech Lahore)

Name: GHI
Function: Project Manager

DOCUMENT REVISION RECORD

Pages not listed remain at the released revision level.

Pages affected by the current ECN reflect the present documents' revision level.

Revision Level	ECN Number	Date	Checked by	CM Release	Pages Affected

Contents

1. Scope

The Design Report (DR) document serves as the Design, Development, and Validation of CNC Milling 5-axis by EDUTECH LAHORE.

2. Introduction

A design report is a document that provides a comprehensive overview of a project's design process, including the objectives, constraints, design decisions, and outcomes.

In the context of engineering, a design report is a critical component of the design process, as it allows designers to communicate their ideas and solutions effectively.

The introduction of a design report should provide a brief overview of the project, including the problem statement, project scope, and objectives. It should also outline the approach taken to design the solution, including any design constraints and assumptions that were made.

The introduction should also briefly describe the design methodology used, such as the design thinking process or the engineering design process. It should also provide an overview of the design team and their roles in the project.

2.1 CNC Milling 5-Axis in Engineering

CNC milling is a type of machining process that involves the use of computer-controlled machines to remove material from a workpiece. 5-axis CNC milling is a specialized technique that uses a CNC milling machine with five axes of motion to achieve complex shapes and contours in a workpiece.

In engineering, 5-axis CNC milling is commonly used in the production of complex parts for industries such as aerospace, automotive, and medical. The ability to produce highly accurate and intricate parts with a high degree of precision makes 5-axis CNC milling a valuable tool in the manufacturing process.

One advantage of 5-axis CNC milling is that it reduces the need for multiple setups and fixtures, which can save time and improve accuracy. The five axes of motion allow the machine to access hard-to-reach areas and to produce complex geometries with fewer operations, reducing the risk of error.

Another advantage of 5-axis CNC milling is that it can produce parts with better surface finishes and tolerances than traditional 3-axis machining. The ability to tilt the tool and workpiece in multiple directions allows for smoother transitions between surfaces and tighter tolerances. Overall, 5-axis CNC milling is a powerful tool in engineering that allows the production of complex parts with a high degree of accuracy and precision.

2.2 Use in Engineering Applications

NC milling 5-axis technology has numerous engineering applications due to its ability to produce complex geometries and parts with high precision and accuracy. Some of the most common engineering applications of CNC milling 5-axis include:

(1) **Aerospace**: The aerospace industry uses CNC milling 5-axis technology to produce complex parts for aircraft and spacecraft, such as turbine blades, engine components, and wing sections.
(2) **Automotive**: CNC milling 5-axis technology is also used in the automotive industry to produce high-precision engine components, transmission parts, and other critical components.
(3) **Medical**: CNC milling 5-axis technology is used in the medical industry to produce complex surgical instruments, implants, and prosthetics.

(4) **Mold Making**: CNC milling 5-axis technology is used in mold making to produce complex and intricate molds used in the manufacturing of products, such as plastic components and metal castings.
(5) **Tool and Die Making**: The ability of CNC milling 5-axis to produce complex geometries with high precision makes it ideal for tool and die making, where precision and accuracy are critical for producing high-quality components.
(6) **Prototyping**: CNC milling 5-axis technology is used in prototyping to quickly produce complex prototypes for testing and evaluation.

CNC milling 5-axis technology is a powerful tool in engineering applications, allowing the production of complex geometries with high precision and accuracy. Its ability to reduce the need for multiple setups and fixtures makes it a cost-effective solution for producing complex parts.

2.3 Design and Model

To manufacture CNC Milling 5-axis, it is easily transferable, various aspects and considerations need to be made. Use of Computer Aided Design (CAD) models using SOLIDWORKS software can be used to develop 3D parts and assembly for an understanding of the model. Calculation of load, right specification for a part used, and various other factors are considered as well through analytical calculation and Analysis report. Selection of the procurement part is also based on the required load. After passing this evaluation it's a final Design for Manufacturing and Assembly (DFMA).

3. Design Specification and Component List

The specification given for the design of CNC Milling 5-axis include the (Table 4.1):

Table 4.1 Component part list

Part No.	Name/ description	Model/supplier part No.	Make	QTY
PTN-100301	Spindle	SZGH10-2-35 5.5/ 7.5-4-1500		01
PTN-100302				01
PTN-100303				01
PTN-100304				01
PTN-100305				01
PTN-100306	Ball screw, x	DFUR-3210-1185	SHAC Motion Technology Co., Ltd.	01
PTN-100307	Ball screw, y	DFUR-3210-1015		01
PTN-100308	Ball screw, z	DFUR-3210-960		01
PTN-100309	Ball nut	DFUR-3210-4		03
PTN-100310	Ball nut support	DSG32H		03
PTN-100311	BK, BF support units	BK25, BF25		06
PTN-100312	Linear rail guide, x	GHW-30HA-1050		02

(continued)

(continued)

Part No.	Name/ description	Model/supplier part No.	Make	QTY
PTN-100313	Linear rail guide, y	GHW-30HA-870		02
PTN-100314	Linear rail guide, z	GHW-30HA-870		02
PTN-100315	Linear rail bearing	GHW-30HA		08
PTN-100316	Linear rail bearing	GHH-30HA		04
PTN-100317	Servo motor 130 DNMA 1-0001ABAM	130DNMA2-0002C	Dorna Technology Co., Ltd.	03
PTN-100318	Driver, servo motor	EPS-B1-0001AA-0000		03
PTN-100319	Timing belt pulleys	ID19, OD32, W19mm	Local	06
PTN-100320	Rotary and tilting table	–	Taiwan	01
PTN-100321	Telescoping bellow cover, x	xx	Cangzhou Shenghao Machine Tool Accessories Co., Ltd.	02
PTN-100322	Telescoping bellow cover, y	yy		02
PTN-100323	Armored bellow cover, z	zz		02
PTN-100324	Leveling pad for main body	D40L150	Local	06
PTN-100325	Leveling pad, for covering sheet	D8L75	Local	04
PTN-100326	Belt timing pulley, x	W18Lx	Local	01
PTN-100327	Belt timing pulley, y	W18Ly	Local	01
PTN-100328	Belt timing pulley, z	W18Lz	Local	01

4. Metal Machining and Cutting Loads

The following calculation methods and procedures for milling operations are intended to be guidelines and not absolute because of the many variables encountered in actual practice.

4.1 Metal-Removal Rates

The metal-removal rate R (sometimes indicated as MRR) for all types of milling is equal to the volume of metal removed by the cutting process each time, usually expressed as cubic inches per minute (in/min). Thus,

$$R = WHf$$

where

R metal removal rate, in^3/min
W width of cut, in
H depth of cut, in
f feed rate, inches per minute (ipm).

4.2 Feed Rate

The speed or rate at which the workpiece moves past the cutter is the feed rate f, which is measured in inches per minute (ipm). Thus,

$$f = F_t N C_{rpm}$$

where

f feed rate, ipm
F_t feed per tooth (chip thickness), in or cpt
N number of cutter teeth
C_{rpm} rotation of the cutter, rpm.

4.3 Feed per Tooth

Production rates of milled parts are directly related to the feed rate that can be used. The feed rate should be as high as possible, considering machine rigidity and power available at the cutter. To prevent overloading the machine drive motor, the feed per tooth allowable Ft may be calculated from

$$F_t = \frac{Kh_{pc}}{NC_{rpm}WH}$$

where

hpc horsepower available at the cutter (80–90% of motor rating), i.e., if motor nameplate states 15 hp, then hp available at the cutter is 0.8 to 0.9 × 15 (80–90% represents motor efficiency)
K machinability factor.

4.4 Cutting Speed

The cutting speed of a milling cutter is the peripheral linear speed resulting from the rotation of the cutter. The cutting speed is expressed in feet per minute (fpm or ft/min) or surface feet per minute (sfpm or sfm) and is determined from:

$$S = \frac{\pi Dn}{12}$$

where

S cutting speed, fpm or sfpm (sfpm is also termed spm)
D outside diameter of the cutter, in
n rotational speed of cutter, rpm.

4.5 Number of Teeth

Cutter. The number of cutter teeth N required for a particular application may be found from the simple expression (not applicable to carbide or other high-speed cutters)

$$N = \frac{f}{F_t C_{rpm}}$$

where

f feed rate, ipm
F_t feed per tooth (chip thickness), in
C_{rpm} rotational speed of cutter, rpm
N number of cutter teeth

$$K = \frac{WHf}{hp_c}$$

K metal removal factor, in^3/min/hpc
hpc horsepower at the cutter
W width of cut, in
H depth of cut, in
f feed rate, ipm

$$K = \frac{WHf}{hp_c}$$

$$h_{pc} = \frac{WHf}{K}$$

Here K is the machinability factor.

$$h_{pc} = \frac{30 \times 12 \times 0.75 \times H}{0.7}$$

$$h_{pc} = 385.7\,H$$

$$N = \frac{f}{F_t C_{rpm}}$$

$$N = 19.5\sqrt{R} - 5.8$$

$$N = 19.5\sqrt{0.36} - 5.8$$

$$N = 6$$

$$F_t = \frac{200}{N C_{rpm}}$$

$$F_t = \frac{200}{6 \times 3200}$$

$$F_t = 0.0104\,\text{m}$$

5. General Parameters

Parameter	Notation	Value	Unit
Pitch of ball screw	pl	0.005	m
Coefficient of friction	meu	0.2	
Weight of lead screw	W	7	kg
Rotation of ball screw	N	893	rpm
Radius of lead screw	r	0.016	m
Modulus of electricity of screw material	E	2.10E+11	pa
Area of cross section of ball screw	A	2.0E−04	m^2
Radius of tool cutter	r tool	8.00	mm
Speed of cutter or spindle motor	rpm	6000	rpm
Cutting speed S as given for MS	S	92	ft/min
Outside diameter of the cutter	D	10.00	mm
Radius of tool cutter	r	5.00	mm
Depth of cut	H	5.00	m

(continued)

(continued)

Parameter	Notation	Value	Unit
Moment of inertia total	I	35.4208896	
Power of motor	P	2.6	hp
Motor efficiency	eta	0.96	%
Linear acceleration	a	2.5	m/s^2
Angular acceleration	alpha	93.47	rad/s^2

6. Z-Axis Calculations

See Fig. 2.59.

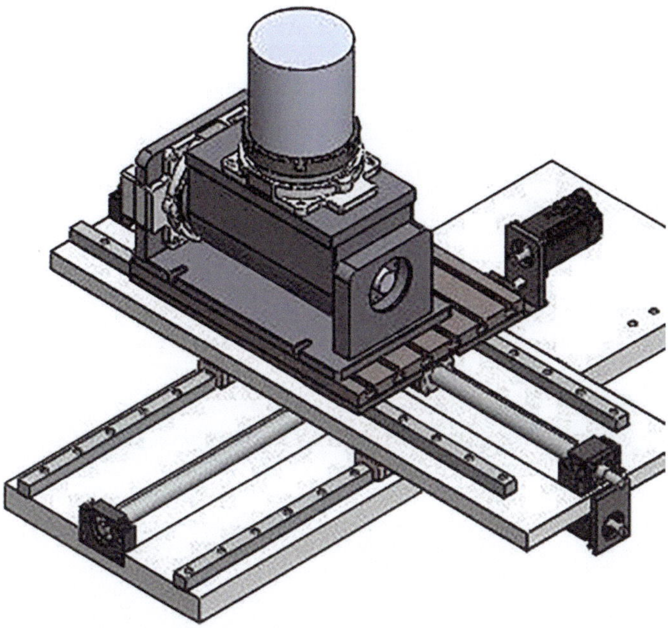

Fig. 2.59 Coordinate calculations

Total mass of z axis	m	**95**	kg
Inertial force	Fi	3546.7	N
Gravitational force	Fg	931.95	N

(continued)

(continued)

Frictional force	Ff	186.39	N
Machining Force	Fm	163.3	N
Thrust force	Fa	3896.4	N
Moment of inertia for load	J load	35.42	kg/m^2
Moment of inertia for screw	J screw	0.0008896	kg/m^2
Torque for z axis	T	*3.2*	Nm

7. Y-Axis Calculations

Mass of specimen		555	
Mass of y axis		75	
Total mass of z axis	m	95	kg
Inertial force	Fi	4875.7	N
Gravitational force	Fg	6180.3	N
Frictional force	Ff	1236.06	N
Machining force	Fm	163.3	N
Thrust force	Fa	6275.1	N
Moment of inertia for load	J load	35.42	kg/m^2
Moment of inertia for screw	J screw	0.0008896	kg/m^2
Torque for z axis	T	*5.2*	Nm

8. X-Axis Calculations

Mass of specimen		555	
Mass of y axis		198	
Total mass of z axis	m	753	kg
Inertial force	Fi	5181.2	N
Gravitational force	Fg	7386.93	N
Frictional force	Ff	1477.386	N
Machining force	Fm	163.3	N
Thrust force	Fa	6821.9	N
Moment of inertia for load	J load	35.42	kg/m^2
Moment of inertia for screw	J screw	0.0008896	kg/m^2
Torque for z axis	T	*5.7*	Nm

8.1 Comparison of Torque

	Designed torque (Nm)	Actual anticipated torque (Nm)	Factor of safety (FoS)
Servo motor with ball screw for x-axis	9.55	5.7	1.67
Servo motor with ball screw for y-axis	9.55	5.2	1.83
Servo motor with ball screw for z-axis	9.55	3.2	2.98

8.2 Linear Guide GHW-30HA

- Static load rating: 69.16 kN = 7060 kg
- Dynamic load rating: 47.17 kN = 4805 kg.

8.2.1 Comparison of Load

	Static load (kg)	Dynamic load (kg)	Actual anticipated load (kg)	Factor of safety w.r.t dynamic load (FoS)
Linear guide for x-axis	7060	4805	700	6.86
Linear guide for y-axis	7060	4805	600	8.00
Linear guide for z-axis	7060	4805	250	19.22

Moment load	MR (Nm)	MP (Nm)	MY (Nm)
Linear guide bearing	880	920	920

8.2.2 Moment Applied

As shown in figure, when the center of gravity of the carried load is overhung from the center of the table of the electric linear actuator, the moment is applied from the center of the table as the support point. The moment applied to a linear guide can be obtained from the following formula (Fig. 2.60).

$$M = m \cdot g \cdot L \quad (1)$$

M moment applied to an electric linear actuator
L overhung distance
m load mass
g gravitational acceleration.

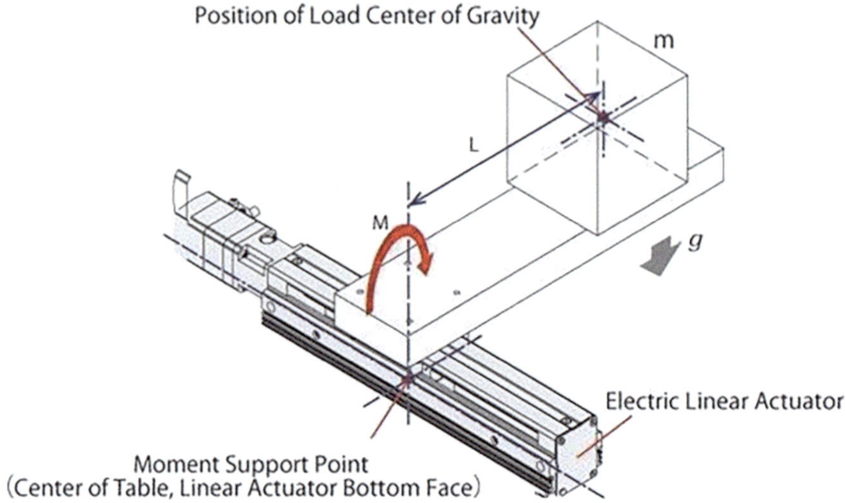

Fig. 2.60 Moment applied to linear actuator

8.2.2.1 Moment Direction

There are three different directions of moments applied to linear guide:

a. pitching direction,
b. yawing direction and
c. rolling direction (refer to figure).

The moment support points are on the bottom face (installation surface) of the linear actuator and the center of the table (Figs. 2.61 and 2.62).

$$M_P = m \cdot \alpha \cdot H$$
$$M_Y = m \cdot \alpha \cdot L$$
$$M_R = m \cdot g \cdot L$$

where

- M_P pitching direction moment
- M_Y yawing direction moment
- M_R rolling direction moment
- L overhung distance in the direction of the Y-axis load center of gravity
- H overhung distance in the direction of the Z-axis load center of gravity
- m load mass
- α operating acceleration
- g gravitational acceleration.

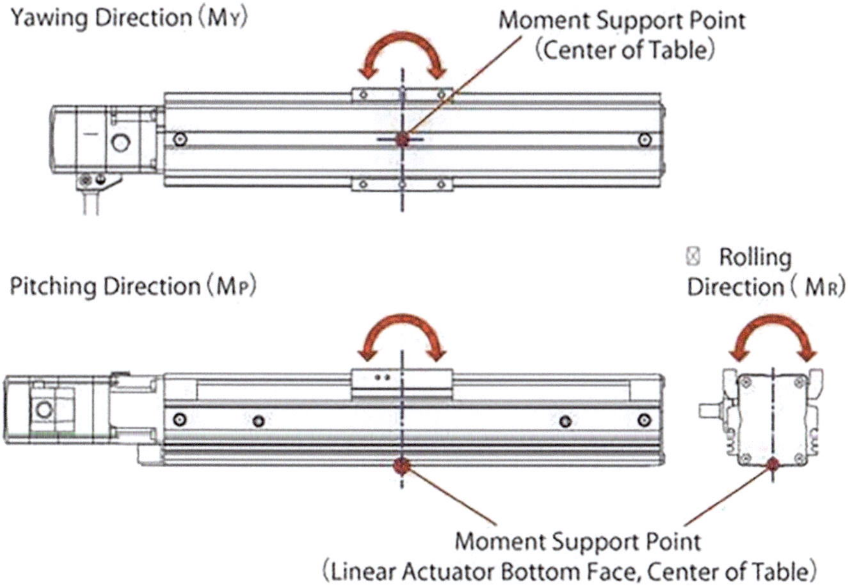

Fig. 2.61 Moment directions applied to electric linear actuators

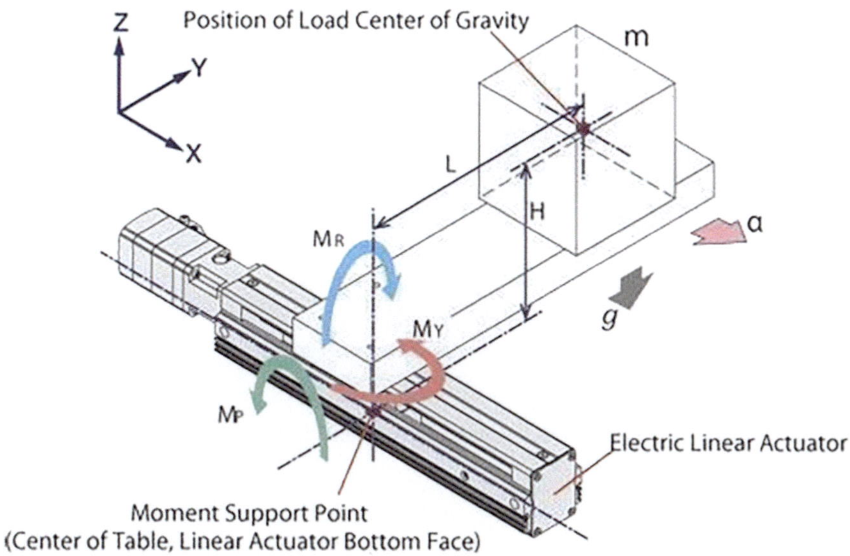

Fig. 2.62 Linear actuators working mechanism

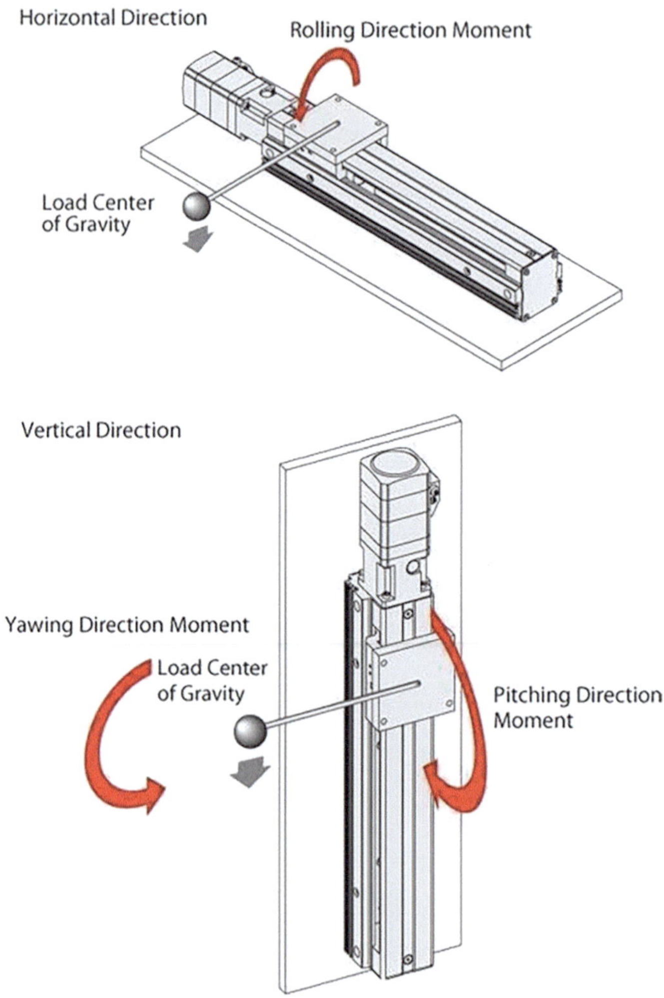

Fig. 2.63 Difference of moments caused by installation directions

8.2.2.2 Moment Direction Determined by Installation Direction

Directions of moments applied to linear guides can change depending on the installation condition of the load and the installation direction of the actuator itself. Figure 2.63 shows how the direction of a moment changes when the installation direction is changed under the same installation condition of the load.

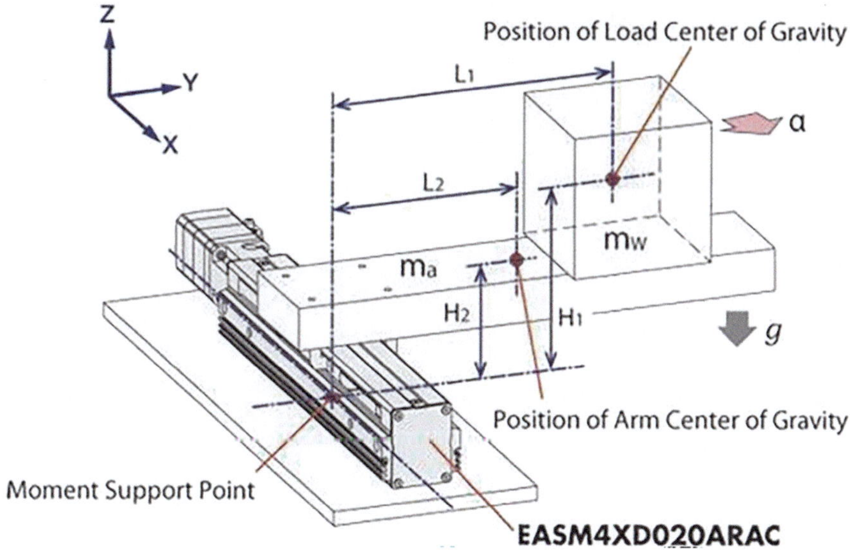

Fig. 2.64 Calculation example of dynamic moment applied to actuator

8.2.3 Moment Applied on x-Axis

No. of linear guide bearing = 4

Mass of maximum material which can hold = 555 kg/4 = 139 kg

Mass of platform maximum = 45 kg/4 = 11 kg

Mass of y-axis placed on x-axis = 100 kg/4 = 25 kg

Total mass on each bearing = 139 + 11 + 25 = 175 kg (Fig. 2.64).

Parameter	Value	Unit
L1	0.47/2 = 0.235	m
L2	0.390/2 = 0.195	m
ma	11	kg
mw	139	kg
g	9.81	m/s^2
a	3.0	m/s^2
H1	0.33	m
H2	0.075	m
MP	140	Nm
MY	104	Nm
MR	341	Nm

8.2.4 Comparison of Moment Load (x-Axis)

	Designed moment load (Nm)	Actual anticipated moment load (Nm)	Factor of safety (FoS)
MP	920	140	6.57
MY	920	104	8.85
MR	880	341	2.58

8.2.5 Moment Applied on y-Axis

No. of linear guide bearing = 4

Mass of maximum material which can hold = 555 kg/4 = 139 kg

Mass of platform maximum = 45 kg/4 = 11 kg

Mass on each bearing = 139 + 11 = 150 kg.

Parameter	Value	Unit
L1	0.39/2 = 0.195	m
L2	0.12	m
ma	11	kg
mw	139	kg
g	9.81	m/s^2
a	3.0	m/s^2
H1	0.255	m
H2	0.08	m
MP	109	Nm
MY	85	Nm
MR	279	Nm

8.2.6 Comparison of Moment Load (y-Axis)

	Designed moment load (Nm)	Actual anticipated moment load (Nm)	Factor of safety (FoS)
MP	920	109	8.44
MY	920	85	10.82
MR	880	279	3.15

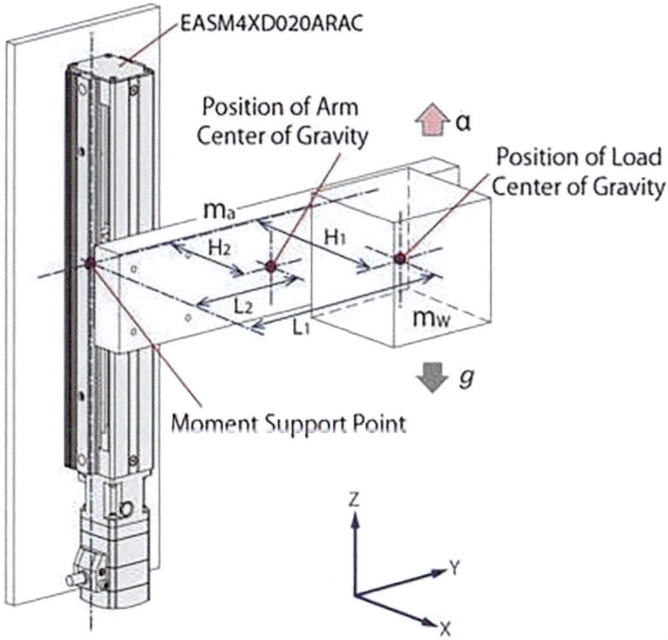

Fig. 2.65 Moment applied on Z-axis

8.2.7 Moment Applied on z-Axis

No. of linear guide bearing = 4

Mass of maximum material z-axis head = 65 kg/4 = 16.5 kg

Mass of spindle = 32 kg/4 = 8 kg

Mass on each bearing = 16.5 + 8 = 24.5 kg (Fig. 2.65).

Parameter	Value	Unit
L1	0.25/2 = 0.125	m
L2	0.125	m
ma	16.5	kg
mw	8	kg
g	9.81	m/s^2
a	3.0	m/s^2
H1	0.6/2 = 0.3	m
H2	0.5	m
MP	136	Nm
MY	39	Nm
MR	30	Nm

8.2.8 Comparison of Moment Load (z-Axis)

	Designed moment load (Nm)	Actual anticipated moment load (Nm)	Factor of safety (FoS)
MP	920	136	6.76
MY	920	39	23.59
MR	880	30	29.33

9. Preliminary FEM Analysis

See Figs. 2.66, 2.67 and 2.68.

10. CE Compliance

CE marking indicates that a product has been assessed by the manufacturer and deemed to meet EU safety, health, and environmental protection requirements. It is required for products manufactured anywhere in the world that are then marketed in the EU (Fig. 2.69).

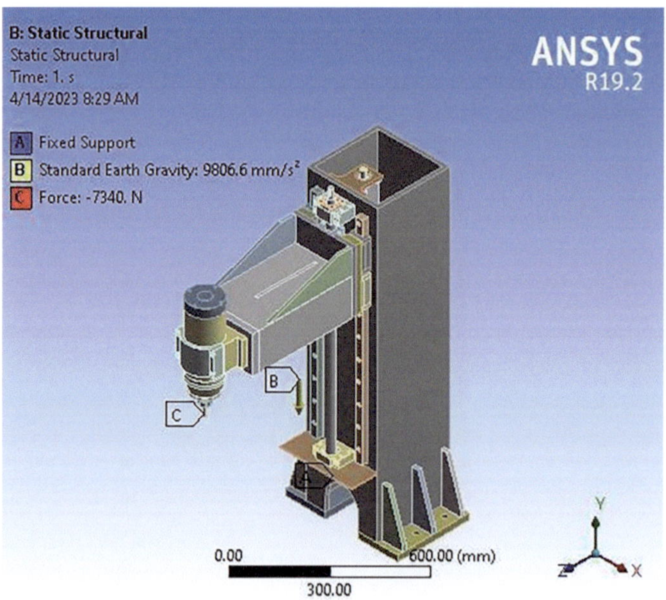

Fig. 2.66 Finite element analysis

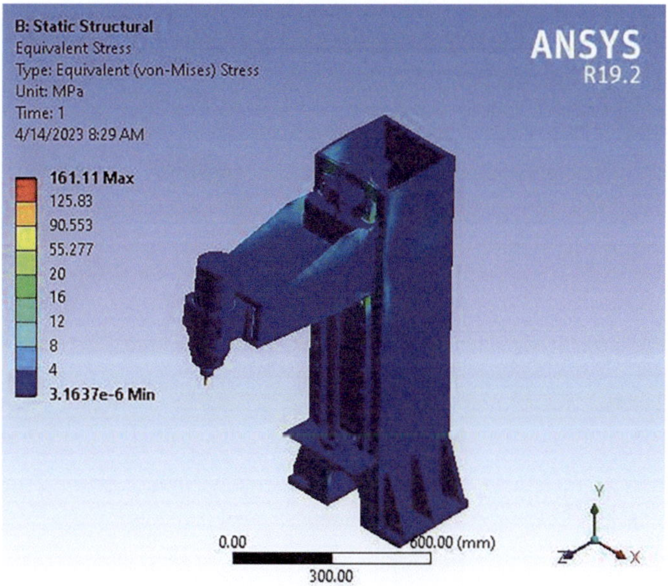

Fig. 2.67 Finite element static structural analysis

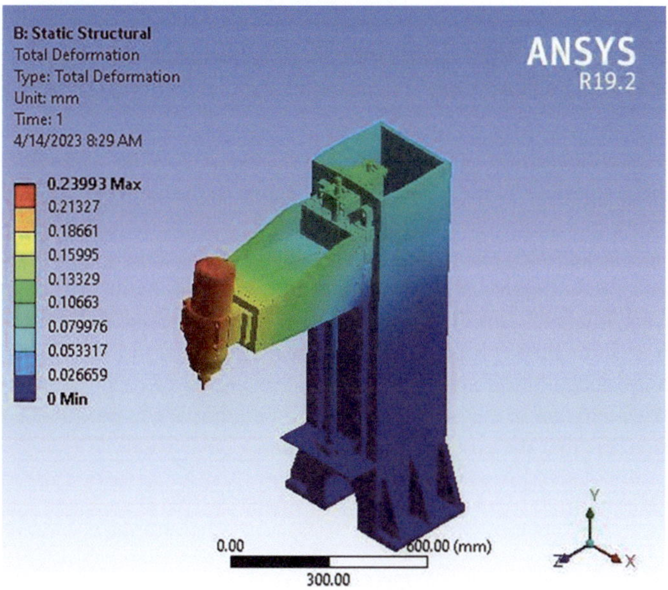

Fig. 2.68 Finite element static structural analysis with deformation

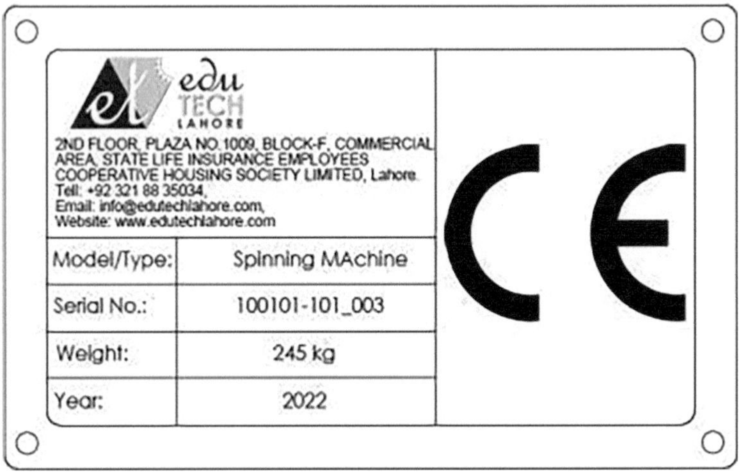

Fig. 2.69 CE compliant

C. Video Link of Proposed Functionality (User eXperience)[31]

https://drive.google.com/file/d/1qWGrodmEG8kRJe_othye_clKYKzjGtqD/view

2.10 Learning Outcomes

The reader after the study of this chapter will understand the difference between tangible and intangible that differentiate the terms manufacturing and production. Manufacturing processes with their conversion into manufacturing equipment are described in a structured manner based on ISO 841: 2001. It gives the basic geometrical concept of the equipment and its characteristics including type, size, accessories, parameters, and control to its actual physical manifestation based on the axes orientation and machine (product) configuration as defined in ISO 841: 2001.

Theoretical formulation and realistic implementation of manufacturing systems based on the inclusion of certain manufacturing equipment to produce a product or range of products in a structured form is explained.

[31] Protected by Pakistan patent number 37/2022 Acceptance Number 144217 and US patent applications number 1786442 (Pending); Logic Design by Wasim Ahmed Khan. Programmed by Aqeel Ahmad. Reproduced with permission.

The level of understanding at this point must be that the user is able to define a product that is coming from a design office, is able to perform manual process planning (explained in Part IV, Chap. 10) and production planning (explained in Part IV, Chap. 11), place the product in manufacturing system defined by the reader, and take it to the state of shipment to the distributor from the manufacturing system within defined lead time using manual sourcing procedure included in this chapter.

2.11 Review Questions

Q1. What is the difference between discrete and continuous manufacturing? Give at least two examples of each type of manufacturing.
Q2. Explain the general classification of manufacturing processes.
Q3. Name at least five of the sheet metal forming processes.
Q4. Differentiate between Long-chip processes and non-chip processes. Please give two examples each of long-chip and non-chip processes.
Q5. What are Manufacturing Systems? Enlist its classification.
Q6. What is the difference between a Job shop and a Project shop? What are two common usages of job shop and project shop?
Q7. What is Cellular Manufacturing? Is cellular manufacturing also referred to as batch manufacturing?
Q8. Explain the characteristics of Reconfigurable Manufacturing Systems. What is maximized in reconfigurable manufacturing and why job setup is relatively given less importance?
Q9. Construct the basic description of virtual manufacturing systems. How digital twining helps control the manufacturing setup remotely?
Q10. Illustrate the typical assembly procedures.
Q11. What is Production Documentation and why it is important to keep a record of it throughout the ages in different forms?
Q12. Name and briefly detail at least seven production philosophies.
Q13. What is the difference between strategic and non-strategic machine tools? Give at least two examples of each type of machine tools used in discrete manufacturing.
Q14. List at least five of the system engineering design aspects.
Q15. Perform basic design steps in the logical form for job 1 chosen in question 11.
Q16. Perform manual sourcing for job 1 chosen in question 11.
Q17. What is the system approach to manufacturing?

Q18. What are the different functions of system engineering design in the production life cycle?
Q19. Perform system engineering design using analytical methods.
Q20. Perform system engineering design using pertinent standards.
Q21. Perform system engineering design using empirical equations.
Q22. Perform system engineering design using numerical methods.

Bibliography

1. Chryssolouris G (2006) Manufacturing systems: theory and practice. In: Mechanical engineering series. Springer
2. Kalpakjian S, Schmid S (2022) Manufacturing processes for engineering materials in SI units. Pearson Education
3. Black JT, Kohser RA (2019) DeGarmo's materials and processes in manufacturing. Wiley
4. Krar SF, Gill AR, Smid P, Gerritsen RJ (2019) ISE technology of machine tools. McGraw-Hill Education
5. C. M. T. Institute (1985) Machine tool design handbook. Tata McGraw-Hill
6. Krar SF, Oswald JW, Amand JES (1977) Technology of machine tools. McGraw-Hill, Gregg Division
7. Kibbe RR, Neely J (1982) Machine tool practices. Wiley
8. Nagpal GR (1999) Machine tool engineering. Khanna
9. Sen GC (2015) Principles of machine tools. New Central Book Agency
10. Ahmed A, Bilal H, Ahmad H, Saifullah. Design and fabrication of small-scale CNC engraving machine. Capstone Design Project Report, Faculty of Mechanical Engineering, GIK Institute of Engineering Sciences and Technology, Topi, Khyber Pakhtunkhwa, Pakistan (advisor: Wasim A. Khan)
11. Moiz A, Taimoor F, Zakir M, Shakeel MZ (2016) Design and fabrication of scaled CNC milling machine. Capstone Design Project Report, Faculty of Mechanical Engineering, GIK Institute of Engineering Sciences and Technology, Topi, Khyber Pakhtunkhwa, Pakistan (advisor: Wasim A. Khan)
12. Ismail Leghari M, Shoaib Butt M, Ali S, Ishfaq S (2017) Design and fabrication of a CNC precision router. Capstone Design Project Report, Faculty of Mechanical Engineering, GIK Institute of Engineering Sciences and Technology, Topi, Khyber Pakhtunkhwa, Pakistan (advisor: Wasim A. Khan, 2016)
13. Tariq F, Afzal S, Ahmed S, Jabran S (2018) Five axis CNC milling machine. Capstone Design Project Report, Faculty of Mechanical Engineering, GIK Institute of Engineering Sciences and Technology, Topi, Khyber Pakhtunkhwa, Pakistan (advisor: Wasim A. Khan)
14. Maqsood A, Mughal MI, Shaharyar, Javed U (2020) Design and development of five axis CNC milling machine. Capstone Design Project Report, Faculty of Mechanical Engineering, GIK Institute of Engineering Sciences and Technology, Topi, Khyber Pakhtunkhwa, Pakistan (advisor: Wasim A. Khan)

15. Hitomi K (1996) Manufacturing systems engineering: a unified approach to manufacturing technology, production management and industrial economics. Taylor & Francis
16. Khan WA, Raouf SI (2005) Standards for engineering design and manufacturing. In: Mechanical engineering. CRC Press
17. ISO, 841: 2001 (2001) Industrial automation systems and integration—numerical control of machines—coordinate system and motion nomenclature. ISO, Geneva, Switzerland

Chapter 3
Remanufacturing

Prelude

The principle of circular economy was theorized in 2002 by the German chemist Michael Braungart and the American designer William McDonough with "cradle to cradle". The principle of cradle-to-cradle speaks of eco-design and the idea of re-using. This is the return of Lavoisier's famous concept from the eighteenth century: "Nothing is lost, nothing is created, everything is transformed" in a global society facing new economic expectations. The concept could be an effective response for several reasons. Scholars, policymakers, and industrialists have highlighted globally that the most successful companies will integrate sustainability, i.e., re-manufacturing, into their core business in the next decade.

The world has tripled natural resource consumption since 1970. The trend of the resource consumption of all materials, from biomaterials to fossil fuels, is going upward. Humans have taken far more advantage of natural resources than they have given to nature. Many of the materials that humans use are non-renewable, so the efficiency in using our resources is critical from a global perspective. The earth has valuable and limited resources, and one way of using them responsibly is through the art of remanufacturing, in which used parts return to their life. Replace any worn-out or broken components, then reassemble, test, and re-sell the product as though it were new or almost new. It is not material recycling but extending the component's or product's life. Repair, overhaul, and refurbishment are terms typically used to denote various types of work involved in bringing the product back to its working condition despite maintaining it again. The broken is maintained and repaired to restore its working condition. It not only saves resources but also decreases downtime.

Supplementary Information The online version contains supplementary material available at https://doi.org/10.1007/978-3-031-77106-4_3.

W. A. Khan et al., *Computer Aided Engineering Design and Manufacturing*, Springer Series in Advanced Manufacturing, https://doi.org/10.1007/978-3-031-77106-4_3

The remanufacturing is a closed-loop system from repair to refurbishment to recycling. It enables us to bring the product back to extract whatever value has been left, and the rest obviously can be directed toward recycling. Remanufacturing of surface is one of the most critical areas to enable circular economy implementation as it has significant margins. In remanufacturing, the product is torn down to the component level, cleaned, and inspected to every part to meet the same specifications as new parts. For example, when an engine block casting is worn, some processes are highly efficient to bring it back to the same condition as new casting by depositing new material from a business perspective. It reduces the need for new raw materials and energy associated with it, lowering the cost through less production waste, and significantly reducing greenhouse gas emissions. It conserves about eighty per cent of the raw material and labor. It offers significant savings in energy and material consumption, resulting in significant environmental impacts.

3.1 Objectives

The world is facing challenges in the supply of materials and energy resources along with the advancement in industrialization. Organizations and companies have been concentrating on waste and energy conservation to mitigate environmental impact globally and reduce carbon footprint, leading to sustainable manufacturing alternatives such as re-manufacturing. Re-manufacturing not only provides the closed-loop system of product/component maintenance in which information on product design such as over-designing, cost and performance are identified for the optimized design measures, but it also re-uses the materials that might help save material cost positively. This chapter introduces state-of-the-art re-manufacturing technologies, standard procedures and regulations and their application in the product life cycle. The re-manufacturing practices in the fourth industrial revolution era are introduced with a case study.

3.2 Industry 4.0 Technologies for Remanufacturing

3.2.1 Introduction to Industry 4.0

Industry 4.0 is the Fourth Industrial Revolution in which digital technologies are incorporated into manufacturing. It represents a paradigm transition from conventional manufacturing processes into smart interconnected manufacturing systems. Industry 4.0 revolutionized the industrial landscape with increased automation, improved communication, self-supervising, and intelligent machines which can evaluate and diagnose issues without human involvement. This transformation is efficient and includes adaptability and optimal resource utilization, which are key measures in a fast-changing industrial era. It has not only presented a substantial opportunity for the conventional manufacturing industry but also brought opportunities for closing material loops and developing remanufacturing [1].

Remanufacturing is an industrial process that would retain the use of the product to at least as good as new with almost the same kind of warranty as the new product. It is the way to return value so that the end of the product will return the value of the end-of-life product rather than scrape to the raw materials. Remanufacturing is considered at the early stage of engineering design and product life cycle because 80% of product cost is determined at this stage. So, it is a much higher value-added process than recycling, extending the component's or product's life. There needs to be more clarity between remanufacturing and refurbishment, reuse, and repair [2]. These value retention processes are differentiated through Fig. 3.1.

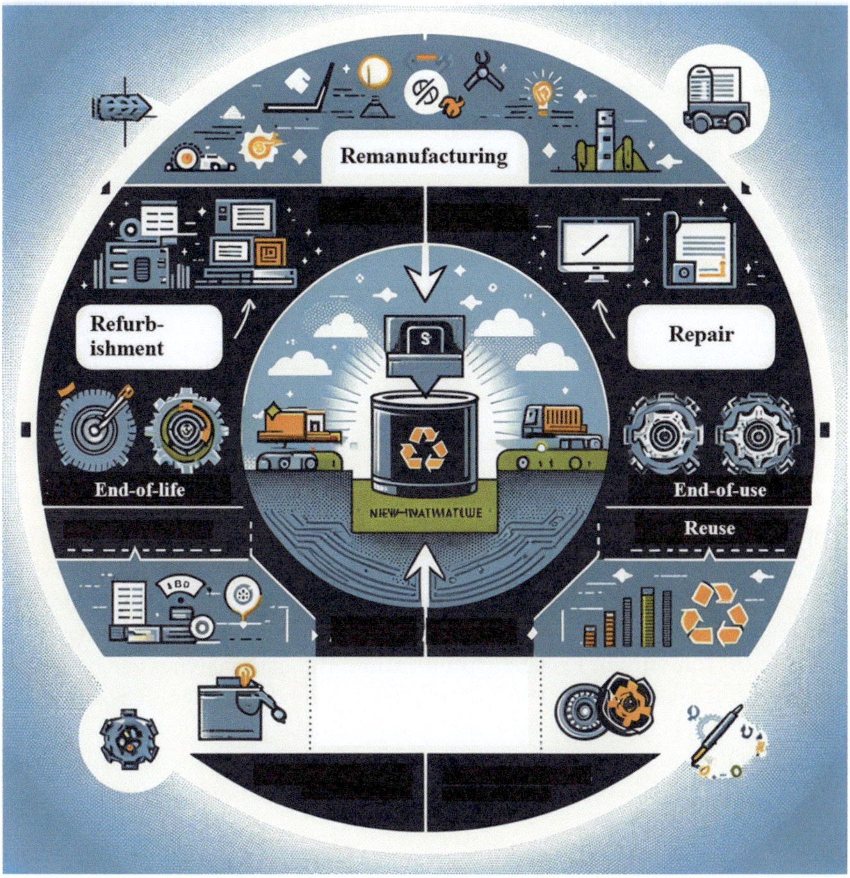

Fig. 3.1 The value retention processes

3.2.2 *Key Technologies for Remanufacturing*

The key technologies for remanufacturing in the context of Industry 4.0 are as follows [3]:

- **Internet of Things (IoT)**: It involves using physical devices embedded with sensors and software to connect with other technologies to exchange data. IoT can monitor and record wear and tear for maintenance requirements and optimize remanufacturing processes.
- **Artificial Intelligence (AI)**: AI comprises machines mimicking human intelligence. In AI-based remanufacturing, the data from IoT can be analyzed to predict failures or maintenance requirements for efficient and cost optimization remanufacturing operations.
- **Big Data Analytics**: It includes the analysis of large volumes of data to characterize hidden patterns/trends to gain insight. Big data in remanufacturing can predict the product life cycles, make the process efficient and make knowledge-based decisions related to material reuse and recycling.
- **Robotics**: Robots can automate repeated tasks, work with hazardous materials, and improve precision in the disassembly and reassembly processes of remanufacturing.

3.2.3 *Technologies Incorporated in Remanufacturing*

The advanced technologies directly incorporated in remanufacturing are as follows [4]:

- **IoT in Life Cycle Tracking**: Internet of Things (IoT) devices are instrumental in examining the conditions and functioning of products throughout their lifecycle. By installing sensors in products, the manufacturer can record the real-time data of wear and tear of products, usage patterns, and potential failures. This data can provide planning insightful and efficient remanufacturing operations, deciding whether components are refurbished or replaced before catastrophic failure and damage to other components occur.
- **Big Data Analytics in Process Optimization**: Big data analytics can be critical in analyzing enormous amounts of data acquired with IoT devices. These analysis measures can help identify the trends and patterns in product usage and failure, which is the informer of the remanufacturing process. For example, data-driven insights can assist in optimizing the disassembly, parts cleaning, and refurbishing techniques.
- **Robotics in Disassembly and Reassembly**: Robots can provide enabling technology for automating disassembly and reassembly tasks in remanufacturing. With machine vision and precision tools, robots can handle complex tasks, decrease labor costs, and minimize human error.

- **AI for the Predictive Maintenance**: Artificial Intelligence (AI) can aid in the predictive maintenance in remanufacturing. AI algorithms can be built on the analysis of IoT sensor data that can predict the expected failure of the component. AI can assist in scheduling the maintenance, refurbishment or remanufacturing measures in advance. The AI-based proactive strategy can significantly decrease downtime and extend the product's life.
- **Machine Learning in Quality Control**: Machine learning models can evaluate product wear and tear patterns to improve the remanufacturing processes' quality control. The models can detect irregularities and predict quality issues before product remanufacturing to ensure higher quality and reliability.
- **Digital Twins for Process Simulation**: Digital twins are virtual models of physical products that can simulate and analyze the remanufacturing process. The technology assists manufacturers in testing and optimizing the remanufacturing process in a virtual environment to decrease the need for actual trials and risks.

The state-of-the-art technologies incorporated in remanufacturing are shown in Fig. 3.2.

3.3 Integrating and Managing the Technologies Through Remanufacturing

Incorporating strategies for integrating and managing the product's life has become progressively more crucial in everyday business operations. Managing and integrating state-of-the-art technologies into remanufacturing can make it much more efficient, high-quality, and sustainable. The remanufacturing system involves both internal processes like the operation and external processes like core collection and product delivery. The challenges and strategies might be faced during the integration and management of the technologies [5].

3.3.1 Compatibility with Existing Systems

- **Technological Mismatch**: Integration of Industry 4.0 technologies into traditional remanufacturing systems is challenging due to compatibility issues. Traditional machinery and software systems may not synchronize with the advanced IoT devices and AI algorithms.
- **Data Management Complexity**: The data from IoT devices can be enormous, particularly for systems not designed to manage big data. Handling, storing, and efficiently processing big data can be a major problem.

Fig. 3.2 Cutting-edge technologies in modern remanufacturing

3.3.2 Workforce Adaptation

- **Skill Gaps**: The workforce may need additional skills to operate and maintain state-of-the-art technologies. There is often a gap between the current skill levels of employees and the skills required to manage new technologies.
- **Resistance to Change**: Employees may be reluctant to adapt, especially when learning new technologies or adapting to new processes.

3.3.3 *Cost and Investment*

- **High Initial Investment**: The cost of procuring, integrating, managing, and maintaining new technologies can be enormous. This not only includes the cost of new technologies but also the cost of employee training and infrastructure upgrades.
- **ROI Uncertainty**: There can be uncertainty in short-term return on investment that can be a limitation for some remanufacturing businesses.

3.3.4 *Managerial Strategies*

3.3.4.1 Incorporating Lean Manufacturing Technologies

- **Process Optimization**: Lean management techniques can be advantageous to minimize waste and improve efficiency in remanufacturing tasks. These include updating operations to decrease downtime and improve workflow.
- **Continuous Improvement**: A culture of continuous improvement in employees is always a proactive approach for process optimization and efficient integration of new technologies.

3.3.4.2 Utilizing Digital Twins and Cyber-Physical Systems

- **Digital Twins**: Digital twins are very helpful in simulating and integrating new technologies into remanufacturing. It assists in identifying potential issues and testing solutions in a virtual environment before implementation.
- **Cyber-Physical Systems**: Cyber-physical systems aid in combining physical and digital processes. The integration can lead to better monitoring control and optimizing remanufacturing processes tasks [6].

3.4 Enhancing Product Life Cycle Through Remanufacturing

Life cycle assessment, also called life cycle analysis, is conducted by product designers to assess their product impacts and improvements associated with all phases of a product's life. The product life cycle can be enhanced through remanufacturing in the following ways [7]:

- **Restoration and Enhancement**: Remanufacturing involves disassembly, cleaning, repairing, or replacement of worn-out parts, and then reassembling products with a standard warranty like a new condition. This can significantly enhance the life of products and increase their performance to meet or exceed original specifications.

- **Modular Upgrades**: Remanufacturing involves integrating new technology into existing products to keep them up-to-date and extend their usability.
- **Preventative Maintenance**: Potential future catastrophic failure can be identified and mitigated to decrease the downtime and increase the product life.

These measures to enhance the product life cycle are shown in Fig. 3.3.

Fig. 3.3 Strategies for extending product life in remanufacturing

3.4.1 Environmental Impact

3.4.1.1 Reduced Waste and Resource Consumption

- **Material Reuse**: Remanufacturing enables the reuse of materials to decrease the need for new raw materials. It is important for metals and plastic whose processing and recycling cause environmental concerns.
- **Waste Minimization**: Refurbishing and reusing components in remanufacturing can significantly decrease materials waste. This is particularly important to decrease the amount of electronic waste, which is a growing environmental concern these days.

3.4.1.2 Lower Carbon Footprint

- **Energy Efficiency**: Remanufacturing usually uses less energy than producing new products from raw materials, causing a decrease in carbon footprint.
- **Transportation Reduction**: Local remanufacturing can significantly decrease the growing need for the long-distance transportation of new products.

3.5 Standards and Regulations for Remanufacturing

3.5.1 Specific Standards and Regulations

Standard procedures aid remanufacturing companies in implementing and maintaining the best environmental health and safety practices and process optimization measures in inter- and intra-industries. The currently available standards encompass the remanufacturing process through various certifications and regulations [8]. For example:

- **ISO 9001**: A widely recognized standard for quality assurance systems across industries, including remanufacturing.
- **ISO 14001**: Focuses on environment management in remanufacturing processes to minimize negative environmental impacts.
- **Automotive Industry**: A standard of IATF-16949 is for remanufacturing automotive parts.
- **Electronic Waste**: Regulations such as the Waste Electrical and Electronic Equipment Directive (WEEE) in Europe and the Resource Conservation and Recovery Act (RCRA) in the U.S.
- **Safety Standards**: OSHA guidelines in the U.S. for workplace safety measures during remanufacturing processes.

3.5.2 Emerging Needs and New Standards

- **Data Security and Privacy**: Since Industry 4.0 is about interconnected devices, standards such as ISO/IEC 27001 are very relevant as they provide information related to system security management.
- **Quality Assurance in AI**: There is growing concern for the availability of standards that can address the reliability and safety of AI-driven decisions in remanufacturing the possible adoption of existing ISO standards related to AI and robotics.
- **Standardizing IoT in Remanufacturing**: Standards such as IEEE 802.15.4 are very important to regulate low-rate wireless personal area networks and interoperability of IoT devices.

3.5.3 Adaptation and Evolution of Standards

- **Sustainability and Circular Economy**: Future regulations may evolve, or new ones might emerge to address sustainability, such as the ISO 14000 series to address the circular economy directly.
- **Global Harmonization**: Significant efforts are required to standardize standards such as ISO 9001 and 14001 globally, especially for industries with extensive international supply chains.
- **Data and Digital Process Standardization**: Developing new ISO standards for digital process standardization, data format, and interoperability in remanufacturing are primarily important.
- **Ethical AI and Automation Standards**: Since AI and automation play a significant role in remanufacturing, ethical guidelines, and standards measures (probably under the ISO/IEC guidelines) will be imperative to guide manufacturers.

The standards and regulations used for remanufacturing are shown in Fig. 3.4.

3.6 Framework and Research Agenda for Remanufacturing

A multitude of technologies facilitate the advent of intelligent remanufacturing. The progress in information and communications technology (ICT) has resulted in the rise of several disruptive technologies, including cloud computing, Internet of Things (IoT), big data analytics, and artificial intelligence. The integration of these technologies into the remanufacturing process is referred to as "Intelligent Remanufacturing." Intelligent Remanufacturing (IRM) or Advanced Remanufacturing is a comprehensive approach to shaping conventional remanufacturing processes into highly advanced, efficient, flexible, and connected operations. In IRM, the complete

Fig. 3.4 Essential standards and regulations for remanufacturing

remanufacturing system is interconnected in such a way so as to provide seamless collaboration, information sharing, and optimization from the suppliers to customers.

In an Industry 4.0 environment, smart networks connect machines, products, ICT systems, stakeholders, and decision-makers throughout the value chain and product life cycle. These technologies include a diverse range of tools that can streamline the process of intelligent remanufacturing, augment the quality of the output, and foster more confidence among customers. There is no three-dimensional framework of Industry 4.0 for intelligent remanufacturing, and its implementation is based on practices and academic research. Most smart remanufacturing systems are built on a framework that integrates concepts from the circular economy approach, machine vision, artificial intelligence, cyber-physical systems, and remote remanufacturing.

To this end, the framework for intelligent remanufacturing is proposed as shown in Fig. 3.5. It consists of two parts: intelligent remanufacturing system (IRS), and intelligent remanufacturing implementation system (IRIS).

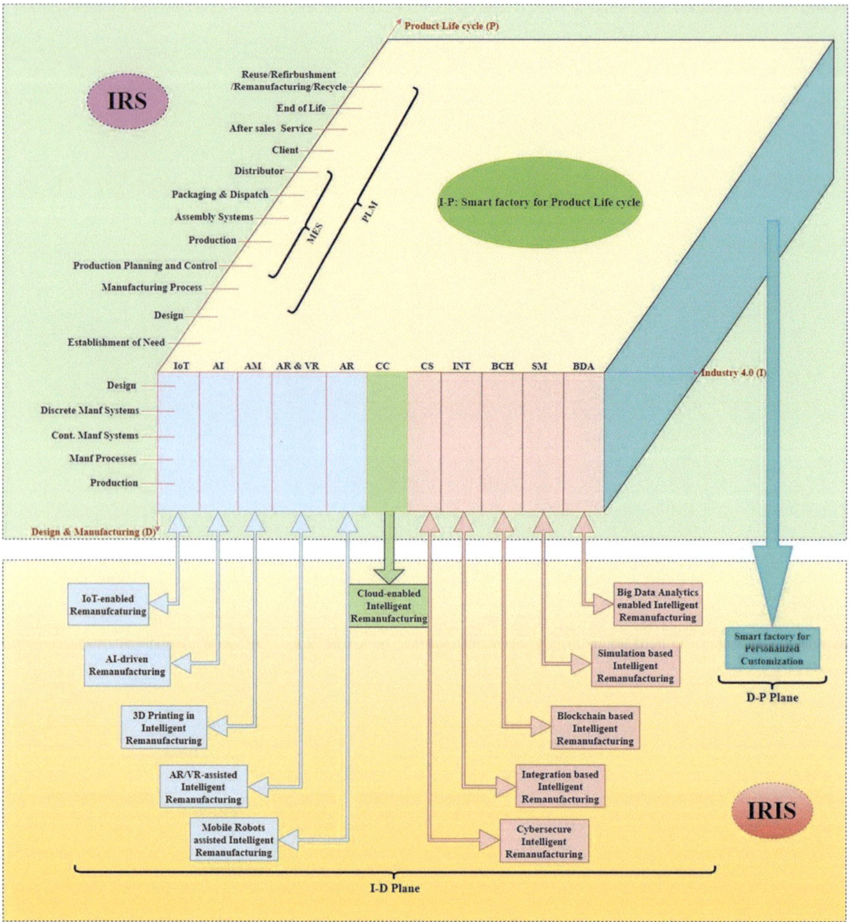

Fig. 3.5 Framework for IRS and IRIS

3.6.1 IRS and IRIS

Product life cycle (P), Industry 4.0 (I), and Design and Manufacturing (D) are the three axes that generate the framework for the Intelligent Remanufacturing System (IRS). Combining any two axes in the deployment context generates the plane in the intelligent remanufacturing implementation system (IRIS). Product life cycle (P) refers to the complete journey passing through the stages from initial design to eventual reuse/remanufacturing/recycling. Industry 4.0 (I) contains the basic pillars that cover the key technologies and concepts that help to define the fourth industrial revolution. The design and manufacturing (D) axis encompass different stages of design activities and manufacturing processes involved in creating and producing goods. For implementation, any two axes of the IRS combine to form the plane of the IRS. For example, the ‘I’ and ‘P’ axes merge to form the I-P plane, the ‘I’ and ‘D’ axes collectively generate the I-D plane, and similarly, the ‘D’ and ‘P’ axes combine to form the D-P plane. The framework for IRS and IRIS is shown in Fig. 3.5.

The description of terms in IRS and IRIS are summarized below. The stages of product life cycle (P) [1] are also as follows:

- **Establishment of Need**: Recognize the need and demand for remanufacturing products, keeping sustainability and cost-effectiveness in mind. For example, a company surveys its customers and finds a growing interest in sustainable practices. Then, they launch a line of remanufactured electronic goods, like smartphones, highlighting their lower environmental impact and more affordable prices than new models.
- **Design**: Adapt design to facilitate disassembly and refurbishment. For example, a furniture manufacturer designs modular sofas in which individual components like armrests and cushions can be easily replaced or refurbished, extending the product’s life and making remanufacturing feasible.
- **Manufacturing Processes**: Select processes that facilitate the remanufacturing of modular component designs. For example, an automotive company designs modular engines, the components of which can be easily replaced or upgraded. The fuel pump and cooling system are replaced rather than replacing the entire engine. Advanced remanufacturing systems and technologies include automation, robotics, and digital control systems. By doing so, productivity and quality can be enhanced while ensuring real-time monitoring and decision-making.
- **Production**: Plan to integrate returned products or components in the production schedule. For example, a consumer electronics company designs a program for customers to return their old laptops. The company schedules the refurbishment of these laptops into their production planning and makes sure they have plans to remanufacture along with new production units. Further, the IRS leverages automation, data analytics, and connectivity to identify any production issue or bottleneck to optimize the production workflows and monitor key performance indicators.
- **Assembly System**: Availability of equipment for handling both the new and remanufactured components. For example, a manufacturing plant of household appliances develops separate assembly lines that can handle both new and remanufactured components. Further, the IRS uses robotics automation and computer vision to increase the assembly process’s speed, flexibility, and accuracy.
- **Packaging and Dispatch**: Ensure the design and materials of packaging can be reused or easily recycled. For example, a beverage company switches to reusable glass bottles with a deposit scheme, i.e. the customers return the bottles after use, which are refilled and redistributed after cleaning, resulting in a decrease in waste and encouraging recycling. Further, the IRS used automation in packaging processes to lessen waste and improve logistics.
- **Distributor**: This involves transporting packed products from the remanufacturing site to the intended distributors or retailers. Further, the IRS facilitates data

analytics and connectivity to manage inventory, demand forecasting, and supply chain optimization to ensure timely delivery and efficient distribution.

- **Client**: Provide awareness and engage stakeholders to enhance the image of remanufactured products. Further, customer satisfaction can be improved by incorporating technologies like IoT-enabled devices and customer feedback systems. A remanufacturer conducts workshops and webinars for distributors and clients about remanufactured products' cost savings and environmental benefits.
- **Reuse/Refurbishment/Remanufacturing/Recycle**: This stage involves the life retention strategies of the product at its end-of-life by focusing on sustainable practices. The following processes, such as recycling, refurbishment, and remanufacturing for reuse, are used for this purpose. For example, a clothing brand introduces a program where customers can return their old garments. The brand then refurbishes the clothes for resale or recycles the material for new garments, decreasing waste.

3.6.1.1 I-D Plane

I-D plane relates the coordination of Industry 4.0 pillars with the design and manufacturing to implement intelligent remanufacturing. The basic pillars of Industry 4.0 (I) in the context of intelligent remanufacturing [9, 10] are considered here as follows:

- **Internet of Things (IoT)**: It tracks the products during the entire life cycle for efficient return and remanufacturing measures. For example, a car manufacturer installs IoT sensors to monitor wear and tear. This real-time data helps anticipate the requirements of component refurbishment or replacement that facilitate efficient returns and remanufacturing processes.
- **AI and Machine Learning**: The use of AI is implemented to improve decision-making in remanufacturing. AI uses technologies such as machine learning, deep learning, and predictive analysis for the required purposes. For example, printer manufacturers employ machine learning algorithms to analyze the performance of various available printer models. Based on the analysis, AI can provide recommendations on whether a particular model should be remanufactured, upgraded, or replaced, increasing the efficiency and effectiveness of the remanufacturing process.
- **Additive Manufacturing**: It is also called 3D printing and involves producing the product by adding materials in layers. It offers a mechanized and highly regulated method to restore the worn parts in remanufacturing. For example, the worn-out gearbox is scanned using 3D scanning technologies to create a digital model of the component. The 3D printer then builds a new gearbox layer by layer using high-strength alloy.
- **Augmented Reality (AR) and Virtual Reality (VR)**: AR and VR aid in product disassembly in remanufacturing.
- **Autonomous Robots**: Automated remanufacturing tasks can increase efficiency, reduce costs, and improve productivity. Autonomous robots can perform various

remanufacturing operations such as disassembly, sorting, cleaning, surface preparation, inspection, and quality measures. For example, autonomous robotic systems are used for welding and joining operations. These robots guarantee accuracy and uniformity in assembling components and the overall quality of the remanufactured products.

- **Cloud Computing**: It involves delivering computing resources, such as storage, processing power, and software applications, over the internet. It promotes secure data storage, real-time data analysis, and stakeholder collaboration. It provides cloud resources to scale remanufacturing operations. For example, a machinery company uses cloud-based software to manage its remanufacturing operations. This approach enables the company to scale up or down based on the influx of machinery requiring refurbishment or remanufacturing and effectively integrate data from various global facilities.
- **Cybersecurity**: It helps to protect the data, systems, and networks from unauthorized access, breaches and cyberattacks. It secures the data involved in the remanufacturing process. For example, a technology firm dealing in remanufacturing smartphones implements advanced encryption and cybersecurity protocols to protect product usage data, repairs, and refurbishment history, ensuring the confidentiality of sensitive information.
- **Integration**: Remanufacturers collaborate to exchange best practices, resources, and information on the acquisition and sourcing of cores and components to optimize operations and reduce expenses. For example, Xerox employs reverse logistics to collect and remanufacture discarded printer cartridges to prolong their lifespan. It maintains a closed-loop system for its consumables that decreases waste and promotes sustainability.
- **Blockchain**: The shared digital databases for storing remanufacturing components/cores information in interconnected blocks using cryptographic techniques. For example, IBM and Lenovo are using blockchain to monitor the electrical product lifecycle and refurbishment of parts. Blockchain enables transparency, traceability of ownership and warranty, and authenticity in remanufacturing and related supply chains.
- **Simulation**: Digital twins replicate the remanufacturing procedures, i.e. the disassembly, cleaning, inspection, and reassembly. This simulation facilitates the optimization of processes, the identification of bottlenecks, and the enhancement of efficiency in the physical remanufacturing environment.
- **Big Data and Analytics**: Analyze data for remanufacturing optimization process measures and predict returns. For example, an appliance manufacturer collects and analyzes data on product usage and failure rates. This information is used to predict which product models are likely to be returned for remanufacturing that enables the company to optimize its resources and inventory for refurbishment processes.

3.6.1.2 I-P Plane

The I-P plane relates the smart factory to the product life cycle. It is a highly advanced and digitized remanufacturing ecosystem where advanced technologies are integrated throughout the entire product life cycle. The design and manufacturing (D) axis consist of the following stages in the context of intelligent remanufacturing are [11, 12]:

- **Concept and Embodiment Design**: This design approach Incorporates remanufacturing measures at the earliest design stages. For example, a bicycle manufacturer designs bikes with a modular concept. This allows each component, like sprockets, frames, or wheels, to be easily replaced or upgraded, which allows for remanufacturing to extend the product life cycle.
- **Detail Design**: Provide materials and component specifications that are durable and suitable for many life cycles. For example, a furniture company designs chairs using high-quality, durable materials that can withstand many refurbishment cycles. The chairs are designed for ease of disassembly, which assists in replacing worn-out parts or repairing them to extend the furniture's life.
- **Discrete and Continuous Manufacturing System**: Incorporate these manufacturing systems in remanufacturing parts and components. For example, an electronic manufacturing line is reconfigured to allow for the incorporation of both new and remanufactured components in their products. Flexibility enables the company to efficiently manage resources and reduce waste by incorporating remanufacturing.
- **Production**: This stage aligns production with the influx of remanufactured components. For example, an automotive company adjusts its production line to allow for the integration of remanufactured engines and transmissions. The production schedules and processes are adapted to blend the assembly of new vehicles seamlessly with the integration of remanufactured parts, optimizing resource use and production facility.

3.6.1.3 D-P Plane

D-P plane relates to the smart factory for Personalized Customization. It focuses on the integration of different state of the art technologies in remanufacturing for its efficient, flexible, and customized operations. The discussion on various aspects essential for remanufacturing operations accompanied by illustrative examples are [13, 14] and are listed as below:

- **Modular Production System**: The intelligent factory production systems are usually modular, providing easy reconfiguring to adapt to various remanufacturing requirements. For example, a modular production system switches from refurbishing automotive components to electronic components by adjusting its setup based on the remanufactured product type.

- **Advanced Quality Control Techniques**: Adopting sophisticated inspection and testing measures; namely, machine vision and automated testing, can ensure that the remanufactured products meet the requirements of the standards. For example, an intelligent factory may employ high-resolution cameras and sensors to inspect electronic boards and detect more minor flaws.
- **Resource Optimization Algorithms**: The intelligent factory uses advanced algorithms for materials and resource optimization measures to minimize waste and improve efficiency. For example, algorithms can calculate the optimal, material-efficient way to refurbish or remanufacture a product or suggest ways to reuse parts from returned products.
- **Customization and Personalization Software**: Software enables customization of products per individual customer preferences. For example, the refurbishment process can be adjusted to upgrade the product features based on customer specifications, such as adding new interfaces to an older smart device model.
- **Supply Chain Integration**: Efficient supply chain integration is critical in remanufacturing. It may involve a software-based system to coordinate customers' returns and used products, manage inventory of refurbished components, and streamline the distribution of remanufactured products.
- **Energy Management Systems**: Incorporation of sophisticated energy management systems in smart factories can decrease consumption and increase sustainability; for example, the adoption of renewable energy resources or the implementation of efficient energy practices in remanufacturing operations.
- **Collaborative Work Environment**: Smart factories blend human expertise and automated systems. Collaborative teams may work alongside advanced machinery to complement each other's capabilities. For example, skilled technicians might efficiently oversee and fine-tune automated systems that disassemble and refurbish complex machinery.
- **Real-Time Data Analysis for Process Improvement**: The ongoing improvement in remanufacturing operations requires continuous monitoring and analysis of its operation. Data analytics can identify the bottlenecks of remanufacturing operations and track the performance of post-sale remanufactured products.

3.6.2 Role of Product Life Cycle Management and Manufacturing Execution Services (MES) in Intelligent Remanufacturing

Product Life Cycle Management (PLM) and Manufacturing Execution Services (MES) are critical elements in the manufacturing sector, mainly when dealing with Intelligent Remanufacturing. PLM is a methodical approach to managing the changes during product development, from its design and development to its end-of-life or disposal strategies. Its elements include concept development, generation and selection, design, engineering, manufacturing, service, and disposal of products. The MES controls and monitors production processes in real time, from raw material handling to finished products. Its elements include production scheduling, quality management, inventory control and performance analysis. The roles of PLM and MES in intelligent remanufacturing (IRS) are [14, 15]:

- **Design for Remanufacturing**: PLM can assist in designing products while easing disassembling, refurbishing, and recycling.
- **Lifecycle Data Management**: PLM can assist in collecting and analyzing data throughout the product's life cycle, which can guide remanufacturing decisions.
- **Resource Optimization**: PLM aids in identifying components that can be reused or recycled, decreasing waste and cost.
- **Sustainability Tracking**: PLM enables tracking of environmental impacts across the product lifecycle, encouraging sustainable practices.
- **Process Optimization**: MES can optimize remanufacturing processes by efficiently utilizing resources and minimizing waste.
- **Quality Control**: MES can ensure standard quality measures for remanufactured products, essential for customers' acceptance of remanufactured products.
- **Real Time Monitoring**: In MES, the real-time status of remanufactured products can be monitored, which can assist in quick adjustments and improvement measures.
- **Data Collection and Analysis**: MES enables the provision of valuable data related to the production process that can assist in improving remanufacturing strategic measures.
- **Seamless Workflow**: Incorporating PLM and MES aids in a seamless workflow from product design to remanufacturing to enhance uniformity and efficiency.
- **Data Sharing**: Sharing data between PLM and MES systems can lead to better decision-making in the remanufacturing process because the information is considered during the entire product lifecycle.
- **Continuous Improvement**: Incorporating PLM and MES in remanufacturing can provide insights which lead to improvement in design and manufacturing strategies, making remanufacturing more efficient and sustainable.

All in all, PLM and MES can play a central role in intelligent remanufacturing using efficient resources, optimizing processes, and maintaining quality. Their incorporation leads to better decision making, and sustainable, and methodical improvements in remanufacturing practices.

3.6.3 Research Agenda

- **Integration of Industry 4.0 Technologies**: A remanufacturing framework based on Industry 4.0 should include integrating emerging technologies like IoT, AI, big data analytics, and robotics. This includes the development of infrastructure for data collection, analysis, and automation of experimental tasks in remanufacturing.

- **Advanced Data Analytics and AI Applications**: The exploitation of advanced data analytics and AI tools as predictive maintenance, quality control, and process optimization measures to improve remanufacturing tasks.
- **Process Optimization and Quality Control**: The framework should consider the optimization measures in the remanufacturing process to increase efficiency and quality. This involves the implementation of lean manufacturing principles and the use of digital twins for optimization in remanufacturing.
- **Economy Models and Market Analysis**: Investigate remanufacturing economic aspects in terms of market dynamics, business models, and cost–benefit analysis to increase industrial feasibility and scalability.
- **Sustainability and Circular Economy Principles**: The adoption of sustainability and circular economy principles can assist in decreasing waste, optimizing resource utilization, and minimizing the environmental impact.
- **Environmental Impact Assessment**: Performing exhaustive research on the environmental impact of remanufacturing practices in terms of decreasing carbon footprint and increasing energy efficiency.
- **Workforce Development and Training**: The strategies of workforce training on emerging technologies and processes to equip the employees with the necessary skills to work with.
- **Technology Development for Disassembly and Reassembly**: Research, development, and deployment of advanced automated remanufacturing systems for efficient disassembly and reassembly operations.
- **Creating a Sustainable Manufacturing System**: The long-term goal in the era of Industrial 4.0 is to make remanufacturing central to develop a sustainable manufacturing ecosystem that can significantly decrease waste, optimize resources, and minimize environmental concerns.
- **Regulatory Compliance and Standardization**: This stage ensures remanufacturing practices are in line with existing standards and regulations and develops guidelines for adopting new standards as they emerge.
- **Integration into Global Supply Chain**: Consideration of remanufacturing as an essential part of global supply chains and a standard practice across industries, where products are designed for re-manufacturability.
- **Technological Leadership and Innovation**: Establishment of leadership for technological innovation in remanufacturing for promoting it as a crucial driver in the next wave of industrial revolution.
- **Materials Science and Sustainability**: Explore new materials and manufacturing processes to increase the sustainability of remanufacturing processes. This involves developing materials with more recyclability and efficient methods of material recovery.
- **Economic Revitalization and Job Creation**: Remanufacturing is promoted for economic growth, including job creation in the new technology sector.
- **Global Standardization and Policy Development**: There is a need for global standardization in remanufacturing practices to influence policy development, support growth, and integrate it into the broader economy.

- **Reuse/Refurbishment/Remanufacturing/Recycle**: The application and selection of sustainable life retention practices, such as reuse, refurbishment, remanufacturing, and recycling, at the product's end-of-life are imperative to retain the product's value. For example, a clothing brand introduces a program where customers can return their old garments. The brand then refurbishes the clothes for resale or recycles the materials for new garments to minimize waste and retain the value.

3.7 Learning Outcomes

The reader can learn the basics of remanufacturing with practical examples. The readers will also be posed with simple examples of integrating and managing Industry 4.0 technologies through remanufacturing, enhancing product life cycle, relevant framework, and research agenda in this area. The reader will also have information related to the relevant standards for remanufacturing. An oblique will be drawn with this newly formalized field with that of pillars of Industry 4.0, as this area is also gaining the research community's attention.

3.8 Real Life Examples

This section describes some examples of everyday remanufacturing products or components for reader inspiration and imagination, including remanufacturing furniture, an increasing and everyday business. The remanufactured chair costs about half of the new chair and is of the same quality. The renowned automobile companies, i.e., Volvo, collect used parts, remanufacture them and sell them as spare parts. The companies bought and remanufactured professional computers, upgrading and selling them. The local bicycle repair shop bought, remanufactured, and sold high-quality second hand bikes.

3.9 Case Study in Remanufacturing

This case study consists of three renowned companies, i.e. Caterpillar, Xerox, and Phillips, which are well known for their advanced remanufacturing procedures and processes. These are leaders in their respective industries and their approaches can provide valuable insights into the state-of-the-art techniques presently used in remanufacturing.

3.9.1 Caterpillar

Caterpillar is a global manufacturer of heavy machinery and equipment. It has a robust remanufacturing program called "reman" to embark on new life to its used machinery and equipment. They bring the used product back to a "like-new" condition with a warranty and quality standard equivalent to a new product. This process involves using state-of-the-art technology to restore and upgrade the performance of equipment and machinery. The company's remanufacturing process starts with collecting end-of-life equipment transported to their remanufacturing facilities. The implementation of the state-of-the-art technologies for remanufacturing is as follows [16]:

- **Computerized Diagnostics and Testing Systems**: The first step of remanufacturing involves using advanced computerized diagnostics to characterize the wear and damage of machine components. This involves using sensors and a computerized system to detect wear and damage at the micro level. This technology allows precise identification of damage and extent of wear, which could not be possible through traditional inspection methods.
- **CNC Machining**: After wear and damage assessment completion, precise manufacturing techniques are used to restore components to their original specifications. Computer numerical control machines (CNC) precisely restore the components to their original specifications and performance standards.
- **Advanced Materials Technology**: Caterpillar integrates advanced materials techniques, such as advanced coatings and heat-treatment cycles, to increase the performance of remanufactured components. These techniques are applied to restore the components with improved performance and extend their lifespan.
- **Laser Scanning**: It is used to digitize tool-path generation for adaptive machining. Adaptive machining analyzes the part's geometry and creates a machining program that is precisely suitable for it. Laser cleaning, blasting for surface preparation and cutting have consistently lowered the cost of remanufacturing processes with improved quality.
- **3D Printing**: Caterpillar exploits 3D printing techniques to improve the performance of worn components and produce new components. It helps manufacture components that have complex geometric features or customized components that are difficult to produce using traditional methods.
- **Upgrade the System**: Caterpillar's production processes are for restoring products and up-gradation. They often integrate up-to-date design changes and technological advancements into remanufacturing machinery and equipment. This guarantees that remanufactured machinery is not only a replica of the original but also the improved version.

3.9.2 Xerox

Xerox is another well-known company for its remanufacturing and refurbishment practices in office printers and copier machines. They use advanced remanufacturing technologies to embark on new life to used machines, with the improvements and updates. The company's approach is based on a "circular economy", in which products are designed and used to maximize the product life cycle and minimize waste.

- **Automated Disassembly Systems**: Xerox employs an automated system for disassembling used printers and copiers to ensure precision efficiency and decrease the likelihood of damage.
- **Advanced Cleaning Techniques**: Xerox cleans the disassembled parts with environmentally friendly and advanced cleaning techniques to ensure sustainability. These processes are essential for further reuse, refurbishment, and remanufacturing of parts.
- **Software Diagnostics and Firmware Updates**: Xerox employs software diagnostics to update the firmware and software to ensure compatibility with advanced digital environments.
- **Sophisticated Testing Rigs**: At the final stage of remanufacturing, Xerox uses advanced testing rigs to replicate actual operating conditions to ensure the intended function of components and machines.

Xerox products undergo rigorous testing and inspection measures to meet the standard of a new product. These quality measures are extended during packing and distribution to ensure that the remanufacturing units are presented and delivered to customers as new [17, 18].

3.9.3 Phillips

Phillips is a medical device maker in the healthcare sector and has a dedicated remanufacturing facility, particularly in imaging systems like Magnetic Resonance Imaging (MRI) and computerized tomography (CT) scanners. They apply advanced technologies to ensure complex devices are as per the specifications of original manufacturers and regulatory requirements [19–21].

- **Specialized Imaging and Diagnostic Technologies**: The initial inspection of medical device components includes using advanced imaging techniques, X-ray, and ultrasound to characterize the internal condition at the microscopic level. This involves a detailed analysis of the critical components for the safety and performance of medical equipment.
- **Hardware and Software Upgrades**: Phillips often upgrades the hardware and software to revitalize medical devices. This involves installing the latest imaging

software and updating hardware to improve performance. For example, a remanufactured MRI machine might be equipped with advanced imaging software to improve image quality and diagnostic capabilities compared to the original version.

- **Regulatory Compliance and Safety Testing Technologies**: Given the critical nature of medical devices, Philips exploits various technologies to ensure that remanufactured devices meet regulatory and safety standards. The rigorous testing and inspection measures verify that devices meet or exceed the original product specification.

A movie on "Waste is food."

https://www.youtube.com/watch?v=4pwCFH1LkCw&ab_channel=vpro documentary (VPRO). http://tegenlicht.vpro.nl/afleveringen/2006-2007/afval-is-voedsel-deel-1.html. An inspiring documentary on the Cradle-to-Cradle design concept of M. Braungart and W. McDonough. Winner of the Silver Dragon at the Beijing International Science Film Festival 2006.

3.10 Review Questions

Q1. Differentiate between remanufacturing and recycling.
Q2. Define and differentiate remanufacturing from repair and reconditioning.
Q3. Describe the benefits of remanufacturing.
Q4. Define the term core used in remanufacturing.
Q5. Define the term remanufacturing.
Q6. Why is a framework and research agenda needed for remanufacturing?
Q7. How will remanufacturing have dropped the existing carbon footprint?
Q8. What are the technologies and their technology readiness level for remanufacturing?
Q9. What is the current state of technology implementation in remanufacturing?
Q10. Why remanufacturing frameworks are needed?
Q11. What is the need for standards and regulations for remanufacturing?
Q12. What are the emerging needs and new standards for remanufacturing?
Q13. What are the IRS and IRIS?
Q14. What are the components of IRS in the remanufacturing framework?
Q15. What are the components of IRIS in remanufacturing frameworks?
Q16. Why are planes such as i-G, i-E, and D-P used to explain the remanufacturing framework?
Q17. What technologies will emerge because of the discussion made in the explanation of the remanufacturing framework through the i-G plane, i-E plane, and D-P plane?

Q18. What is the role of TLM and MES in remanufacturing? Does the definition of remanufacturing framework enhance the definition of TLM and MES; and how?

Q19. What is the current research agenda for remanufacturing and how would you see it being enhanced in coming years?

Q20. Give at least two real-life examples of remanufacturing not listed in this chapter.

Bibliography

1. Li K, Liu J, Fu H, Zhao N (2021) An integrated system with multiple product lifecycles for remanufacturing (IS-MPLR): new opportunities and challenges. Int J Comput Integr Manuf 34(1):20–40
2. Zacharaki A et al (2021) RECLAIM: toward a new era of refurbishment and remanufacturing of industrial equipment. Front Artif Intell 3:570562
3. Östlin J, Sundin E, Björkman M (2009) Product life-cycle implications for remanufacturing strategies. J Clean Prod 17(11):999–1009
4. Yang S, Aravind Raghavendra MR, Kaminski J, Pepin H (2018) Opportunities for industry 4.0 to support remanufacturing. Appl Sci 8(7):1177
5. Tolio T et al (2017) Design, management and control of demanufacturing and remanufacturing systems. CIRP Ann 66(2):585–609
6. Ridley SJ, Ijomah WL, Corney JR (2019) Improving the efficiency of remanufacture through enhanced pre-processing inspection—a comprehensive study of over 2000 engines at Caterpillar remanufacturing, UK. Prod Plan Control 30(4):259–270
7. Ragampeta P (2022) Industry 4.0 remanufacturing: a novel approach towards smart remanufacturing. Missouri University of Science and Technology
8. Pratapa P, Subramoniam R, Gaur J (2022) Role of standards as an enabler in a digital remanufacturing industry. Sustainability 14(3):1643
9. King A, Barker S, Cosgrove A (2007) Remanufacturing at Xerox: evaluating the process to establish principles for better design. In: DS 42: proceedings of ICED 2007, the 16th international conference on engineering design, Paris, France, 28–31 July 2007, pp 743–744
10. ben Khedher A, Henry S, Bouras A (2009) Interaction between product life cycle management and production management: PLM-MES integration. In: 2nd doctoral workshop IFIP WG 5.7, p 13
11. Kerr W, Ryan C (2001) Eco-efficiency gains from remanufacturing: a case study of photocopier remanufacturing at Fuji Xerox Australia. J Clean Prod 9(1):75–81
12. Kerin M, Pham DT (2020) Smart remanufacturing: a review and research framework. J Manuf Technol Manag 31(6):1205–1235
13. Kerin M, Pham DT (2019) A review of emerging industry 4.0 technologies in remanufacturing. J Clean Prod 237:117805
14. Jensen JP, Prendeville SM, Bocken NMP, Peck D (2019) Creating sustainable value through remanufacturing: three industry cases. J Clean Prod 218:304–314
15. Hatcher GD, Ijomah WL, Windmill JFC (2011) Design for remanufacture: a literature review and future research needs. J Clean Prod 19(17–18):2004–2014
16. Goodall P, Sharpe R, West A (2019) A data-driven simulation to support remanufacturing operations. Comput Ind 105:48–60
17. Fofou RF, Jiang Z, Wang Y (2021) A review on the lifecycle strategies enhancing remanufacturing. Appl Sci 11(13):5937

18. Erboz (2017) How to define industry 4.0: main pillars of industry 4.0. In: Managerial trends in the development of enterprises in globalization era, vol 761, pp 761–767
19. Charter M, Gray C (2008) Remanufacturing and product design. Int J Prod Dev 6(3–4):375–392
20. Bernard S (2011) Remanufacturing. J Environ Econ Manag 62(3):337–351
21. Abbey JD, Meloy MG, Blackburn J, Guide VDR Jr (2015) Consumer markets for remanufactured and refurbished products. Calif Manage Rev 57(4):26–42

Part II
Computer Aided Engineering Design and Manufacturing

Prelude

The History of CAD/CAM/CAE and CIM

Computer aided design and manufacturing is the cornerstone to design and build things. Its historical milestones can be traced back to the nineteenth century and are closely related to the development of photography and computers. At the start of 1861, a French chemist Alphonse Louis Poitevin discovered a process involving sunlight and substance found in gum which enables architectural drawing to be reproduced. The design can be replicated and distributed for everyone to work from the same design without fear of errors. In 1936, British scientist Alan Turing invented the Turing machine which forms the basis of modern computers. During early 1950, the main frame computer was introduced and up till 1955, transistors started replacing vacuum tubes, and interactive computer graphics were introduced. In the 1960s, with the advancement in interactive computer graphics, Ivan Sutherland developed a sketchpad system which established the possibility of creating drawings and change of objects interactively on a cathode ray tube (CRT). In 1961, renowned computer scientist Patrick J. Hanratty at General Motor research laboratory developed Design Automated by Computer (DAC). Later, W. T. Ross, pioneer of computerized machining, coined the term CAD. During the 1970s, the industry, government, and academia realized that improving productivity could be done through computer drafting. Intel introduced the microprocessor to the World and Patrick J. Hanratty introduced the software Automated Drafting and Machinery (ADAM). An estimated 90% of today's commercial drafting software based their roots back to ADAM. The 1980s was the era of new algorithms development for expanding the use of design and manufacturing in engineering applications. IBM released the desktop PC and John Walker introduced AutoCAD software.

The CAD/CAM/CAE is now in the evolution stage to meet the demand of industrial digitization since it was built for distinct industrial requirements. The new challenge for CAD, CAM, and CAE technologies is to play a fundamental role to fulfill smart production requirements.

In the 1980s, British research funding agency ESPRIT and other national research agencies in Western Europe and in Northern America started research on 'Integration' leading to production philosophy Computer Integrated Manufacturing—CIM. Integration today (2023) is a pillar of Industry 4.0 and has a much wider scope than what was envisaged in the 1980s.

Programming Languages[1]

Programming languages have evolved significantly over the years, giving rise to five different generations of high-level languages. Each generation represents a paradigm shift in programming, bringing in new features and capabilities. We explore the generations of high-level programming languages in the following discussion, with a specific focus on the current generation and its impact on the world of software development.

1. *First Generation: Machine Language and Assembly Language*
   The first generation of programming languages emerged during the early days of computing with machine language and assembly language being the first programming languages. These languages were complex, tiresome, and required significant time, e.g., the machine language consisted of binary code directly operated on the hardware. Assembly language provided a slight improvement, allowing for symbolic representation of machine code. However, these languages were cumbersome, error-prone, and closely tied to the underlying hardware.
2. *Second Generation: Assembly Language and Procedural Languages*
   Procedural languages like COBOL and FORTRAN constituted the second generation. These languages improved upon the user-friendliness of the first generation by abstracting the hardware details and therefore, made programming more accessible. Programmers could work with statements that resembled natural language, improving code readability and maintainability. Procedural languages also introduced the concept of structured programming, which enhanced code organization and control flow.
3. *Third Generation: High-Level Languages and Object-Oriented Programming*
   High-level languages were the main product of the third generation of programming languages. Typical examples of programming languages of this time are C, Pascal, and Lisp. These languages further simplified programming by introducing more built-in features, libraries, and tools. Additionally, this era saw the advent of object-oriented programming, with languages like C++ and Smalltalk, which allowed for more complex and modular code structures. Object-oriented programming emphasized reusability, data encapsulation, and inheritance, making it easier to develop large and complex software systems.

[1] Courtesy: Dr. Raja Hashim Ali, Associate Professor, Faculty of Computer Science and Engineering, GIK Institute of Engineering Sciences and Technology, Topi—23640, Khyber Pakhtunkhwa, Pakistan. Reproduced with permission.

4. *Fourth Generation: Domain-Specific and Visual Programming Languages*
   The fourth generation marked a shift towards domain-specific languages and visual programming. Some examples of languages in this category are SQL for database management and MATLAB for mathematical computation. These languages aimed to address specific application domains and increase the productivity of programmers. Visual programming tools, such as Scratch and LabVIEW, made it possible for users with varying levels of programming experience to create applications through graphical interfaces.
5. *Fifth Generation: The Current State of High-Level Languages*
   The current generation of high-level languages, often referred to as the fifth generation, is marked by several trends and developments.

(a) **Multifaceted Language Ecosystem:** There is a wide array of programming languages catering to diverse needs. Languages like Python, JavaScript, and Ruby are popular for web development, while C and C++ remain essential for system-level programming. Functional programming languages like Haskell and Scala are gaining popularity for their elegant and expressive syntax. High-level languages like Python have become the de facto choice for artificial intelligence, machine learning, and data science. Libraries like TensorFlow and PyTorch have contributed to Python's dominance in this field.
(b) **Abstraction and Productivity:** Modern high-level languages emphasize abstraction and productivity. They provide extensive libraries and frameworks, enabling developers to build complex applications more efficiently. Integrated development environments (IDEs) offer features like code completion and debugging tools, further enhancing productivity.
(c) **Cross-Platform Development:** With the rise of mobile devices and the importance of cross-platform compatibility, languages like Swift for iOS and Kotlin for Android have gained prominence. Additionally, technologies like JavaScript's Node.js enable server-side development using the same language as front-end web development.
(d) **Open Source and Community Collaboration:** The current generation of high-level languages is characterized by extensive community collaboration and open-source development. This collaborative approach has led to rapid language and library improvements, with developers actively contributing to the ecosystem.

Scientific Laws, Standards and Databases

The day primitive humans started using logic to overcome their basic problems, the process of guided development started taking place. The need was there, and observation and experimentation were the only tools. Accumulation of knowledge was started and was being forged in principles many of those modern humans still use today.

Scientific laws, technical standards, technical specification, codes of practice, framework, benchmarking, regulations are modern names of continuously accumulating knowledge from the days of primitive human beings to today's modern person.

The accumulated knowledge was stored from stone to animal skin to paper to Random Access Memory (RAM—Cloud) to permanently storing mediums such as Read Only Memory (ROM) and other forms of repositories to save the knowledge from all human inflicted damage or the force majeure conditions.

Information Transformation

Today's digital technology uses information (processed or unprocessed data) that can be utilized for multiple purposes. It is important to understand the word 'convention' as per a certain standard that is accepted by a large community. In the context of this book, it is "by Engineers" that is widely accepted in all its forms. To exemplify this concept, take the multiple information transformation of English letter 'A' manufacturing using the Computer Numerical Control process. For a literate human, it is the first letter of the English alphabet. To make this letter 'A' understandable by the computer and to process it, for say a CNC part program for a five-axis milling machine, the letter may be expressed in Binary Coded Decimal (BCD), American Standard Code for Information Interchange (ASCII), or Unicode. Today, ASCII is the widely accepted convention in English speaking/understanding communities while for different languages the Unicode will be chosen. Assuming ASCII 'A' is adopted, it will have DATA BUS wide (8, 16, 32, 64 bits, etc.) binary representation meaning that rotation about X-Axis represented by letter 'A' is being programmed. With each binary high (1) voltage level, the transistor will give one step to a stepper motor or one signal from shaft encoder in case of servo motor. Therefore, rotation of A-Axis motor will be the least count of motor shaft rotation with magnitude that gives the rotation of the workpiece as programmed. In the case of X-Axis, it will be the least count of the linear motion in X direction. It can be seen from this example that the layers of information transformation are multiple and are widely accepted.

Such information transformation as per widely accepted convention is embedded in many equipment and other applications. The logic is to operate the equipment to do the productive work efficiently.

Communication Protocols

Physical and cellular communication protocols are in common use and have reached a technology Readiness level that data communication is performed at a sheer speed of fiber optic internet, clocking in at 1 Gbps (Gigabits per second), providing speeds nearly 20 times faster than conventional cable connections. Cellular networks (5G) operate at 150–200 Mbps. 6G cellular networks will soon hit the market.

Modern data communication has given rise to the application of the cyber security pillar of Industry 4.0.

Physical and cellular data communication and cyber security are detailed in Part VI Chap. 13.

Multidisciplinary Product Development

A product becomes multi-discipline by virtue of use of different materials, mechanical features, electrical features, control using electronics, user interface (UI) through mechanical means or through a graphical user interface based on computer software, digital control using computer hardware, ergonomics, and use of any other field of engineering specific to the use of product in a particular environment. All this combined with reliability, availability, maintainability, and safety over a period forms the user experience (UX) of the person with the multidisciplinary product in comparison to the same or similar product manufactured by any other manufacturer.

Manufacturing Execution System (MES)

Manufacturing Execution System (MES) are all the tangible and intangible services used in tangible value addition of a product from raw material to finished product.

Product Lifecycle Management (PLM)

Product Lifecycle Management (PLM) involves the establishment of need for a product, complete design cycle, pilot production, production, packaging, and dispatch to distributor and after sales services to Client. MES is the tangible value addition and all intangible services required for this tangible value addition and is an embedded part of PLM.

Product Life Cycle

Product Life Cycle is the cradle-to-grave life span of the product. It starts from production of raw material for product, spans over PLM embedding MES, change of hand of product if any, maintenance, end of useful life, reuse through remanufacturing, recycling, and waste disposal.

Chapter 4
Computer-Aided Design

Prelude
Computer Aided Design today (2024) is not just producing lines with some annotations. It has reached to a level where carbon footprint, sourcing, costing with optimized design alternatives for reliability, availability, maintainability, and safety at different manufacturing sites with optimal supply chain of raw material along with interface to PLM and MES software is possible. Product Life Assessment (PLA) is available and complete product life cycle (from Cradle-to-Grave) with change of hands even among countries (e.g. planes and automobiles) is possible through complete record keeping with possibility to use Blockchain, a pillar of Industry 4.0.

4.1 Objectives

The learning objectives of this chapter are the same as that of Chap. 1 except that the ergonomics is supplemented by User Interface (UI) and User Experience (UX). The design cycle is explained using a system approach to engineering design where every element starting from need analysis to design documentation is demarcated and is performed by a Team (Process). The advancement in Mathematical Technology implemented on CAD packages is used here and is explained in detail in Part III, Chap. 7. The same is the case with the Technology Readiness Level of Computing introduced in the prelude of Part I- and Part II. Multidisciplinary aspects of the design of a product are normally based on software and electronics hardware. Other aspects render a design multidisciplinary like the use of different materials which is explained in Part I, Chap. 1. Concept generation and Concept Selection methods are

---

Supplementary Information The online version contains supplementary material available at https://doi.org/10.1007/978-3-031-77106-4_4.

W. A. Khan et al., *Computer Aided Engineering Design and Manufacturing*, Springer Series in Advanced Manufacturing, https://doi.org/10.1007/978-3-031-77106-4_4

explained here. The use of databases to have a systematic approach to design fully implemented is explained. The importance of the use of standards and time factors for product launches is also addressed herein.

4.2 User Interface (UI) and User Experience (UX)

User Interface (UI) as defined in the prelude to this part is the comparative convenience with which a tangible or intangible product can be used mechanically (Man Machine Interface—MMI), through Graphical User Interface—GUI (Human Computer Interaction—HCI) or through Cellular Signals (from Bluetooth to control an air conditioner to microwave signals to control e.g. Mars Rover).

User Experience (UX) is the comparative convenience with which same or similar products manufactured by the same or different manufacturer can be used through the provided User Interface (UI) over a certain span of time.

4.3 Concept Modeling Using CAD Packages

Concept Modelling is defined in Chap. 1, Part I. Use of CAD packages expedites the concept generation specifically in the case of original design. Generally, Concept modeling efforts are much less in case of adaptive design and variant design. An example of modular concept modeling using BS for different types of spanners is provided in Fig. 4.1. These concepts go through a concept selection procedure as described in Chap. 1 Part I. The use of standards minimizes the concept selection effort, and the product is available as pictures and can be used for marketing after costing with appropriate profit margins.

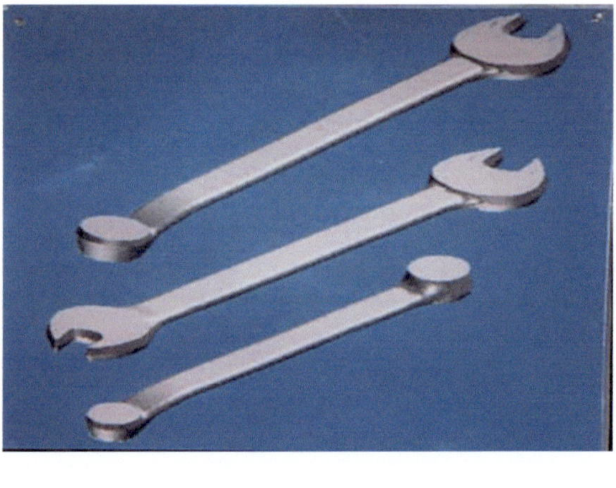

Fig. 4.1 A modular concept modeling using BS for different types of spanners

4.4 Embodiment Design Using CAE Packages

The selected Concept Model needs analysis depending on the type of exposure it will be subjected to during its lifespan to convert it to embodiment design that performs analysis at component, structure, assembly, and mechanism level. All types of matter need to be considered including solids, liquids, gases, and plasma. The specifications of the product were raw (not in above case as Standard information is being used) in case of Concept Modelling and it gets firmer after Embodiment Design using Analytical, Numerical and Experimental methods of Analysis. For type and depth of analysis required the sensitivity of use of product is involved. A balance must be brought in as analysis adds to the cost of design. No compromise can be made in applications where life of living beings is involved alone or collectively. Examples of Analytical, Numerical and Experimental analysis are available in case studies provided in Part IV Chap. 6.

4.5 Detail Design Using CAE Packages

The detailed design encompasses analysis of the assembly of components, structures, assemblies, and mechanisms forming a machine. It also considers the type of matter and its interaction with other types of matter. In detail design, the design analysis ranges from the foundation requirement of a machine tool to modal analysis for avoiding natural frequency during machine operation, safe exit in case of emergency, and saving the machine in the best possible manner. The detail design stage has a frozen specification whereby the product is allowed to be used by customer after all possible safety measures while design office keeps on working on improving the product as per technology readiness level for maximum innovation while optimizing the design for safety, cost, ease of operation and ease of production as customized product or for mass production. The case studies related to this topic are placed in Chap. 1 Part I and Chap. 2 Part I.

4.6 Design Documentation Using CAD

All currently available CAD packages can produce design documentation with appropriate annotations in first angle projection, third angle projection and/or isometric view as well as the exploded view of the assembly of the machine. An example of design documentation is provided in Chap. 2 Part I.

4.7 Case Study[1]: Design and Manufacturing of Industrial Scale 6-Spindle Spinning Machine[2]

1. Data Acquisition (DAQ) System

Industrial Scale 6-Spindle Spinning Machine Control Specification.

1.1 DAQ Includes:

Electric Panel (tower type).

Programmable Logic Controller (PLC)

Central Processing Unit (CPU)

Touch Screen HMI Data Monitoring and saving Module

Emergency stop function

1.2 Component to Integrate with the Control System

Motor for spindle rotation, 1hp, 2800 rpm

Inverter for spindle rotation

3 × stepper motors with drives for drafting rollers, 3.5 A

1 × stepper motors with drives for Lifting of ring rail, 3.5 A

Vacuum pump-800 W, 0–18,000 rpm

DC Power supply 20 A, Volt:

Industrial Scale 6-Spindle Spinning Machine
Design Report (DR)
Prepared by:
Faculty of Mechanical Engineering, GIK Institute & Edutech
Approved by: FME, GIK Institute
Recommendation: Edutech, Lahore
GIK Institute of Engineering Sciences and Technology
Approving Authority:
Directorate of Science and Technology, Government of Khyber Pakhtunkhwa
Document Revision Record

Pages not listed remain at the released revision level.
Pages affected by the current ECN reflect the present documents' revision level.

[1] Work performed using Grant under ADP Scheme No. 1286-160565 (2016-17) Directorate of Science and Technology, Department of ST&IT, Government of Khyber Pakhtunkhwa, Pakistan.

[2] Pakistan Patent Applications in favor of GIK Institute in process.

Revision Level	ECN Number	Date	Checked by	CM Release	Pages Affected

Contents

Tables

Figures

1. Scope

The Design Report (DR) document serves as the Design, Development, and Validation of 6-Spindle Spinning Machine by EDUTECH LAHORE.

2. Introduction

Six spindle spinning machine is a ring-spinning machine that draws, rewinds, or rolls half-finished roving or sliver into a fine yarn. The roving is made through the drafting of the fine gauze machine and by the drawing and holding of the roving machine. The roving can be made into fine yarn by the drafting and tenting of the fine gauze machine; the roving and tenting of the roving yarn can be made into fine yarn by the drawing and tenting of the fine gauze machine; the combing machine can comb the fibers into fine comb strips; the needle combing machine can combine the strips.

A Ring spinning machine is the main machine of spinning, and the output and quality of spinning is a comprehensive reflection of the advantages and disadvantages of each spinning process.

The basic function of the ring-spinning frame is drafting. Drafting is carried out to such an extent as to achieve the desired fineness of the yarn.

The final yarn produced after drafting and twisting is wound on special ring bobbins also called cops. The building of the package kept is such that it is suitable for storage, transportation, and further processing. The isometric view of the ring frame is given in Fig. 2.1.

This is the most widely used method of staple-fiber yarn production. The fibers are twisted helically around each other to impart strength to the yarn with the help of a spindle, ring, and traveler as the main components of twist insertion and winding.

2.1. 6-Spindle Spinning Machine in Engineering

Ring spinning is the most versatile spinning process. It can spin yarns of a wide range of counts (from very fine to coarse) from different types of fibers (short as well as long-staple fibers). The quality of ring-spun yarns has been a benchmark against which the quality of yarns produced on other spinning systems is judged.

The basic principle of ring spinning has not changed much since its invention by Thorpe in 1828. But there have been numerous developments of the ring spinning system, particularly in the short-staple sector. These developments include:

- Automatic doffing of full cops (bobbins)
- Linkage to roving frame
- Linkage to an automatic winding machine.

The recent developments in ring spinning are mainly concerned with improvements in drive systems, drafting arrangement, design aspects of ring and traveler, machine monitoring, and automation that includes roving stop motion, roving bobbin transfer, doffing, and transfer to winding. Improved the versatility of ring spinning without change in the very basic concept of ring spinning.

In ring spinning, twist insertion requires the rotation of the whole yarn package on the spindle. About 95% of the power used in ring spinning is consumed by rotating the yarn package to insert a twist. This leads to relatively high-power consumption for the ring-spinning systems. The package size is limited in ring spinning due to the need to reduce the balloon height and yarn tension, as discussed in the section on the physics of ring spinning. Perhaps the most serious limitation of ring spinning is its low production rate. We already know the relationship between yarn twist level, spindle speed, and yarn production or delivery speed. According to its end-use, a ring spun yarn needs to have a certain level of twist, which is determined before spinning is started. From the production equation, we know that the only way of speeding up the yarn delivery speed or production rate is to increase the spindle rotation speed. We already know that any increase in spindle speed will lead to a significant increase in yarn tension, hence the possibility of ends down.

In addition, with the increase in spindle speed, the traveler's speed increases. This increases the friction between the traveler and the ring. Considerable heat is generated because of this friction, which may result in the traveler running during spinning. Because of these, spindle speed cannot be increased at will, and yarn production is limited consequently.

2.2. Use in Engineering Applications

The ring spinning machine is used in the textile industry to simultaneously twist staple fibers into yarn and then wind it onto a bobbin for storage. The yarn loop rotating rapidly about a fixed axis generates a surface referred to as a balloon.

2.3. Design and Model

To manufacture 6-Spindle Spinning Machine, is easily transferable, various aspects and considerations need to be made. Use of Computer Aided Design (CAD) models using SOLIDWORKS software can be used to develop 3D parts and assembly for an understanding of the model. Calculation of load, right specification for a part used, and various other factors are considered as well through analytical calculation and Analysis report. Selection of the procurement part is also based on the required load. After passing this evaluation it's a final Design for Manufacturing and Assembly (DFMA).

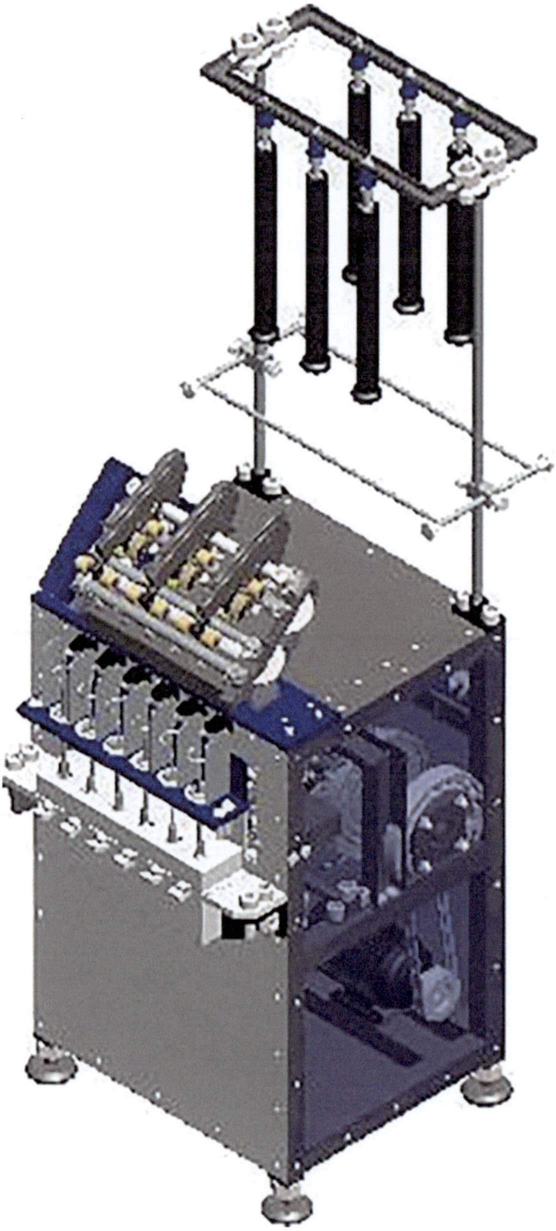

Fig. 2.1 6-Ring Spinning Machine CAD Model

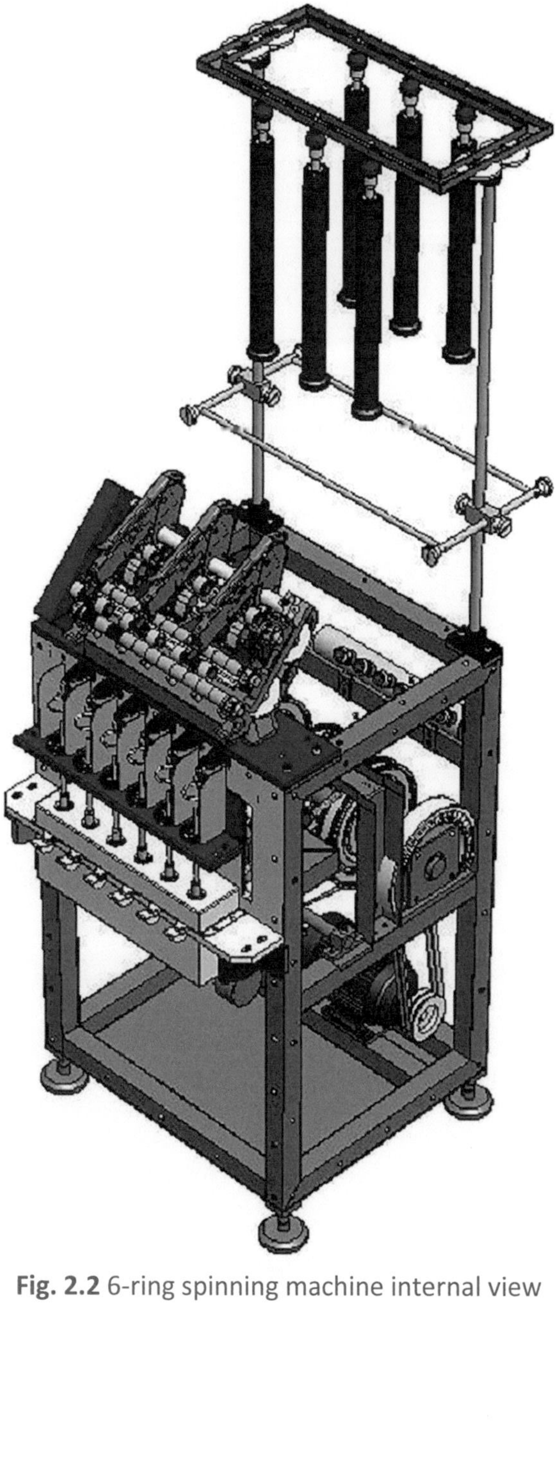

Fig. 2.2 6-ring spinning machine internal view

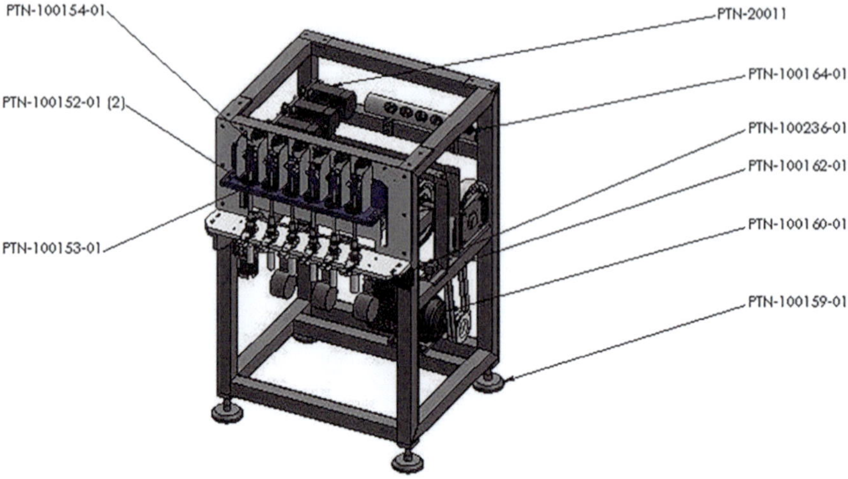

Fig. 2.3 Base Frame of 6-Spindle Spinning Machine

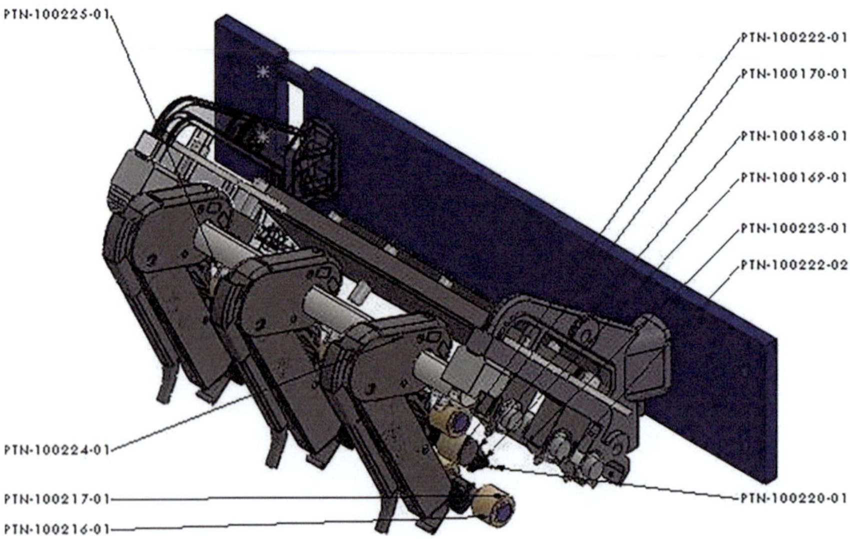

Fig. 2.4 Assembly (100111-101) of 6-SSM

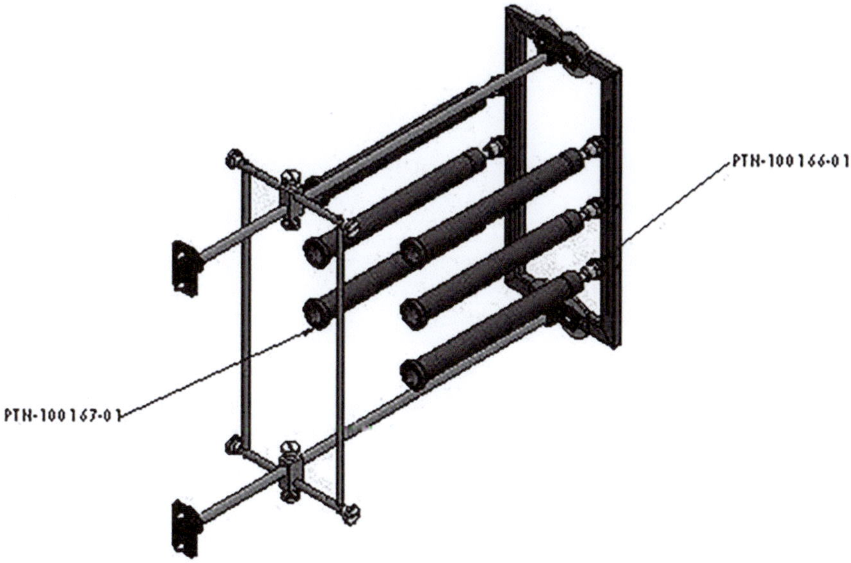

Fig. 2.5 Top Frame of 6-SSM

3. Theoretical Calculations

Diameter of the 1st Pulley $d_1 = 80$ mm Diameter of the 2nd Pulley $d_2 = 800$ mm Diameter of the 3rd Pulley $d_3 = 250$ mm Diameter of the 4th Pulley $d_4 = 22.5$ mm.

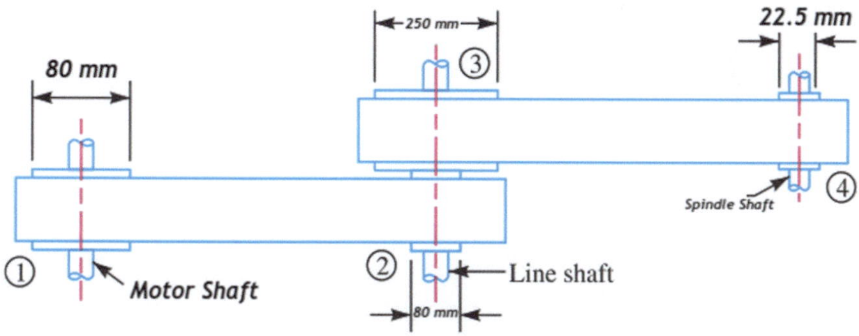

All dimensions in mm.

As

$$N_1 = 2800 \text{ rpm}$$
$$\mathrm{P}_1 = 1hp = 746 \text{ W}$$

So,

$$\frac{N_3}{N_4} = \frac{d_4}{d_3}$$
$$\frac{N_3}{N_4} = \frac{22.5}{250}$$
$$\frac{N_3}{N_4} = 0.9527$$
$$\boldsymbol{\frac{N_4}{N_3} = 10.5}$$

Self-weight of Spindle = m = 1 kg

$$T = I\alpha$$
$$I = \frac{1}{2}mr_4^2$$
$$I = \frac{1}{2}(1)\left(\frac{22.5 \times 10^{-3}}{2}\right)^2$$
$$\boldsymbol{I = 6.328 \times 10^{-5}\ \mathrm{kg\ m^2}}$$

$$\omega_4 = \frac{2\pi N_4}{60}$$

where, $N_4 = 19{,}000$ rpm (spindle speed)

$$\omega_4 = \frac{2\pi(19000)}{60}$$
$$\boldsymbol{\omega_4 = 1990\ \mathrm{rad/s}}$$

$$\omega_4 = \omega_0 + \alpha t$$
$$1990 = 0 + \alpha(1)$$
$$\alpha = 19690\ \frac{\mathrm{rad}}{\mathrm{s}}$$

$$T_4 = (1.2656 \times 10^{-4})(1990)$$
$$T_4 = 0.125\ \mathrm{Nm}$$

3.1. Torque Required for Lifting the Ring Rail

Stepper motor	57HY57-100
Torque	3 Nm
Speed	250 rpm
Current	3.5 A

(continued)

(continued)

Phase	2
Step	1.8°
Mass of ring rail	5.1 kg
External force	50.03 N

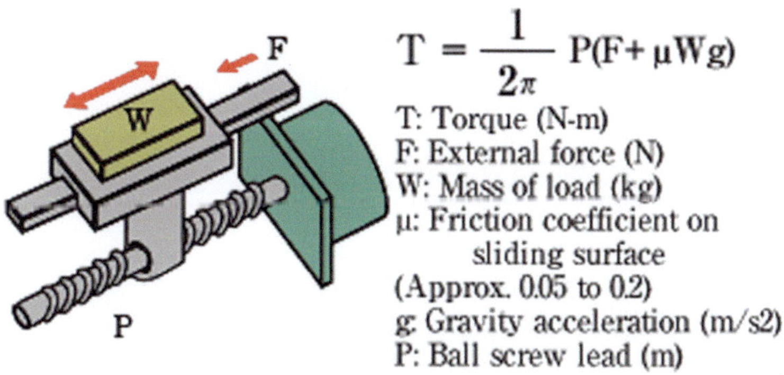

$$T = \frac{1}{2\pi}P(F + \mu Wg)$$
$$T = \frac{1}{2\pi}0.003(50.03 + (0.2 * 5.1 * 9.81))$$
$$T = 0.18\ \text{Nm}$$

4. Design Specification and Component List

The specification given for the design of 6-Spindle Spinning Machine include the:

Table 4.1 Component part list

Approved part list (6-spindle spinning machine)					
Sr. #	Name	Part No	Model/Supplier Part No	Make	QTY
1	Spindle with brake	PTN-100151-01	D4203C	Grand Peace Enterprise Limited, China	6
2	Ring	PTN-100152-01	GT2000	TEXParts, Germany	6
3	Ring adapter	PTN-100153-01	R. Adapter GT2000	TEXParts, Germany	6
4	Ring adapter lock ring	PTN-100155-01	Lock Ring GT2000	TEXParts, Germany	6

(continued)

(continued)

Approved part list (6-spindle spinning machine)					
Sr. #	Name	Part No	Model/Supplier Part No	Make	QTY
5	Eyelet ASY	PTN-100156-01	Eyelet	VD Textile, China	6
6	Traveler	PTN-100157-01	Flat	Tongda, China	100
7	Pulley timing chain for motor	PTN-100158-01	Timing pulley belt	Local	6
8	Leveling pad	PTN-100159-01	M16	Local	4
9	Motor for spindle rotation, 1hp, 2800 rpm	PTN-100160-01	M2VA80B-2	ABB	1
10	Belt for spindles	PTN-100161-01	Flat belt	Local	3
11	The belt for the motor to the shaft	PTN-100162-01	V-belt type-B	Local	2
12	The belt for drafting rollers	PTN-100163-01	Timing pulley belt	Local	3
13	Control ring	PTN-100165-01	Balloon C. Ring 34.5 mm	Tongda, China	6
14	Bobbin hanger	PTN-100166-01	DD2-22 KHK	Sichron International Trading China	6
15	Roving bobbin	PTN-100167-01	Roving Bobbin	Qingdao Guoao Machinery, China	6
16	Mid-top roller	PTN-100168-01	LP1003	TEXParts, Germany	3
17	Cradle for Appron, 1247888 PA6.6-GF50	PTN-100169-01	OH 2022	TEXParts, Germany	3
18	Top apron	PTN-100170-01	37 × 27.8 × 1.1 mm 0009 11	Tongda Textile, China	6
19	Manifold	PTN-100171-01	1/2in air manifold 6 outlet	Local	1
20	VDF/Inverter	PTN-100172-01	ED51	Invent, Pakistan	1
21	Power Supply	PTN-100173-01	NDR-480–24	Meanwell, Taiwan	1
22	Top rollers	PTN-100216-01	LP 1002	TEXParts, Germany	9
23	Top Rubber rollers	PTN-100217-01	LP 1002	TEXParts, Germany	18
24	Driver for stepper motor	PTN-100218-01	DM542	Dorna, China	4
25	Weighting arm	PTN-100219-01	PK2025	TEXParts, Germany	3

(continued)

(continued)

Approved part list (6-spindle spinning machine)					
Sr. #	Name	Part No	Model/Supplier Part No	Make	QTY
26	Appron guide pin	PTN-100220–01	OH 2022	TEXParts, Germany	3
27	Bearing for rollers	PTN-100221-01	LZ3224	Tongda, China	6
28	Front bottom rollers	PTN-100222-01	Bottom Rollers	Saurer, Germany	2
29	Middle bottom rollers	PTN-100223-01	Bottom Appron Rollers	Saurer, Germany	1
30	T slide for bottom appron	PTN-100224-01		Saurer, Germany	1
31	Roving guide	PTN-100225-01		D Textile, China	6
32	Spindle brake	PTN-100226-01	D4203C, brake	Grand Peace Enterprise Limited, China	6
33	Bearing and locksbore35mm, t = 35, OD = 80, speed = 5600 rpm	PTN-100227-01	3307 2RS	MBC bearing, China	2
34	BF12 unit	PTN-100228-01	BF-12 unit	Werb, china	2
35	BK12 unit	PTN-100229-01	BK-12 unit	Werb, china	2
36	Bal screw	PTN-100230-01	SFU1605-400	Werb, china	
37	Ball Nut	PTN-100231-01	SFU1605-3	Werb, china	2
38	Coupling	PTN-100232-01	D25L30_8-10	Werb, china	2
39	Stepper motor	PTN-100233-01	57HY57-100	Dorna, China	1
40	Bottom apron	PTN-100234-01	83 × 30x1.1 0020 12	Tongda Textile, China	6
41	Timing chain pulley for ball screw	PTN-100235-01	16XL037F	Local	2
42	The belt for the ball screw	PTN-100236-01	1timing belt, 1/2in	Local	1
43	V-Pulley B-type for shaft	PTN-100237-01	Bore40, OD78	Local	1
44	V-Pulley B-type for motor	PTN-100238-01	Bore 19, OD78	Local	1
45					

4.1. Spinning Geometry

To achieve good spinning results, the geometrical ratios described below must be complied with as precisely as possible:

d:D:	If the value of d:D is too small, this leads to extreme traveler strain. Traveler wear and end breakages will increase as a result If the value of d:D is too large, it will result in low productivity, however, this does not have an adverse effect on the spinning conditions
H:	If the bobbin or spindle selected is too long (e.g., $H = 5, 5 \times D$), the yarn balloon will touch the tip of the bobbin. Besides frequent end breakages and inferior yarn quality is registered
D and t:	When choosing the ring Ø (D) the spindle gauge (t) must be considered In comparison to the spindle gauge, the maximum ring diameter must be no more than 25 mm smaller than the gauge This ensures that the traveler and yarn balloon have the same flexibility. The insertion of the traveler, the elimination of end breakage, and cop changes are less problematic
BE:	The balloon control ring should be 2–3 mm larger than the ring diameter. If the balloon control ring is too large, it will not provide any relief for the yarn balloon

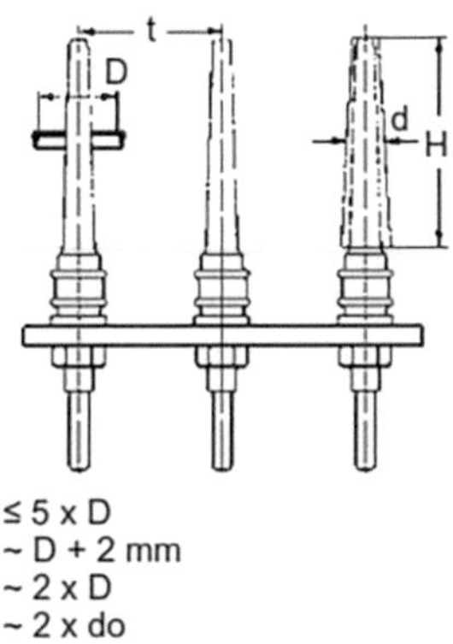

Fig. 4.1 Function of ring spinning frame

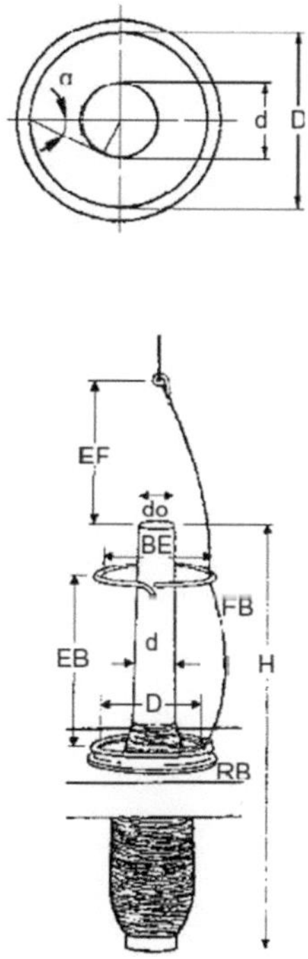

Fig. 4.2 Ring spinning frame

Symbols:

t = spindle gauge D = inside ring Ø
d = mean bobbin Ø do = top bobbin Ø H = bobbin length
BE = balloon control ring
EB = setting distance ring/balloon control ring FB = yarn balloon
RB = ring rail
EF = setting distance top of bobbin/yarn-guide eyelet (measures in mm)

Ratio values

D = t − 25 mm.
d:D in spinning:
0.48–0.5 bzw. α 29°–30°
(Not less than 0.44 or α 26°)

4.2. Functions of the Ring Spinning Frame

The basic function of the ring-spinning frame is drafting. Drafting is carried out to such an extent as to achieve the desired fineness of the yarn. The attenuated yarn formed by drafting is weak and lacks cohesion. To give strength to the final yarn, a twist is inserted. The twist is inserted at the ring- spinning frame by using the popular traveler method. The amount of twist inserted in the yarn varies with the count of the yarn. To make finer yarns, greater TPI is given, and vice versa.

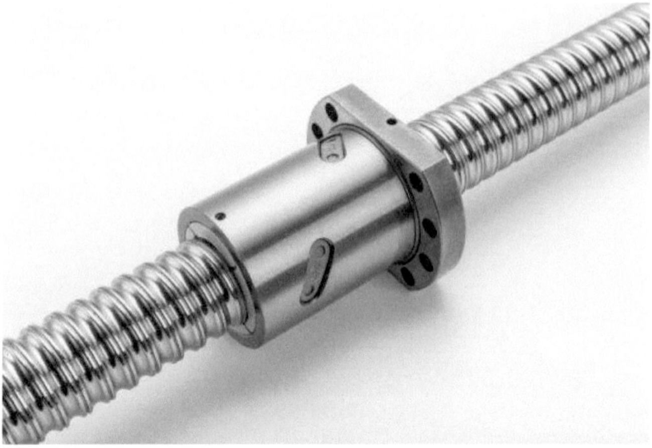

Fig. 4.3 Ball screw

4.2.1. Using Trumpet Guides

The strand of roving has few numbers of fibers in a cross-section and upon drafting even fewer fibers will remain in a yarn. These fibers have hardly any cohesion between themselves. Special fiber guiding elements are therefore essential to carry out drafting without breaking the strand. For this purpose, specially designed trumpet guides are used in the drafting zone to help keep the fiber mass together and to avoid tearing apart the strand. Three trumpet guides or condensers are used, one just before the back pair of rollers, the second in between the back and middle rollers, and the last just before the front pair of rollers. These guides are mounted on a reciprocating bar that oscillates slowly so that the wear of the rollers may spread out gradually throughout the entire roller width.

4.2.2. An Overhang of the Front and Middle Top Roller

As the twist is given to the yarn with the help of a traveler, the twist must run back as close as possible to the nip line of the front rollers. But it never penetrates completely to the nip because at the nip the fibers are diverted inwards and are made to wrap around the surface of the roller. So, at the exit point of the front pair of rollers, there is always a small triangular mass of fibers that remains without twist, this is called the spinning triangle. This triangle bundle of fibers will remain without twist during twisting and create a weak point in the yarn hence breakage rate of the yarn during spinning increases. The effect of the spinning triangle is enhanced at the ring-spinning frame where the greatest draft along with the greatest twist is applied.

4.2.3. Twisting Zone

The yarn leaving the front pair of drafting rollers is threaded through a guide also called the lappet guide that is placed directly over the spindle axis. The yarn then passes through a traveler onto the yarn package (cop). The traveler is mounted on the ring encircling the spindle. The cop is mounted on the spindle and rotates with the spindle. When the cop rotates the tension on the yarn pulls the traveler to rotate around the ring. Due to the friction created between the traveler and the ring and also due to the air drag created by the balloon formation at the winding section, the rotary and surface speed of the traveler is slightly less than that of the spindle and is given by:

$$RPM\ \ of\ \ Traveller = RPM\ \ of\ \ Spindle - Delivery\ Speed\ \pi\ \times Pachage\ Dia.$$

So, in one rotation of the traveler one twist is inserted in the yaRPMrn. Since the difference between traveler and spindle rotary speed is very small, this difference is ignored. Thus, calculating the twist rotary speed of the spindle is considered, mathematically:

$$T.\,P.\,I. = Rotary\ Speed\ \ of\ \ Spindle\ \text{(rpm)}/Delivery\ Spped\ \text{(inches/min)}$$

The ring/traveler arrangement and the spindle are also very important elements of the ring-spinning frame because it not only helps to insert a twist in the yarn but also helps in the winding of the package.

So, in the ring-spinning drafting arrangement, the front top roller is deliberately kept 2–4 mm forward relative to the bottom front roller while the middle top roller is kept 2–4 mm backward relative to the middle bottom roller. This type of arrangement gives smooth running of the top roller and helps to reduce the spinning triangle and ultimately the breakage rate during the spinning process is reduced.

The ring/traveler arrangement and the spindle are also very important elements of the ring-spinning frame because it not only helps to insert a twist in the yarn but also helps in the winding of the package.

4.3. Ball Screw

A ball screw is a mechanical linear actuator that translates rotational motion to linear motion with little friction. A threaded shaft provides a helical raceway for ball bearings which act as a precision screw. As well as being able to apply or withstand high thrust loads, they can do so with minimum internal friction. They are made to close tolerances and are therefore suitable for use in situations in which high precision is necessary. The ball assembly acts as the nut while the threaded shaft is the screw. In contrast to conventional lead screws, ball screws tend to be rather bulky, due to the need to have a mechanism to recirculate the balls.

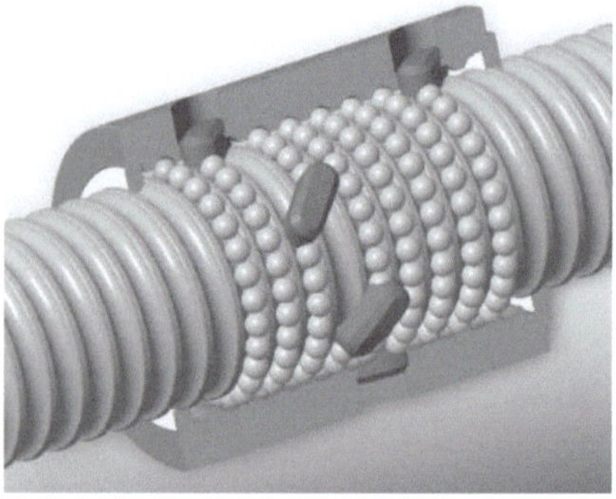

Fig. 4.4 Ball screw—internal view

Fig. 4.5 Induction motor

We select the ball screw of nominal dia. 16 mm. In our design, we use two of the same specification.

4.4. Stepper Motors

The DM542 is a fully digital stepper drive developed with an advanced DSP control algorithm based on the latest motion control technology. It has achieved a unique level of system smoothness, providing optimal torque and null mid-range instability. Its motor auto-identification and parameter auto-configuration features offer quick setup to optimal modes with different motors. Compared with traditional analog drives, DM542 can drive a stepper motor at much lower noise, lower heating, and smoother movement. Its unique features make DM542 an ideal choice for high-demand applications.

4.4.1. Features

- Anti-Resonance provides optimal torque and nulls mid-range instability.
- Motor auto-identification and parameter auto-configuration technology offer optimal responses with different motors.
- Multi-Stepping allows a low-resolution step input to produce a higher micro- step output, thus offering smoother motor movement.
- 15 selectable micro step resolutions including 400, 800, 1600, 3200, 6400, 12,800, 25,600, 1000, 2000, 4000, 5000, 8000, 10,000, 20,000, 25,000.
- Soft start with no "jump" when powered on
- Input voltage 18–50 VDC
- 8 selectable peak current including 1.00 A, 1.46 A, 1.91 A, 2.37 A, 2.84 A, 3.31 A, 3.76 A, 4.20A
- Pulse input frequency up to 200 kHz, TTL compatible and optically isolated input
- Automatic idle-current reduction
- Suitable for 2-phase and 4-phase motors
- Support PUL/DIR and CW/CCW modes
- Over-voltage, over-current protections

4.4.2. Applications

Suitable for a wide range of stepping motors, from NEMA size 17–34. It can be used in various kinds of machines, such as X–Y tables, engraving machines, labeling machines, laser cutters, pick-place devices, and so on. Particularly adapts to the applications desired with low noise, low heating, high speed, and high precision.

4.4.3. Power Supply Selection

The DM542 can match medium and small size stepping motors (from NEMA frame size 17–34) made by Lead shine or other motor manufacturers around the world. To achieve good driving performance, it is important to select supply voltage and output current properly. Supply voltage determines the high- speed performance of the motor, while output current determines the output torque of the driven motor (particularly at lower speed). The higher supply voltage will allow higher motor speed to be achieved, at the price of more noise and heating. If the motion speed requirement is low, it's better to use a lower supply voltage to decrease noise, heating and improve reliability.

4.4.4. Regulated or Unregulated Power Supply

Both regulated and unregulated power supplies can be used to supply the driver. However, unregulated power supplies are preferred due to their ability to withstand current surges. If regulated power supplies (such as most switching supplies.) are indeed used, it is important to have a large current output rating to avoid problems like a current clamp, for example using a 4 A supply for 3 A motor-driver operation. On the other hand, if an unregulated supply is used, one may use a power supply of a lower current rating than that of a motor (typically 50–70% of motor current). The reason is that the driver draws current from the power supply capacitor of the unregulated supply only during the ON duration of the PWM cycle, but not during the OFF duration. Therefore, the average current withdrawn from the power supply is considerably less than the motor current. For example, two 3A motors can be well supplied by one power supply of 4A rating.

4.4.5. Multiple Drivers

It is recommended to have multiple drivers share one power supply to reduce costs if the supply has enough capacity. To avoid cross-interference, DO NOT daisy-chain the power supply input pins of the drivers. (Instead, please connect them to the power supply separately.)

4.4.6. Selecting Supply Voltage

The power MOSFETs inside the DM542 can operate within + 20 to + 50 VDC, including power input fluctuation and back EMF voltage generated by motor coils during motor shaft deceleration. The higher supply voltage can increase motor torque at higher speeds, thus helpful for avoiding losing steps. However, the higher voltage may cause bigger motor vibration at a lower speed, and it may also cause over-voltage protection or even driver damage. Therefore, it is suggested to choose only sufficiently high supply voltage for intended applications, and it is suggested to use power supplies with a theoretical output voltage of + 20 to + 45 VDC, leaving room for power fluctuation and back-EMF.

4.4.7. Micro Step Resolution Selection

Table 4.2 Micro step resolution selection

Micro step	Steps/rev. (for 1.8°motor)	SW5	SW6	SW7	SW8
2	400	OFF	ON	ON	ON
4	800	ON	OFF	ON	ON
8	1600	OFF	OFF	ON	ON
16	3200	ON	ON	OFF	ON
32	6400	OFF	ON	OFF	ON
64	12,800	ON	OFF	OFF	ON
128	25,600	OFF	OFF	OFF	ON
5	1000	ON	ON	ON	OFF

(continued)

(continued)

Micro step	Steps/rev. (for 1.8°motor)	SW5	SW6	SW7	SW8
10	2000	OFF	ON	ON	OFF
20	4000	ON	OFF	ON	OFF
25	5000	OFF	OFF	ON	OFF
40	8000	ON	ON	OFF	OFF
50	10,000	OFF	ON	OFF	OFF
100	20,000	ON	OFF	OFF	OFF
150	25,000	OFF	OFF	OFF	OFF

4.5. Induction Motor

The AC induction motor is a rotating electric machine that is designed to operate on a phase supply. Motor is also called an asynchronous motor. The principle of operation of this motor is based on the production of a rotating magnetic field.

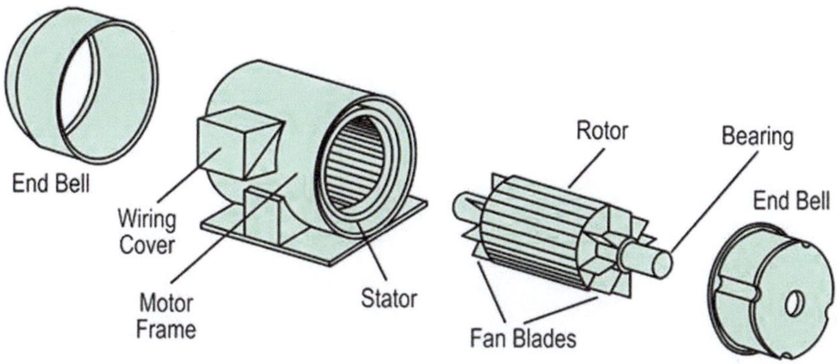

Fig. 4.6 3-phase induction motor

When the motor is excited with a three-phase supply, the three-phase stator wind produces a rotating magnetic field with 120 displacements at a constant magnitude which rotates at synchronous speed. This changing magnetic field cuts the rotor conductors and induces a current in them according to the principle of Faraday's laws of electromagnetic induction. As these rotor conductors are shorted, the current starts to flow through these conductors.

In the presence of the magnetic field of the stator, rotor conductors are placed, and therefore, according to the Lorenz force principle, a mechanical force acts on the rotor conductor. Thus, all the rotor conductors' force, i.e., the sum of the mechanical forces produces torque in the rotor which tends to move it in the same direction as the rotating magnetic field.

This rotor conductor's rotation can also be explained by Lenz's law which says that the induced currents in the rotor oppose the cause for its production, here this opposition is a rotating magnetic field. This results in the rotor starting to rotate in the same direction as the stator rotating magnetic field. If the rotor speed is more than the stator speed, then no current will induce in the rotor because the reason for rotor rotation is the relative speed of the rotor and stator magnetic fields. This stator and the rotor field difference are called slips. This is how a 3-phase motor is called an asynchronous machine due to the relative speed difference between the stator and the rotors.

As we discussed above, the relative speed between the stator field and the rotor conductors causes them to rotate the rotor in a particular direction. Hence, for producing the rotation, the rotor speed Nr must always be less than the stator field speed Ns, and the difference between these two parameters depends on the load on the motor.

5. Analysis of 6-Spindle Spinning Machine

5.1. 6-Spindle Spinning Machine Base Frame

5.1.1. Static Structural Analysis

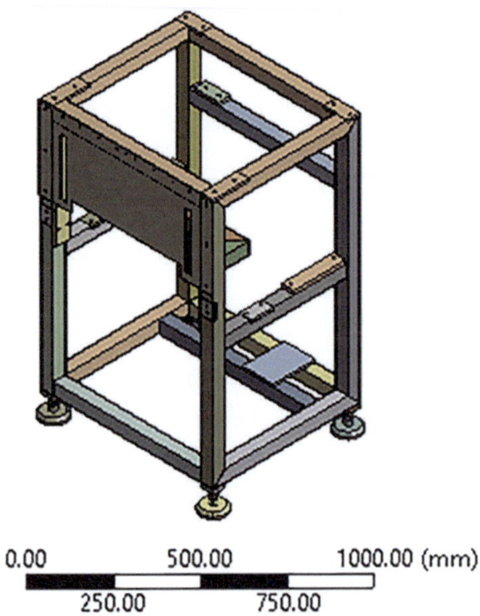

Fig. 5.1 Geometry of base frame of 6-SSM

5.1.2. Mesh

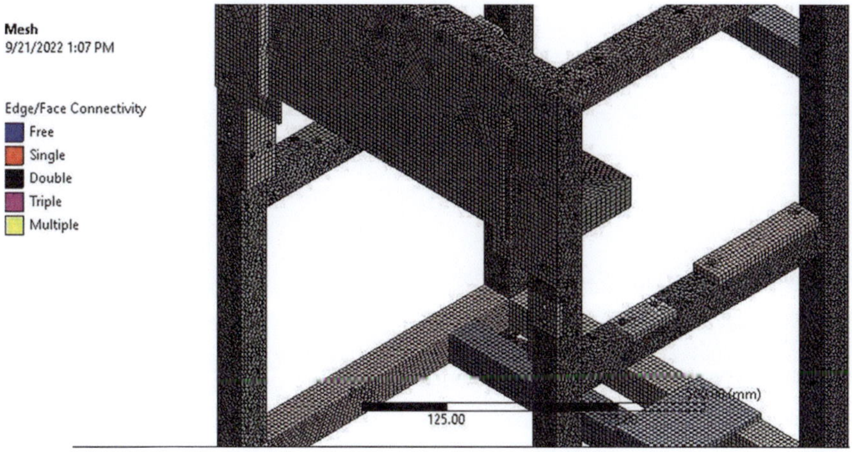

Fig. 5.2 Mesh of base frame of 6-SSM

5.1.3. Forces

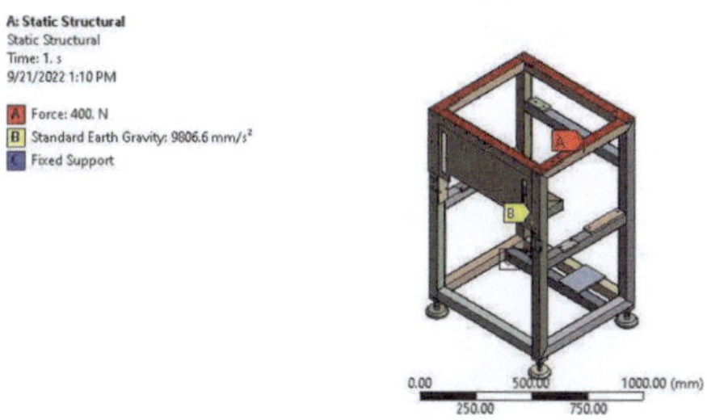

Fig. 5.3 6-SSM base frame with forces

5.1.4. Total Deformation and Equivalent Stress

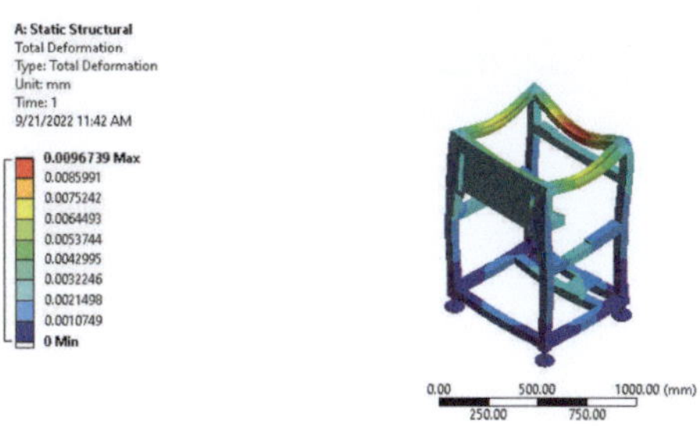

Fig. 5.4 Total deformation of base frame of 6-SSM

5.1.5. Modal Analysis

5.1.5.1 Mode 01

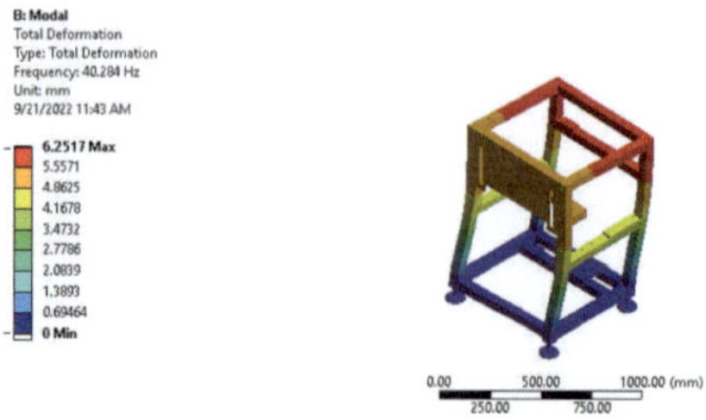

Fig. 5.5 Total deformation at mode 01 (6-SSMBF)

5.1.5.2 Mode 02

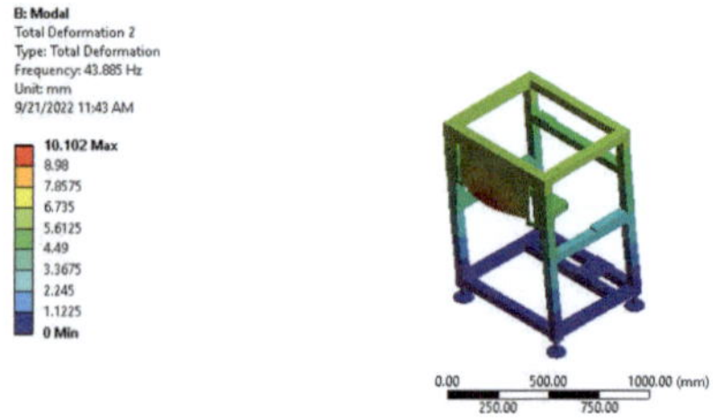

Fig. 5.6 Total deformation at mode 02 (6-SSMBF)

5.1.5.3 Mode 03

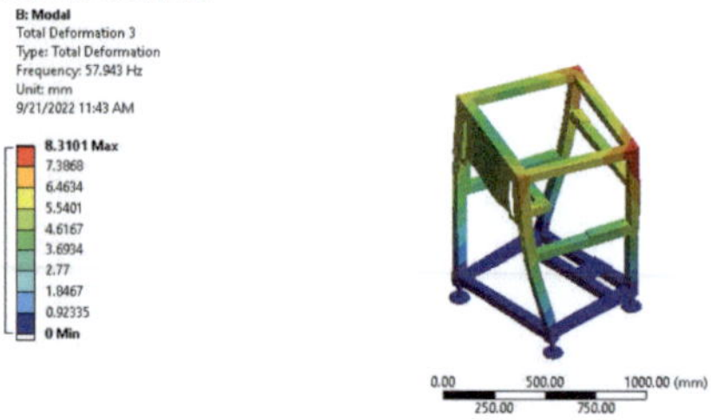

Fig. 5.7 Total deformation at mode 03 (6-SSMBF)

5.1.5.4 Mode 04

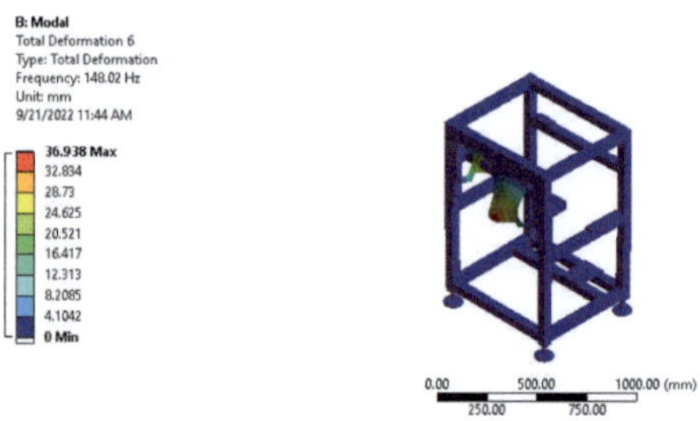

Fig. 5.8 Total deformation at mode 04 (6-SSMBF)

5.1.5.5 Mode 05

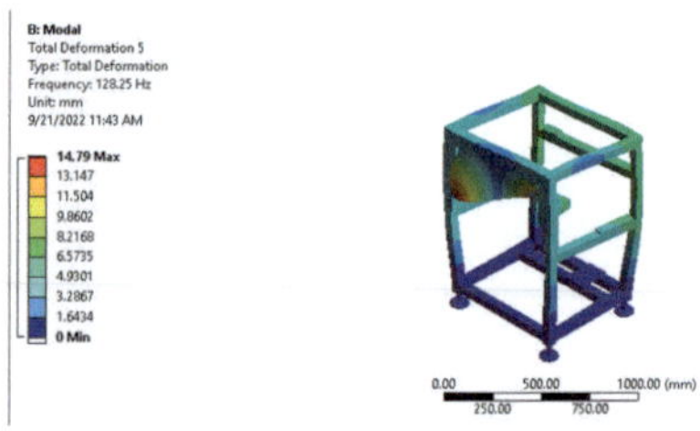

Fig. 5.9 Total deformation at mode 05 (6-SSMBF)

5.1.5.6 Mode 06

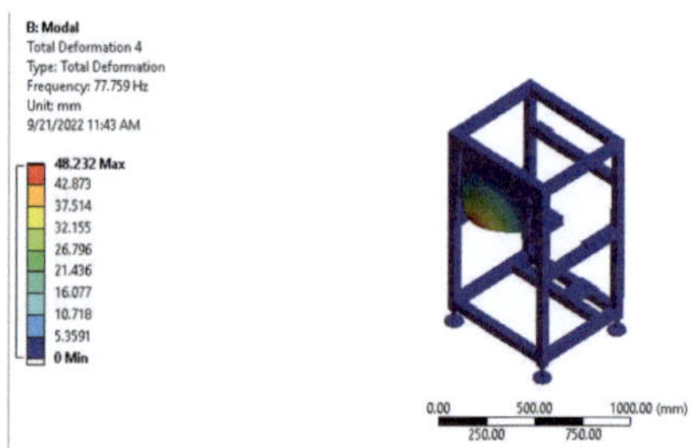

Fig. 5.10 Total deformation at mode 06 (6-SSMBF)

5.1.6. First 6-Natural Frequencies of 6-SSM Base Frame

Table 5.1 First 6-mode and total deformation of 6-SSMBF respectively

Mode	Total deformation (mm)	Natural frequencies (Hz)
1	6.2517	40.284
2	10.102	43.885
3	8.3101	57.943
4	48.232	77.759
5	14.79	128.25
6	36.938	148.02

5.1.7. Summary

In this section, only the base frame is analyzed. Fixed support and gravitational force are applied to the base of the frame and at point B, whereas the force of 400 N is applied at the upper surface of the base frame. Hence the total deformation is maximum is 0.0096739 mm.

5.2. 6-Spindle Spinning Machine Top Frame

5.2.1. Static Structural Analysis

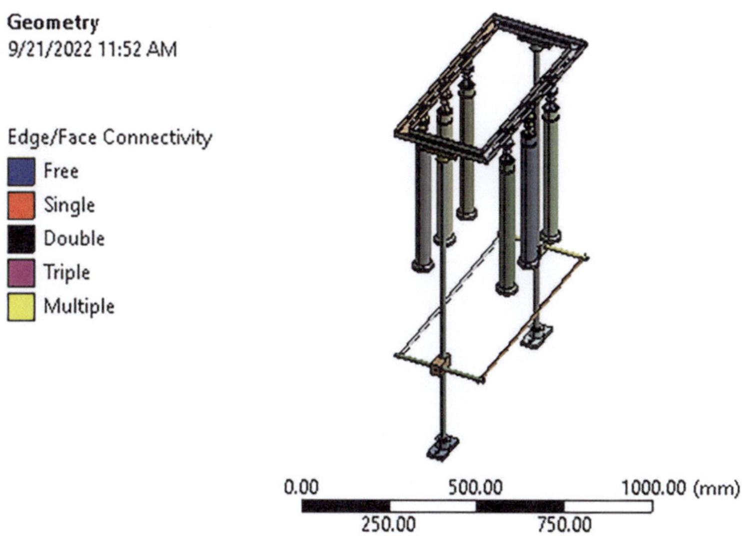

Fig. 5.11 Geometry of top frame of 6-SSM

5.2.2. Mesh

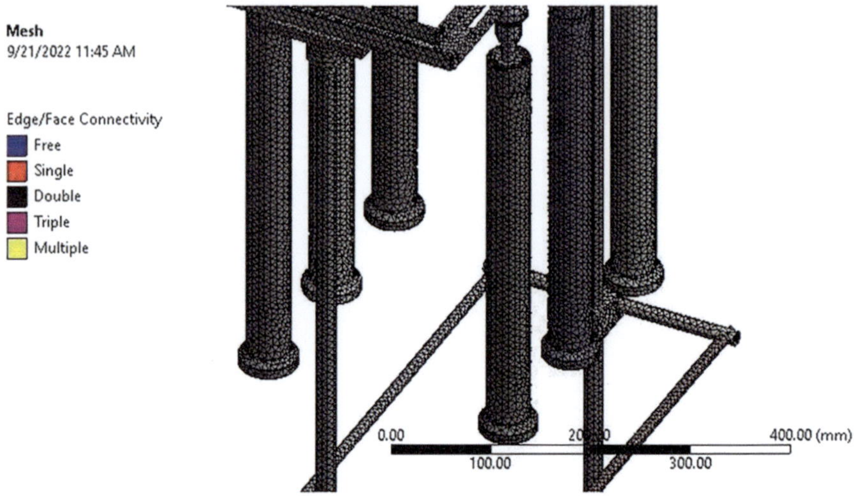

Fig. 5.12 6-SSM top frame after performing meshing

5.2.3. Forces

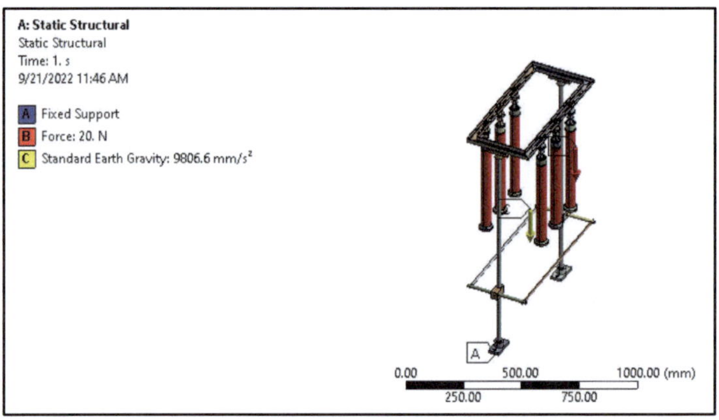

Fig. 5.13 Forces on top frame of 6-SSM

5.2.4. Total Deformation and Equivalent Stress

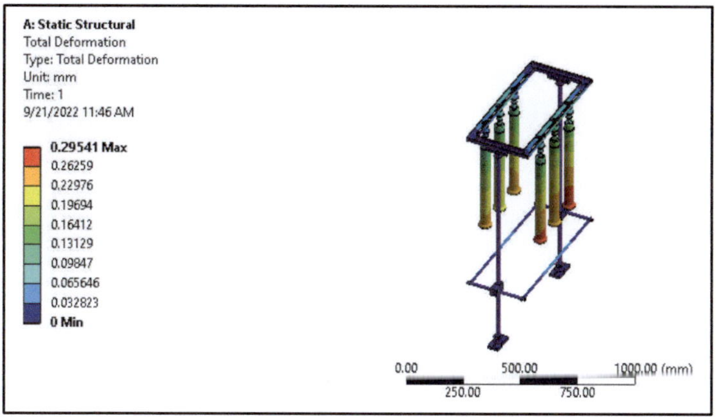

Fig. 5.14 Total deformation of top frame of 6-SSM

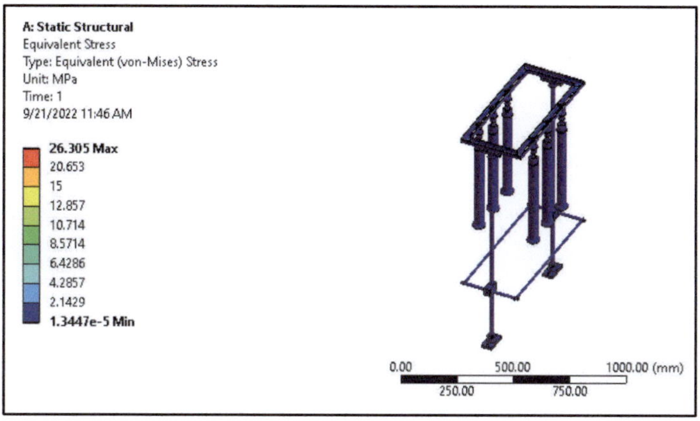

Fig. 5.15 Equivalent stress of 6-SSM top frame

5.2.5. Modal Analysis

5.2.5.1Mode 01

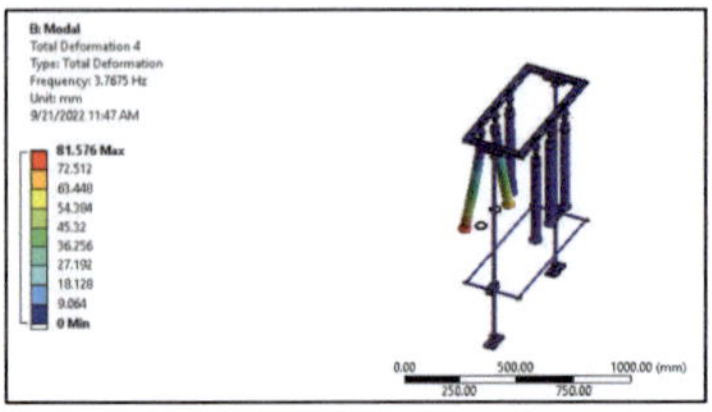

Fig. 5.16 Total deformation at mode 1 (6-SSMTF)

5.2.5.2 Mode 02

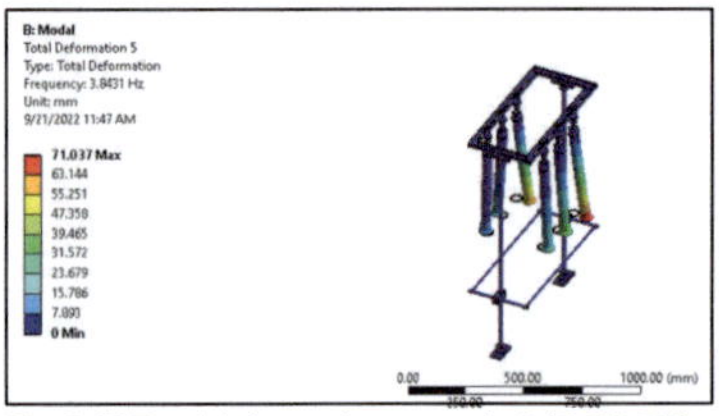

Fig. 5.17 Total deformation at mode 2 (6-SSMTF)

5.2.5.3 Mode 03

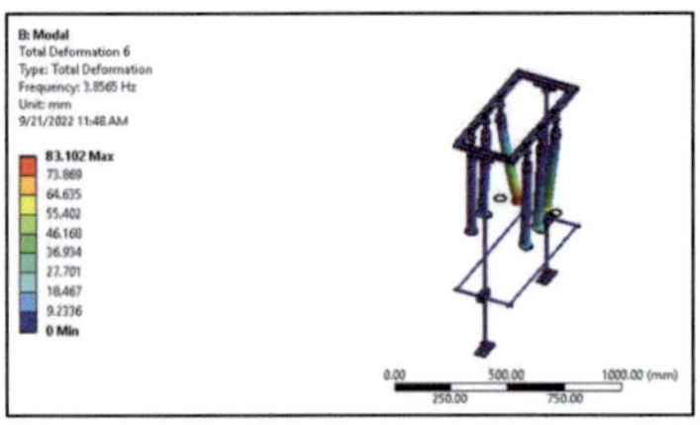

Fig. 5.18 Total deformation at mode 3 (6-SSMTF)

5.2.5.4 Mode 04

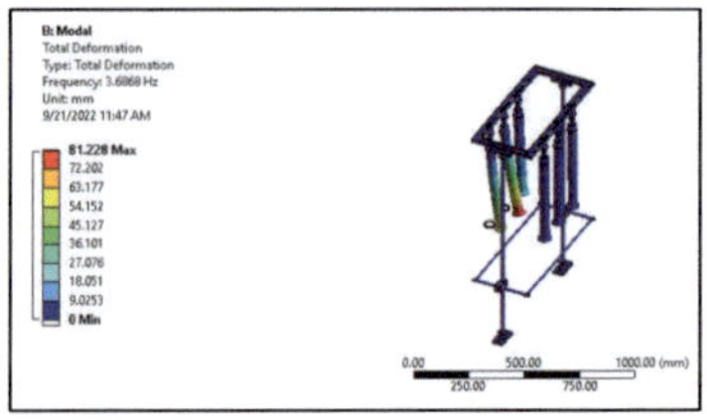

Fig. 5.19 Total deformation at mode 4 (6-SSMTF)

5.2.5.5 Mode 05

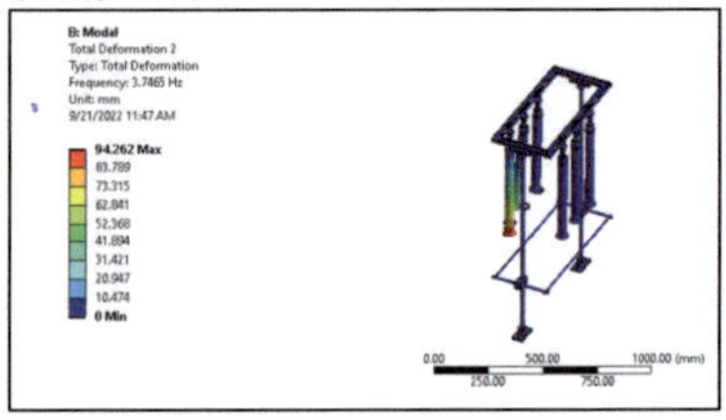

Fig. 5.20 Total deformation at mode 5 (6-SSMTF)

5.2.5.6 Mode 06

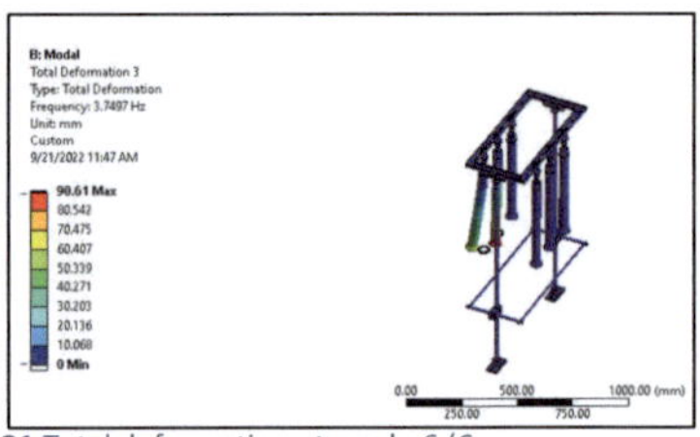

Fig. 5.21 Total deformation at mode 6 (6-SSMTF)

5.2.6. First 6-Natural Frequencies of 6-SSM Top Frame

Table 5.2 First 6-mode and total deformation of 6-SSMTF respectively

Mode	Total deformation (mm)	Natural frequencies (Hz)
1	81.228	3.6868
2	94.262	3.7465
3	90.61	3.7497
4	81.576	3.7675
5	71.037	3.8431
6	83.102	3.8565

5.2.7. Summary

In this section, only the top frame of 6-SSM is analyzed. Fixed support and gravitational force are applied to the base of the frame and at point C, whereas the force of 20 N is applied to the hangers. Hence the total deformation is maximum at 0.29541 mm and maximum stress at 26.305 MPa.

5.3. Modal Analysis of 6-Spindle Spinning Machine (100110-101)

5.3.1.1 *Mode 1*

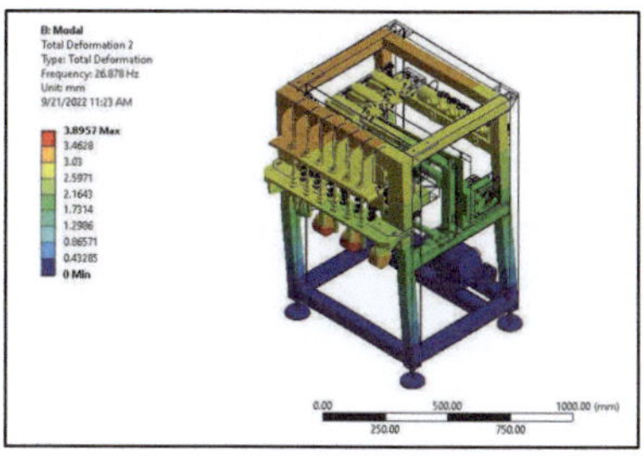

Fig. 5.22 Total deformation at mode 1

5.3.1.2 *Mode 2*

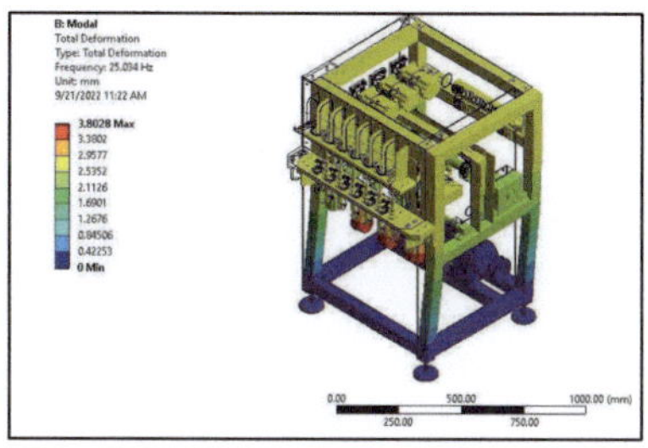

Fig. 5.23 Total deformation at mode 2

5.3.1.3 Mode 3

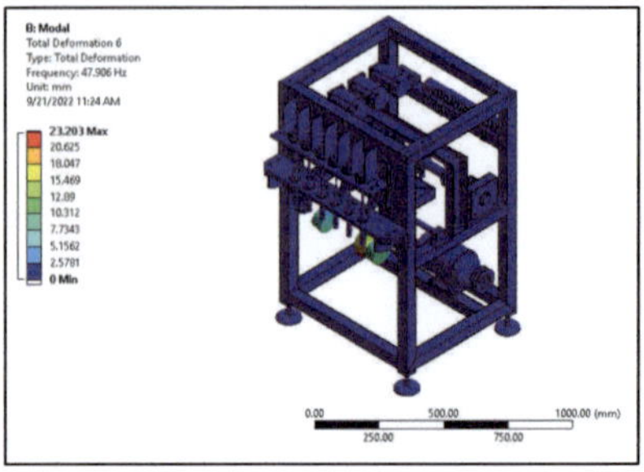

Fig. 5.24 Total deformation at mode 3

5.3.1.4 Mode 4

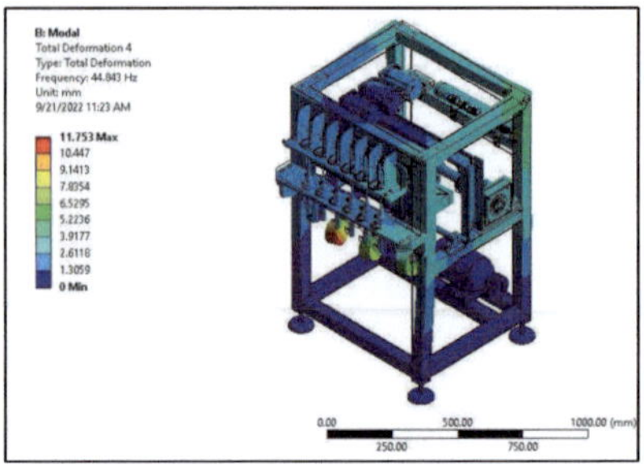

Fig. 5.25 Total deformation at mode 4

5.3.1.5 Mode 5

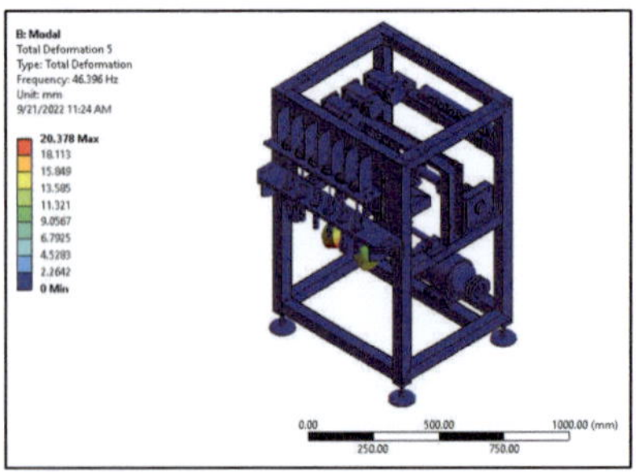

Fig. 5.26 Total deformation at mode 5

5.3.1.6 Mode 6

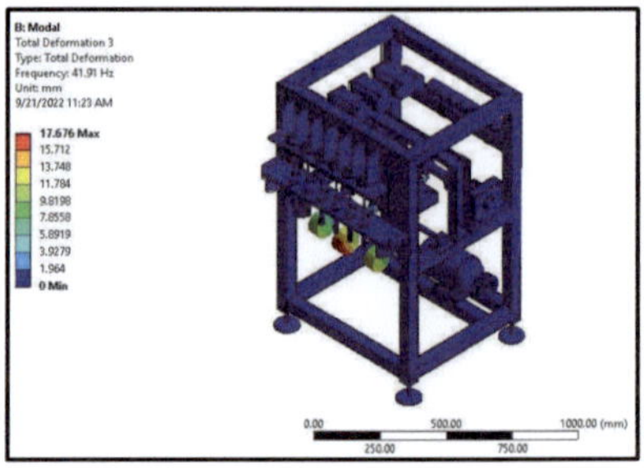

Fig. 5.27 Total deformation at mode 6

5.3.1.7 Mode 7

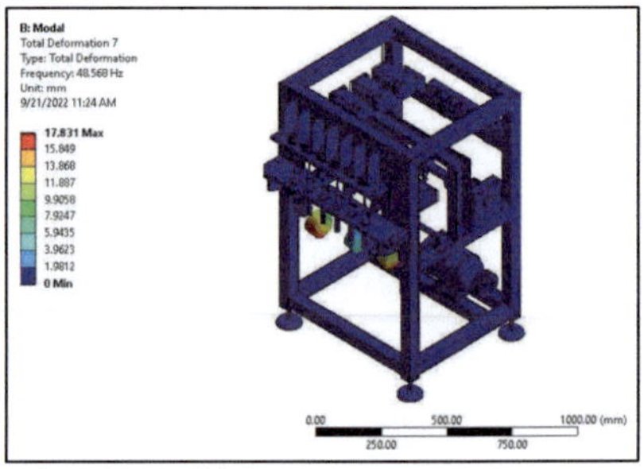

Fig. 5.28 Total deformation at mode 7

5.3.1.8 Mode 8

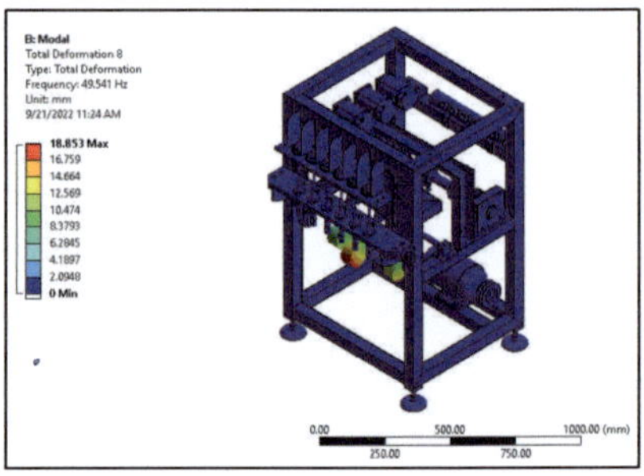

Fig. 5.29 Total deformation at mode 8

5.3.1.9 Mode 9

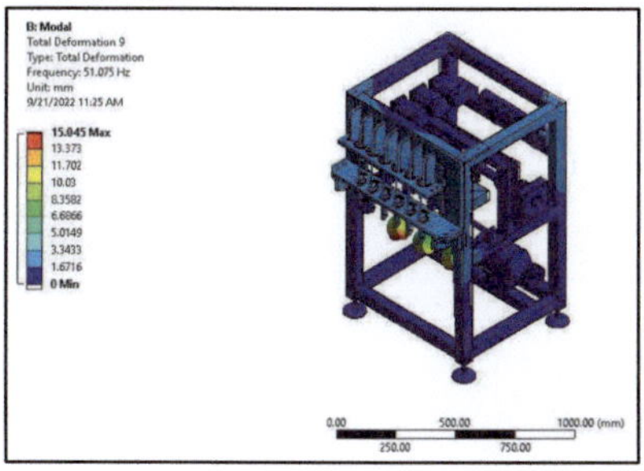

Fig. 5.30 Total deformation at mode 9

5.3.1.10 Mode 10

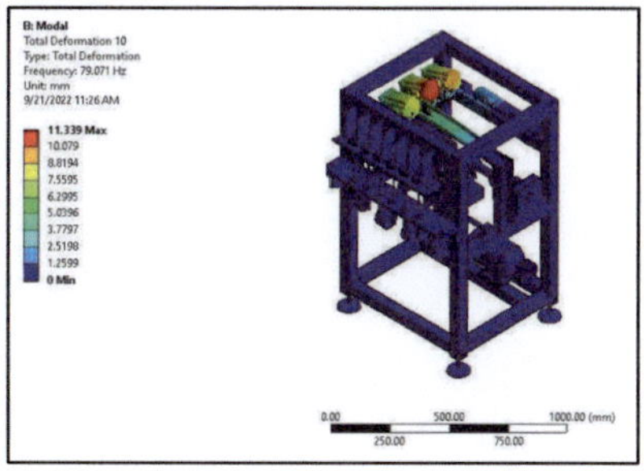

Fig. 5.31 Total deformation at mode 10

5.3.2. First Ten Natural Frequencies

Table 5.3 First ten mode and total deformation of 6-SSM (100110-101) respectively

Mode	Total deformation (mm)	Natural frequencies (Hz)
1	3.8028	25.034
2	3.8957	26.878
3	17.676	41.910
4	11.753	44.843
5	20.378	46.396
6	23.203	47.906
7	17.831	48.568
8	18.853	49.541
9	15.045	51.075
10	11.339	79.071

5.3.3. Summary

In this section, the subassembly of the base frame is analyzed. Fixed support and gravitational force are applied to the base of the frame and at point B, whereas the force of 40 N is applied at the upper surface of the base frame.

5.4. 6-Spindle Spinning Machine (100103-101)

5.4.1. Static Structural Analysis

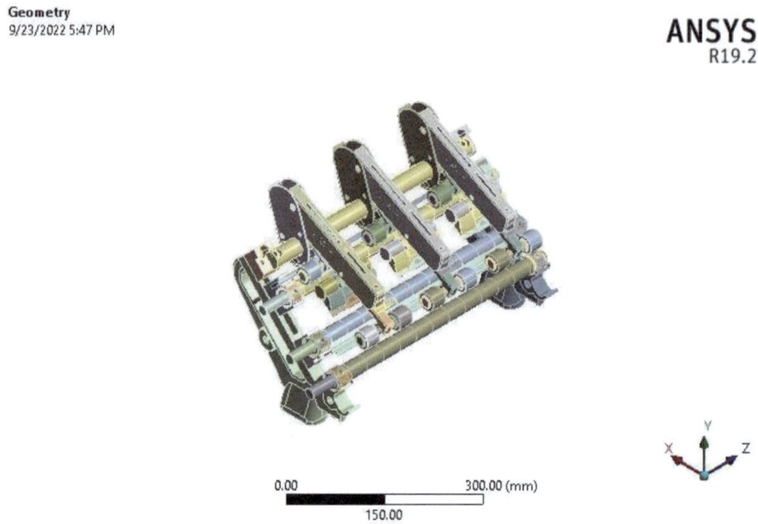

Fig. 5.32 Geometry of top frame of 6-SSM

5.4.2. Mesh

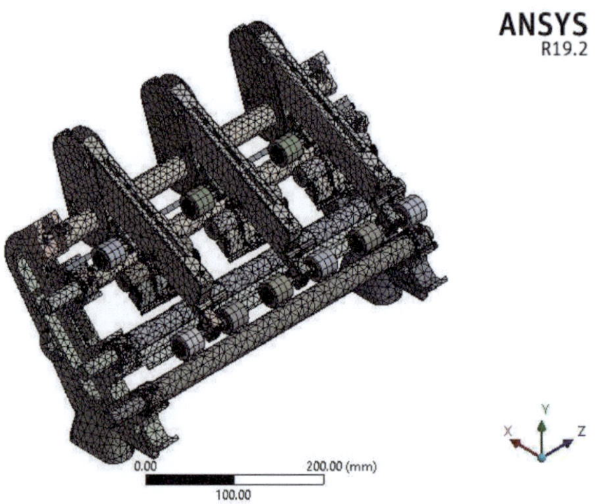

Fig. 5.33 6-SSM top frame after performing meshing

5.4.3. Forces

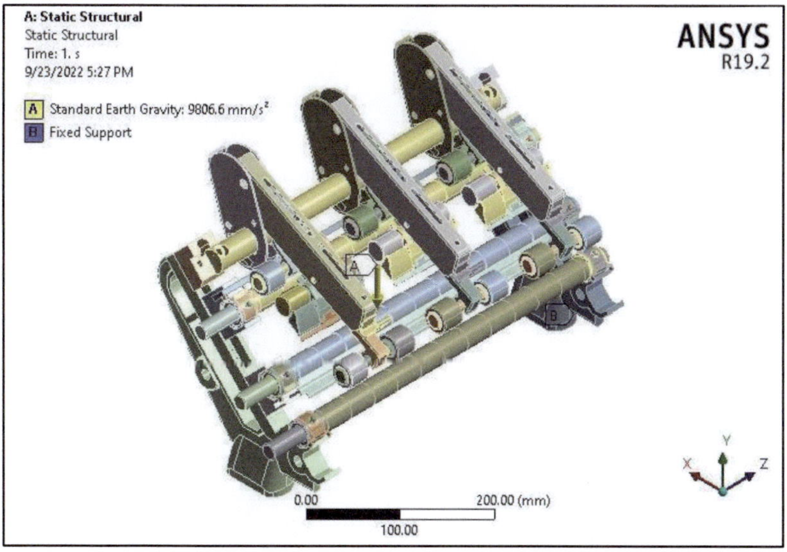

Fig. 5.34 Forces on Top Frame of 6-SSM

5.4.4. Total Deformation and Equivalent Stress

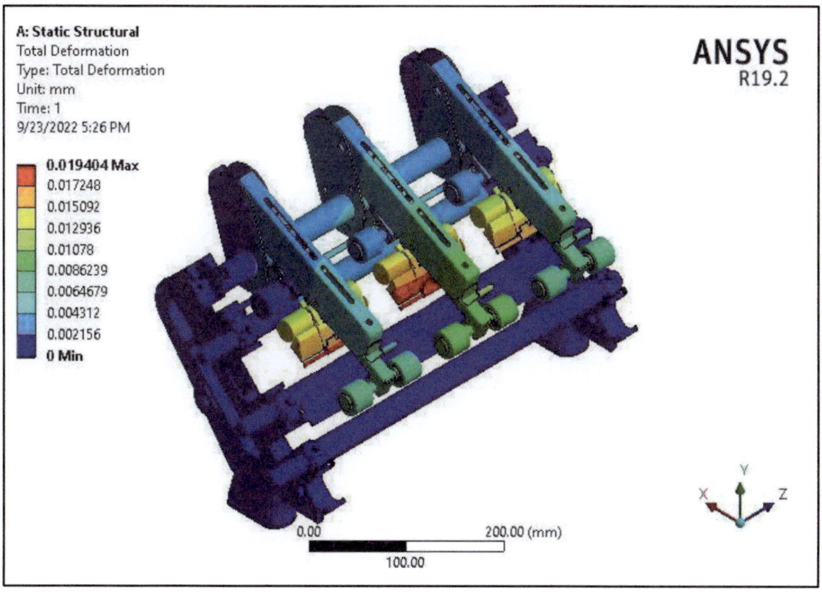

Fig. 5.35 Total deformation of top frame of 6-SSM

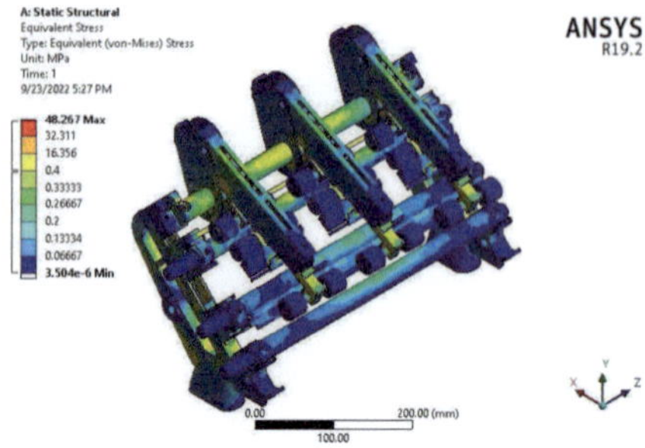

Fig. 5.36 Equivalent stress of 6-SSM top frame

5.5. Modal Analysis of 6-Spindle Spinning Machine (100103-101)

5.5.1.1 Mode 1

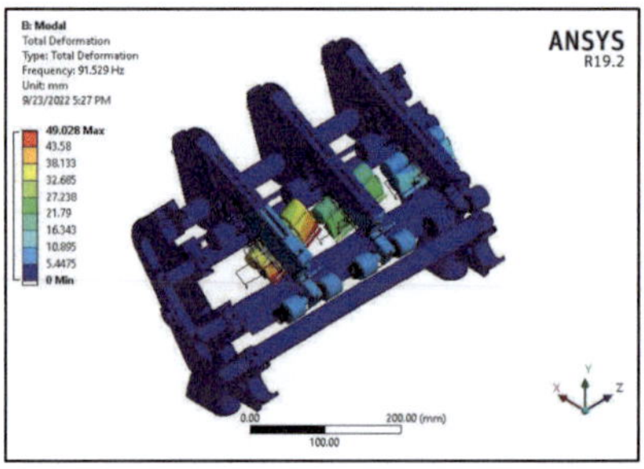

Fig. 5.37 Total deformation at mode 1

5.5.1.2 Mode 2

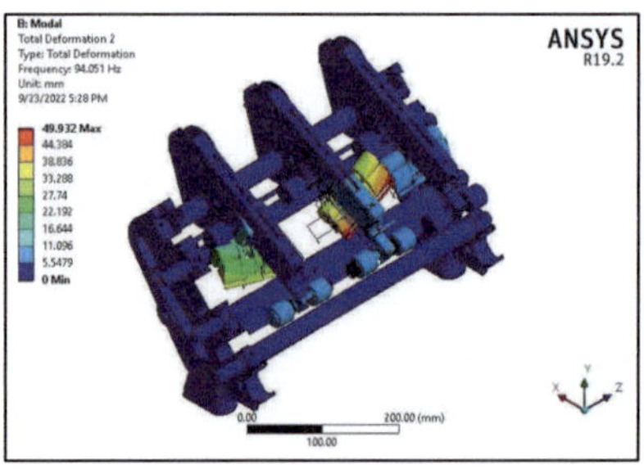

Fig. 5.38 Total deformation at mode 2

5.5.1.3 Mode 3

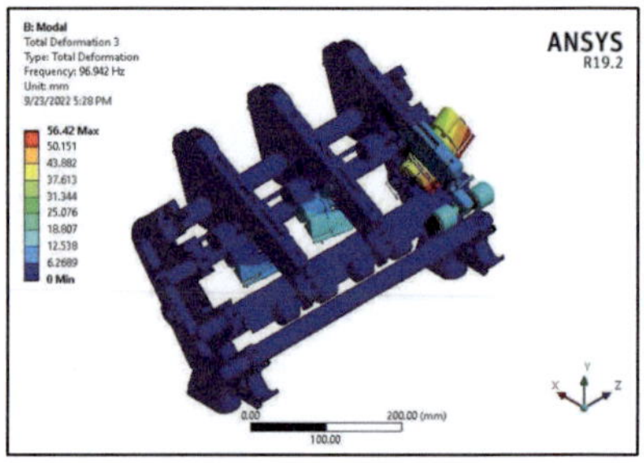

Fig. 5.39 total deformation at mode 3

5.5.1.4 Mode 4

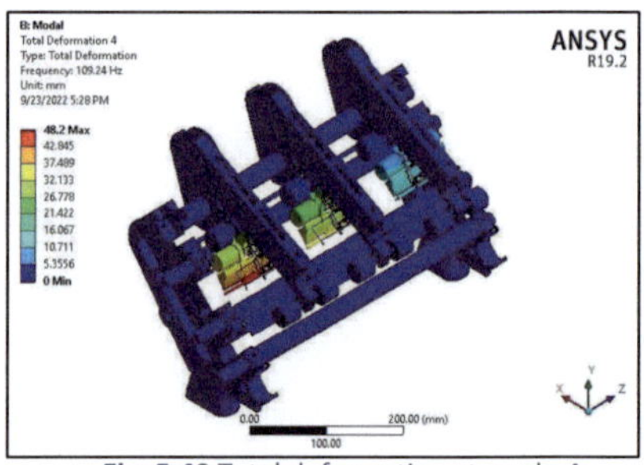

Fig. 5.40 Total deformation at mode 4

5.5.1.5 Mode 5

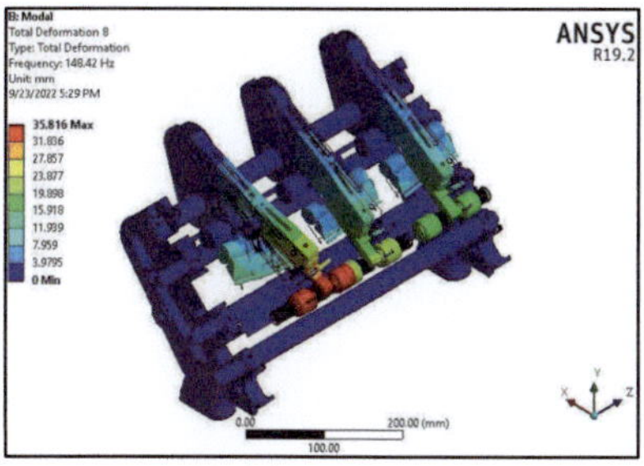

Fig. 5.41 Total deformation at mode 5

5.5.1.6 Mode 6

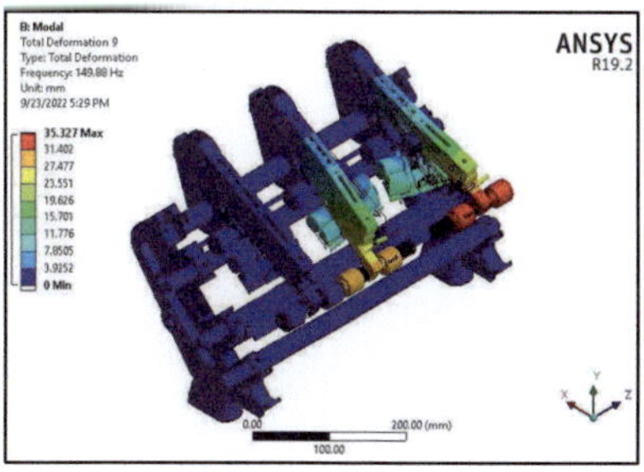

Fig. 5.42 Total deformation at mode 6

5.5.1.7 Mode 7

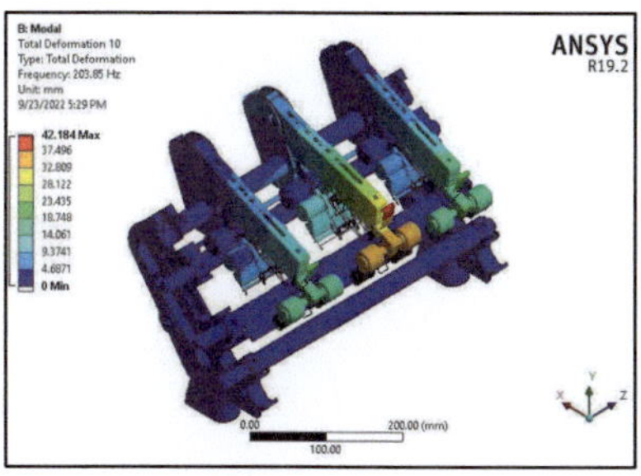

Fig. 5.43 Total deformation at mode 7

5.5.1.8 Mode 8

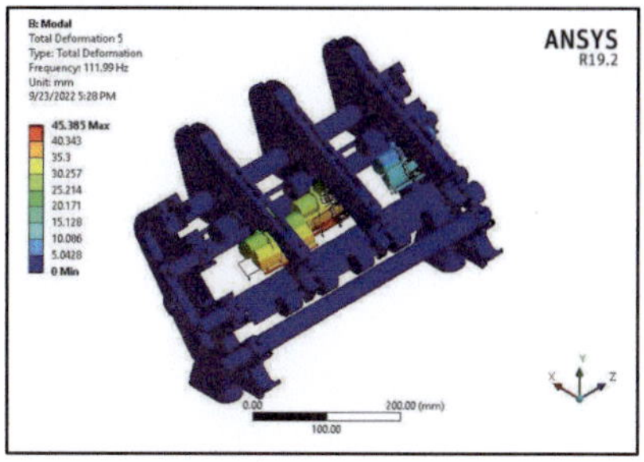

Fig. 5.44 Total deformation at mode 8

5.5.1.9 Mode 9

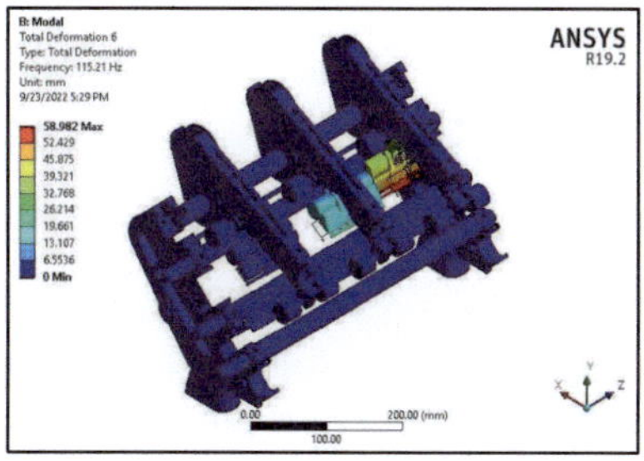

Fig. 5.45 Total deformation at mode 9

5.5.1.10 Mode 10

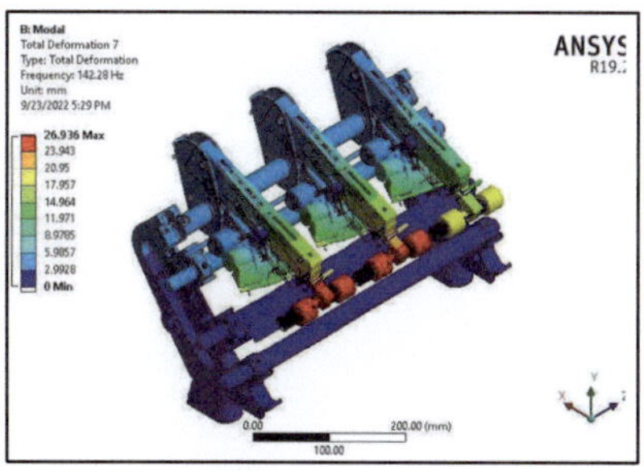

Fig. 5.46 Total deformation at mode 10

5.5.1. First Ten Natural Frequency

Table 5.3 First ten mode and total deformation of 6-SSM (100110-101) respectively

Mode	Total deformation (mm)	Natural Frequencies (Hz)
1	49.028	91.529
2	49.932	94.051
3	56.42	96.942
4	48.2	109.24
5	45.385	111.99
6	58.982	115.21
7	26.936	142.28
8	35.816	148.42
9	35.327	149.88
10	42.184	203.85

5.5.2. Summary

In this section, only the base frame is analyzed. Fixed support and gravitational force are applied to the base of the frame and at point B. Hence the maximum value of total deformation is 0.019404 mm highlighted in red color, whereas, maximum stress has a value of 48.267 MPa.

5.6. 6-Spindle Spinning Machine

5.6.1. Static Structural Analysis

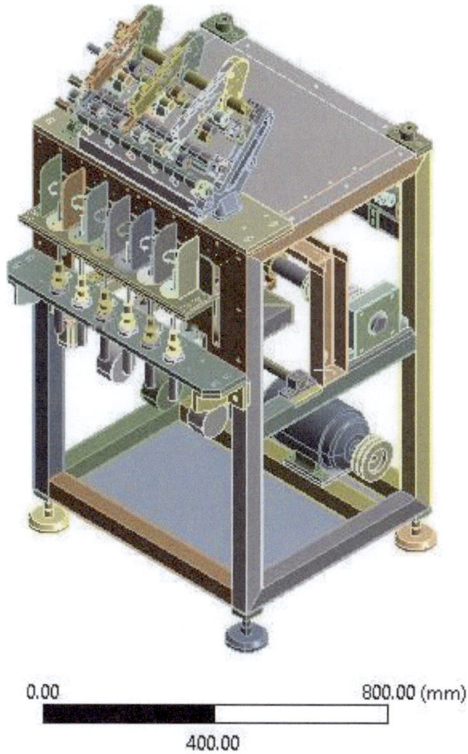

Fig. 5.47 Geometry of 6-SSM (100) assembly

5.6.2. Mesh

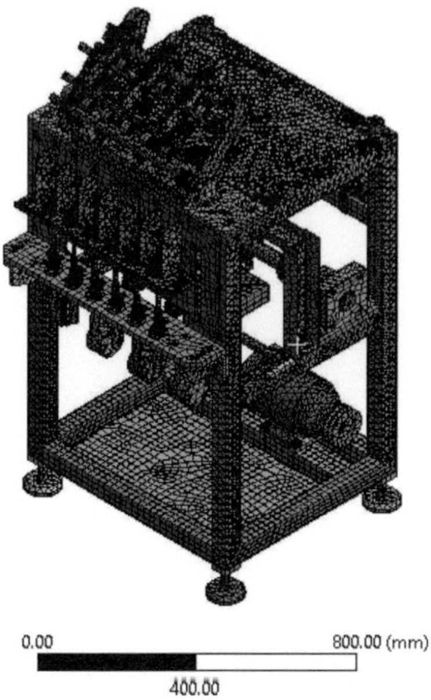

Fig. 5.48 Meshing of 6-SSM (100) assembly

5.6.3. Forces

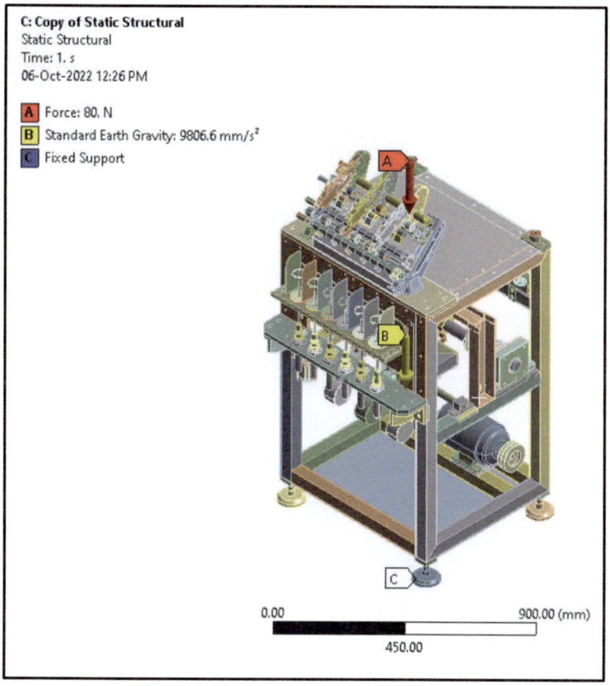

Fig. 5.49 Forces on 6-SSM (100) assembly

5.6.4. Total Deformation and Equivalent Stress

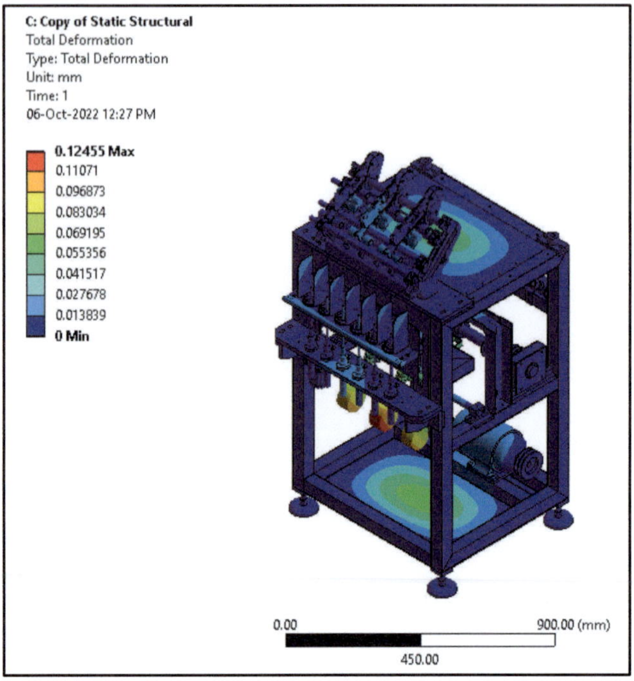

Fig. 5.50 Total deformation of 6-SSM (100) assembly

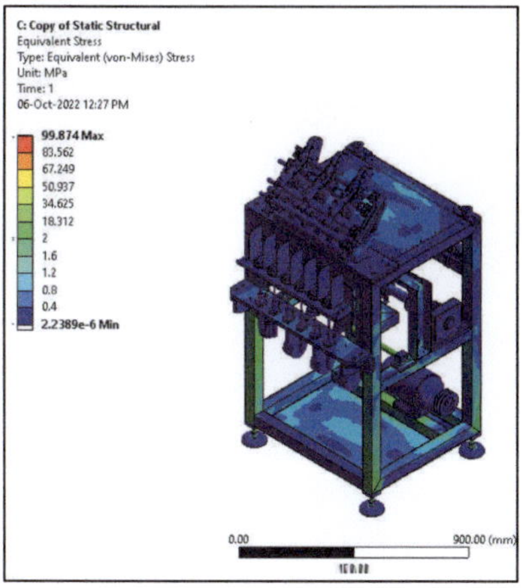

Fig. 5.51 Equivalent stress of 6-SSM (100) assembly

5.7. Modal Analysis of 6-Spindle Spinning Machine (100103-101)

5.7.1.1 Mode 1

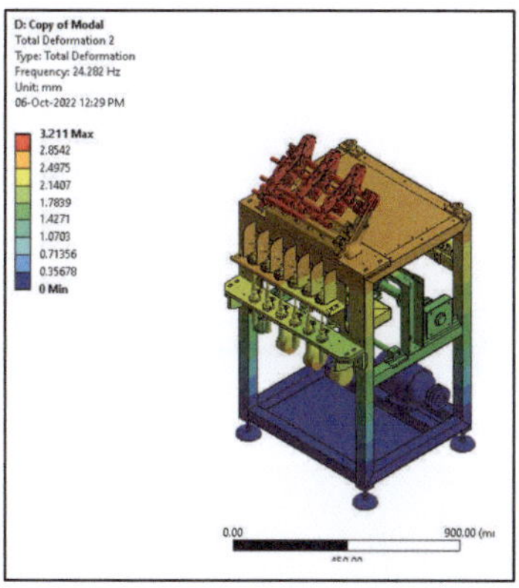

Fig. 5.52 Total deformation at mode 1 (6-SSM-100)

5.7.1.2 Mode 2

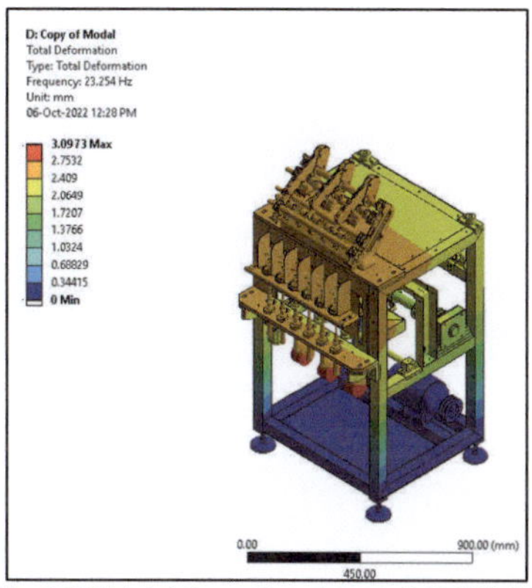

Fig. 5.53 Total deformation at mode 2 (6-SSM-100)

5.7.1.3 Mode 3

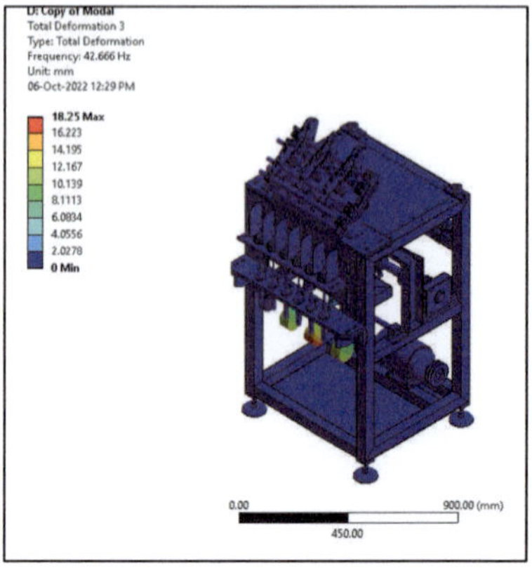

Fig. 5.54 Total deformation at mode 3 (6-SSM- 100)

5.7.1.4 Mode 4

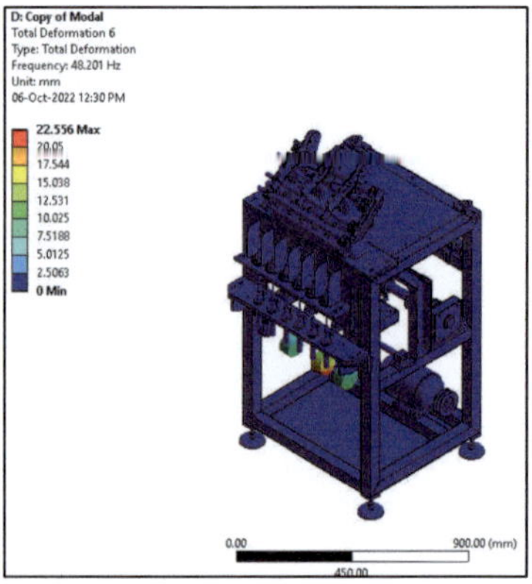

Fig. 5.55 Total deformation at mode 4 (6-SSM-100)

5.7.1.5 Mode 5

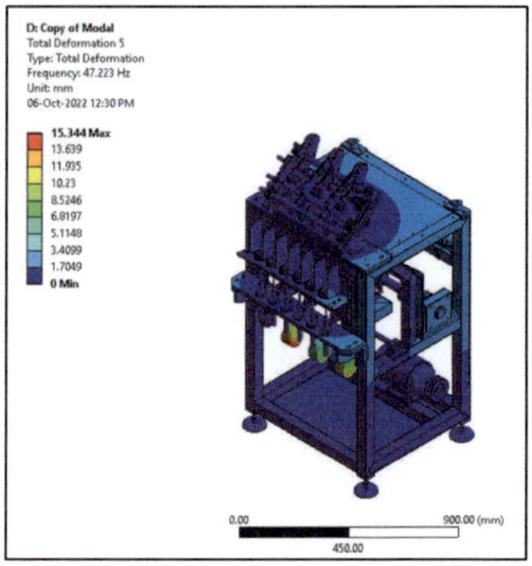

Fig. 5.56 Total deformation at mode 5 (6-SSM-100)

5.7.1.6 Mode 6

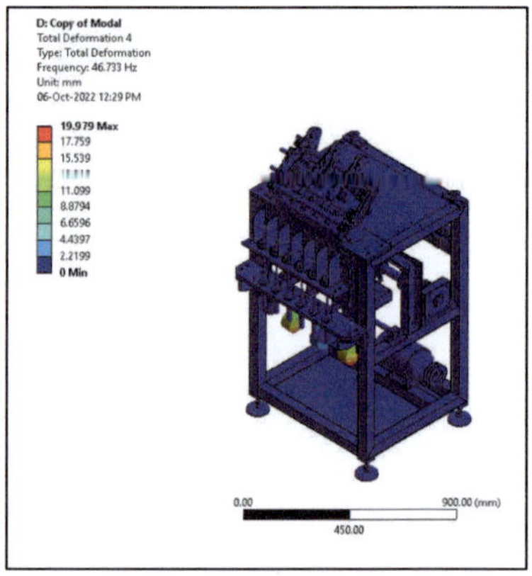

Fig. 5.57 Total deformation at mode 6 (6-SSM-100)

5.7.1.7 Mode 7

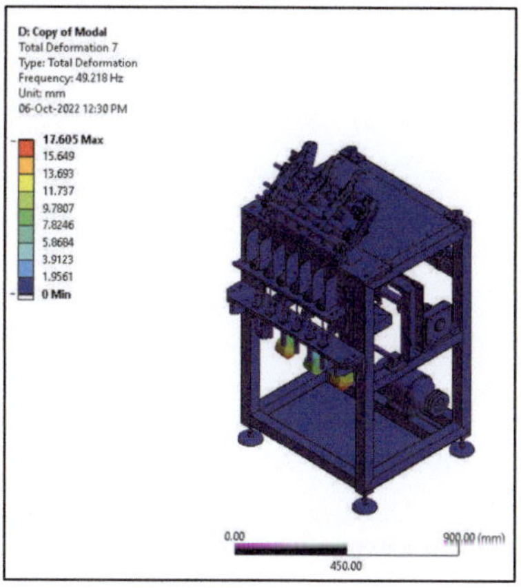

Fig. 5.58 Total deformation at mode 7 (6-SSM- 100)

5.7.1.8 Mode 8

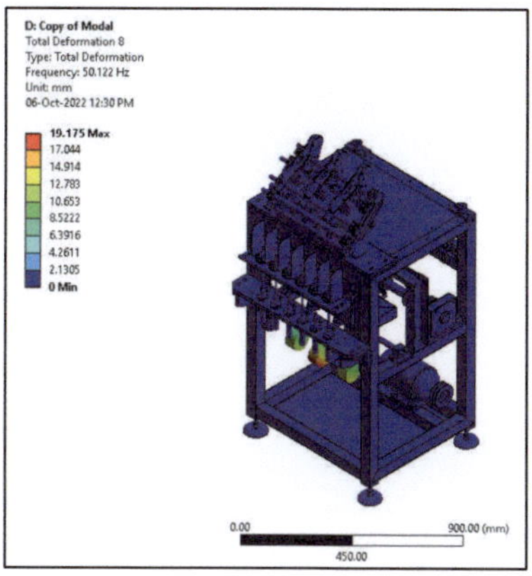

Fig. 5.60 Total deformation at mode 8 (6-SSM- 100)

5.7.1.9 Mode 9

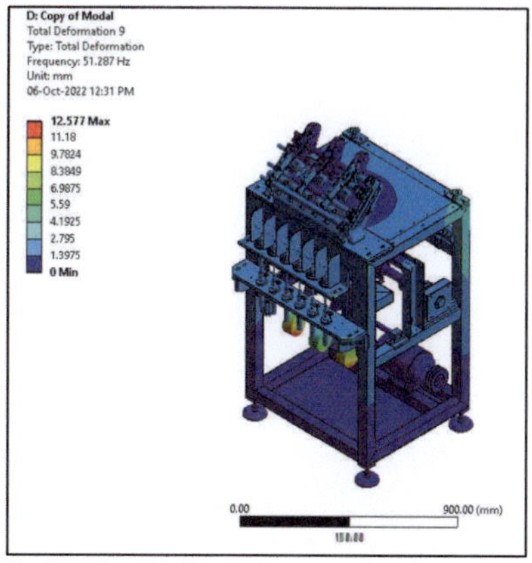

Fig. 5.59 Total deformation at mode 9 (6-SSM- 100)

5.7.1.10 Mode 10

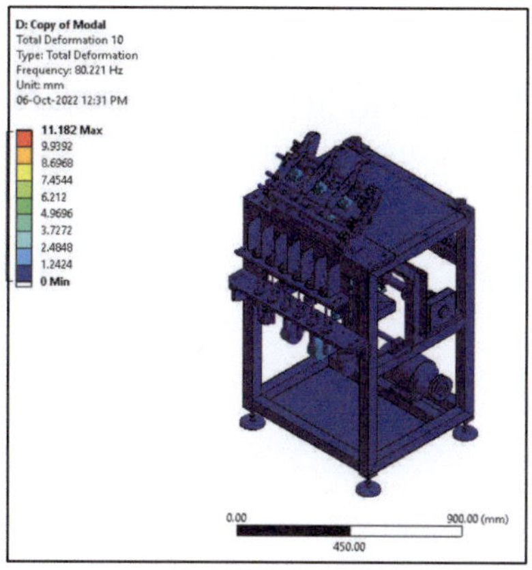

Fig. 5.61 Total deformation at mode 10 (6-SSM- 100)

5.7.2. First Ten Natural Frequency

Table 5.4 First ten mode and total deformation of 6-SSM (100110-101) respectively

Mode	Total deformation (mm)	Natural frequencies (Hz)
1	3.0973	23.254
2	3.211	24.282
3	18.25	42.666
4	19.979	46.733
5	15.344	47.223
6	22.556	48.201
7	17.605	49.218
8	19.175	50.122
9	12.577	51.287
10	11.182	80.221

5.7.3. Summary

Sub-assembly of 6-SSM has been evaluated in this section. Fixed support and gravitational force are applied to the base of the frame and at point B, whereas the force of 80 N is applied at the upper surface labeled with the letter A. Hence the total deformation is maximum at 0.12455 mm and the maximum stress has the value of 99.874 MPa shown in the respective figures.

5.8. 6-Spindle Spinning Machine (100101-101)

5.8.1. Static Structural Analysis

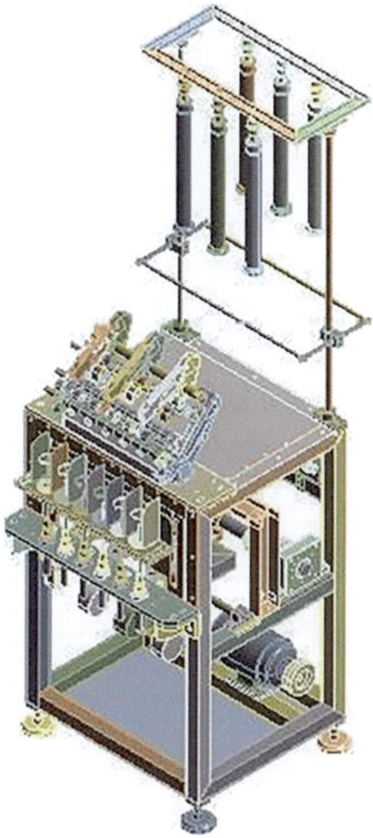

Fig. 5.62 Geometry of 6-SSM (100101-101) assembly

5.8.2. Mesh

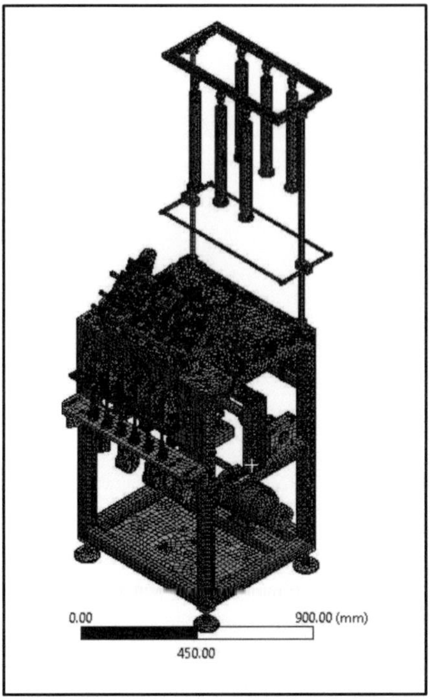

Fig. 5.63 Meshing of 6-SSM (100101-101) assembly

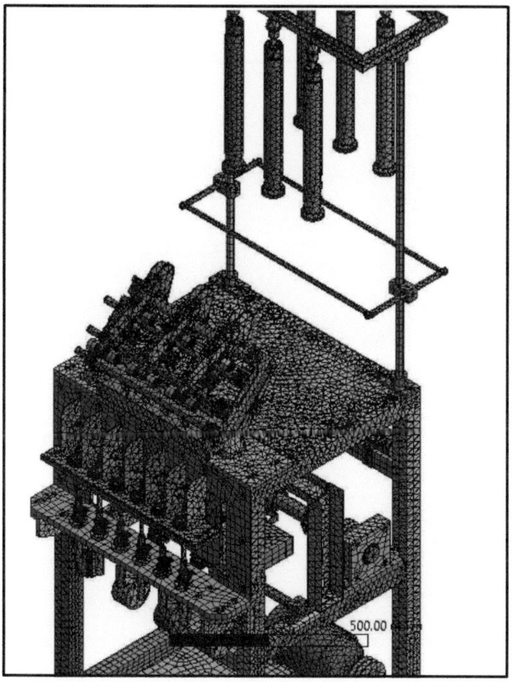

Fig. 5.64 Meshing of 6-SSM (100101-101) assembly (zoom)

5.8.3. Forces

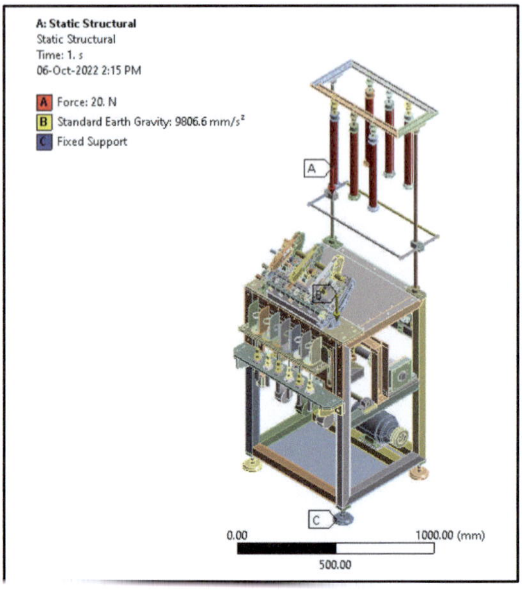

Fig. 5.65 Forces on 6-SSM (100101-101) Assembly

5.8.4. Total Deformation and Equivalent Stress

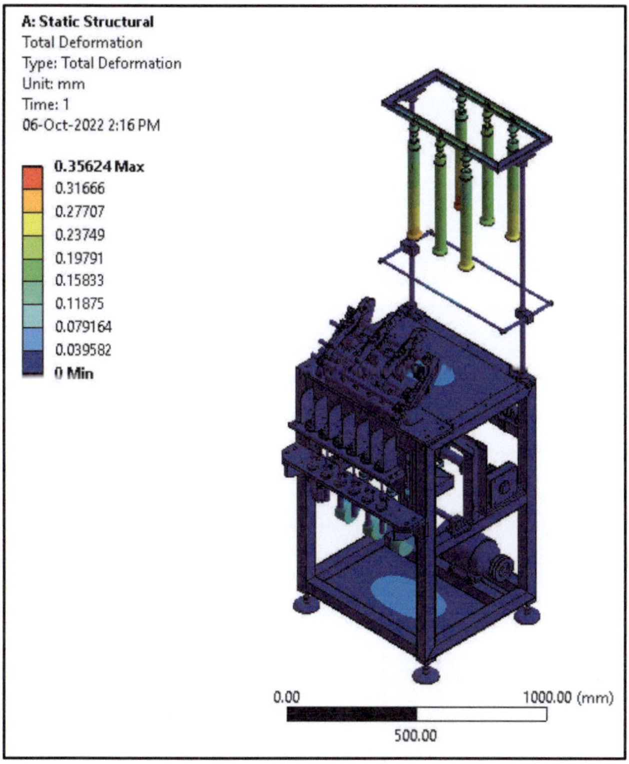

Fig. 5.66 Total deformation of 6-SSM (100101-101) assembly

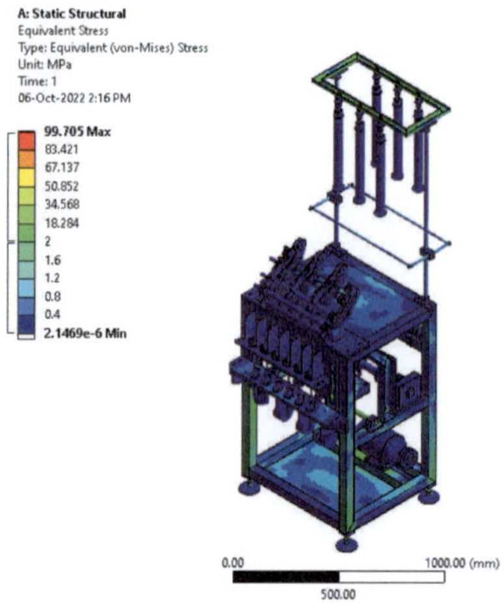

Fig. 5.67 Equivalent stress of 6-SSM (100101-101) assembly

5.9. Modal Analysis of 6-Spindle Spinning Machine (100103-101)

5.9.1.1 Mode 1

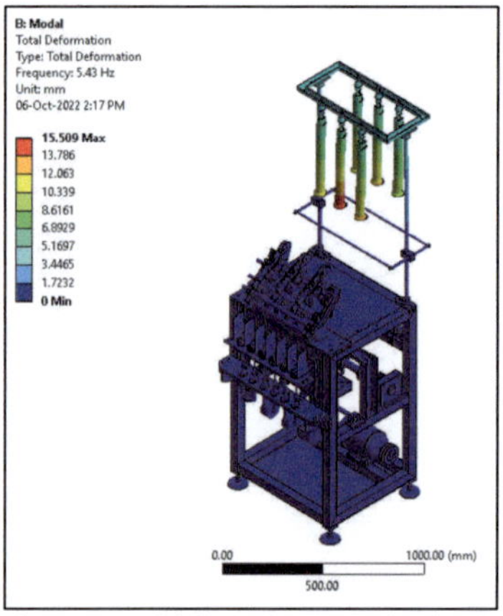

Fig. 5.68 Total deformation at mode 1 (6-SSM-100101-101)

5.9.1.2 Mode 2

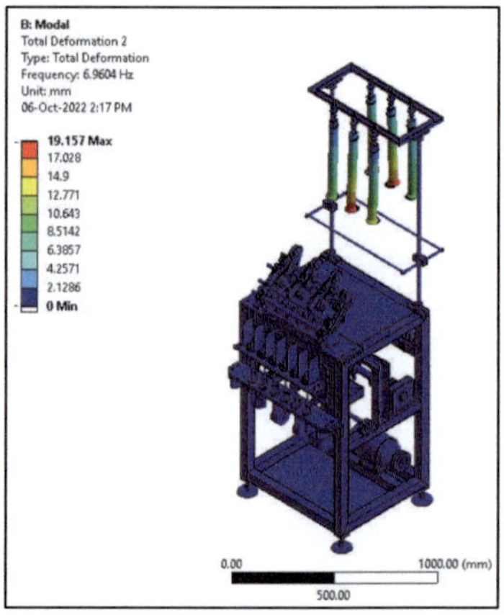

Fig. 5.69 Total deformation at mode 2 (6-SSM-100101-1

5.9.1.3 Mode 3

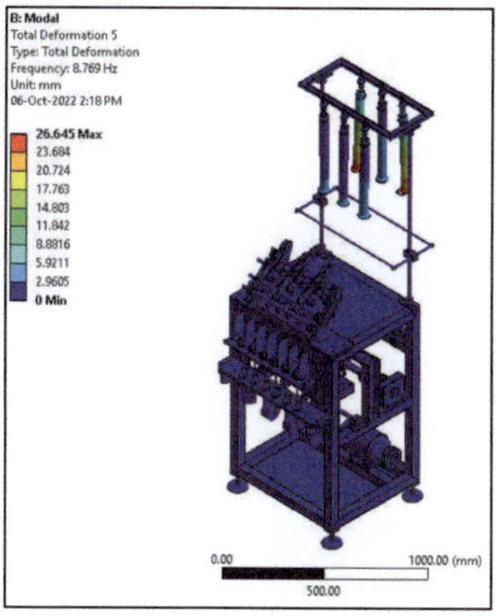

Fig. 5.70 Total deformation at mode 3 (6-SSM- 100101-101)

5.9.1.4 Mode 4

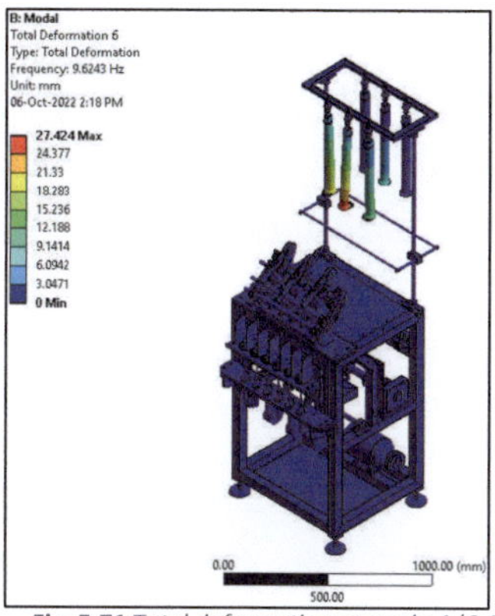

Fig. 5.71 Total deformation at mode 4 (6-SSM-100101-101)

5.9.1.5 Mode 5

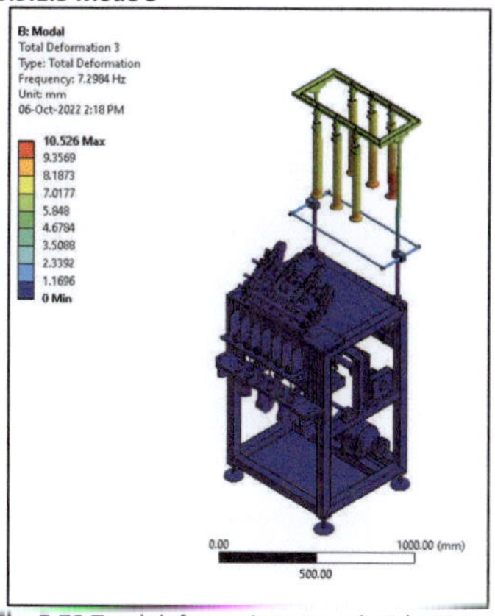

Fig. 5.72 Total deformation at mode 5 (6-SSM-100101-101)

5.9.1.6 Mode 6

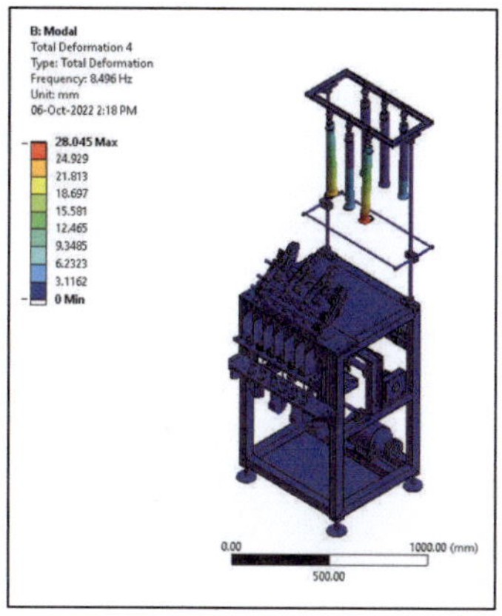

Fig. 5.73 Total deformation at mode 6 (6-SSM-100101-101)

5.9.1.87Mode 7

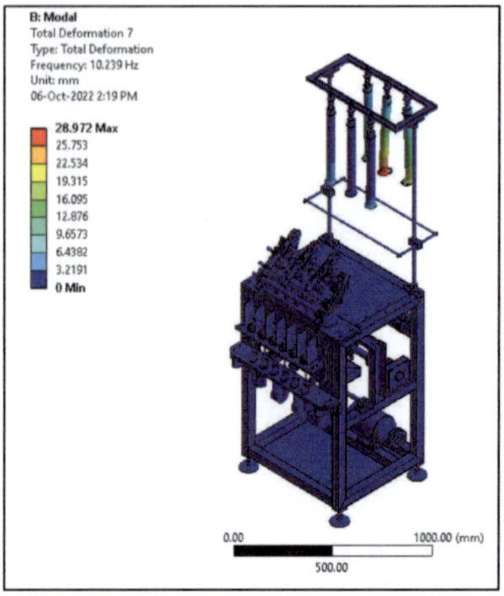

Fig. 5.74 Total deformation at mode 7 (6-SSM-100101-101)

5.9.1.8 Mode 8

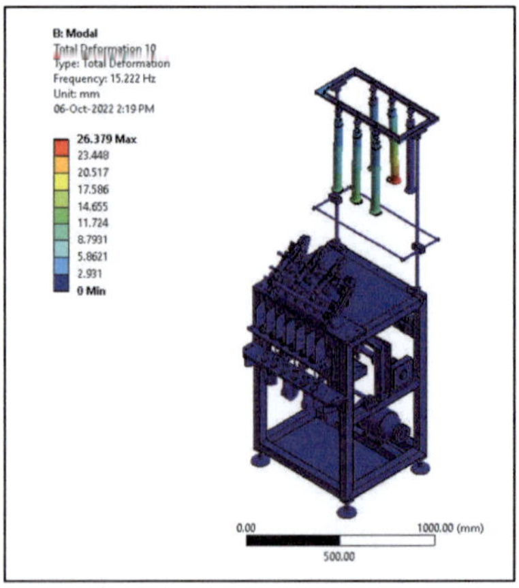

Fig. 5.75 Total deformation at mode 8 (6-SSM-100101-101)

5.9.1.9 Mode 9

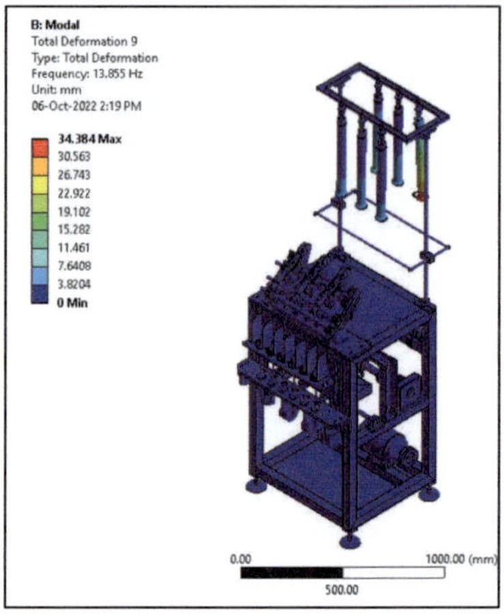

Fig. 5.76 Total deformation at mode 9 (6-SSM-100101-101)

5.9.1.10 Mode 10

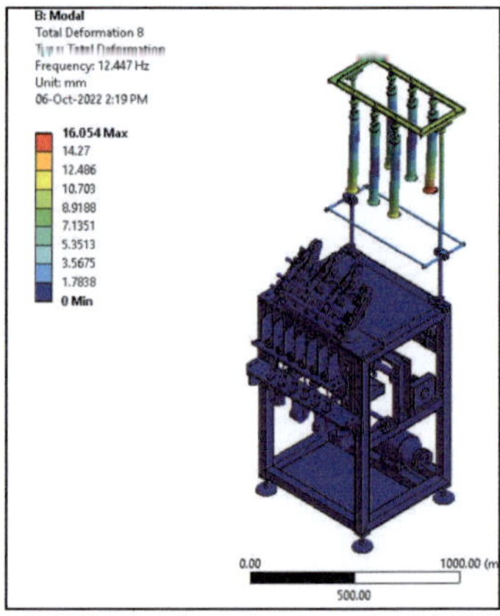

Fig. 5.77 Total deformation at mode 10 (6-SSM-100101-101)

5.9.2. First Ten Natural Frequencies

Table 5.5 First ten mode and total deformation of 6-SSM (100103-101) respectively

Mode	Total deformation (mm)	Natural frequencies (Hz)
1	15.509	5.43
2	19.157	6.9604
3	10.526	7.2984
4	28.045	8.496

(continued)

(continued)

Mode	Total deformation (mm)	Natural frequencies (Hz)
5	26.645	8.769
6	27.424	9.6243
7	28.972	10.239
8	16.054	12.447
9	34.384	13.655
10	26.379	15.222

5.9.3. Summary

The complete assembly of 6-SSM has been analyzed. Fixed support and gravitational force are applied to the base of the frame and at point B, whereas the force of 20 N is applied at the hangers of the top frame. Hence the total deformation has a maximum value of 0.35624 mm, and the maximum stress has a value of 99.705 MPa.

6. CE Compliance

CE marking indicates that a product has been assessed by the manufacturer and deemed to meet EU safety, health, and environmental protection requirements. It is required for products manufactured anywhere in the world that are then marketed in the EU.

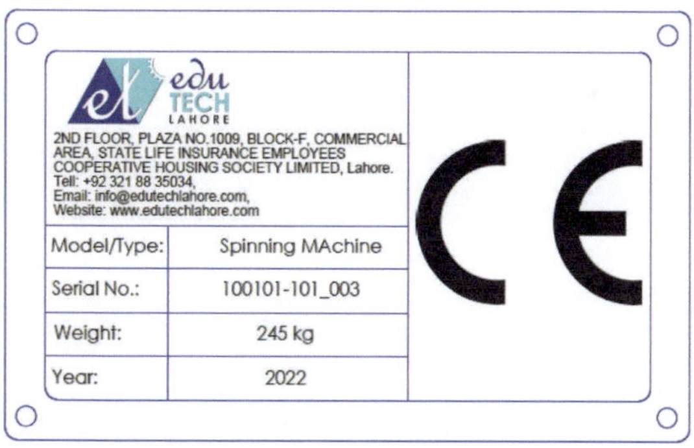

Fig. 6.1 CE compliance

4.8 Learning Outcomes

The readers will understand the role of Ergonomics, and modern concepts: User Interface (UI), and User Experience (UX), as elaborated and exemplified in the case study here. The use of a system approach to design to perform computer-aided design (from concept design to working detail design) with concept Generation and Concept

Selection methods and a reference to Computer Aided Engineering (CAE) methods to perform design analysis must be understood before we move on to further chapters. Design documentation is exemplified in the case study here through an example of a real product by providing a real design of a Six Spindle Spinning Machine in part.

4.9 Review Questions

Q1. What is the basic definition of Engineering Design?

Q2. What converts modeling into design?

Q3. Differentiate between User Interface (UI) and User Experience (UX).

Q4. How to quantify level of satisfaction with UI through a metric?

Q5. How to quantify level of satisfaction with UX through a metric?

Q6. Why disregarding all efforts at the embodiment design stage is difficult?

Q7. What is meant by concept design? Is it a qualitative exercise?

Q8. How many methods are there in concept generation and concept selection? Give at least two examples of concept generation and selection of discrete products of your own choice?

Q9. What is meant by an embodiment design? Is embodiment design is performed at components, structures, assemblies, and mechanisms level?

Q10. What techniques are used to convert concept design into an engineering design through analysis of components, structures, assemblies, and mechanisms level?

Q11. What is meant by detail design? Is detail design is performed at assembly, mechanism, and machine level as a whole entity?

Q12. What analysis techniques are used to perform engineering design analysis of assembly, mechanism, and machine level?

Q13. Is it pertinent to verify engineering design using design through standards, design through empirical equations, and design through real life models?

Q14. Is it satisfactory to perform engineering design using only numerical methods? Give examples.

Q15. Is it essential to use all design methodologies to ensure components, structures, assemblies, mechanisms, and machines will work as per their functionality when human life is involved?

Q16. Is time having importance in today's competitive marketplace to have engineering design and its versions available whenever needed for maximum profitability—the very purpose of engineering design and manufacturing?

Q17. Where concept design is applicable in engineering design exercise and why?

Q18. Where embodiment design is applicable in engineering design exercise and why?

Q19. Where detail design is applicable in engineering design exercise and why?

Q20. Do the standards, both legacy and industry 4.0 and industry 5.0 help reduce design time as sourcing becomes simpler?

Q21. Is today's design being innovative, multidisciplinary, customized, and sustainable needs a broad base of knowledge through having a design team from manufacturing shop to design office?
Q22. What are programmable languages and why are they needed in engineering design?
Q23. What are the different generations of programmable languages and what is their utility?
Q24. What are the trends and developments of the fifth-generation programmable language? Does it lead to more automation for Cyber Physical Systems (CPS)?
Q25. Discuss the multifaceted language ecosystem and where it is used.
Q26. Elaborate on Multidisciplinary product development. Give at least two examples.
Q27. What is Product Lifecycle Management (PLM)?
Q28. What is the Manufacturing Execution System (MES)?
Q29. What is Product Life Cycle Assessment (PLA)? Explain through examples.
Q30. What is the Product Life Cycle?

Bibliography

1. Shigley JE, Mischke CR (eds) (2004) Standard handbook of machine design, 3rd edn. Hardcover, McGraw Hill
2. Smid P (202) Technology of machine tools ISE. McGraw-Hill Education
3. Gupta MP (2011) Quantitative techniques for decision making. PHI Learning Pvt. Ltd.
4. CMTI (2000) Machine tool design handbook. Tata McGraw Hill, New Delhi
5. Chang TC, Wysk RA (1984) An introduction to automated process planning systems. Prentice Hall Professional Technical Reference
6. Kibbe RR, Neely JE, Meyer RO, White WT, Bonkoski M, Bradshaw P (1982) Machine tool practices. Wiley
7. Krar SF, Oswald JW, Amand JES, Krar SF (1990) Technology of machine tools. Gregg Division, McGraw-Hill New York
8. Nagpal GR (1999) Machine tool engineering. Khanna
9. Norton RL (2010) Machine design: an integrated approach, Hardcover. Pearson
10. Pahl G, Wallace K, Blessing LTM, Beitz W, Bauert F (2013) Engineering design: a systematic approach. Springer, London
11. Pahl G, Beitz W, Feldhusen J, Grote KH (2019) Engineering design: a systematic approach, Hardcover. Scholars Portal. Accessed: 28 Aug 2024
12. Norton RL (2010) Machine design: an integrated approach, Hardcover. Pearson
13. Pahl G, Wallace K, Blessing LTM, Beitz W, Bauert F (2013) Engineering design: a systematic approach. Springer, London
14. Coticchia ME, Crawford GW, Preston EJ (1993) CAD/CAM/CAE systems: justification, implementation, productivity measurement. CRC Press
15. Dwivedi SN, Verma AK, Sneckenberger JE (2012) CAD/CAM robotics and factories of the future'90: volume 2: flexible automation 5th international conference on CAD/CAM, robotics and factories of the future (CARS and FOF'90) proceedings. Springer Science & Business Media
16. Raouf A, Ben-Daya M (1995) Flexible manufacturing systems: recent developments

17. Mahon M, Browne J (1993) CAD/CAM from principles to practice. Addison-Wesley, UK, Suffolk
18. Zeid I (1991) CAD/CAM theory and practice. McGraw-Hill series in mechanical engineering. McGraw-Hill
19. Teicholz E (1985) CAD/CAM handbook. Computer communications series. McGraw-Hill
20. McMahon C, Browne J (1993) CADCAM: from principles to practice. Addison-Wesley Publishing Company
21. Zeid I (2005) Mastering CAD/CAM. McGraw-Hill Higher Education
22. CAD/CAM: computer-aided design and manufacturing. Pearson Education (2013)
23. Goetsch DL (1983) The CAD/CAM workbook. South-Western Publishing Company
24. Rao PN (2004) CAD/CAM: principles and applications. Mechanical engineering series. McGraw-Hill Education (India) Pvt Limited

Chapter 5
Computer Aided Manufacturing

Prelude

The repeatability of producing parts, structure, assembly, mechanism, and machine based on automation leading to higher productivity with less dependency on workers is the key factor that led to the invention of the Numerical Control Machine and Special Purpose Machine. Today, discrete manufacturing is being advanced by pillars of Industry 4.0 and Industry 5.0. Manufacturing for SMEs is the backbone of any industrial economy, and it encourages setting up small manufacturing units with local job creation and contribution to local, provincial, and federal government through paying taxes and dropping the pressure on foreign exchange reserves by reducing imports. This socio-economic benefit leads to less muscle and brain drain.

5.1 Factors Influencing Industrial Activity

Industrial activity is triggered when an enterprise (SME or LSE) starts production for a client or business. It generates dependent activities like the establishment of waste disposal for a product covering value addition in MES embedded in PLM with a wider scope that is further encapsulated in the Product Life Cycle or cradle-to-grave phase for a Product. All allied tangible activities and intangible services are thus part of the whole production scheme working under one or multiple philosophies to enhance productivity, quality, and profitability.

Supplementary Information The online version contains supplementary material available at https://doi.org/10.1007/978-3-031-77106-4_5.

W. A. Khan et al., *Computer Aided Engineering Design and Manufacturing*, Springer Series in Advanced Manufacturing, https://doi.org/10.1007/978-3-031-77106-4_5

If the product is competitive (innovative, multidisciplinary and/or customized) it generates demand for a certain period for which it remains competitive. During the 4th Industrial Revolution, apart from products that are needed for public consumption, other products have a shorter life span with a better model round the corner with new catching features.

The design office caters for this type of market and manufacturing is modular and done for a very competitive price with vendors doing the job as per a persistent quality standard.

This industrial activity is based on a well thought out plan that considers the time factor, potential market, financial capability of the customer with lucrative leased based purchasing schemes involving banks, media, and a complex structure of taking the product to the masses.

The designer of the product, normally from the modern world, ensures that the product design fetches significant profits over a long period of time and once the vendor's deletion plan reaches a stage when the product is of no significance in terms of profit a new product is launched with re-organization of the business entity.

New organizations based on new products, mergers, bankruptcy are common phenomena being governed by large banks and DFIs and for relatively poor countries companies producing the product seek sovereign guarantees to operate in such countries.

This is a complete capital culture that is sweeping its way into the countries that had a long political stand to not to adapt to this scheme of business.

5.2 Objectives

In succession to logical thinking and knowing Engineering Design and Manufacturing are done for wealth creation, the authors open Computer Aided Manufacturing (CAM) which converts ordinary manufacturing into a repetitive high-productivity activity.

Contemporary manufacturing analysis techniques are introduced assuming this function is performed by consultants or the business entity itself leading to the selection of a discrete product design and manufacturing enterprise to do the job.

The reader is further introduced to today's continuous and discrete manufacturing process controllers: Programmable Logic Controllers (PLC) and Computer Numerical Control (CNC). The Ladder Diagram and Part Program are briefly introduced here.

As more emphasis is given to CNC, the anatomy of these controllers is provided for manufacturing of discrete products on machine tools through system engineering aspects of these controllers including but not limited to Mechanical, Electrical, Electronics, Computer Software, Computer Hardware, and control engineering aspects.

Part programming using standards (ISO 6983-1:2009 and STEP-NC [ISO 14649]) with their history of evolution and from position-based part programming based on G&M code to feature-based programming based on STEP-NC is introduced. Details about these standards are provided in Part III Chap. 8.

The concepts are further extended by taking a single CNC machine to multiple CNC machines forming a manufacturing enterprise by introducing Manufacturing Engineering Services (MES) and Product Life Management (PLM), please also refer to the objectives and learning outcomes of Part IV, Chap. 11.

Emphasis is given to the Product Life Cycle of both the machines at the manufacturing enterprise and to products that these machines produce. The concept is explained using a case study on Product Life Cycle Code that can keep a record of every single product that is produced on Earth from its cradle-to-grave cycle.

The reader is expected to understand the logical progression and the concepts provided so that further discussion on these topics and new topics where these topics are prerequisite can take place.

5.3 Contemporary Manufacturing Analysis Technique

The manufacturing analysis technique encompasses analysis of the PLM and the MES cycle for SMEs and LSEs. These software packages operate on a UI that describes the product range and processes used to manufacture these product ranges. This software is either open source or has established brand names such as SAP and Oracle etc. In either case, be it open source or branded source, the package needs to be configured. On many occasions the SME or LSE must adopt the procedures defined by the expensive branded software. The UX for such software is the level of acceptance of processes and procedures defined by the branded software company to technicians, engineers, and other non-technical staff.

The depth of manufacturing process analysis for parameters estimation at branded software has not touched the level of Virtual Manufacturing System for discrete products described in this book.

5.3.1 Graphical Programming Techniques for Discrete Manufacturing

Graphical programming techniques for discrete manufacturing are limited to use at large manufacturing enterprises having established brands such as automotive manufacturing enterprises and airplane manufacturing enterprises. This software is not for public use and is developed and used at the individual company level e.g. Toyota, Ford, Boeing, Airbus etc. Some Open-source software such as OpenGL can be configured to be used at SMEs with established business of some product range.

5.3.2 Graphical Programming Techniques for Continuous Manufacturing

Use of Graphical programming techniques for continuous manufacturing is less common at SMEs and more common at LSEs. Established purpose built commercial software is available for graphical programming for continuous manufacturing such as ASPEN and NX. Two examples, water supply control through SCADA using VMS developed by the authors and production of Ammonia using ASPEN are provided in the following sections:

5.3.2.1 Water Supply Control Through SCADA (VMS)[1]

https://drive.google.com/file/d/1qMRkhGYiYTDpQzeTy81PTzpVl5yoAieo/view?usp=sharing.

5.3.2.2 Aspen Plus Simulation Problem Statement: Ammonia Production Process[2]

The ammonia production process combines nitrogen and hydrogen to produce ammonia. The process starts with an inlet stream (FEED) containing 100 kmol/h of nitrogen (N_2) and 300 kmol/h of hydrogen (H_2) at 25 °C and 1.013 bar, which is mixed in a mixer before entering the reactor.

In the reactor, the following reaction takes place:

$$N_2 + 3H_2 \rightarrow 2NH_3 \quad \text{(25\% conversion of nitrogen per pass)}$$

After the reaction, the mixture of product and unreacted reactants (PROD-MIX) is sent to a separator, where ammonia is separated from the unreacted nitrogen and hydrogen (Fig. 5.1).

The product stream ([3]PRODUCT) from the separator contains 98% NH_3, with small amounts of N_2 and H_2 (0.05% each). The remaining stream from the separator (RECYCLE) is recycled back to the mixer, where it is combined with fresh feed for reentry into the reactor.

---

[1] Virtual Manufacturing System (VMS), Logic Design by Wasim A. Khan, Programmed by Amir Hussain. Protected by Pakistan patent number 37/2022 Acceptance Number 144217 and US patent applications number 1786442 (Pending). Reproduced with permission.

[2] Designed by Ms. Ramesha Tariq, Lecturer, Department of Chemical Engineering, GIK Institute, Topi-23640, Khyber Pakhtunkhwa, Pakistan.

[3] Produced on Licensed Aspen VII.

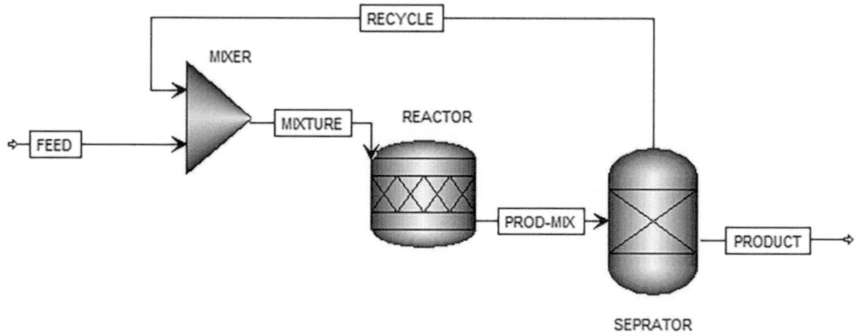

Fig. 5.1 Ammonia production process

URL

https://drive.google.com/file/d/1nVdZeF3VSiAx4OWtshykG365knb-jMGs/view?usp=drive_link.

5.4 Modern Microprocessors and Microcontrollers[4]

A processor is at the heart of modern communication, performing all types of computation. It's made up of transistors which form logic gates which do the computations. One of the first microprocessors was Intel 4004 which comprised 2300 transistors fabricated at 10 μm. The latest offerings by the same company, which are much more powerful, have billions of transistors fabricated at 5 nm. Apart from Intel, other companies which make processors include AMD, Apple, ARM, and IBM, etc.

A modern processor's circuitry can broadly be divided into control and data path. These two works in conjunction with storage (RAM) and I/O to make a full computation unit. RAM is usually external, and its purpose is to hold data as well as instructions needed for the computation at hand. The Central Processing Unit (CPU) brings in instructions from RAM one by one and executes them.

The two major design philosophies when designing processors are CISC (Complex Instruction Set Computer) and RISC (Reduced Instruction Set Computer). The former can execute complex instructions at the cost of a very convoluted design whereas the latter proposes to keep the instructions simple so that the design is simple. Intel and AMD processors are based on CISC whereas ARM and IBM follow the RISC path. RISC-V is another rising processor architecture which follows the RISC philosophy.

[4] Courtesy: Dr. Taj Muhammad Khan, Assistant Professor, Faculty of Computer Science and Engineering, GIK Institute of Engineering Sciences and Technology, Topi-23640, Khyber Pakhtunkhwa, Pakistan. Reproduced with Permission.

5.4.1 Microcontrollers

These are simpler and low-cost variants of processors with notable architectural differences. They are much less powerful than our typical processor but also cost much less. They have their storage as well as a lot of I/O peripherals on chip and due to their simpler architecture require much less energy to operate. They are typically used in embedded settings to control various peripherals or get data from diverse sensors. Manufacturers like Atmel, STMicroelectronics, Espressif, and ARM, etc., offer different ranges of microcontroller families with varying capabilities targeting a variety of tasks.

5.4.2 Processors and Microcontrollers in Manufacturing Applications

A typical mechanical engineering project will involve the use of both processors as well as microcontrollers, typically connected in a hub fashion with the processor at the center. The microcontrollers will typically interface with the outer world and receive data from various peripheral devices. They would transmit this data to the central processor which does the heavy lifting of performing complex computations and analyses on the data received. Based on its conclusions, the processors would transmit new data and instructions to the microcontrollers according to which the microcontrollers will readjust their peripheral machines. An autonomous robot will be a good example where the microcontrollers receive data about the balance from different limbs and transmit it to the processor. The processor, after doing analyses, instructs each microcontroller to readjust its limb in a new position to restore the balance. This cycle happens continuously resulting in a feedback loop.

5.5 Programmable Logic Controller (PLC)

A Programmable Logic Controller, PLC, is a durable, shock- and vibration-resistant, reliable small computer that is employed especially in industrial facilities that adopt automation. A PLC embodies inputs and outputs that help the device transfer data, through which an internally embedded logic is processed. In the case of CNCs, PLC is physically located within the machine and protected by electrical junction boxes. Using a PLC, a CNC manufacturer benefits from a well-developed and established versatile technology that is easy to program, relatively cheap to procure, and simple to troubleshoot and maintain.

5.6 Computer Numerical Control (CNC)

Humans retard effort therefore they need to be working in a framework whether it is services industry or discrete or continuous manufacturing. In 1206 AD, Al-Jazari and Shooshtari, used Cams in their automata, The practice continued in more formal forms of industrial activities meaning industrial revolutions. There was another behavior of the families taking industrial actions since they had the knowledge of operating the machines used for discrete or continuous manufacturing leading to production stoppage. The first action the owners of the industrial concern, especially related to discrete manufacturing, was to use Cams to automate discrete manufacturing processes and to lessen the threat of families of workers to take industrial action.

The journey of automation reached its peak towards the start of the third industrial revolution when Massachusetts Institute of Technology, USA, demonstrated the first Numerically controlled (NC) machine in the 1950s. The major characteristic of the NC machine was repeatability and continuous operation using a part program.

The inception of first NC machine led to the development of up to 12 axes Computer Numerical Control (CNC) machining centers (2023) using input medium such as Part Programs at cloud, produced through Conventional or Modern CAD/CAM cycle (See Sects. 5.7.1 and 5.7.2), producing parts or structures with most difficult combination of surfaces that would otherwise be requiring many setups and would increase the lead time.

The difference between Numerically Controlled (NC) and Computer Numerical Control (CNC) is that the CNC machine has onboard computer with the provision of Random Access Memory (RAM) for the storage of part program, Read Only Memory (ROM) for provision of stored programs such as part program interpreter or compiler, software and hardware drivers for peripheral devices such as Visual Display Unit (VDU), Keyboard, and; Input/output ports for data communication as per certain serial or parallel data transfer protocol allowing CNC machine communication with design office and for e-Commerce.

Today's Computer Numerical Control (CNC) Machine has following aspects that provides an effective UI and UX:

- Mechanical Engineering Aspects
- Electrical Engineering Aspects
- Electronic Engineering Aspects
- Computer software Aspects
- Computer hardware Aspects
- Computer Communication Aspects
- Control Engineering Aspects of CNC [Open Loop and Feedback Control through Transfer function]
- Ergonomics
- Reliability, Availability, Maintainability and Safety.

Controller for 3-Axes Milling Machine

https://drive.google.com/file/d/1yokCtFiq7gu-bDZSXxLU9LkenzxRxU-R/view.

5.7 Part Programming

Part programming is an art to program tool movement to produce a part. It is meant for subtractive manufacturing or for any other machine performing motion in a full or partial cartesian coordinate system (refer to Section on ISO-841:2001) for more detail. The part programs have been written and stored using different methods. At the time of inception, the part program was input to the controller using a punched card. Afterward paper tape, magnetic tape, floppy disk, and Random Access Memory (RAM) were used. Now transfer of part programs to CNC controllers is possible through industrial ethernet. SERCOSE IV and WI-FI with cloud being the storage medium.

The programming languages used to program a part included binary coding, hexadecimal coding, microprocessor instruction set and G&M Mnemonics using EIA 274 (Electronic Industries Association standard) and ISO 6983 (International Standards Organization ISO standard). High level languages including APT (Automated Programming Tool) and NELAPT (National Engineering Laboratory APT) were also used. CNC software package that supports Python integration, can use Python to create scripts that control the CNC machine. This typically involves using Python libraries and modules specifically designed for CNC programming, such as pygcode and cnc-gcode.

STEP-NC is a modern method of part programming. As STEP-NC carries high-level information of machining processes and features, it is an efficient format for controlling the machine tool compared to G-M codes. With its rich set of information, STEP AP238 (AIM format of STEP-NC) is a more sustainable and reliable format than ISO 14649 (ARM format of STEP-NC).

The operation of ISO 6983-1:2009 is limited to the machine tool. It is post-processed while STEP-NC has wide adaptability over the same type of machines. STEP-NC is applicable to a limited number of machines such as woodworking machinery and subtractive manufacturing machinery while ISO 6983-1:2009, refers to all machines that can be defined using ISO 841-2001. ISO 841-2001 was reviewed in 2023 by ISO and is a valid standard.

Machine tool technicians still prefer using G&M mnemonics based on ISO 6983-1:2009 that is a derivative of other versions of ISO 6983 that themselves were driven from different versions of EIA 274. ISO 6983-1:2009 was reviewed by ISO in 2020 and is currently (2023) a valid standard. All machines defined by ISO 841-2001 and all other machines derived from Right Hand Rule (The Basic Concept of ISO-841) are programmable using ISO 6983-1:2009. These machines include subtractive manufacturing machinery, Additive Manufacturing Machinery, Measurement machines and scanning machines.

Interoperability between ISO 6983-1:2009 a legacy standard that is relied upon heavily even today and STEP-NC that has the potential for capturing the industry because of its integration capability in Manufacturing Engineering Services (MES) and Product Life Management (PLM) is intensively researched.

Producing part programs using ISO 6983-1:2009 is conventional CAD/CAM Cycle as provided in above section while producing part programs using STEP-NC is modern CAD/CAM cycle as lamented in the following section (Fig. 5.2).

Raw Material, Tools, Measurement Devices, Jigs and Fixtures and coolants and lubricants are defined through process plan (see Chap. 10) and the technician acquires these items through requisition forms in manual sourcing from store or through Computer Integrated Manufacturing (CIM) (Or Integration as defined in Industry 4.0) via electronic forms (See Chap. 10) from store. Keeping record of inventory of items and tied up capital for above listed items manually is a difficult job while a software-based dashboard displays the status of inventory of each item and the tied-up capital with just the click of buttons.

Part III Chap. 9 provides how databases are used to store all the information mentioned in the above paragraph and how alerts can be set up to maintain an optimum level of inventory.

Fig. 5.2 Conventional CAD/CAM Cycle

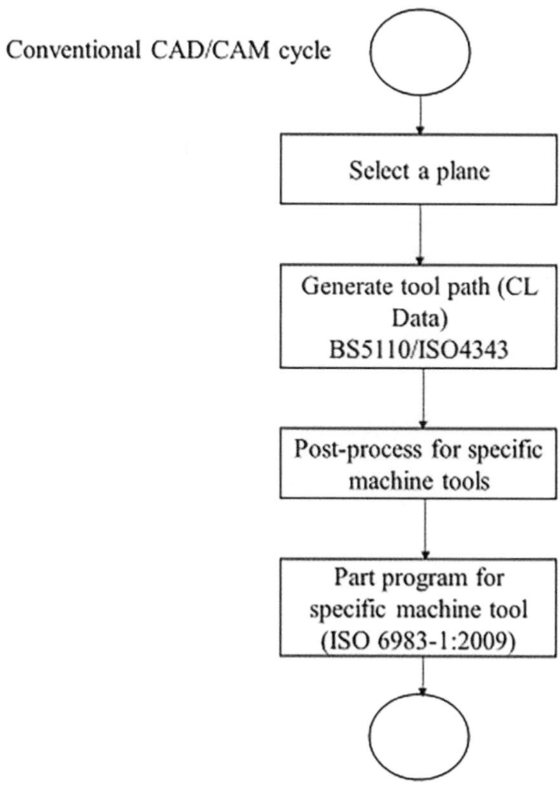

Several methods of part programming existed since its time of inception:

5.7.1 EIA RS 274-D:1979 Standard

Electronic Industries Association (of USA—today named as Electronic Industries Alliance Recommended Standard) EIA RS 274D:1979 standard is one of most adopted G&M Part Programming standards of its time. It was later adopted by the International Standards Organization and evolved into ISO 6983-1:2009 standard that is currently (2023) a valid ISO standard. EIA RS 274-D:1979 standard can be obtained from libraries.

5.7.2 ISO 6983-1:2009 Standard

ISO 6983-1:2009 after various revisions is now a stable standard and is widely used in industry (2023). It has Normative References of following standards and is handled by the Technical Committee ISO/TC 184.

ISO 6983-1:2009 operates on ISO 841:2001 axes orientation thus has movement as the basis of its operation. It is applicable to all machines defined based on Right Hand Rule.

It provides G codes (Preparatory Functions) and M codes (Miscellaneous Functions) for operating a computer numerical control machine. Preparatory Functions prepare the machine for operation and include simple movement (positioning) to interpolated movement (for machining analytic and synthetic curves and surfaces). It has 1000 preparatory functions (G00 to G999). Three-digit preparatory functions are used to program special features such as mirror, translate and rotate a feature. The standard has 9 classes of Miscellaneous functions that are applicable to different categories of machines defined under ISO 841:2001.

The standard uses all English alphabets to denote an 'Address Letter' that combines with its numerical value forms an 'Address Word'. Combination of Address Words separated by Address Word separator 'Space' forms a 'Block' that is terminated by a 'Carriage Return' known as 'End of Block or EOB' as per convention used in this book. The part program is a collection of 'Blocks' and is terminated by an End of File 'EOF' inserted by the text editor such as Microsoft 'Notepad'. Every visible and non-visible character has an ASCII representation and is stored in RAM as a binary equivalent of the character used.

Advanced programming techniques are also available including Cycle (Fixed parametric subroutines), Loops, Nested Loops, Subroutine, and Macro.

The subset of 'Address Letters' including Axes Used, Preparatory Function Used, Miscellaneous Functions Used, Invisible Characters Used, and Symbols Used as per a defined 'Format Classification' defines the scope of the controller of Computer Numerical Control Machine Tool.

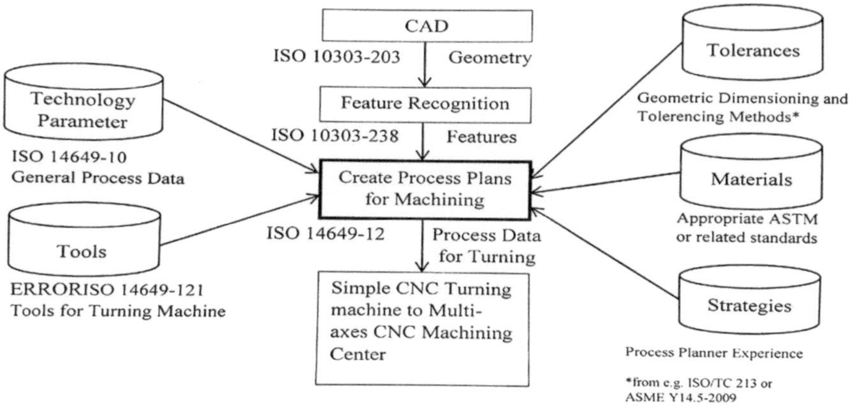

Fig. 5.3 The modern CAD/CAM cycle

The case study given in this chapter provides pseudo coding using flowcharts to define an interpreter for a specific knee and column type Milling Machine Tool.

5.7.3 STEP NC Standard

Integration, a pillar of Industry 4.0, is difficult to implement if rules and methods as per the principles used in ISO 6983-1:2009 are implemented. Contrary to use of movement as principal criterion to operate the CNC machines STEP-NC uses features such as holes, pockets, chamfer etc. to operate a Computer Numerical Control machine eliminating the post processing step from the modern CAD/CAM cycle given in Fig. 5.3.

The modern CAD/CAM cycle utilizes standard information except for features that cannot be unique, and the resulting strategies worked out by the process planner. The process planner is normally the most experienced person with exposure to management, therefore can interact with personnel at the design office, personnel at production planning and control and the personnel at the shop floor.

STEP-NC is expected to prevail in future years as its automation prone capabilities leads it to be adopted in 4th Industrial Revolution technologies especially in the 'Integration' pillar.

Development of interpreters/compilers can be facilitated by information available at the National Institute of Standards and Technology (NIST, USA).

A case study related to development of an interpreter for G&M code as per ISO 6983-1:2009 for all machines that can be defined using ISO 841:2001 is given in Sect. 8.5—Chap. 8 Part III.

5.8 Case Study—Extended Product Life Cycle Code—ePLCC

Product life cycle code from original publication (refer to Bibliography) is extended and activities related to Manufacturing Execution System (MES), Product Lifecycle Management are embedded in Product Life Cycle (PLC); please refer to Fig. 5.4. New entities are defined here in each management layer (i.e. MES, PLM and PLC).

The procedure to develop the Product Life Cycle Code remains the same as per mentioned research paper that can easily be downloaded from given DoI. The code is now called the extended Product Life Cycle Code (ePLCC).

ePLCC has new definitions for Attribute Category Name, Pre-Defined Code Bits, and; User Defined Code Bits. Rest of the procedure to workout ePLCC remains the same as given in the cited research paper.

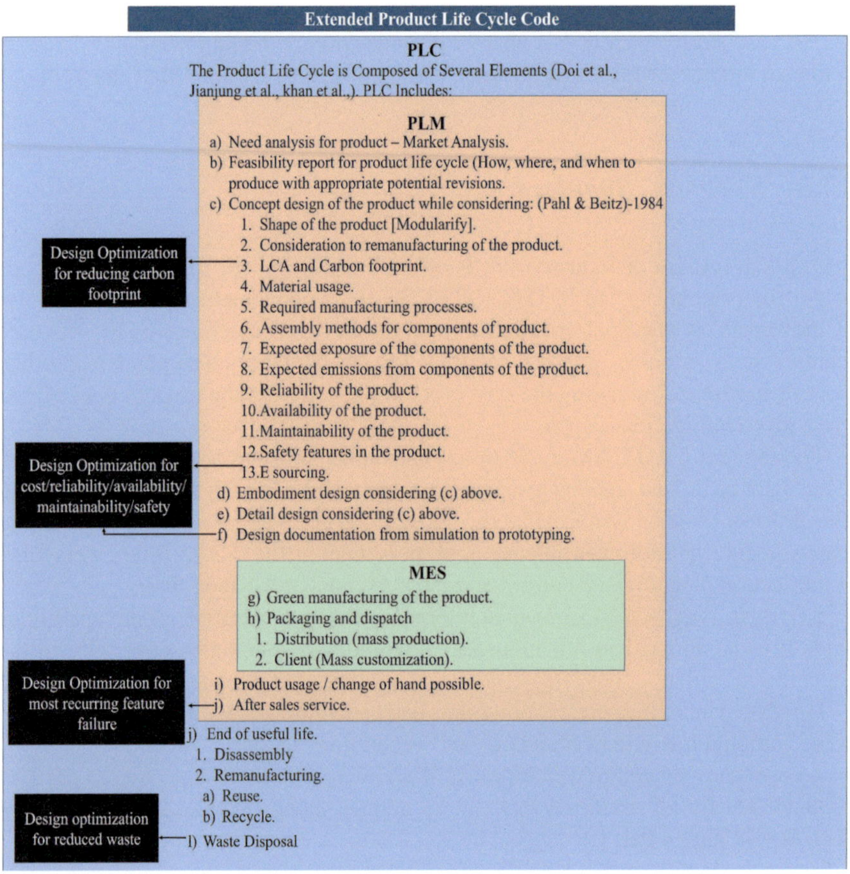

Fig. 5.4 Entities for extended product life cycle code (ePLCC)

5.9 Learning Outcomes

By now the reader must be able to understand the relationship between the CAD and CAM and is able to implement these functions at least in a rudimentary manner. This includes the ability to operate any CAD/CAM package and the ability to produce part programs using the basic machining module of the CAD/CAM package excluding the production Planning and control (PPC) aspect.

The reader is also required to understand the concepts related to MES, PLM, Product Life Cycle, and Product Life Cycle Code (PLCC).

5.10 Review Questions

Q1. What are the key economic, environmental, and social factors influencing industrial activities in the context of sustainable development? How do these factors interact and influence decision-making in industrial processes?

Q2. Analyze the advantages and limitations of contemporary manufacturing analysis techniques, such as digital twin technology and use of Artificial Intelligence in optimizing production processes.

Q3. Discuss the architecture, performance benchmarks, and applications of modern microprocessors and microcontrollers in manufacturing machinery. How do advances in these technologies impact the development of industrial automation?

Q4. Compare and contrast the principles, operational complexities, and application areas of Numeric Control (NC) and Computer Numeric Control (CNC) systems. How have advancements in CNC technology influenced precision engineering?

Q5. Identify and evaluate the key functional aspects of a Computer Numerical Control (CNC) machine, including but not limited to machine kinematics, tool path optimization, User Interface (UI) design, and feedback control systems. How do these aspects contribute to the overall performance and accuracy of CNC machining?

Q6. What is Part Programming in CNC machining? Discuss the different types of part programming methods and their impact on production efficiency and error reduction.

Q7. Explain STEP-NC and its significance in the evolution of CNC machining. How does STEP-NC improve upon traditional G-code, and what are its implications for the future of smart manufacturing?

Q8. Draw and critically analyze the Conventional CAD/CAM cycle. How does it compare to modern integrated CAD/CAM/CAE systems in terms of efficiency, accuracy, and design flexibility?

Q9. What constitutes a modern CAD/CAM cycle, and how do advanced technologies such as AI, machine learning, and cloud computing integrate into this cycle? Discuss the impact of these technologies on the product design and manufacturing process.

Q10. Describe the basic types in the classification and coding of Product Life Cycle Codes (PLCC). How do these classifications aid in the management and sustainability of product lifecycles?

Q11. What are the three basic types of 3D modeling, and how do they differ in terms of their application in engineering design and analysis?

Q12. List and evaluate at least five different manufacturing attributes in PLCC. How do these attributes contribute to the optimization of product design and manufacturing processes?

Q13. List and analyze at least five different exposure (usage) attributes in PLCC. How do these attributes influence product lifecycle management and sustainability considerations?

Q14. Discuss the various techniques for using binary numbers in generating product life cycles. How do these techniques facilitate the encoding and management of complex product lifecycle data?

Q15. Explain and critically assess the design attribute bit mapping process in PLCC. How does this mapping influence design decisions and lifecycle management?

Q16. Define disposal attribute bit mapping in the context of PLCC. Discuss its role in enhancing the sustainability and environmental impact of products at the end of their life cycle.

Q17. Illustrate and discuss the object-oriented design approach for product lifecycle coding. How does this approach improve the modularity, reusability, and scalability of lifecycle management systems?

Q18. List at least five different ISO standards relevant to discrete manufacturing and discuss their significance in ensuring quality, safety, and sustainability in industrial processes.

Q19. How do advancements in additive manufacturing and 3D printing technologies influence the classification and coding systems within PLCC? Provide examples of how these technologies are reshaping lifecycle management.

Q20. Discuss the impact of Industry 4.0 on the evolution of CAD/CAM systems. How does smart manufacturing integrate into the modern design and manufacturing cycle?

Q21. Develop Pseudo code for generating ePLCC for two different makes of refrigerator?

Bibliography

1. Bartelt TLM (2002) Industrial control electronics: devices, systems, and applications
2. Gupta VP, Arora AK (2022) Automation in healthcare services. In: Research anthology on cross-disciplinary designs and applications of automation. IGI Global, pp 285–303

3. Gaonkar R (2010) Fundamentals of microcontrollers and applications in embedded systems
4. Berger A (2001) Embedded systems design: an introduction to processes, tools, and techniques. CRC Press
5. Pack DJ, Barrett SF (2008) Microcontroller theory and applications: HC12 and S12. Prentice Hall
6. Onwubolu G (2005) Mechatronics: principles and applications. Elsevier
7. Groover MP (2015) Automation, production systems, and computer-integrated manufacturing, Global Edition. In: Law express. Pearson Education
8. Groover MP (2012) Introduction to manufacturing processes. Wiley
9. Harris S, Harris D (2015) Digital design and computer architecture. Morgan Kaufmann
10. Modlen G (1989) Manufacturing engineering and technology. By Serope Kalpakjian. Addison-Wesley
11. Ogata K (2020) Modern control engineering
12. Pabla BS, Adithan M (1994) CNC machines. New Age International
13. Ray AK, Bhurchandi KM (2006) Advanced microprocessors and peripherals. Tata McGraw-Hill
14. Smid P (2003) CNC programming handbook: a comprehensive guide to practical CNC programming. Industrial Press Inc.
15. Webb JW, Reis RA (1994) Programmable logic controllers: principals and applications. Prentice-Hall, Inc.
16. Kalpakjian S, Schmid S (2022) Manufacturing processes for engineering materials. Pearson
17. Black JT, Kohser RA (2019) DeGarmo's materials and processes in manufacturing. Wiley
18. Kibbe RR, Neely JE, Meyer RO, White WT, Bonkoski M, Bradshaw P (1982) Machine tool practices. Wiley
19. Khan WA, Hussain A (2014) Automatic object-oriented coding facility for product life cycle management of discrete products. Int J Comput Integr Manuf 27(1):60–84
20. Krar SF, Gill AR, Smid P, Gerritsen RJ (1985) ISE technology of machine tools. McGraw-Hill Education, 2019. C. M. T. Institute, Machine tool design handbook. Tata McGraw-Hill

Part III
Computer-Aided Engineering

Prelude

In the Industry 4.0 era, computer-aided engineering (CAE) plays a more focal, and vital role in implementation of computer and software in modeling, simulation, and analysis of engineering and basic sciences problems that interest both academia and industry. CAE is an umbrella concept that embodies the use of software for design, analysis, simulation, optimization, validation, and manufacturing efforts of parts, processes, and systems. CAE establishes a broad platform that includes numerical techniques such as *Finite Difference Method* (FDM), *Finite Element Method* (FEM), *Finite Volume Method* (FVM), and *Boundary Element Method* (BEM); *Multi-body Dynamics* methods for kinetic and kinematic analysis of mechanisms; control systems design and analysis; Mechanical Event Simulations (MES); other (analytical, etc.) engineering calculations, and process simulations for various manufacturing processes.

As mentioned, CAE has become more important than ever before in the current age of Industry 4.0. Efficient and comprehensive use of CAE certainly brings advantages such as collecting swift feedback and data from the ongoing processes, rapid implementation of design changes, effective performance monitoring, minimizing the need for experimental testing, rapid prototyping, and improved risk management.

In Part III of the book, the reader will have a good understanding of the most common computer-aided numerical analysis techniques, mathematical basics and standards of CAD/CAM/CAE, and use of databases developed for these computer-aided techniques.

Design Analysis

As highlighted in Chap. 1; engineering design, materials selection and substitution, and manufacturing are profoundly co-dependent endeavors. In this respect, CAE is the fundamental modality in Industry 4.0 for design conceptualization, engineering design, assembly, and analysis testing. CAE utilizes software through the use of one or more techniques to speed up design and analysis efforts, enabling effective optimization and validation.

The Behavior of Solids, Liquids, and Gases (SLG)

The physical and chemical properties of three of the states of matter, namely solids, liquids, and gasses (SLG), as well as their behaviors under certain given conditions require meticulous efforts in expressing the governing differential/integral equations for the engineering/basic science problems to be investigated. In this context, the vibration and free movement of gas particles, vibration and sliding past each other for liquid particles, and vibratory tightly packed arrangement of solids need to be asserted within these equations reasonably. The nature of these problems directs the engineers and scientists towards choosing an appropriate method, which thereafter leads to careful selection of CAE software that can potentially model and simulate the phenomena realistically and effectively. As an example, amongst CAE methods, FEM is a suitable choice for analyzing solids, whereas FVM integrated computational fluid dynamics (CFD) software can efficiently simulate liquids and gasses. It should be mentioned that some CFD software also utilizes the FEM approach. A combination of methods is a possibility to benefit from the strengths of the individual methods.

Applications Using SLGs—Independently and in Multiple Medium

CAE software has been improving rapidly; some are becoming more dedicated to investigating a certain state of matter, whereas some others can now handle Multi-physics problems, which are physics phenomena that co-exist. Therefore, the Multi-physics approach warrants a deeper investigation of the problem and a much more sophisticated numerical platform to resolve the targeted co-existing phenomena. Comprehensive lists of software that handle a problem for a selected independent state of matter are provided within the following chapters.

Software Packages for Independent SLG Analysis

As mentioned, comprehensive lists of software for independent SLG analysis are given within the following chapters.

Software Packages for SLG Applications

COMSOL Multiphysics, ANSYS, and *MATLAB* can be mentioned as some of the widely used Multiphysics software packages.

Mathematics for SLGs

Part III of the book aims to provide mathematical relations for the topics within its relevant chapters as well as in the examples and case studies provided. These relations can be found in the relevant chapters, where a typical mathematical relation can help understand the method more thoroughly. Obviously, the book is intended and developed as a CAD/CAM textbook in the era of Industry 4.0; therefore, delving into the very details of the methodologies is not a scope of the book. However, various further reading and references are also provided to inform the reader of quality material on the topics.

Chapter 6
Computer-Aided Numerical Analysis Techniques

Prelude

This chapter is dedicated to discussing the four prominent numerical approximation methods; namely, Finite Difference Method (FDM), Finite Element Method (FEM), Finite Volume Method (FVM), and Boundary Element Method (BEM). Many powerful, widely used CAE software adopt one or more of these methods as their fundamental computational approach in solving the differential or integral equations that govern the engineering problem at hand. Finally, application of computer-aided engineering methods in solid, liquid, and gas-based products is reviewed, which is followed by case studies.

6.1 Objectives

Computational methods such as Finite Difference Method (FDM), Finite Element Method (FEM), Finite Volume Method (FVM), and Boundary Element Method (BEM) have dominated the design and analysis efforts in the engineering field since their theoretical emergence around 1700s (FDM), and particularly in the second half of the 1900s (FEM, BEM, FVM). Especially with the exponential developments in computer technology, numerical analysis techniques have been empowered with the computational capability they require, turning simulation and analysis of complicated real-life problems into reality. These methods have not only curtailed the need for excessive, time-consuming, and potentially expensive experimental studies but also laid the groundwork for a combined approach of the two methods employed together efficiently. Incorporation of numerical analysis tools along with CAD/CAM

Supplementary Information The online version contains supplementary material available at https://doi.org/10.1007/978-3-031-77106-4_6.

W. A. Khan et al., *Computer Aided Engineering Design and Manufacturing*, Springer Series in Advanced Manufacturing, https://doi.org/10.1007/978-3-031-77106-4_6

software forms effective computer-aided engineering (CAE) solutions that can potentially constitute one of the backbones of smart factories in the era of 4th Industrial Revolution.

One of the challenges addressed in the literature appears to be how these computational methods can be integrated into the smart factory concept with the capability of real-time solutions. Depending on the computational power as well as the size of the problem at hand, analysis time can take longer than desired which may cause troublesome lags and setbacks in production. Powerful and computationally fast analysis tools can undoubtedly contribute to the decision-making for design alterations and effectiveness of the smart factory; however, technical details and feasibility of this integration remain to be an arduous matter that warrants further research and development.

This chapter briefly introduces the techniques, the hardware and software required, and various solutions for different phases of the materials of the products. Choice of method adequate to the problem and guidelines for successful analysis along with common pitfalls to be avoided are provided. The challenges and the opportunities are discussed to equip the reader to think about and generate ideas for the successful amalgamation of these techniques in the realization efforts for the effective smart factory.

6.2 Finite Difference Method (FDM)

The Finite difference method is a fundamental mathematical approximation scheme that is used to numerically evaluate the derivative of a boundary value problem at a specified point. The methodology is based on expressing a function through a Taylor series expansion, where a selected function g (of a single independent variable x) is varied with a sufficiently small increment, h, as can be seen in Eq. 6.1.

$$g(x_0 + h) = g(x_0) + g'(x_0)h + \frac{1}{2!}g''(x_0)h^2 + \frac{1}{3!}g'''(x_0)h^3 + \cdots \tag{6.1}$$

In a similar fashion, it can also be written as:

$$g(x_0 - h) = g(x_0) - g'(x_0)h + \frac{1}{2!}g''(x_0)h^2 - \frac{1}{3!}g'''(x_0)h^3 + \cdots \tag{6.2}$$

- **Forward Difference Approximation**

Using Eq. 6.1, neglecting the terms with second and higher order derivatives (as the exponential values of a too small h in those Taylor series expansion terms shall get even smaller), a forward difference approximation is constructed as follows:

$$g'(x_0) \approx \frac{g(x_0 + h) - g(x_0)}{h} \tag{6.3}$$

- **Backward Difference Approximation**

Using Eq. 6.2, again neglecting the terms with second and higher order derivatives, this time a *backward difference approximation* is formulated as follows:

$$g'(x_0) \approx \frac{g(x_0 - h) - g(x_0)}{-h} \tag{6.4}$$

- **Centered (Central) Difference Approximation**

If the arithmetic mean of Eqs. 6.3 and 6.4 are taken, a *centered (*or also known as *central) difference approximation* is achieved:

$$g'(x_0) \approx \frac{g(x_0 + h) - g(x_0 - h)}{2h} \tag{6.5}$$

In all these finite difference methods above, selection of a sufficiently small increment, h, plays an important role in the relatively closer estimations of the numerical derivations taken.

Example 6.1 For the function $g(x) = x^2 \cos(x)$, estimate the value of $g'\left(\frac{\pi}{6}\right)$ by using all three finite difference approximations, and compare against the exact solution by providing the absolute percentage errors.

Solution

The derivative of the function $g(x) = x^2 cos(x)$ can be found as below:

$$g'(x) = x(2\cos(x) - x\sin(x))$$

First derivative $g'(x)$ is to be evaluated at $x_0 = \frac{\pi}{6}$ with two different h values (i.e., $h = 0.005$ and $h = 0.01$) to demonstrate the importance of choosing a significantly small increment, h. The exact solution as well as the estimations through use of all three finite difference approximations, and the associated absolute percentage errors are given in Table 6.1.

6.2.1 Finite Difference Approximations for Partial Derivatives

Employing a centered difference approximation, a sample partial derivative at given points x_0 and y_0 can be evaluated as follows:

$$\frac{\partial u}{\partial x}(x_0, y_0) \approx \frac{u(x_0 + h, y_0) - u(x_0 - h, y_0)}{2h} \tag{6.6}$$

Table 6.1 Exact and approximate solutions for $g'(\pi/6)$

for h = 0.005	**Solution**	**Absolute error**
Exact	0.769822	
Forward diff. approx	0.770917	0.142%
Backward diff. approx	0.768680	0.148%
Centered diff. approx	0.769799	0.003%
for h = 0.01	**Solution**	**Absolute error**
Exact	0.769822	
Forward diff. approx	0.771966	0.278%
Backward diff. approx	0.767492	0.303%
Centered diff. approx	0.769729	0.012%

$$\frac{\partial u}{\partial y}(x_0, y_0) \approx \frac{u(x_0, y_0 + h) - u(x_0, y_0 - h)}{2h} \tag{6.7}$$

6.2.2 Finite Difference Approximations for Second Derivatives of Functions

Utilizing the Taylor series expansion terms up to and including the term containing the second derivative of the function $g(x)$, a centered difference approximation yield the following equation:

$$g''(x_0) \approx \frac{g(x_0 + h) - 2g(x_0) + g(x_0 - h)}{h^2} \tag{6.8}$$

With use of Eq. 6.8, the second partial derivative of the function $u(x, y)$ can be approximated. As an example, second partial derivative of u with respect to the independent variable x can be shown as follows:

$$\frac{\partial^2 u}{\partial x^2}(x_0, y_0) \approx \frac{u(x_0 + h, y_0) - 2u(x_0, y_0) + u(x_0 - h, y_0)}{h^2} \tag{6.9}$$

6.3 Finite Element Method (FEM)

Finite element method (FEM) is one of the most common, highly efficient, powerful, and cost-effective numerical methods which can be applied to many engineering and basic sciences problems ranging from structural analysis of any system or product (such as vehicles, buildings, machines, mechanisms, etc.) to thermo fluidic, heat transfer, mass transport, and electromagnetism. FEM is based on discretizing the medium of the problem at hand into several finite elements, within each the elemental governing differential equation is defined. An assembly of the elemental equations provides an overall mathematical model of the problem over the whole domain and solved numerically in an approximate manner providing detailed results at every infinitesimal point as necessary.

Typical steps in FEM algorithm consist of three distinct stages:

i. Preprocessing: following the conceptualization of the problem, involving all modeling steps from geometry and mesh generation to material properties and boundary conditions
ii. Solver: solution of the assembled matrix equation, and obtaining the necessary unknowns
iii. Post Processing: containing predictions, results, plots, etc.

A schematic of the typical steps in the FEM algorithm is illustrated in Fig. 6.1.

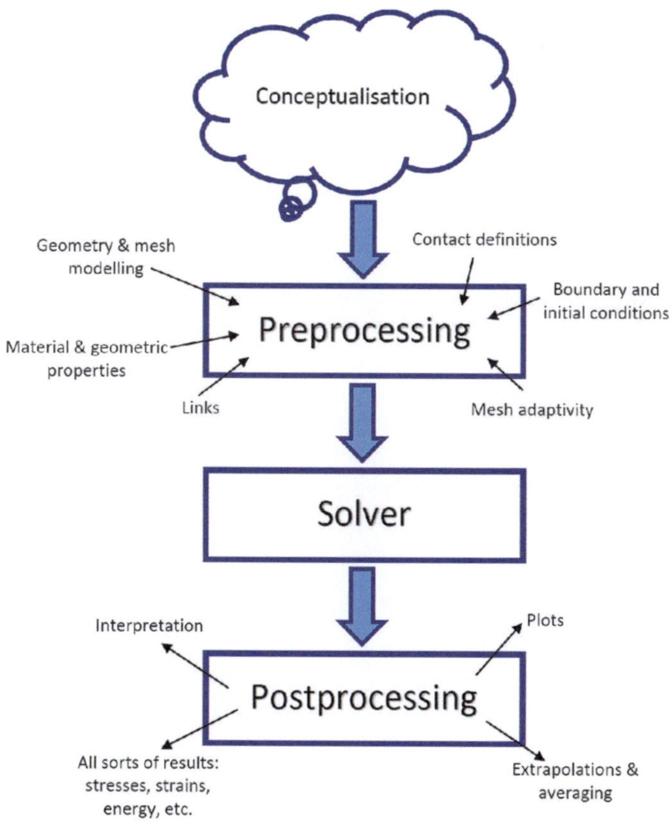

Fig. 6.1 Modeling and analysis process in FEM: the three stages

When FEM is practically applied to real-life problems, it is referred to as *finite element analysis* (FEA). There are several commercially available FEA software available that are used vigorously both in industry and academia. Some of these widely preferred FEA software packages can be listed as *ANSYS, Abaqus, Hexagon MSC Marc/Mentat, COMSOL Multiphysics, Hexagon MSC Nastran, Hexagon MSC Patran, LS-DYNA, Altair HyperWorks, Femap, OpenFoam, Autodesk Inventor Nastran, Autodesk Fusion 360, and SimScale.*

In FEM, discretization of the domain is called a mesh. Mesh is composed of *elements* and *nodes*. Elements are the subdomains that seamlessly constitute the whole physical domain when combined like the pieces of a puzzle. Corners of the elements are the nodes, at which the primary unknown of the problem (i.e. $u(x)$ in a one-dimensional (1D) problem) are solved. Some advanced element classes can contain additional nodes in the middle of the edges, faces, or volumes, too. The method employs interpolation (shape) functions to figure out the results within the element via interpolation using the found nodal unknowns. Common 1D and two-dimensional (2D) element classes including linear, triangular, and quadrilateral geometries with various node configurations are given in Fig. 6.2. In Fig. 6.3, some three-dimensional (3D) element classes of hexagonal and tetragonal geometries with several node modalities can be seen. Mesh definition is complete when element types are assigned to the chosen element classes. Generally, FEA software prefers to designate numbers that define certain element types. An element type uses a relevant element class and can simulate certain realistic behavior for the selected material.

In FEA, 1D, 2D, and 3D meshes can be generated using the necessary number of elements for all selected domains in all areas of basic sciences and engineering. Examples can range from a mechanical system such as a railway bogie, or a biological organ or tissue such as the human brain as can be seen in Fig. 6.4.

After completion of preprocessing, provided that a successful model devoid of any errors is created, the assembled matrix equations are solved at the solver stage, working out the nodal unknowns. These unknowns are then used to predict all the rest of the desired results either from the matrix equations, if possible, or from the mathematical/physical definitions. As an example, prediction of strains and stresses from the primary unknowns of displacements/deformations in different states of a bending problem is given in Fig. 6.5.

6.3.1 FEM Theory: Procedure for a Sample 1D Problem

It is a good idea to start learning FEM with its simplest form; namely, through 1D problems. The procedure can easily be applied to 2D and 3D problems by carefully adapting the necessary parameters. Developing a FEM procedure for a 1D problem starts with mesh generation. The physical domain of the problem (i.e. $0 < x < L$) is discretized into a finite number of elements with associated number of nodes. At this step, necessary geometric properties (such as cross-sectional areas, coordinates, etc.) can also be determined.

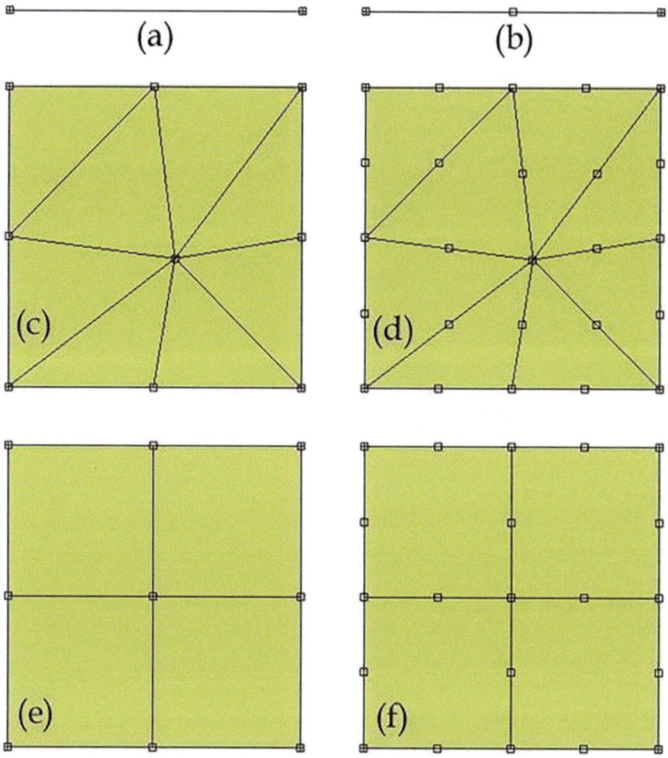

Fig. 6.2 Common 1D and 2D element classes; **a** linear—2 nodes, **b** linear—3 nodes, **c** triangular—3 nodes, **d** triangular—6 nodes, **e** quadrilateral—4 nodes, and **f** quadrilateral—8 nodes

Typical FEM elemental equations can be derived in three different ways, via; direct equilibrium methods, work/energy methods, or variational (weighted residual) methods. Here, a derivation procedure adopting a variational (weighted residual) approach is provided. The following model differential equation in Eq. 6.10 is assumed for our selected boundary value problem (such as deformation of elastic bars, heat transfer, electrostatics, etc.) at hand.

$$-\frac{d}{dx}\left(b\frac{du}{dx}\right) + mu - g = 0 \quad \text{for } 0 < x < \mathrm{L} \tag{6.10}$$

Parameters b, m, and g are data of the problem. For example, in the deformation of elastic bars problem, b is axial stiffness, and g is axial force. Boundary conditions are:

$$u(0) = u_0 \quad \text{and} \quad b\left(\frac{du}{dx}\right)_{x=L} = Q_0 \tag{6.11}$$

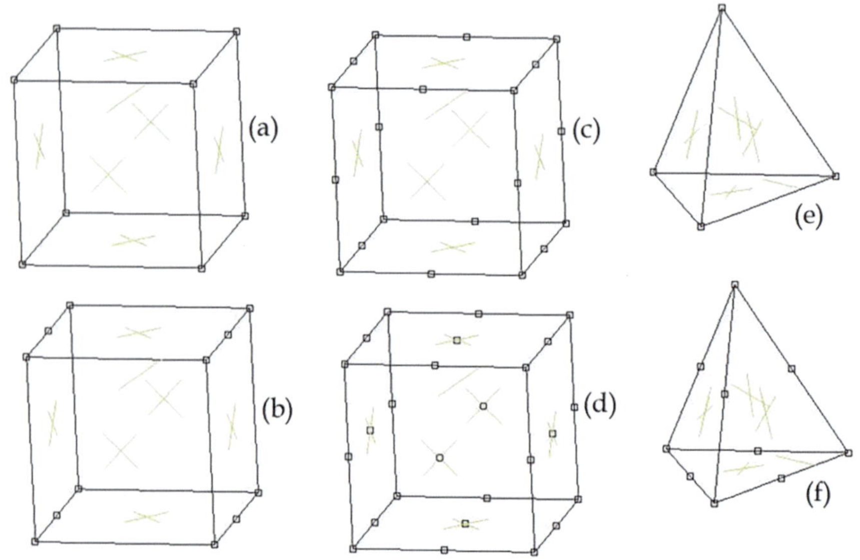

Fig. 6.3 Common 3D element classes; **a** hexagonal—8 nodes, **b** hexagonal—12 nodes, **c** hexagonal—20 nodes, **d** hexagonal—27 nodes, **e** tetragonal—4 nodes, and **f** tetragonal—10 nodes

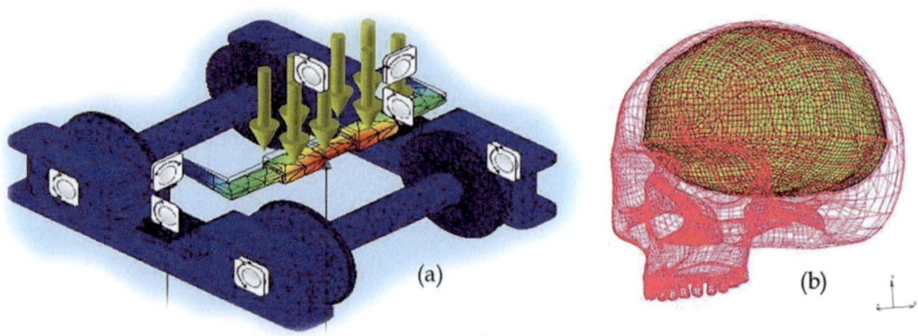

Fig. 6.4 Mesh examples formed with 3D brick elements: **a** a railway bogie, and **b** human brain inside the skull

Next step is to build the variational formulation of the equation above over a typical element. Generating the variational form, an equivalent weighted integral form is created utilizing mathematical methods such as integration-by-parts as in Eq. 6.12:

$$\int_{x_a}^{x_b} \left(b\frac{dw}{dx}\frac{du}{dx} + mwu - wg \right) dx - wx_a Q_1 - wx_b Q_2 = 0 \tag{6.12}$$

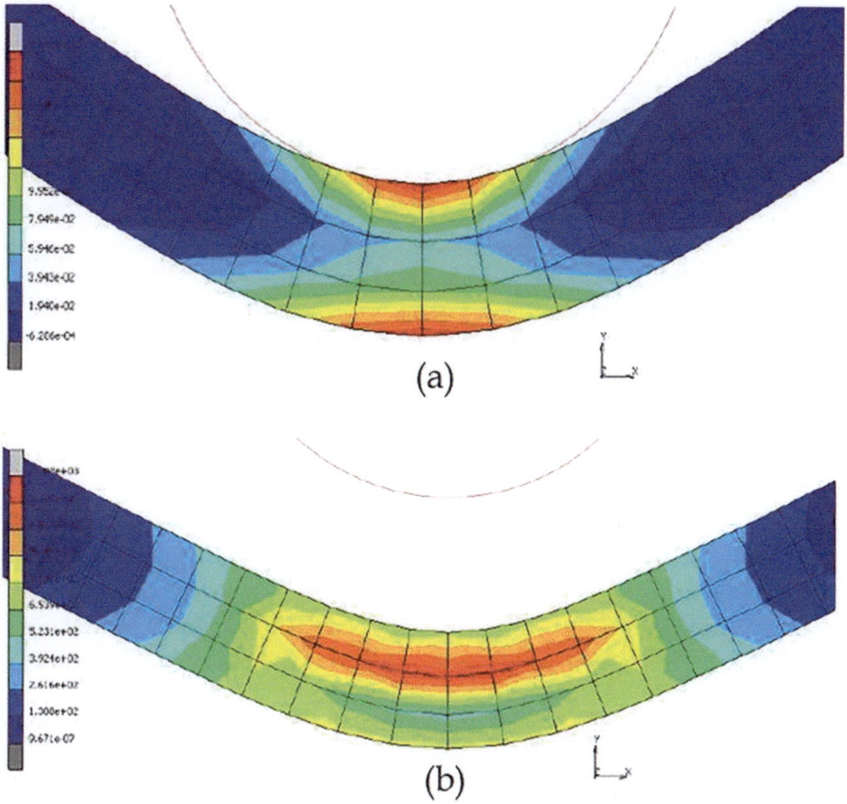

Fig. 6.5 Post Processing carried out on a sheet under air bending; **a** total equivalent plastic strains at the fully-loaded state; and **b** equivalent von Mises stresses at the unloaded state

where w is an introduced weight function, u is the primary unknown/dependent variable/solution, x_a and x_b are the nodal coordinates of a typical element, and Q_1 and Q_2 are the secondary variables such as point force, heat, flux, etc. The elemental solution u_h^e is approximated as follows:

$$u \approx u_h^e = \sum_{j=1}^{n} u_j^e \psi_j^e(x) \tag{6.13}$$

where ψ are systematically selected interpolation functions, and superscript e means elemental. In the next step, u's in Eq. 6.12 are substituted with u_h^e, and w's with ψ's in order to form the elemental FE equations. In the developed equations, relevant terms are grouped together which eventually provide the elemental matrix equation:

$$[K^e]\{u^e\} = \{g^e\} + \{Q^e\} \tag{6.14}$$

where K^e is the coefficient matrix, u^e is the solution (dependent variable), g^e is the source vector, and Q^e is the secondary variable. Satisfying inter-element continuity and equilibrium conditions at the nodes, and combining the right-hand side of Eq. 6.14 adequately, elemental equations are assembled into a single matrix equation representing the whole problem as given in Eq. 6.15:

$$[K]\{u\} = \{F\} \tag{6.15}$$

Boundary conditions are then reflected into the necessary cells corresponding to the relevant nodes in Eq. 6.15. Solution of Eq. 6.15 gives us the primary unknowns, u, after which secondary unknowns, Q, can be determined. Using the found unknowns, all desired results can be computed in the postprocessing stage.

6.4 Finite Volume Method (FVM)

Finite Volume Method, in short FVM, is one of the most widely and frequently utilized methods in several areas of computational physics, arguably first and foremost, fluid mechanics. Just as with the variational methods employed in FEM, FVM can tackle all differential equations written in divergence form, meaning containing mathematical divergence operators. The next step is to integrate the equation over the volume, hence the name FVM. The following step is generally the application of Gauss's theorem which transforms the volume integral over the divergence into a surface integral on the boundaries. The discretized subdomains in FVM are generally called *cells contrary* to the term *elements* in FEM. Therefore, the integral converts from integrating the differential of the dependent variable in the cells into surface integrals representing the fluxes of the dependent variable on the boundary of the cells.

Generation of mesh, or grid as widely used in the FVM context, is one of the essential and vital stages of the whole process. The quality of the grid, from volume element class and type selection to the number of cells used, determines the accuracy of the predictions when compared to the exact solution, if possible, at all. A coarse, rough grid may also cause instabilities during solving. In FVM, the created discrete volumes; namely, the *cells*, should stay stationary in space. Two-dimensional (2D) cells are generally in the form of quadrilaterals (e.g. square, rectangle, trapezoid, etc.) or triangles. In three-dimensions (3D), tetrahedrons and hexahedrons are the main cell types with either planar or curved surfaces.

There are many commercially available, highly powerful software that employ FVM solvers particularly for modeling and analyzing Multiphysics phenomena, including computational fluid dynamics (CFD) problems. A selective list of most widely used CFD software utilizing FVM solvers includes *ANSYS Fluent, ANSYS CFX, ANSYS Chemkin-Pro, Phoenics, Autodesk CFD, SOLIDWORKS Flow Simulation, SimScale, OpenFOAM, CFD Module, Siemens Simcenter, COMSOL Multiphysics, Flowsquare, ParaView,* and *Altair CFD*.

FVM operates on the principle of estimating the variables at distinct locations in the computational domain just as in FEM. Again, like FEM, FVM partitions a random geometric domain into a finite number of subdomains in the form of a structured or unstructured mesh/grid, followed by constructing control volumes known as a dual mesh/grid. The discretization of the random domain into subdomains and control volumes is carried out via two means: a cell-centered approach, and a vertex-centered approach. In a cell-centered approach, control volumes exactly match with the cells, whereas in vertex-centered approach, each node of the grid constitutes the center of a finite control volume, whose boundaries are attained by joining the centroids of each cell. These two approaches are shown in Fig. 6.6 for the 2D case. Each one of the commercial software mentioned above generally tends to choose and adopt one of these approaches.

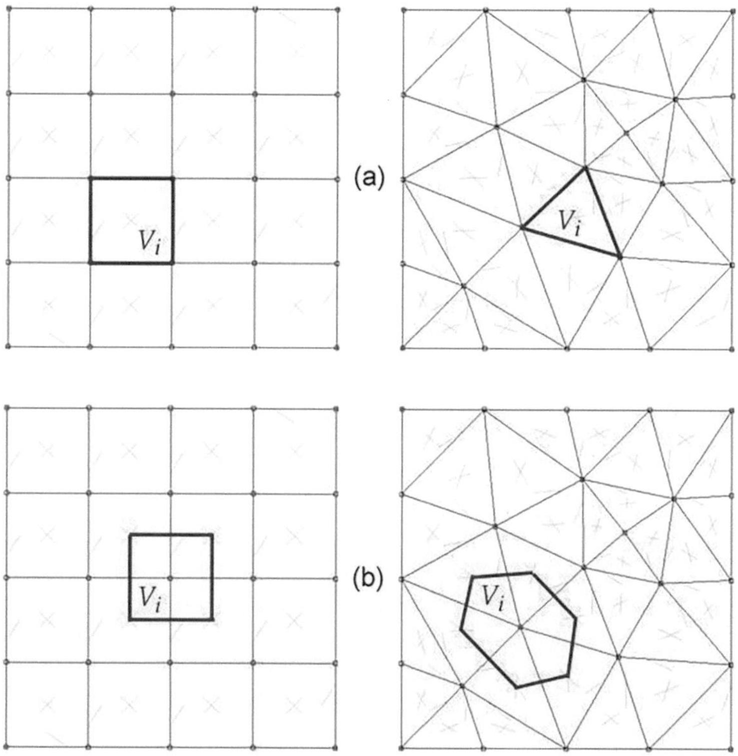

Fig. 6.6 Two-dimensional mesh/grid discretization created over the domain with quadrilateral and triangular cells, Vi's representing typical control volumes for: **a** cell-centered, and **b** vertex-centered configurations

Following the formation of the grid, each control volume requires an integral formulation of the balance equations. These equations are then approximated via numerical integration. As a model equation, the mass transport balance for an incompressible fluid can be utilized with the conditions of a fixed non-deformable grid. For further simplification, source terms can also be neglected as in Eq. 6.16 [1]:

$$\frac{\partial \rho}{\partial t} + \nabla \cdot \rho u = 0 \tag{6.16}$$

where ρ is the density, and u is the velocity vector. FVM tries to impose Eq. 6.17 for a typical small control volume with a volume of V_i, and a boundary surface S_i within the computational grid:

$$\int_{V_i} \frac{\partial \rho}{\partial t} dV + \int_{V_i} \nabla \cdot \rho u dV = 0 \tag{6.17}$$

As a fundamental difference when compared to FEM which uses weak forms of the governing differential equations, FVM employs discretization on the balance equation directly, imposing conservation properties that can be observed as the flux going into a given volume being the same as that leaving the bordering volume. Due to this characteristic, FVM has been widely accepted across academia and industry as a highly convenient platform for modeling and analyzing fluid mechanics, and heat and mass transfer phenomena. Equation 6.17 can be converted to the following form in Eq. 6.18 through taking the average of volume getting rid of the integral, and thus, obtaining the first term; and employing the Gauss theorem to the second term:

$$\frac{d\rho_i}{dt} + \frac{1}{V_i} \oint_{S_i} \rho u \Delta n dS = 0 \tag{6.18}$$

where n represents the normal vector to the boundary S_i, and reaches outwards from V_i. A finite difference scheme can be applied to approximately evaluate the function values and derivatives for the first term in Eq. 6.18:

$$\frac{\partial \rho_i}{dt} \approx \left(\frac{\rho - \rho^{n-1}}{\Delta t} \right) \tag{6.19}$$

where Δt is the time increment, and the index $(n - 1)$ is pointing to the previous increment. As used typically in integrations, the second term in Eq. 6.18 can be handled as a sum over all integration points of the control volume. As in FEM, assembly is carried out combining the equations written for each control volume in a systematic way, which then can be solved via numerical techniques.

6.4.1 Comparison of FVM with FDM and FEM

- An advantage of FVM reveals itself in flexible grid generation. In FVM, the shape of the cells can conform to the geometry of the domain, whereas in FDM, the elements (control volumes) must be equally shaped. Due to this fact, FVM appears to be a superior choice for more complex geometries with curved surfaces.
- Another advantage of FVM, with respect to FDM, appears in directly employing and satisfying the governing conservation of mass, momentum, and energy equations over the whole computational domain.
- FVM can be frequently called the *control volume finite element method*, the *cell-centered finite difference method*, or the *generalized finite difference method* due to its inherently existing similarities with FDM and FEM [1].
- The main difference between FVM and FDM approaches is that FDM is used for the flux rather than for the operator itself [1].
- The numerical methods of FDM and FEM are based on solving governing equations in the form of partial differential equations (PDEs), whereas FVM tackles the integral form of the conservation laws rather than their differential form. This fundamental difference makes FVM more suitable over FEM especially in CFD analysis, bringing more accuracy and stability to the numerical approximation of the fluid mechanics problem.

6.5 Boundary Element Method (BEM)

Boundary value problems for partial differential equations that govern physical phenomena can also be converted and resolved through boundary integral equations. Boundary element method (BEM) embodies a collection of methods that provide approximate computational solutions to the mentioned boundary integral equations. The approximate solution to the boundary value problem at hand provided by BEM, which is based on dividing the boundary into a limited number of elements, represents an exact solution of the given differential equation within the domain.

The BEM, also known as the boundary *integral equation method* (BIEM) is reported to be an effective tool particularly for free surface problems. BEM can be classified into two [2]:

- Direct Boundary Element Method (DBEM): based on a closed boundary for which physical variables (such as stress, normal velocity, etc.) can be placed only on one side of the surface (either interior or exterior)
- Indirect Boundary Element Method (IBEM): Both sides of the boundary surface can be utilized.

6.5.1 Comparison of BEM with FDM and FEM

Some of the strengths and shortcomings of BEM when compared to FDM and FEM are provided below [3]:

- As an advantageous aspect, only the boundary of the domain is required to be discretized. In other words, no need to discretize the interior domain, which is, on the contrary, a precondition in FEM. Due to this fact, mesh generation is less arduous, resulting in smaller scale system matrices.
- It is reported that exterior problems containing unbounded domains with bounded boundaries are resolved as efficiently as interior ones.
- In some cases, solutions of boundary integral equations can directly produce physically relevant data for the problem rather than by the solution through the interior of the domain. In these cases, on contrary to BEM, FEM often cannot provide highly precise results.
- In BEM, it is claimed that a greater convergence rate is achieved in estimating the solution within the interior of the domain.
- A disadvantage of pure BEM appears in not being attainable essentially for problems with inhomogeneities or nonlinear differential equations. However, at times, a coupling of FEM and BEM may turn out to be practical.
- BEM exhibits the requirement to treat singularities such as corners and edges on the boundary, or boundary conditions which are discontinuous, more straightforwardly than FEM.

6.6 Application of Computer-Aided Engineering Methods in Solid, Liquid, and Gas-Based Products

The methods discussed above have mostly been turned into practical design and analysis software, commercial or not, enabling the users to model, simulate, and analyze many basic sciences and engineering problems. FEM is often used for solid based products, whereas FVM integrated CFD software are preferred for liquid- and gas-based products. Some CFD software also utilize the FEM approach. FDM and BEM techniques are applied wherever necessary, depending on the physical nature of the problem, the governing equations most suitable defining the problem, and its boundary conditions. Some case studies are provided in this chapter for the reader to study to further comprehend the computer-aided numerical analysis techniques.

6.7 Case Studies

6.7.1 *Case Study—1[1]: Innovative Microfluidic Device for Enhanced Quality Control and Analysis in the Food Industry*

6.7.1.1 Introduction

Microfluidics is a system that refers to the manipulation of a small volume of fluids utilizing tiny channels that range in size from ten to hundreds of micrometers. Molecular analysis, molecular biology, biodefense, and microelectronics are all included in this interdisciplinary subject. To analyze and manipulate fluids in microscale devices known as microfluidic chips or lab-on-a-chip systems, it integrates ideas from physics, engineering, chemistry, and biology.

A microfluidic chip or gadget is a molded or etched pattern of microchannels. A few holes carved out through the microfluidic chip of various sizes connect this network of microchannels to the macroenvironment. Fluids are introduced into and removed from the microfluidic chip via these channels. To achieve multiplexing, automation, and high-throughput systems, fluids are directed, mixed, segregated, or manipulated. To accomplish the necessary characteristics (lab-on-a-chip, pathogen detection, electrophoresis, DNA analysis, etc.), the microchannel network architecture must be carefully designed.

Environmental sensing, medical diagnosis, drug discovery, drug administration, microscale chemical manufacturing, combinatorial synthesis and assays, artificial organs, micro propulsion, and microscale energy systems are application areas of microfluidics.

Microfluidic devices are widely used in the food and chemical sectors and have had a significant influence on clinical analytical procedures. Usually, the pre-treatment and injection of samples, species mixing, polymerase chain reaction, cell/particle separation, and counting are operations carried out using microfluidic chips. The methods for creating microfluidic devices are broken down into two categories: high-volume production (hot embossing, injection molding, and film or sheet operations) and low-volume production (casting, laminating, laser ablation, and 3D printing).

[1] Developed by Hafiz Muhammad Rizwan under guidance from Dr. Taqi A. Cheema using Licensed COMSOL Version. Reproduced with permission.

6.7.1.2 Design of Microfluidics Device

For several reasons, the design of microfluidic devices is important. Small fluid quantities (usually in the microliter to nanoliter range) are controlled and maneuvered using these devices inside of microchannels or chambers. For accurate fluid control, increased effectiveness, the integration of numerous functions, improved sensitivity and detection limits, high throughput, and customization, microfluidic device design is essential.

6.7.1.3 Governing Equations

Fluid flow in a rectangular channel is basically a Poiseuille flow and can be modeled by the incompressible Navier–Stokes equation.

$$\nabla \vec{v} = 0$$

$$\rho \cdot \frac{\delta \vec{v}}{\delta t} + \rho(\vec{v} \cdot \nabla)\vec{v} = -\nabla P + u\nabla^2 \vec{v} + \rho \vec{g}$$

where "ρ" is the mass density, "V" is the velocity of fluid, "P" is the pressure and "η" is the viscosity. By solving the Navier stokes equation the velocity field for the rectangular channel is given as,

$$v_x(y, z) = \frac{nh^2 \Delta P}{\pi^3 \eta L} \cdot \sum_{n,odd}^{\infty} \frac{1}{n^3} \left[1 - \frac{\cos\left(\frac{n\pi y}{h}\right)}{\cos\left(\frac{n\pi w}{h}\right)} \right] \sin\left(\frac{n\pi z}{h}\right)$$

For a rectangular channel, the Flow rate Q induced by the pressure difference ΔP is given by the following equation,

$$v_x(y, z) = \frac{h^3 w \Delta_P}{12\eta L} \cdot \left[1 - \sum_{n,odd}^{\infty} \frac{1}{n^5} \cdot \frac{195}{\pi^5} \cdot \frac{h}{w} \cdot \tanh\left(\frac{n\pi w}{2h}\right) \right]$$

Very useful approximate results can be obtained in the $\lim_{h/w \to 0}$ of a flat and wide channel, for which $\frac{h}{w} \cdot \tanh\left(\frac{n\pi w}{2h}\right) \to \frac{h}{w} \tanh(\infty) = \frac{h}{w}$, and Q becomes,

$$Q = \frac{h^3 w \Delta_P}{12\eta L} \left[1 - 0.0630 \frac{h}{w} \right], \ for\ h < w.$$

6.7.1.4 Hydraulic Circuit Analysis

Hydrodynamic resistance

An essential design factor for microfluidic devices that must reliably and accurately manage the transfer of fluid is hydraulic resistance, which is a measurement of a channel's resistance to the passage of a fluid flow. Flow rate Q in a channel is proportional to the applied pressure drop.

$$\Delta P = R_h \cdot Q$$

With "R_h" the hydrodynamic resistance. This expression is formally the analog of Ohm's law relation,

$$V = R.I$$

The following is the formulation for the hydraulic resistance for a rectangular cross-sectional channel,

$$R_h = \frac{12\eta L}{wh^3\left(1 - 0.630\frac{h}{w}\right)}, \quad h < w$$

Calculations

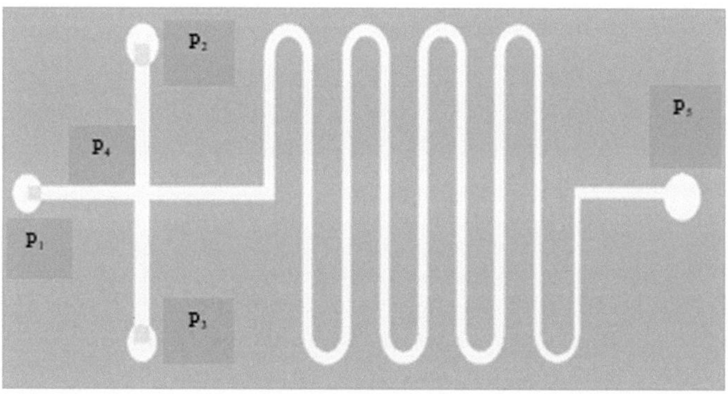

$$\mathrm{P_1} = \mathrm{P_2} + \mathrm{P_3} = 400000\ \mathrm{Pa}$$
$$\mathrm{P_5} = 100000\ \mathrm{Pa}$$

From the conservation of mass,

$$\mathrm{Q_1} + \mathrm{Q_2} + \mathrm{Q_3} = \mathrm{Q_4}$$
$$\Delta P = R_h \cdot Q$$
$$P_1 - P_4 = R_{h1}Q_1$$
$$P_2 - P_4 = R_{h2}Q_2$$
$$P_3 - P_4 = R_{h3}Q_3$$
$$P_4 - P_4 = R_{h4}Q_4$$

These relations can be written as one matrix equation,

$$\begin{bmatrix} 1\,0\,0 & 0 & 0 & 0 & 0 & 0 & 0 \\ 0\,1\,0 & 0 & 0 & 0 & 0 & 0 & 0 \\ 0\,0\,1 & 0 & 0 & 0 & 0 & 0 & 0 \\ 0\,0\,0 & 1 & 0 & 0 & 0 & 0 & 0 \\ 0\,0\,0 & 0 & 0 & 1 & 1 & 1 & -1 \\ 1\,0\,0 & -1 & 0 & -Rh_1 & 0 & 0 & 0 \\ 0\,1\,0 & -1 & 0 & 0 & -Rh_2 & 0 & 0 \\ 0\,0\,1 & -1 & 0 & 0 & 0 & -Rh_3 & 0 \\ 0\,0\,0 & 1 & -1 & 0 & 0 & 0 & -Rh_4 \end{bmatrix} \begin{bmatrix} 400000 \\ 400000 \\ 400000 \\ 100000 \\ 0 \\ 0 \\ 0 \\ 0 \\ 0 \end{bmatrix} = \begin{bmatrix} P_1 \\ P_2 \\ P_3 \\ P_4 \\ P_5 \\ Q_1 \\ Q_2 \\ Q_3 \\ Q_4 \end{bmatrix}$$

$$R_h = \frac{12\eta L}{wh^3(1 - 0.630\frac{h}{w})}$$

By putting values in the above relation of hydraulic resistance,

$$\mathrm{Rh_1} = \mathrm{Rh_2} = \mathrm{Rh_3} = 2.5 \times 10^{-5}\ \mathrm{Pa\,s/\mu m^3}$$
$$\mathrm{Rh_4} = 3.7 \times 10^{-4}\ \mathrm{Pa\,s/\mu m^3}$$

MATLAB Code: -

```
clc
clear
P1= 400000 %Pressure in Pascal.
P2= 400000 %Pressure in Pascal.
P3= 400000 %Pressure in Pascal.
P5= 100000 %Pressure in Pascal.
Rh1= 2.5e-5 %hydraulic resistance in Pa-s/μm^3.
Rh2= 2.5e-5 %hydraulic resistance in Pa-s/μm^3.
Rh3= 2.5e-5 %hydraulic resistance in Pa-s/μm^3.
Rh4= 3.7e-4 %hydraulic resistance in Pa-s/μm^3.

X=[1 0 0 0 0 0 0 0 0;
   0 1 0 0 0 0 0 0 0;
   0 0 1 0 0 0 0 0 0;
   0 0 0 0 1 0 0 0 0;
   0 0 0 0 0 1 1 1 -1;
   1 0 0 -1 0 -Rh1 0 0 0;
   0 1 0 -1 0 0 -Rh2 0 0;
   0 0 1 -1 0 0 0 -Rh3 0;


   0 0 0 1 -1 0 0 0 -Rh4]
Y=[400000;400000;400000;100000;0;0;0;0;0]
Z = linsolve(X,Y)
```

Result

$$Z = \begin{bmatrix} 0.004 \times 10^8 \text{ Pa} \\ 0.004 \times 10^8 \text{ Pa} \\ 0.004 \times 10^8 \text{ Pa} \\ 0.0039 \times 10^8 \text{ Pa} \\ 0.0010 \times 10^8 \text{ Pa} \\ 2.60 \times 10^8\ \mu\text{m}^3/\text{s} \\ 2.60 \times 10^8\ \mu\text{m}^3/\text{s} \\ 2.60 \times 10^8\ \mu\text{m}^3/\text{s} \\ 7.80 \times 10^8\ \mu\text{m}^3/\text{s} \end{bmatrix} = \begin{bmatrix} P1 \\ P2 \\ P3 \\ P4 \\ P5 \\ Q1 \\ Q2 \\ Q3 \\ Q4 \end{bmatrix}$$

6.7.2 *Fabrication of Microfluidic Device*

For microfluidic devices, microchannels are crucial components. The fabrication of the microchannel can be done in a variety of ways, including employing the silicon-based organic polymer polydimethylsiloxane (PDMS), femtosecond laser micromachining and inscription, or deep reactive ion etching (DRIE).

6.7.2.1 Fabrication Procedure

- The microchip was fabricated on inexpensive and commercially available microscope glass.
- The glass slides were initially subjected to a 4-h annealing process at 400 °C to relieve any remaining internal stresses.
- Hexamethyldisilane was employed as an adhesive primer and a 3 mm thick layer of positive photoresist (PR) served as the etching mask instead of metal layers for glass etching in the HF-based etchant.
- The substrate was then hard baked (150 °C, 10 min) after being soft baked (100 °C, 3 min) and UV patterned to create the desired pattern.
- Buffered oxide etchant (BOE) etchant solution was then applied to the patterned substrates.
- With BOE etching, channels with a depth of 40 could be produced.
- The glass substrates were then submerged in a warm, diluted KOH solution for a few seconds to remove the PR etch mask.
- In the meantime, fluid through holes were bored into the second glass slide with the identically printed microchannels which was then carefully aligned with the patterned glass.
- The two plates were then fused together using a thermal fusion bonding procedure that took place in a sintering oven at 670 °C for 10 min with a 5 °C/min ramp rate.
- Finally, Teflon tubing was attached to the fluid intake holes using epoxy (Fig. 6.7).

3D isometric view of device

The CAD model of microfluidic device is created on PTC Creo software (Figs. 6.8 and 6.9).

6.7.3 *Simulation Analysis*

Computational fluid dynamics simulation

CFD simulation of the pressure and velocity fields in the microchannels was performed in COMSOL Multiphysics. Physics of Laminar flow and stationary study

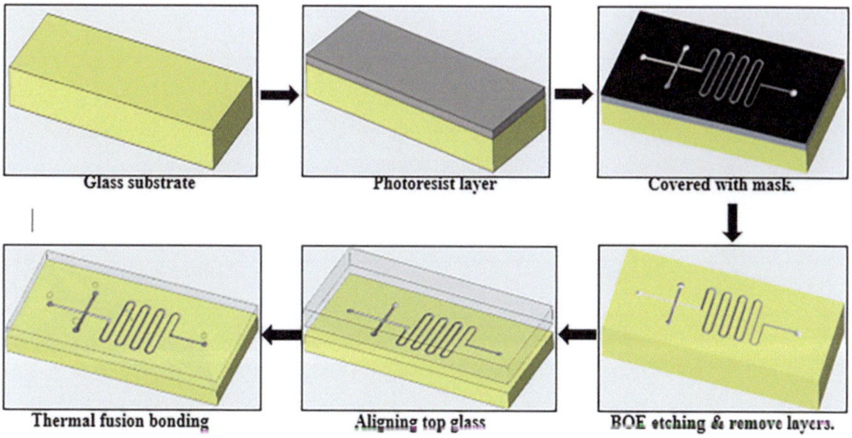

Fig. 6.7 Fluid Intake using epoxy

3D isometric view of device: -

The CAD model of microfluidic device is created on PTC Creo software.

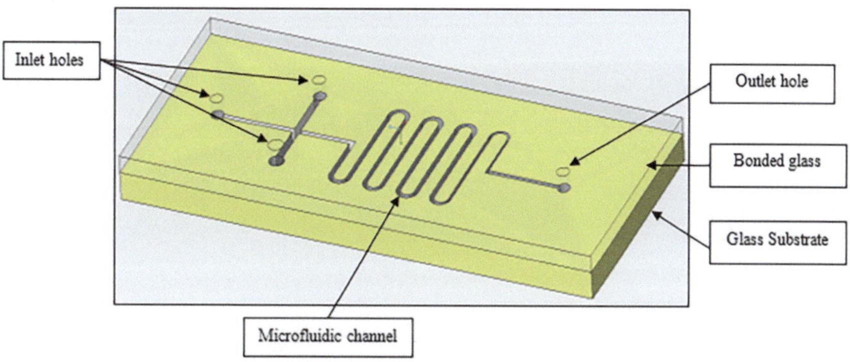

Fig. 6.8 3D Isometric view of device

is used to simulate the model to predict the pressure and velocity profiles of flow inside the microchannels.

CAD model and Meshing

Design of Microfluidic device is generated in COMSOL design builder and hexagonal meshing is used.

2D schematic; -

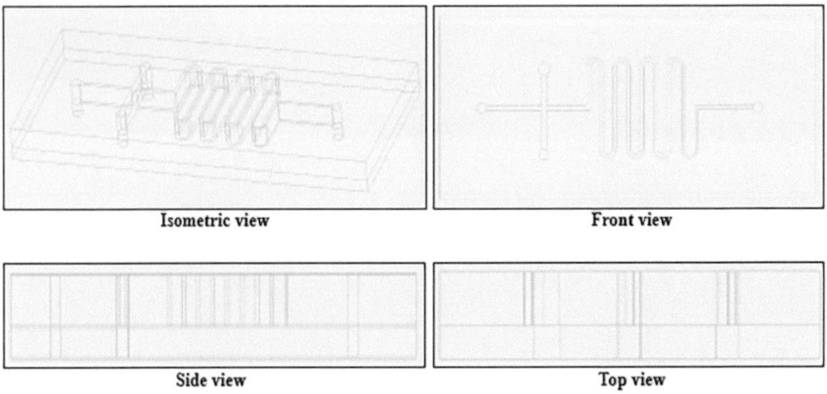

Fig. 6.9 Different schematic views

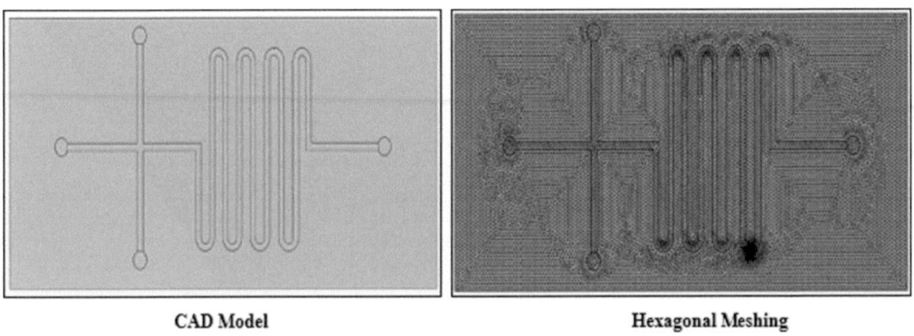

6.7.3.1 Velocity Contours

In the boundary conditions of laminar flow physics, the surfaces of inlet holes are selected, and a velocity magnitude is provided and low-pressure conditions are provided to the outlet hole surface (Figs. 6.10 and 6.11).

6.7.3.2 Pressure Contours

Pressure contour shows that at inlet holes there is high pressure, and that pressure distributes towards out let hole (Fig. 6.12).

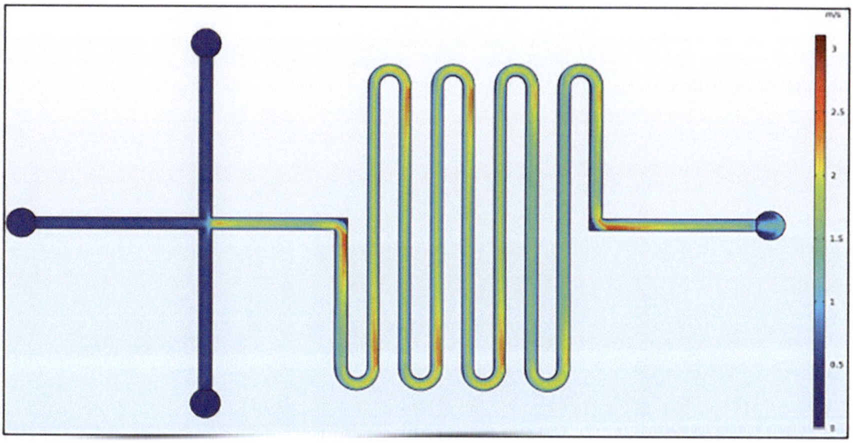

XY plane

Fig. 6.10 Description of velocity contours (XY plane)

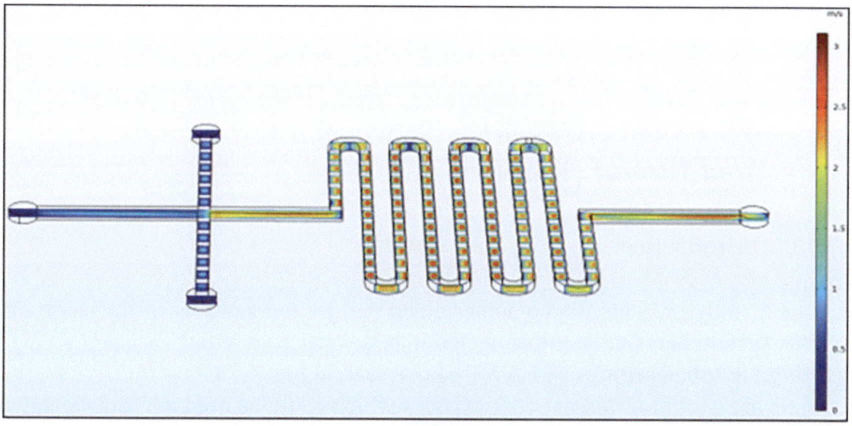

YZ plane

Fig. 6.11 Description of velocity contours (YZ plane)

Pressure Contours: -

Pressure contour shows that at inlet holes there is high pressure, and that pressure distributes towards outlet hole.

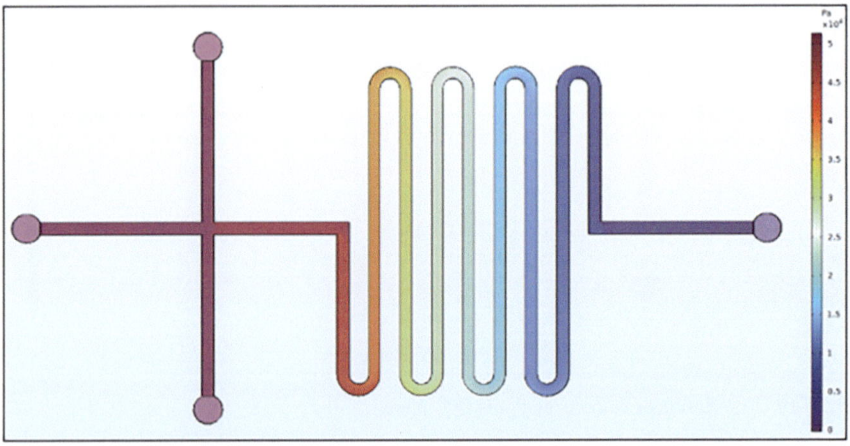

Fig. 6.12 Pressure contours description

6.7.4 Case Study—2[2]: Computer-Aided Design and Analysis of the Shaft and Gearbox System of a Vertical Axis Wind Turbine (VAWT)

6.7.4.1 Introduction

This case study presents the computer aided design and analysis of the shaft and gearbox system that can be implemented in a Vertical Axis Wind Turbine (VAWT) to be used in urban settings. A VAWT can be seen in Fig. 6.13.

There are a lot of different configurations which could be used when designing a VAWT shaft and gearbox. In this section, different implementations used by manufacturers, and the ideal conditions for each of the mentioned implementations are discussed.

As a general overview, there are two main types of VAWT designs: namely, Savonius and Darrieus. Savonius uses 2–3 blades which extend down like scoops and generate rotation with drag. They generally have an S-shaped cross-section. Darrieus uses the lift-type design with split blades which are either straight or curved. Of the two, Darrieus has a higher power coefficient.

[2] Produced by Zuhayr Amin, Talal Amin, Walid K. Badwi Albitar, Melisa Altuntas at Middle East Technical University, Northern Cyprus Campus, Türkiye, under the guidance of Volkan Esat. Reproduced with permission.

Fig. 6.13 A Darrieus type VAWT

Some commercial VAWT products are listed in Table 6.3. The rated wind speed is the speed at which the turbines generate their desired output of power. The survival wind speed is the maximum speed the turbines can handle before damage occurs. These values are used in calculations for shaft design. The Hoyi 300 and Vision Air 3 utilizes a helical Darrieus configuration, and the DS-700 uses a combination of both Savonius and Darrieus rotor types in a hybrid setup [6.CS2.2, 6.CS2.3, 6.CS2.4].

Gearbox systems used in VAWTs are more specifically known as Right Angle Gearboxes. This is because the generators for VAWTs are placed on the ground, while the shaft rotates perpendicularly with respect to the ground. There are multiple gear systems that can be used to change the direction of torque. According to Shigley's Mechanical Engineering Design [6.2.5], such an effect can be observed by bevel gears, spiral bevel gears, worm gears, hypoid and spiroid gears.

6.7.4.2 Theory and Methodology

Before designing any mechanical device, it is imperative to first understand the conditions it will be used in and the loads it will have to endure. This will not only ensure that the designed device can fulfill its purpose, but that it can do so at the minimum required cost.

In this design effort, a vertical shaft with two bearings is constructed. In addition to the bearing at the bottom, the shaft is connected to a bevel gear, which will transfer torque to the gearbox. The second bearing is placed just above this bevel gear. Calculations are done for different gear configurations to determine the optimum setup necessary for this application.

Since the goal is to design a VAWT system which can be applied in a wide variety of urban settings, data on the average height of buildings in some of the largest cities in the world are found as can be seen in Table 6.4.

Based on the data displayed in Table 6.2, it would be appropriate to assume an average wind height of 10m in the calculations on the shaft and gearbox system. Engineering-based methods for measuring the flow of air report errors in the range of 5–20%. Compared to CFD methods which report errors of around 10%, the engineering methods are quite unpredictable; therefore, researchers prefer the CFD methods in developing models and are working on making the predictions of such software more accurate [6.CS2.7].

Table 6.2 Parameters dimensions

Dimension table	
Parameters	Dimensions
Channel width	100 μm
Channel height	40 μm
Inlet/outlet holes diameter	0.3 cm
Length (L1)	1 cm
Length (L2)	1 cm
Length (L3)	1 cm
Length (L4)	15 cm
Substrate width	2.6 cm
Substrate length	5.5 cm

Table 6.3 Commercial VAWT products

VAWT model	Weight (kg)	Blade material	Rated wind speed (RPM)	Rated power (W)	Survival speed (m/s)
Hoyi 300	49	Fiberglass	300	300	50
Vision Air 3	274	Fiberglass	200	1500	50
DS-700	60	Anodized aluminum	405	700	60

Table 6.4 Number of buildings and average height of buildings in various cities

City	Number of buildings	Average height (m)
San Francisco, CA	151,787	9.7
Boston, MA	80,409	9.0
New York City, NY	1,066,354	8.4
Austin, TX	452,439	7.0
Vancouver, BC	111,024	6.8

Table 6.5 Average wind speed in multiple cities at a height of 10m

City	Average Wind Speed (m/s)
San Francisco, CA	4.20
Boston, MA	4.93
New York City, NY	5.55
Austin, TX	4.14
Vancouver, BC	3.64

Having determined that the optimum location for a wind turbine in an urban setting is the roof of a building, the wind speeds that can be expected by the VAWT shall be figured out. An online directory known as the Global Wind Atlas [6.CS2.8] is used which has compiled wind data from around the world at different heights: 10 m, 50 m, 100 m, and 200 m. The wind data for all the cities mentioned in 6.CS2.2 shows that the average wind speed is similar throughout the different cities as in 6.CS2.3 (Table 6.5).

6.7.4.3 Forces on Shaft

The shaft is responsible for holding the weight of the blades. However, it is also responsible for withstanding the torque generated by the rotation of the blades. Finally, the force exerted by the wind directly on the shaft is also considered; however, it is expected to be negligible when compared to the other operating loads.

Before calculating the force exerted by the wind, air mass, A_m, of the system is determined by using Eq. 6.20:

$$A_m = A\rho \tag{6.20}$$

where A is the surface area in contact with the air, and ρ is the air density.

Using the air mass, wind speed is converted into a force using Eq. 6.21:

$$F = A_m * v^2 \tag{6.21}$$

where, v is the wind speed and A_m is the air mass [6.9].

There are two wind forces which need to be considered in the analysis. The first wind force acts directly on the surface of the shaft, whereas the second wind force causes the blades to rotate and generate torque around the shaft.

The possibility of improper installation must also be acknowledged. This would lead to eccentricity, which causes centrifugal force deflections. These deflections are resisted by the flexural rigidity of the shaft's material; however, it gives us a critical speed value for the designed shaft. According to [6.5], the shaft is considered safe, if the critical speed is at least twice as large as the operating speed. The critical speed can be determined using Eq. 6.22:

$$\omega = \left(\frac{\pi}{l}\right)^2 \sqrt{\frac{EI}{m}} \tag{6.22}$$

6.7.4.4 Forces on Gearbox

The stresses caused by the wind in the gearbox must be estimated; however, a negligible effect on the gearbox is expected. The main consideration in the gearbox is the inter-gear forces and the stresses that they generate. The gear ratios need to be adapted to ensure that the gearbox can generate the torque necessary to power up the generator.

The stresses to be firstly considered are the *Fundamental Contact Stresses* using Eq. 6.23:

$$\sigma_H = Z_E \left(\frac{1000W^t}{bdZ_1} K_A K_v K_{H\beta} Z_x Z_{xc} \right)^{\frac{1}{2}} \tag{6.23}$$

where all the *K and Z* values represent different AGMA Equation Factors [6.5].

The *Permissible Bending Stresses* are also considered via Eq. 6.24:

$$\sigma_{FP} = \frac{\sigma_{F\,\lim} Y_{NT}}{S_F K_\theta Y_z} \tag{6.24}$$

Using the various tables and catalogs available in the 9th Edition of Shigley's Mechanical Engineering Design [6.5], optimum design for a gearbox can be carried out which can handle the applied loads and torques.

Since the various AGMA factors will change based on design decisions, the exact values used in calculations will be mentioned in the Results and Discussion Section. This allows us to compare the effects of different values used in different design concepts.

6.7.4.5 Modelling and Simulation

When initially setting the shaft diameter, it is based on the ratio of shaft diameter to blade diameter of 5%. The blade diameter is retrieved from a VAWT company catalog. After the initial simulations, the shaft is found to be more than sufficient; thus, to reduce weight and material usage, a 70 mm hole is extruded in the shaft. This allows the structure to maintain its structural integrity and reduce material usage.

In the initial shaft analysis, a free-body diagram of the shaft is made conducting force analysis at the two bearing locations used on the shaft. The forces considered are:

- Weight of the shaft and blade
- Direct force on the shafts and the blades
- Torque by the blades on the shaft.

Using a free-body diagram, the total axial and radial forces are determined at each bearing location. Knowing the angular speed of the shaft, all the information to make use of the bearing design tool in Autodesk Inventor are collected. This tool allows the engineer to select a class of bearing, following which, the tool goes through every bearing in Inventors catalog, at the chosen diameter, and determines which bearings are suitable. This tool also calculates the expected bearing life values. The bearing chosen by Inventor is the optimum bearing, meaning the one with the lowest factors of safety at the expected life. However, the decision here is made for the bearing that has a reasonably large life value.

Static analysis is carried out on the shaft using 3 different materials: AISI 4340, AISI 1022, and ASTM A34 steels. All three attained infinite life in the simulations as the maximum Von Mises stresses turned out to be quite low. Next, the design of bevel gears with a spur gear gearbox are accomplished. Based on the calculations, it is determined that the gear ratios between the gear sets should be kept at 5 (which is close to the upper limit for spur gears) to maintain the number of gears used at a minimum. Keeping the minimum number of gears allows the engineer to minimize the points of failure in the system.

The bevel gear on the shaft, used for torque conversion, requires a ratio of 3.2 to have the output speed to be sufficient for the generator. The spur gears are designed, based on the torque and speed of the shaft, using different materials and running multiple simulations. Again, the material with the highest life is chosen along with increasing the face width of the gear due to the high gear ratios.

To compare different configurations, bevel gears are then used in the gearbox using the same analysis to check its performance. For all the bevel gears used, a helix angle of 22° is chosen to reduce the contact stresses in between the gears. Various angles are tested; however, 22° appears to be the optimum selection for the current design. Like the spur gearbox design, Autodesk Inventor's Bevel Gear Calculator tool is utilized to determine the optimum material for each gear pair.

6.7.4.6 Results and Discussion

Based on the city data mentioned in Tables 6.2 and 6.3, it is decided to use New York as the model city. Based on data published by MeteoBlue [6.10], the strongest wind speeds in New York are modeled to be close to 61 km/h. Using the air pressure data for New York [6.11], it is assumed that the average pressure is like the standard air pressure of 101.3 kPa. Therefore, the standard value for air density is used. Based on these conditions, operating conditions are mentioned in Table 6.6.

The diameter and swept area of VAWT can vary widely based on the blade design. A 3-blade Darrieus turbine design is considered assuming adequate values for the rotor diameter, height, and swept area as tabulated in Table 6.7. The reason to prefer a Darrieus configuration is because they generate electricity at a lower cost and with slower wind speeds, compared to a Savonius setup [6.12].

Based on the turbine parameters, the shaft parameters are decided. The height will be the same as the turbine; however, the recommended 5% ratio is used for the diameter. The shaft parameters can be seen in Table 6.8.

Using the data mentioned in Tables 6.6 and 6.8, the forces acting on the shaft are determined. The moments caused by the rotation of the blades are considered for all three blades. The free-body diagram can be seen in Fig. 6.14.

The loads under normal operating conditions are mentioned in Table 6.9, whereas the loads under extreme conditions are provided in Table 6.10. Based on the wind speeds mentioned in Table 6.6, the operational angular speed of the low-speed shaft can be estimated. These values can be seen in Table 6.11:

Table 6.6 Operational Parameters

Parameter	Value
Air density	1225 kg/m^3
Operational height of turbine	10 m
Average wind speed	5.55 m/s
Highest wind speed	17 m/s

Table 6.7 Turbine parameters

Parameter	Value
Height	3.5 m
Diameter	2.8 m
Swept area	10.5 m^2
Blade weight	110 kg

Table 6.8 Shaft design parameters

Parameter	Value (m)
Height	3.5
Diameter	0.14

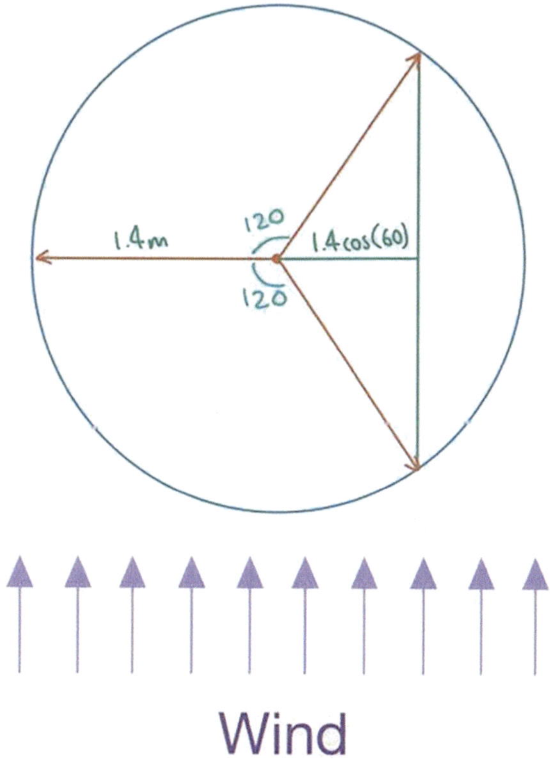

Fig. 6.14 Schematics for the top view of VAWT

Table 6.9 Loads under normal operating conditions

Load	Value
Moment (by blades)	1109.352 N m
direct Load	29.04 N

Table 6.10 Loads under extreme conditions

Load	Value
Moment (by blades)	10,408.335 N m
Direct load	272.490 N

Table 6.11 Average and maximum values for the rotational speed of the low-speed shaft

Rotational speed	Value (rpm)
Average	37.86
Maximum	115.96

After determining these parameters, the free body diagram of the shaft is drawn where two bearings are assumed to support the shaft (Fig. 6.15). These bearings are placed at the base of the shaft and just above a bevel gear. This bevel gear is used to transfer torque to the gearbox. The bevel gear design is discussed in further detail when designing the gearbox.

For the design of the bearings, the Bearing Design Calculator tool in Autodesk Inventor is employed. This tool allows to enter the axial and radial loads at a given angular speed and would check all the bearings in Inventor's library to find the most suitable ones. An example prompt can be seen in Fig. 6.16.

The first bearing designed is the one to be used at the base of the shaft. According to the calculations, none of the tapered thrust bearings, roller thrust bearings, or thrust ball bearings could be used. At a diameter of 140 mm, there are only 2–3 options for each category, and none of them are sufficient to withstand the expected loads. However, it is found that angular contact bearings are sufficient for this application. Table 6.12 details three angular contact bearings.

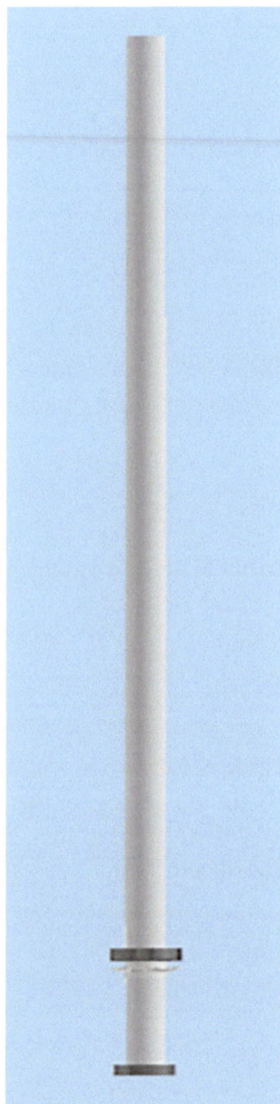

Fig. 6.15 Shaft after adding the bearings and driving bevel gear

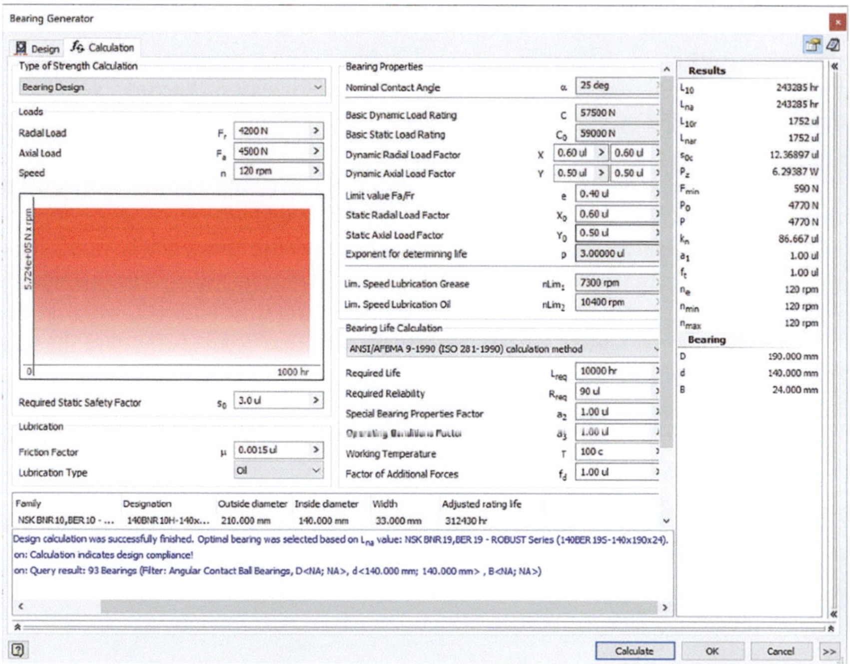

Fig. 6.16 Inventor's bearing design calculator tool

Table 6.12 Angular contact bearings to be used at the base of the shaft

Bearing	Expected life (years)
NSK 79—7928C	4
NSK 70—7028C	19
BS 292: Part (II) Metric—7028	36

Based on this table, the BS 292 bearing is chosen due to its considerably longer life. It should be noted that the prices for any of the bearings could not be determined at the time; therefore, the selection is based purely on performance.

Running a similar simulation for the bearing above the bevel gear, a deep groove ball bearing is preferred, since the angular contact bearing is expected to hold all the axial loads generated by the weights of the shaft and blades. For this application, the ideal gear is determined to be JIS B 1521-6228. This gear is preferred as its life is also in the order of 30 years. This would mean that both gears on the shaft would need to be changed around the same time, thus reducing the maintenance downtime of the system over its life. The JIS bearing is placed at 365 mm from the base of the shaft, giving it a little clearance above the bevel gear.

When performing shaft analysis, the results of the materials listed below are compared: A36 Steel, AISI 1022, and AISI 4340. These are applied to the shaft along with the axial and radial loads to study three specific properties: von Mises stresses experienced by the shaft, factor of safety, and failure time. These results can be seen in Figs. 6.17, 6.18, and 6.19.

The data of the strength-life fatigue curves are imported into the Inventor Nastran environment as in Fig. 6.20.

For the vibrational analysis, twenty modes are calculated. Fortunately, none of the vibrational modes coincided with the frequencies of wind under the given operational conditions. The frequencies caused by wind range from 0.5 to 2.0 Hz, whereas, for all materials, the first natural frequency occurs at around 0.00009 Hz, and the next

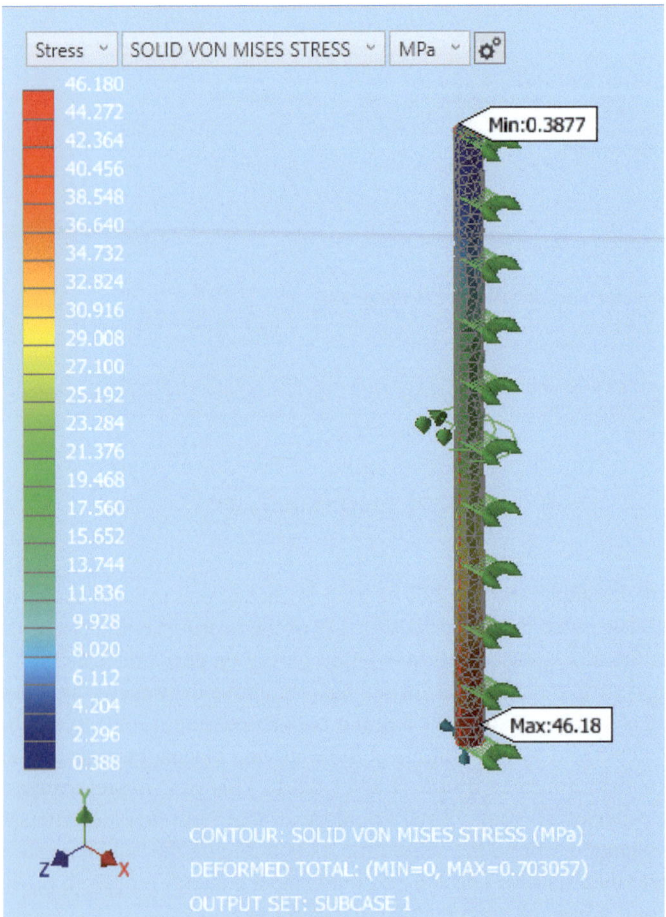

Fig. 6.17 A36 steel static stress analysis values

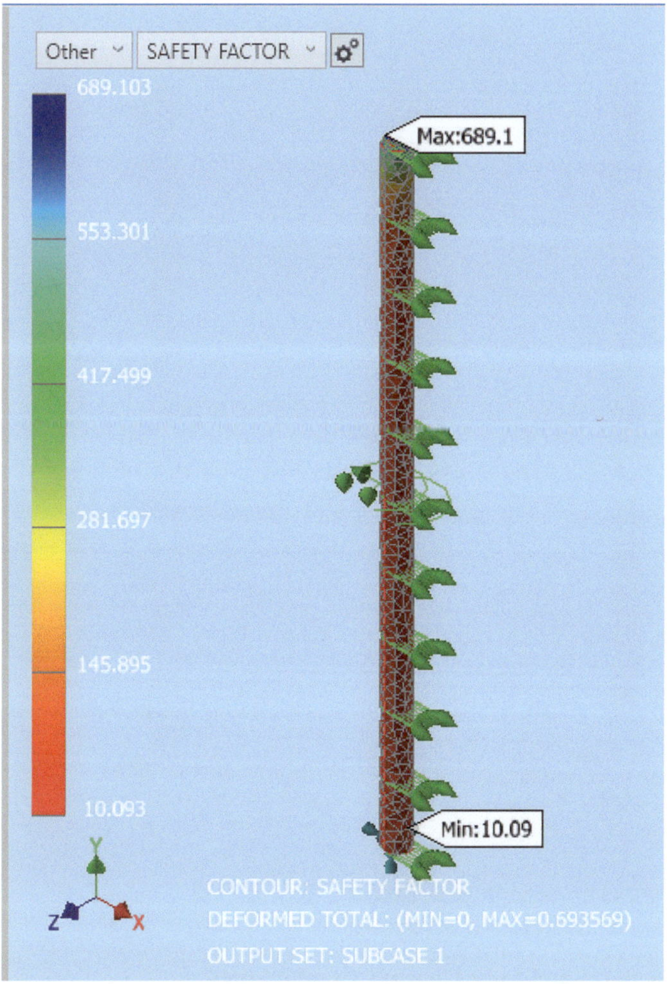

Fig. 6.18 AISI 1022 factor of safety values

mode occurs around 9 Hz (Fig. 6.21). Therefore, resonance should not be an issue under the current operational conditions.

As can be seen in Table 6.13, von Mises stresses are similar across all shafts; however, factors of safety for the AISI steels are significantly larger. The fatigue life of all materials is infinite since the maximum von Mises stress is well below the endurance limit for all materials. The natural frequencies for all three materials are also very similar. The frequencies are significantly larger than the expected frequency of the wind in New York City.

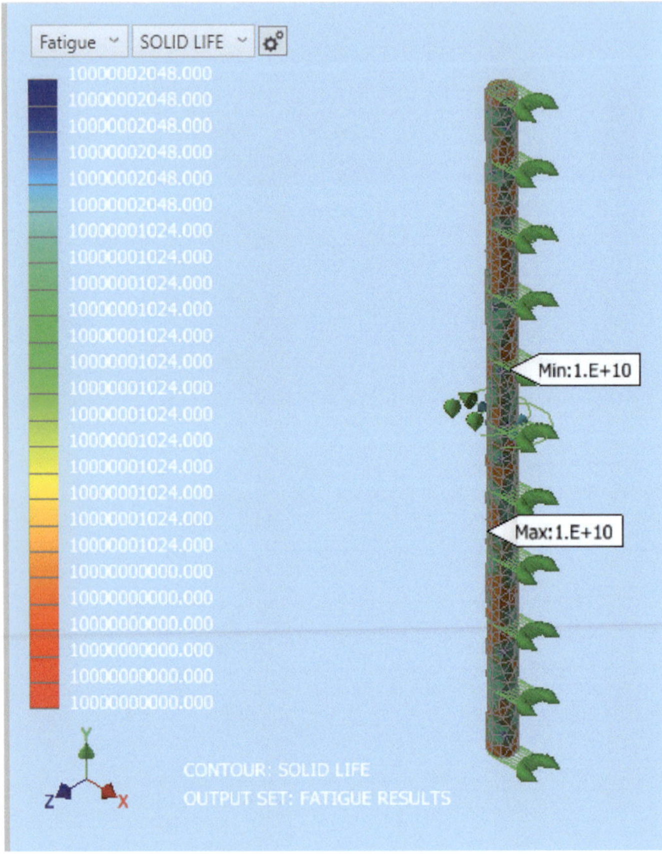

Fig. 6.19 AISI 4340 fatigue life predictions

6.7.4.7 A Bevel-and-Spur Gearbox Combination

For this combination the first gear on the shaft bevel gear is made of tooth face hardened carbon steel and the second shaft bevel gear is made of 14 Nickel Chromium 18 Steel. For both spur gear sets, nodular cast iron is used. The materials chosen offered the highest factors of safety. All the materials in Inventor's library were utilized for the gear design, and based on the results, the best performing material is chosen. Using a prompt like the one seen in Fig. 6.22, the input data are added and run a calculation for each material in Inventor's library.

Along with these material adjustments, to support the shaft holding the gear pairs, Inventor's Bearing Designer Tool is employed resulting in the CSN 02 4630 (16016) bearings being suitable. All critical findings are tabulated in Table 6.14.

The final assembly for the shaft with the spur gearbox can be seen in Fig. 6.23.

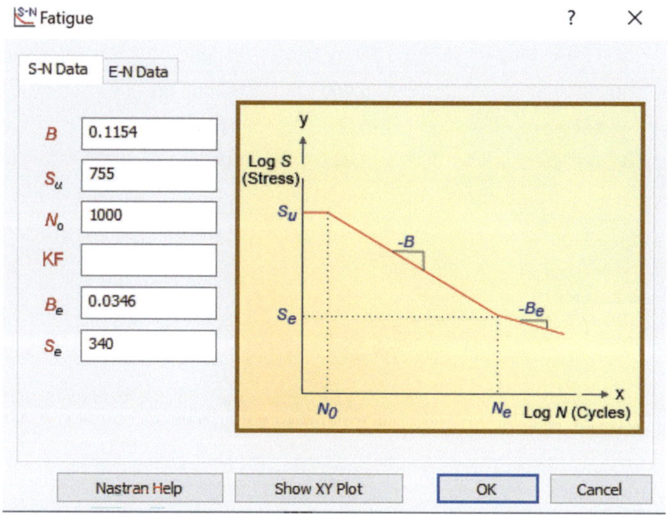

Fig. 6.20 Strength-life data for AISI 4340 steel alloy

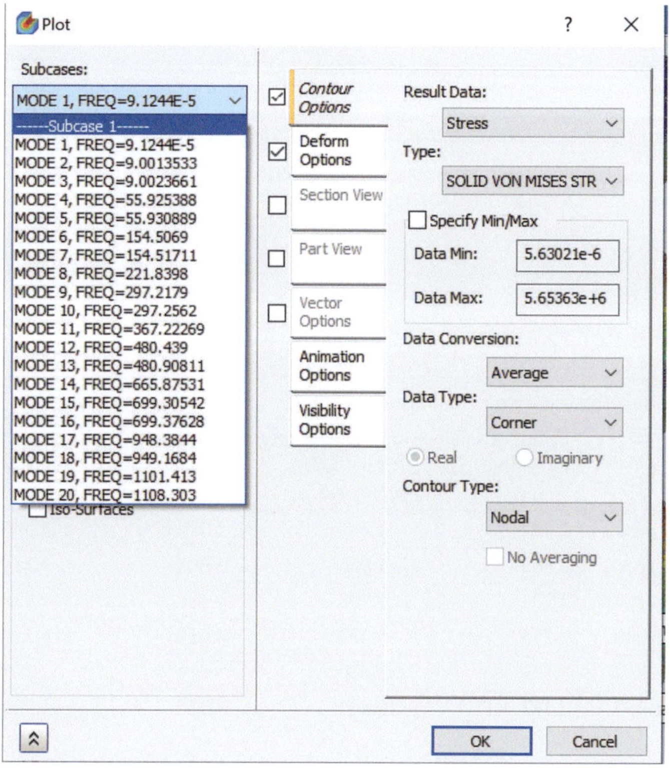

Fig. 6.21 Frequencies for 20 modes for the AISI 4340 steel alloy

Table 6.13 Shaft analysis results

Material	Von Mises stresses (MPa)	Safety factor	Fatigue	Natural frequency (Hz)
ASTM A36	46.180	6.960	Infinite life	8.32
AISI 1022	46.189	10.093	Infinite life	8.75
AISI 4340	46.180	10.443	Infinite life	8.47

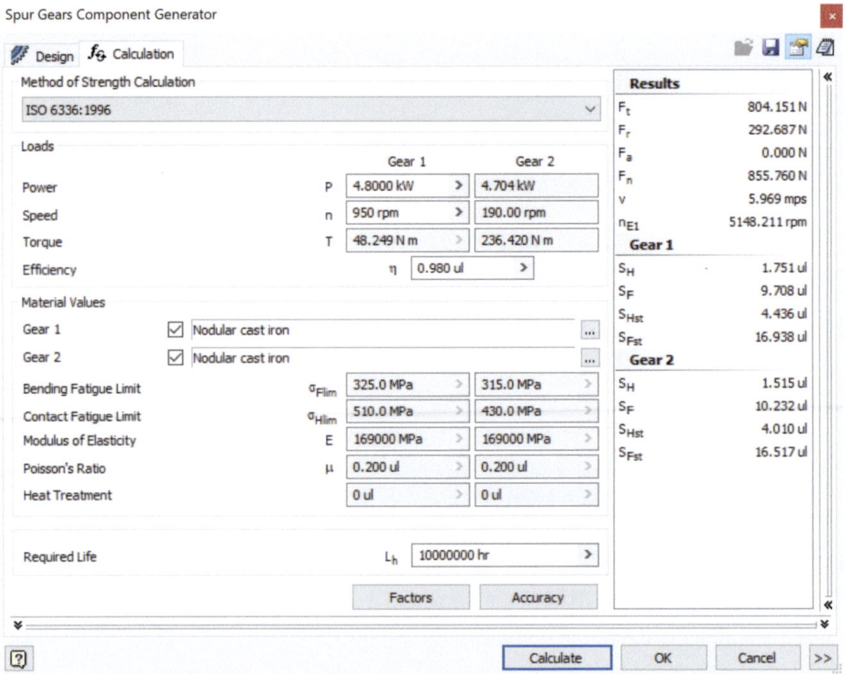

Fig. 6.22 Example for inventor's spur gear design calculator tool

Table 6.14 Bevel and spur gear combination results

Modality	Factor of safety for pitting, S_H	Factor of safety from tooth breakage, S_F	Static safety in contact, S_{Hst}	Static safety in bending, S_{Fst}
Shaft bevel gear 1	2.137	1.485	1.715	3.090
Shaft bevel gear 2	2.493	1.656	5.308	3.138
Spur set 1 gear 1	1.751	9.708	4.436	16.938
Spur set 1 gear 2	1.515	10.232	4.010	16.517
Spur set 2 gear 1	2.529	18.670	5.914	32.575
Spur set 2 gear 2	2.183	19.604	5.335	31.909

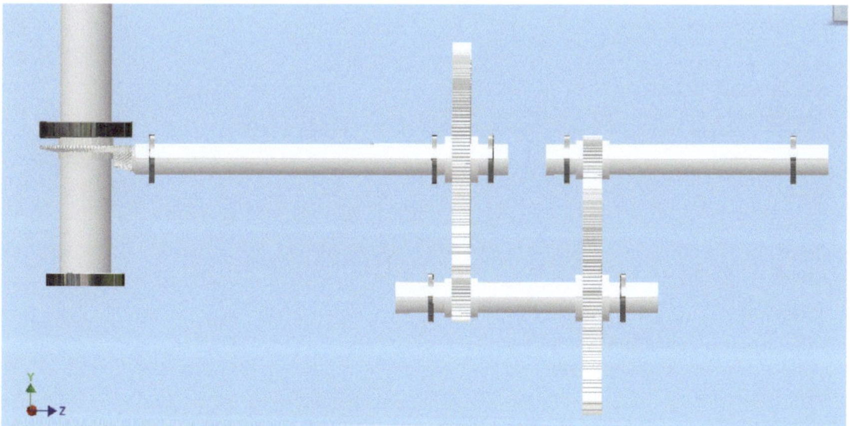

Fig. 6.23 Full assembly with spur gearbox

6.7.4.8 B Bevel-and-Bevel Gearbox Combination

For this configuration the material used for the shaft bevel gear is tooth face hardened carbon steel and the second shaft bevel gear is made of 14 Nickel Chromium 18 (14NiCr18) Steel. For the first bevel gear set, case hardened EN C10E material is selected. Finally, for the second bevel gear set heat treated nodular cast iron is employed. These materials are compared with different combinations such as with the gear sets using heat treated nodular steel, and cast/gray cast iron to compare their factor of safety. The material which provided the highest factor of safety is the heat-treated nodular cast iron. Analysis details are shown in Table 6.15.

An example of the prompt used to determine the factor of safety for the tested loading conditions can be seen in Fig. 6.24.

Table 6.15 Bevel and bevel gear combination results

Modality	Factor of safety for pitting, S_H	Factor of safety from tooth breakage, S_F	Static safety in contact, S_{Hst}	Static safety in bending, S_{Fst}
Shaft bevel gear 1	2.137	1.485	1.715	3.090
Shaft bevel gear 2	2.493	1.656	5.308	3.138
Bevel set 1 gear 1	2.287	3.596	5.243	7.113
Bevel set 1 gear 2	2.287	2.337	5.243	4.367
Bevel set 2 gear 1	2.061	7.637	4.944	12.991
Bevel set 2 gear 2	2.355	6.390	4.944	9.822

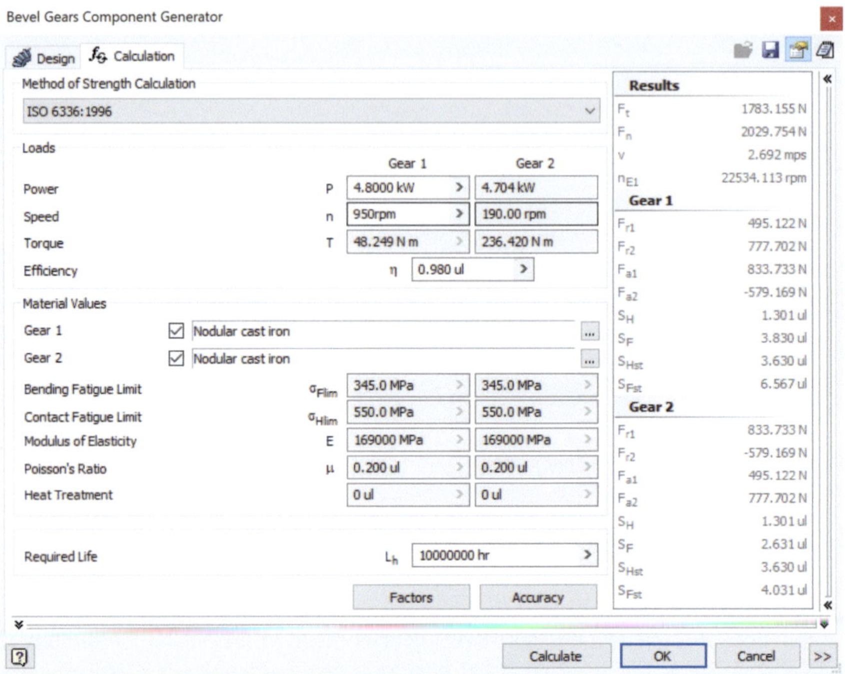

Fig. 6.24 Example for inventor's bevel gear design calculator tool

To support the bevel gearbox, different bearings must be used when compared to the spur gearbox. This is because, due to the size of the smaller gear, the diameters of the shafts must be altered in the gearbox to match the diameter. This resulted in a diameter of 50 mm for the internal shafts of the bevel gearbox. Therefore, the ideal bearing is determined as JIS B 1521 (6210 50 × 90 × 20). The full assembly including the bevel gearbox can be seen in Fig. 6.25.

Comparing the two gearboxes, the values for factor of safety for pitting and static safety in contact are similar across the gearboxes; however, the values for factor of safety for tooth breakage and static safety in bending are significantly larger for the spur gearbox. Based on these predictions, the spur gear set combination provides a better overall factor of safety and is the preferred choice for the current VAWT assembly.

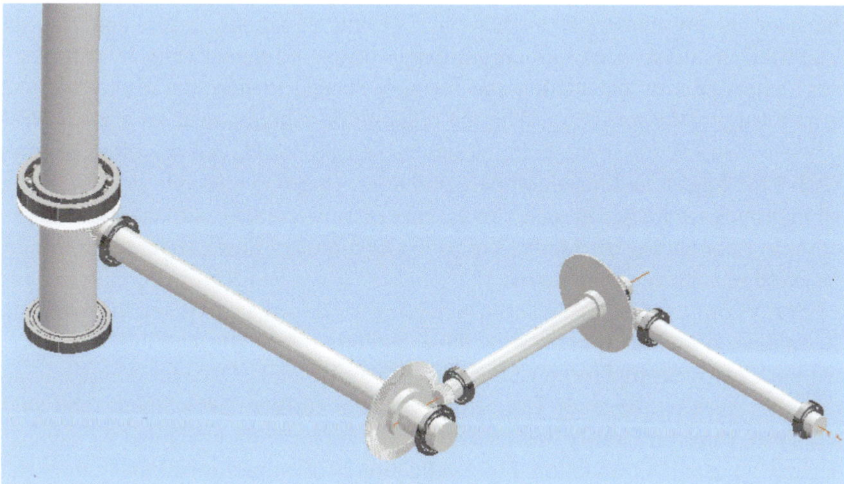

Fig. 6.25 Full assembly with the bevel gearbox

6.7.5 Case Study—3[3]: CFD Analysis of Wind Flow Over Car

6.7.5.1 Computational Methods in Fluid Mechanics

In real life, many applications are faced with where there is a flow of a particular fluid such as air flow over aircraft and water flow inside the pipe. To design those systems, it is needed to understand how flow and the solid bodies, or two different flows interact. The interactions could be as of force, energy, and mass diffusion. However, in classic Fluid Mechanics due to the complexity of governing equations i.e. conservation laws for mass, momentum, and energy, there is now analytical solution for those partial differential equations unless many terms in equations are neglected as simplification. Computational Fluid Dynamics (CFD) tries to solve the equations numerically which gives a scattered approximation solution of the flow field. The governing equations of fluid flow in integral form reads [1]:

$$\frac{\partial}{\partial t}\iint_{\Omega} \mathbf{W}dA + \oint_{\partial\Omega} \mathbf{F_N}dL = \oint_{\partial\Omega} (\mathbf{R}dy - \mathbf{S}dx) \tag{6.25}$$

where $\mathbf{W}$ is the vector of conserved variables, $\mathbf{F_N}$ is the vector of inviscid fluxes, and $\mathbf{R}$ and $\mathbf{S}$ viscous fluxes normal to the volume surface. These equations hence are solved numerically by methods such as FDM and FVM methods. CFD as a branch of fluid mechanics uses data structures to analyze problems and is highly utilized in many fields of Fluid Mechanics such as Aerodynamics, Propulsion, and Turbomachinery.

[3] Courtesy: Dr. Ali Atashbar Orang, Middle East Technical University, Northern Cyprus Campus, Türkiye, Reproduced with permission.

The same as any numerical methods in CFD the equations, space, and time are discretized then converted into algorithms to be solved numerically by computers with higher speed in calculation and memory needed to store the discrete results. As an example, the result could be the pressure distribution over an aircraft. With high-speed supercomputers, better solutions can be found, and are often required to solve the largest and most complex problems. Ongoing research yields software that improves accuracy and speed. In the case of more complex simulation scenarios with heavy computing load such as transonic or turbulent flows, supercomputers are employed to achieve the solutions.

For any computational simulation validating the results are required to be reliable. The validation is done by comparing the computational results with those of experiments already measured by apparatus such as wind tunnels. Once the computational results are verified the study can be extended for further developments and analyses. Also, some previously obtained analytical and/or verified computational analysis may be used for validation. Final validation is often performed using full-scale testing, such as flight tests.

In general, a CFD study consists of standard steps namely sketching the geometry, grid generation (meshing), input data such as fluid properties and time step, and applying consistent boundary conditions such as inlet and solid wall boundary conditions. CFD study of a problem could be performed by elaborating direct CFD codes written in programming languages such as Python or using CFD packages and software like ANSYS-FLUENT within which the CFD codes are developed and embedded already. To learn the basics of a CFD simulation during the next sections the air flow over a moving vehicle is chosen to be studied using the ANSYS-FLUENT package.

Here, the air flow over a car is investigated and modified. For simplicity the mid-plane of the actual three-dimensional car is taken as the outer entity of the car model such that the flow can be solved in two-dimensional domain. The modified configurations on the car lead to achieving better stability of the vehicle along with reducing the fuel consumption by drag reduction. This drag reduction can be monitored by the drag coefficient which reads.

$$c_D = \frac{D}{0.5\rho V^2} \tag{6.26}$$

where D is drag force, ρ and V are free stream density and velocity, respectively. C_L, lift coefficient, is defined in the same manner. The model is constructed and analyzed in the ANSYS FLUENT software to understand the aerodynamic properties of the selected car.

6.7.5.2 Primary Steps of CFD Analysis

The following steps depict the pre-processing stage that needs to be configured before running the solution.

Creating the model

The car model A180 AMG Mercedes is chosen as the study case. Planar geometry is created using the Inventor 2020 software as shown in Figs. 6.26 and 6.27.

Flow domain and mesh generation

The created entity is plugged into the ANSYS-FLUENT software. The next step is to define the flow domain as seen in Fig. 6.28. For the proposed problem the dimensions of the **flow domain** are taken as 25m in length and 5m in height as it is large enough to

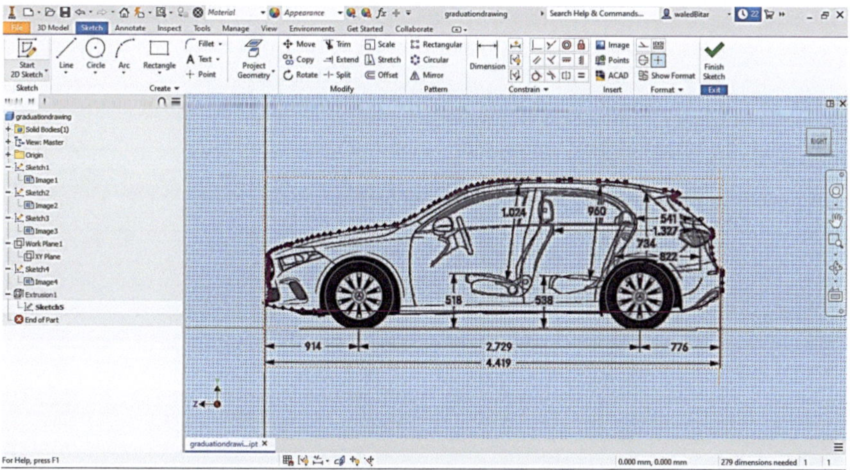

Fig. 6.26 Creating the model using Inventor 2020 [3]

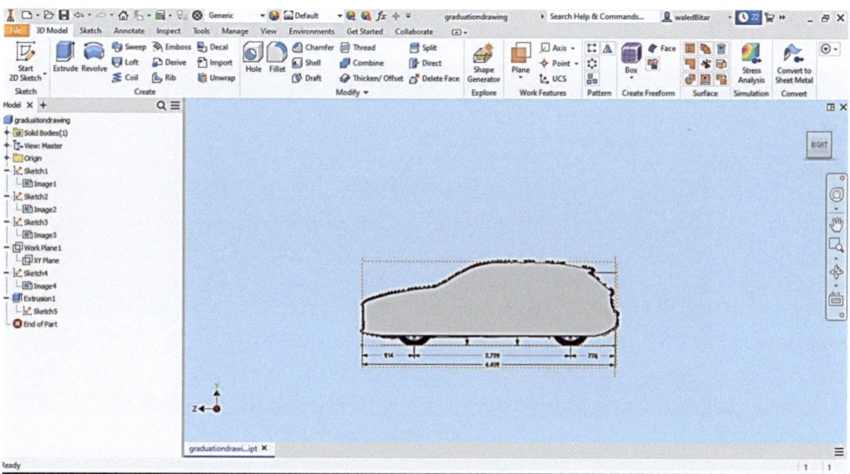

Fig. 6.27 Final two-dimensional model using Inventor 2020

avoid reflection of any bias boundary conditions into the computational domain. Since high gradients occur behind the car, the back far-field boundary is larger than that of the frontal. As shown in Fig. 6.29 the unstructured triangular meshes are created with element size of 0.025 m. Moreover, the inflation of the mesh is considered to have a finer grid close to the solid wall of the car where there are high gradients as well.

Input data and flow properties

The type of fluid, the turbulence modeling, and other properties are determined as it is shown in Fig. 6.30.

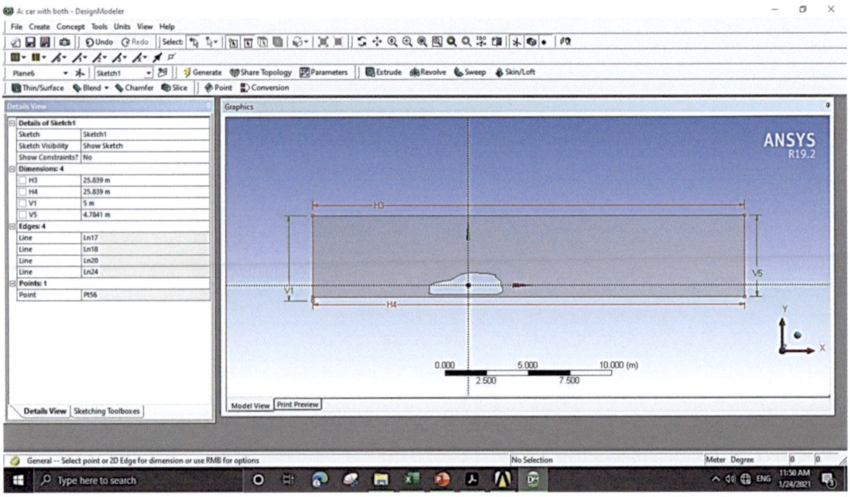

Fig. 6.28 Computational domain

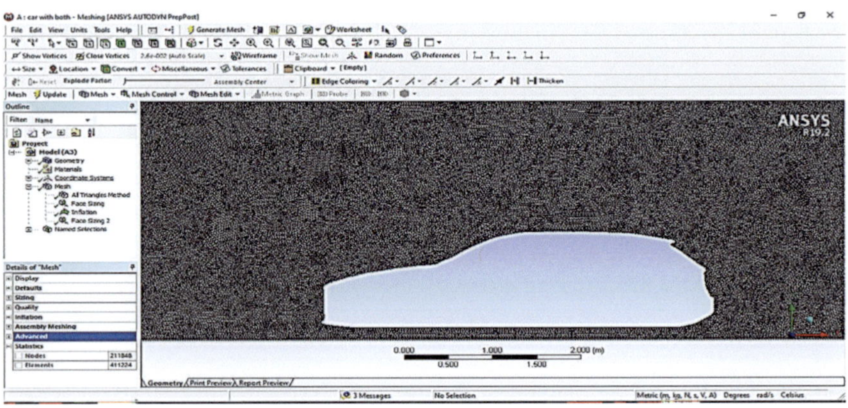

Fig. 6.29 The creation of the mesh

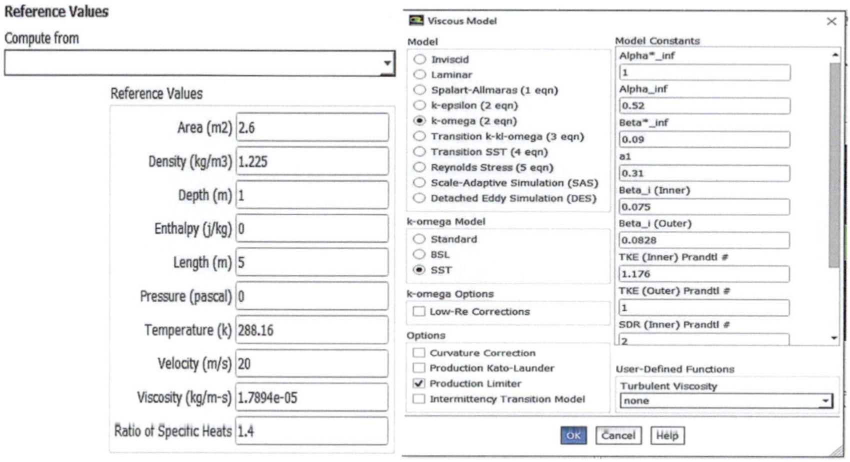

Fig. 6.30 Input reference values and properties

In this part the type of numerical scheme and its order of accuracy is chosen as well. The time step needs to be prescribed such that the scheme is stable according to the CFL (Courant Friedrichs Lewy) number [1]. Time integration method such as Runge-Kutta method is set in this step as well.

Boundary conditions

For the proposed problem there are three boundary conditions namely inlet, outlet, and solid boundary conditions. As shown in Fig. 6.31, at the inlet the far-field velocity equal to 20 m/s is set whereas at the outlet the atmospheric pressure is set. On the car surface the no slip solid boundary condition is imposed which is valid for any viscous flow.

Once the input configurations are set the program is run then time iteration starts and continues till reaching the steady state solution.

6.7.5.3 The Results

When the solution converges to the steady state the results are extracted to be discussed in the post processing stage. Here the results for two cases with different geometry are reported.

Case 1: Car with actual geometry

The first study case is the car with actual geometry. In Fig. 6.32 the velocity and pressure contours are illustrated. As seen the maximum pressure occurs at the frontal area of the car that is stagnation point while the maximum velocity appears on the upper part of the car with the small pressure. This is expected in accordance with the

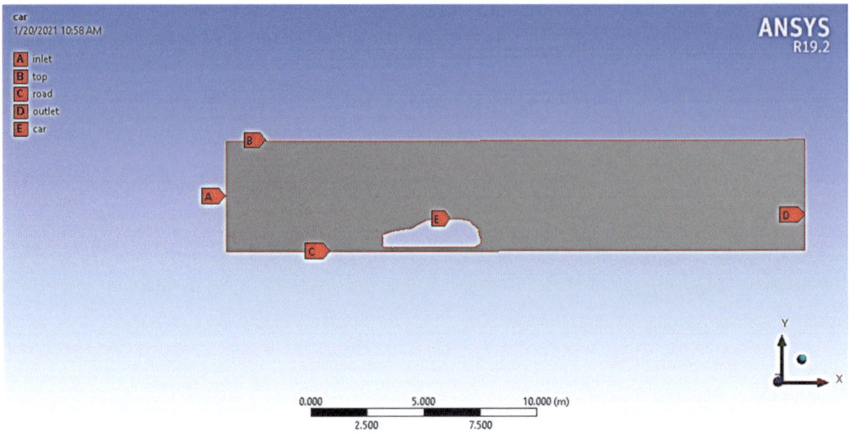

Fig. 6.31 The boundary conditions

Bernoulli equation [4]. The drag and lift coefficients of the case are found as 0.27 and 0.89, respectively.

Case 2: Car with spoiler and diffuser

The second study case is the car with modified geometry by installing the spoiler and diffuser at the end of the car. These two additional features are used to reduce the drag coefficient of the car, hence increasing the performance. As seen in Fig. 6.33a, spoiler is a device attached to the end of the car and used for straightening the stream and avoids developing the possible vortices leading to lessened wake region hence drag reduction. The other performance enhancement device is a diffuser which is attached to the bottom of the car as shown in Fig. 6.33b. This device increases the backside pressure hence reducing the drag force by lessening the pressure difference between up and down streams.

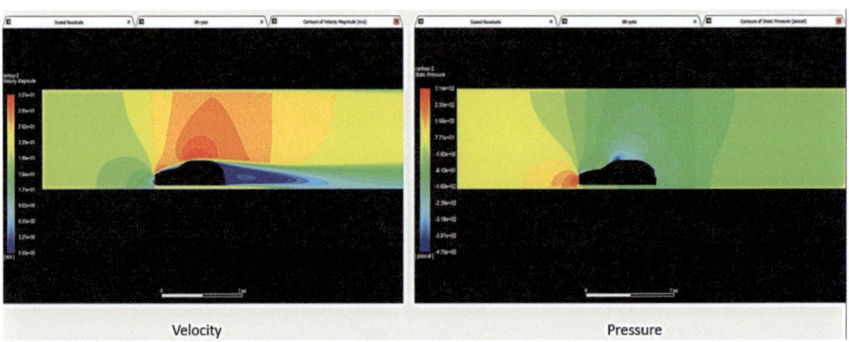

Fig. 6.32 The velocity and pressure contours, case 1

(a) (b)

Fig. 6.33 **a** Car with spoiler, and **b** car with diffuser

In Fig. 6.34 the velocity and pressure contours are illustrated. The drag and lift coefficients of the case are found as 0.22 and − 0.85, respectively. As it was expected the drag coefficient is dropped whereas the lift coefficient is negative that means the normal force is downward leading to better stability especially in high-speed rides.

Figure 6.35 depicts the streamlines of case 2 where the wake region can be seen at the backside of the car.

The comparison of the drag and lift coefficients of two cases is shown in Table 6.16.

In summary, this case study provides the introduction of the computational methods with applications in Fluid Mechanics. However, CFD is a very vast area including simple problems as well as more complex flows which require application of advanced techniques.

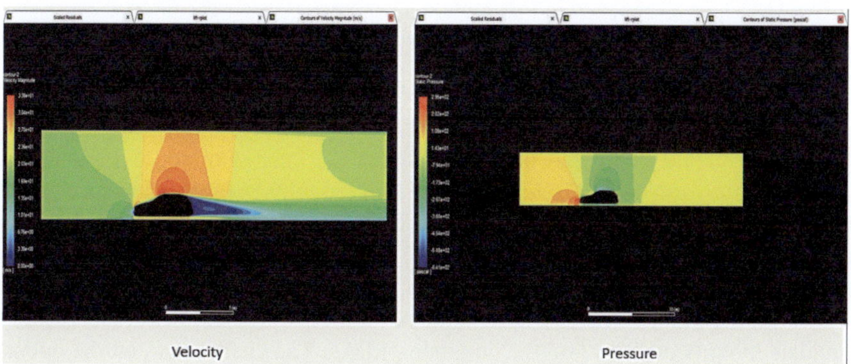

Fig. 6.34 The velocity and pressure contours, case 2

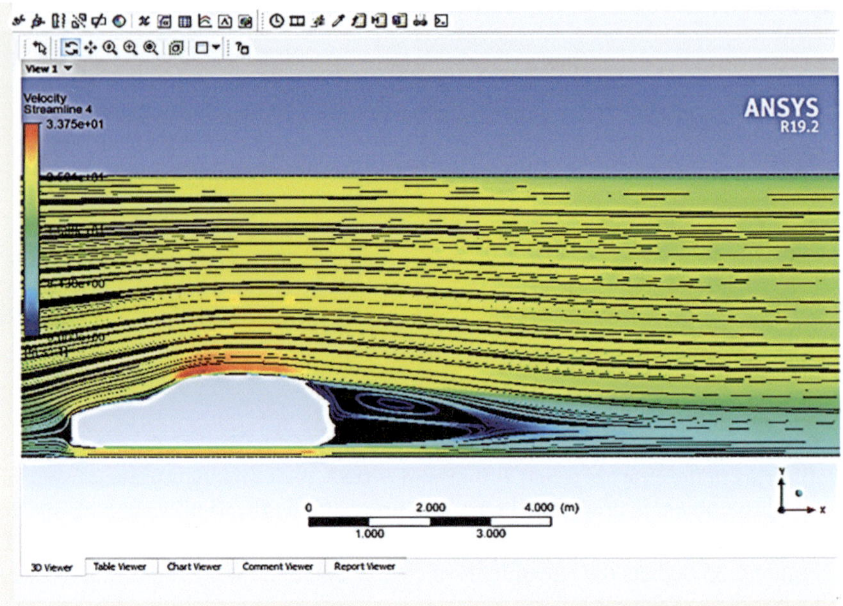

Fig. 6.35 The wake region for case 2

Table 6.16 The summary of the study cases

Case	Drag coefficient	Lift coefficient
Case 1	0.27	0.89
Case 2	0.22	– 0.85

6.8 Learning Outcomes

The reader should know and be able to select the most common computational methods in the form of the Finite Difference Method (FDM), Finite Element Method (FEM), Finite Volume Method (FVM), and Boundary Element Method (BEM) suitable to the problem or process at hand. The basics of the numerical techniques are grasped and recollected. The importance of hardware and commercially available software is highlighted. The learner should follow the given guidelines to successfully simulate the process avoiding the common pitfalls. The challenges and the opportunities are recognized, and knowledge for successful integration of the numerical techniques into the CAE required by the smart factory must be utilized.

6.9 Review Questions

Q1. For the function, estimate the value of by using all three finite difference approximations, and compare against the exact solution by providing the absolute percentage errors. Use two different increment values of h = 0.005 and h = 0.01.

Q2. Evaluate different FDM approaches in terms of accuracy and computational efficiency. Which approach provides the most accurate results with the least percentage error, particularly for complex, nonlinear boundary conditions? Discuss the trade-offs between accuracy and computational cost.

Q3. Name and explain the three distinct stages in the FEM algorithm. How do these stages interact with each other, and what are the challenges in optimizing each stage for solving large-scale, multi-physics problems?

Q4. Discuss the common 1D, 2D, and 3D element classes in FEM. How do these elements vary in terms of shape functions, interpolation methods, and application to specific types of engineering problems? Provide examples of scenarios where higher-order elements would be preferable.

Q5. Write down the final assembled matrix equation in FEM. What do the parameters in this equation represent, and how do boundary conditions influence the solution? Discuss the implications of using different solution techniques (direct vs. iterative) on the accuracy and stability of the results.

Q6. Compare and contrast FDM, FEM, BEM, and FVM in terms of their discretization techniques, stability, and applicability to different types of physical problems. Which method would be more suitable for solving a coupled heat transfer and fluid flow problem, and why?

Q7. Explore the differences between FDM, BEM and FEM. How do the boundary-focused nature of BEM and the domain-focused nature of FEM influence their respective strengths and limitations in solving engineering problems? Discuss the scenarios where BEM would outperform FEM and FDM, particularly in terms of computational resources and problem setup.

Q8. Discuss the impact of element type and mesh density on the convergence and accuracy of FEM solutions. How can adaptive mesh refinement techniques be employed to enhance solution accuracy in regions with high gradients?

Q9. Analyze the role of preconditioning in iterative solvers for FEM. How do different preconditioning techniques affect the convergence rate, especially for large-scale, ill-conditioned problems?

Q10. Discuss the extension of FEM to nonlinear problems, such as plasticity and large deformations. What are the challenges associated with nonlinear FEM, and how do different approaches (e.g., Newton–Raphson, arc-length methods) address these challenges?

Q11. Explain the process of integrating multi-physics phenomena within FEM simulations. What are the computational and modeling challenges associated with coupling different physical fields, such as thermal and structural analysis?

Bibliography

1. Aleksendric D, Carlone P (2015) Soft computing techniques. In: Soft computing in the design and manufacturing of composite materials, vol 4, pp 39–60
2. Amin Z, Amin T (2022) Vertical axis wind turbine (VAWT) shaft & gearbox design. Graduation Design Project, 2022, Mechanical Engineering Program, Middle East Technical University Northern Cyprus Campus, Türkiye
3. Aryasomayajula A, Bayat P, Rezai P, Selvaganapathy PR (2017) Microfluidic devices and their applications. In: Springer handbook of nanotechnology, pp 487–536
4. Badwi Albitar WK, Altuntas M (2020) CFD analysis of wind flow over car. Graduation Design Project, 2020, Mechanical Engineering Program, Middle East Technical University Northern Cyprus Campus
5. Blaettler K (2020) Wind speed vs. air pressure, internet, sciencing
6. Kirby BJ (2010) Micro-and nanoscale fluid mechanics: transport in microfluidic devices. Cambridge University Press
7. Budynas R, Nisbett K (2010) Shigley's mechanical engineering design. McGraw-Hill Education
8. Casini M (2016) Small vertical axis wind turbines for energy efficiency of buildings. J Clean Energy Technol 4(1):56–65
9. Lin C-H, Wang Y-N, Fu L-M (2012) Integrated microfluidic chip for rapid DNA digestion and time-resolved capillary electrophoresis analysis. Biomicrofluidics 6(1)
10. CNYWeather.com (2022) Barometric pressure. Cnyweather.com
11. Costabel M (1987) Principles of boundary element methods. Comput Phys Rep 6(1–6):243–274
12. Florida R, The relationship between skyscrapers and great cities. Bloomberg.com
13. Fu L, Tsai C, Lin C (2008) A high-discernment microflow cytometer with microweir structure. Electrophoresis 29(9):1874–1880
14. Global Wind Atlas (2022) GlobalWindAtlas.info
15. Bruus H (2007) Theoretical microfluidics, vol 18. Oxford University Press
16. Hi-Vawt Technology Corp (2012) DS-700 vertical axis wind turbine
17. https://drive.google.com/file/d/1Z7eG3IFuqHXWNiolBSAMxC4lOFNKNiFX/view?usp=sharing.
18. https://www.wikiwand.com/en/VAWT
19. Reddy JN (2005) An introduction to the finite element method, vol 3. McGraw-Hill New York
20. Egab L (2020) Computer-aided engineering findings on the physics of tire/road noise. In: Automotive tire noise and vibrations. Elsevier, pp 217–243
21. Mestrich Palau P, Sellart Garcia MT (2022) Design and development of an integrated renewable energy centre in Cervià de les Garrigues, Catalonia
22. Micallef D, van Bussel G (2018) A review of urban wind energy research: aerodynamics and other challenges. Energies 11(9):2204
23. Munson BR, Rothmayer AP, Okiishi TH (2012) Fundamentals of fluid mechanics. In: Blackwell handbooks in linguistics, 7th edn. Wiley
24. Yunus NAM (2013) Microfluidic devices fabrication for bioelectrokinetic system applications. In: Electrochemistry. IntechOpen
25. Zienkiewicz OC, Taylor RL (2005) The finite element method set. Elsevier
26. Zienkiewicz C, Taylor RL, Zhu JZ (2005) The finite element method: its basis and fundamentals. Elsevier Science
27. Razavi SE, Orang AA (2012) A comparative investigation of hydrofoil at angles of attack. Int J Numer Meth Fluids 68(9):1087–1101

28. Patankar S (2018) Numerical heat transfer and fluid flow. CRC Press
29. Sarkar S, Thummar K, Shah N, Vagrecha V (2019) A review paper on aerodynamic drag reduction and CFD analysis of vehicles. In: Proc. A rev. paper aerodyn. drag reduction CFD anal. vehicles, pp 231–235
30. Chandrupatla T, Belegundu A (2021) Introduction to finite elements in engineering. Cambridge University Press
31. Simulated historical climate & weather data for New York—meteoblue, Internet, MeteoBlue
32. Vision Air Wind Technologies (2020) V-Air Products
33. Vision Air Wind Technologies (2020) Vision Air 3

Chapter 7
Mathematics for CAD/CAE/CAM

Prelude
Inference from numbers using computers leads to productive results for humans. Today's discrete products originate from some need that refers to one or more concepts amalgamated in one coherent body. This is done through CAD/CAE/CAM working on principles based on mathematics with number crunching done by the Turing machine. This chapter attempts to explain this puzzle.

7.1 Introduction

Mathematics for CAD/CAE/CAM started taking root in the 1970s with a specific name called Computational Geometry. This is a branch of mathematics that allows the handling of geometric primitives based on Euclidean Mathematics[1] like lines, arcs, surfaces, and solids presented in primary, secondary, rotational, and tertiary axes. A number of algorithms were developed since the inception of computational geometry and the art is still practiced today to amalgamate it with other requirements of CAD/CAM/CAE such as storage and access of geometric primitives in databases, combining it with other types of information such as machine elements primitives, control elements primitives, pneumatic elements primitives, hydraulic elements

[1] In contrast to Computational Geometry based on Euclidean principles Fractal Geometry based on non-Euclidean principles also emerged in the 1970s that has roots in past centuries but was formalized in this era. Fractal geometry has the distinction of using time, color, and perspective to define objects with self-similarity.

Supplementary Information The online version contains supplementary material available at https://doi.org/10.1007/978-3-031-77106-4_7.

W. A. Khan et al., *Computer Aided Engineering Design and Manufacturing*, Springer Series in Advanced Manufacturing, https://doi.org/10.1007/978-3-031-77106-4_7

primitives, etc. in order to have complete design of part, structure, assembly, mechanism and of machine at one place. The User Interface (UI) performance and User Experience rating with a particular software package also matters that incorporates the above-mentioned auxiliaries.

Different types of databases are accessed as company-provided databases through authenticated access to company databases that may also have a small price tag. The use of semantic databases and bots takes importance here. This is still the duty of the engineer to identify the object that needs sourcing and select it from different databases including but not limited to above mention databases and databases like material databases, machining parameters databases, manufacturing tolerance databases, etc. specific to a particular product that is associated with a national or a regional or a company or a professional institution. The databases are discussed in Part III Chap. 9.

The reader is guided through the mathematics of geometric primitives and their manipulation using mathematical operators as these functionalities are transparent to the user when a CAD/CAM/CAE API is operational. Today it takes a fraction of time to compute that is not even noticeable when the equations are computed on a basic Turing machine.

The reader is provided with the solution methodology of Bezier Curves and Surfaces as a case study after the basic introduction to the elements of computational geometry.

Parametric equations are introduced, and resources are identified from where the basic solution methodology of the working of these equations can be understood.

7.2 Basic Terminology

A CAD model can either be a wireframe model, a surface model, or a solid model. These models are composed of:

A. Vertex (Point, Coordinates, Association)
B. Edges (Line, Curve)
C. Wireframe Model (Connectivity of Vertex and Edges)
D. Patches (Skin of model based on vertex and Edges)
E. Surface Model (Connectivity of Patches to form a Shape)
F. Solid Model (result of Boolean operators on Geometric Primitives and/or User Defined Primitives).

Items B–F above have a mathematical representation that can either be implicit or parametric while A is defined in terms of coordinates in an appropriate coordinate system. Item F is based on Boolean operations at Geometric Primitives (Figs. 7.1, 7.2 and 7.3) or User Defined Primitives (See Sect. 7.3). Depending on the type of industry, the user can define 'User Defined Primitives' to make tasks specific to a particular industry easier.

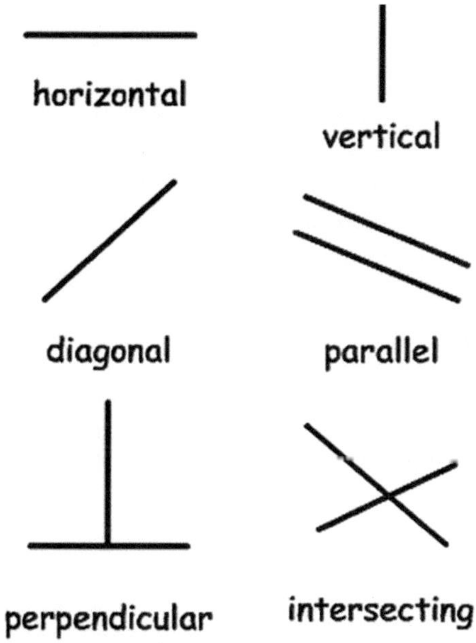

Fig. 7.1 Types of straight lines. https://www.youtube.com/watch?v=CfwHKtMY2C4

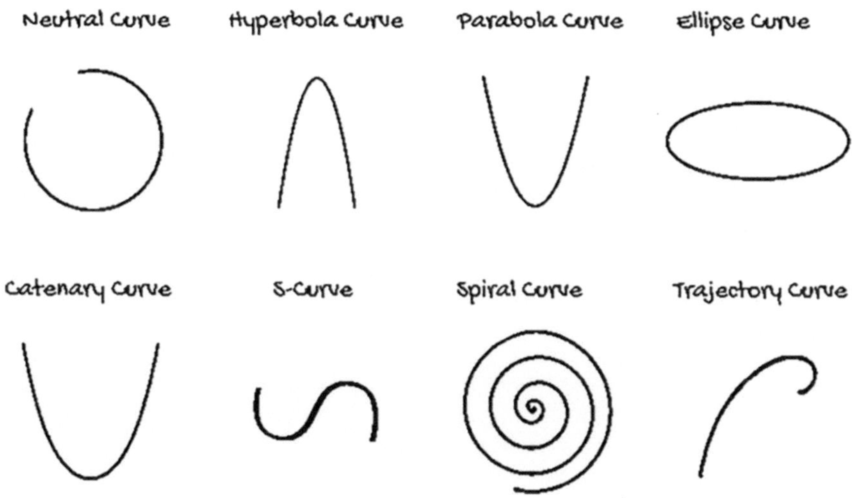

Fig. 7.2 Types of curves. Image Source Unknown: Googled on 17th February 2024

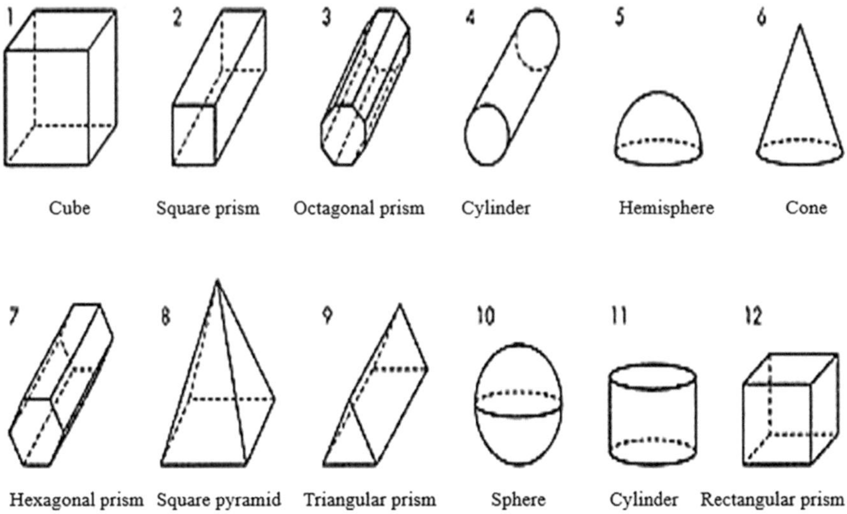

Fig. 7.3 Solid geometric primitives. https://nap.nationalacademies.org/read/10050/chapter/26

7.3 Anatomy of Curves and Solids

Starting from very basic, some of the types of lines and curves are:

Types of Straight Line are:

Types of curves are illustrated in Fig. 7.2:

Some solid geometric primitives are given in Fig. 7.3:

The basis of curve formation is:

Analytic Curves are points, lines, arcs and circles, filets and chamfers, and conics (ellipses, parabolas, and hyperbolas).

Synthetic curves include various types of splines (cubic spline, B-spline, Beta-spline) and Bezier curve.

Analytical Curves

Analytical curves can be represented by an implicit equation that needs to be transformed in parametric form to draw lines or curves between two points with a step that has a least count. The smaller the step, smoother the curve with extra computational cost, from manufacturing perspective a better surface finish will be obtained with smaller step size, but it has mechanical limitations including cutter size and available lead time.

Higher the order of the equation, problems associated with square roots and derivatives are encountered, and a need for synthetic curves arises. Curve fitting is necessary with analytical curves if the shape of the curve is complex and often leads to discontinuity where the derivative may be zero or undefined. Curve fitting itself may be cumbersome and may not give the exact desired shape.

Synthetic Curves

The types of lines and other analytical curves can normally be defined by implicit and parametric equations. These lines and curves are embedded between vertices that ultimately result in patches (Surfaces) with an established hierarchy for storage and relation between vertices in 3D space.

The 3D geometric or user defined primitives are used in 3D modeling and operation is specifically known as Constructive Solid Geometry (CSG) where Boolean operations are used among 3D primitives (Standard Geometric or User defined Primitives) to achieve the desired shape.

The geometric data related to machine elements, control elements and other specific elements of design can be imported from the Company database after due authentication (and may be after paying a small amount as fee) in a data exchange format normally DXF (AutoCAD) or IGES (to facilitate most of the CAD APIs) resulting in complete engineering design in digital format that can be worked upon for multi-faceted engineering analysis. Essentially, the exercise also supports sourcing and life cycle assessment.

Following paragraphs provide storage information related to lines and curves and some of the Boolean Operators:

Basic database information

Vertices

- Coordinate values for vertices.

Edges

- Vertices associated with edges (endpoints).

Faces

- Loops (faces) formed by edges (Fig. 7.4).

Basic database structures

Relational

- A set of lists, using arrays for storage, is given in Table 7.1.

Hierarchical

- A tree structure, think of a company's executive structure (Fig. 7.5)

Network

- Use data pointers (Figs. 7.6 and 7.7)

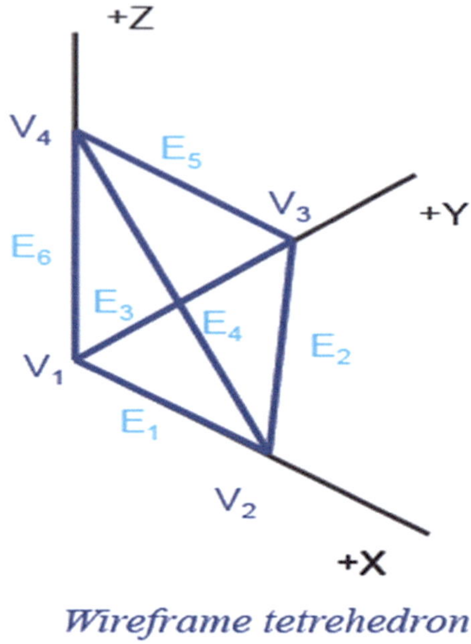

Fig. 7.4 A wireframe tetrahedron. Image Source Unknown: Googled on 17th February 2024

Table 7.1 A set of lists containing arrays for storage

Vertex list	Edge list
$V_1(0, 0, 0)$	$E_1[V_1, V_2]$
$V_2(1, 0, 0)$	$E_2[V_2, V_3]$
$V_3(0, 1, 0)$	$E_3[V_3, V_1]$
	$E_4[V_2, V_4]$
	$E_5[V_4, V_3]$
	$E_6[V_1, V_4]$

Image Source Unknown: Googled on 17th February 2024

7.4 Implicit and Parametric Equations for Geometric Primitives

Taking the example of a circle, it is a type of analytical curve. In mathematics, an **implicit equation** is a relation of the form:

$$R(X1, \ldots, Xn) = 0,$$

where *R* is a function of several variables (often a polynomial). For example, the implicit equation of the unit circle is:

$$X^2 + Y^2 - 1 = 0.$$

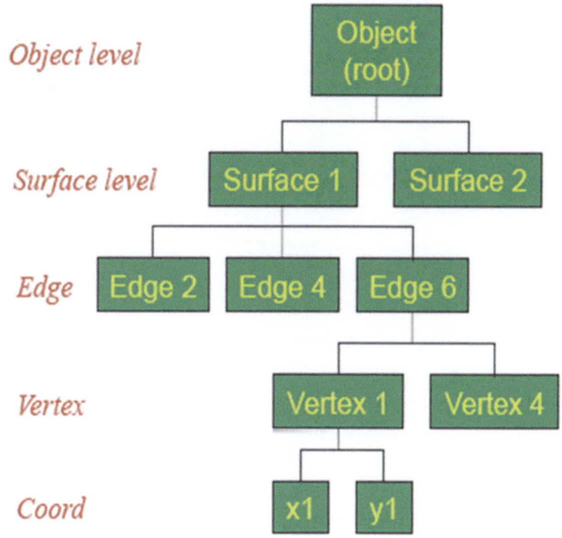

Fig. 7.5 A tree structure. Image Source Unknown: Googled on 17th February 2024

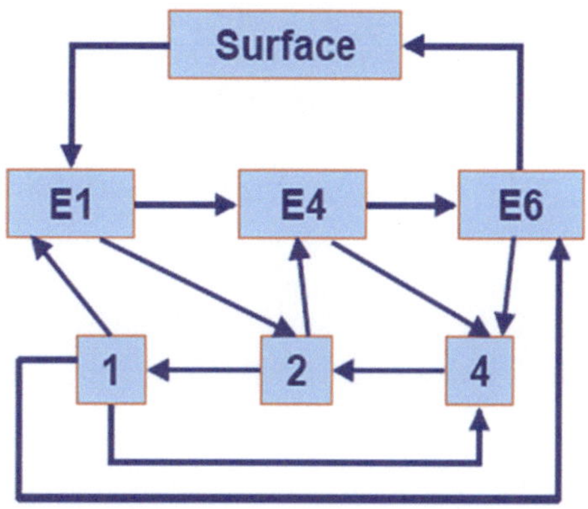

Fig. 7.6 Network with data pointers. Image Source Unknown: Googled on 17th February 2024

An **implicit function** is a function that is defined by an implicit equation that relates one of the variables, considered as the value of the function, with the others considered as the arguments. For example, the equation

$$X^2 + Y^2 - 1 = 0$$

The unit circle defines Y as an implicit function of X if $-1 \leq X \leq 1$, and Y is restricted to non-negative values.

The implicit function theorem provides conditions under which some kinds of implicit equations define implicit functions, namely those that are obtained by equating to zero multivariable functions that are continuously differentiable.

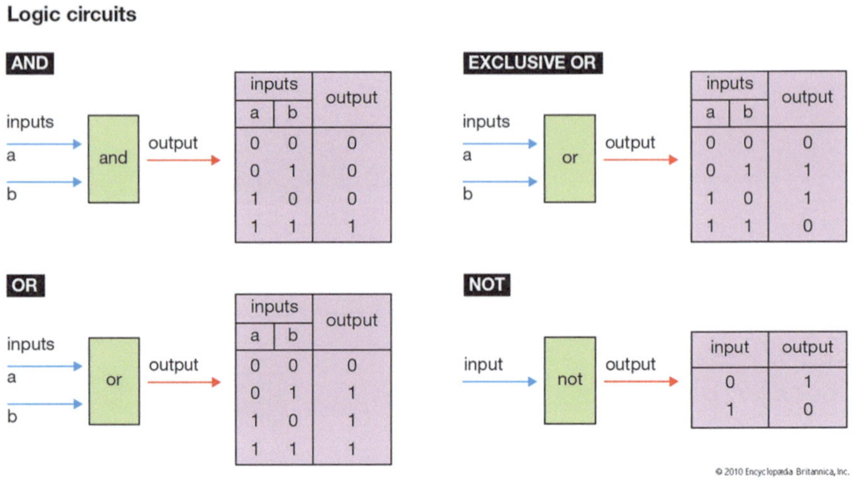

Fig. 7.7 Basic Logic circuits. https://www.britannica.com/technology/logic-design

Parametric equations are commonly used to express the coordinates of the points that make up a geometric object such as a curve or surface, in which case the equations are collectively called a parametric representation or parameterization (alternatively spelled as parametrization) of the object.

In CAD, It is desired to find a mathematical form for representing curved surfaces. Reasons are:

(a) Smooth Contours
(b) Easy to manipulate and manufacture
(c) Follows a prescribed shape

Following the URL below, a recorded lecture about analytical and synthetic curves will start playing:

https://drive.google.com/file/d/1xLZdoi3IeKVoCJ6g5e1HGvjWd6DahjvY/view?usp=share_link.

Following the URL below, a recorded lecture about surfaces is provided (Fig. 7.8):

https://drive.google.com/file/d/1ZYMRECTFuUuHxzL7DCF6IuA91_2vLhmA/view?usp=drive_link.

Following URL provides a recorded lecture about solid modeling using Constructive Solid Geometry:

https://docs.google.com/presentation/d/1IL-AyY6KmYCUKzqv2PasUfN5Lb1SG6aA/edit?usp=drive_link&ouid=113180720659833781020&rtpof=true&sd=true.

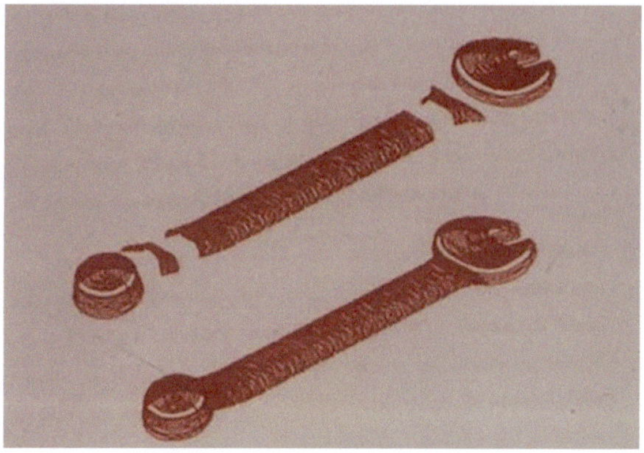

Fig. 7.8 Tool path generation for blended surfaces

7.5 Case Study: Bezier Curve[2]—In 5-Axis CNC Milling Machine Interpreter Development

Calculating Toolpath from Parametric Equations
Python Code for Bezier Curves (Cubic)

Summary

- The code defines the x, y, and z values for the starting and ending positions.
- The code extracts the x, y, and z values for the control points from the input argument.
- The number of steps for the Bezier curve is calculated based on the number of control points.
- The x, y, and z values for the toolpath are calculated using the parametric equation of a cubic Bezier curve and stored in a list.

The difference between the two methods is seen in the way the toolpath is calculated. In the Bezier curve code, the toolpath is calculated directly using the parametric equations of the Bezier curve, using control points. In the cubic spline code, the coefficients of the cubic spline are computed first, and then the toolpath is calculated using the cubic spline equations.

Bezier Curves

```
import numpy as np
def bezier_curve(start_pos, end_pos, control_points):
# Define the control points for the spline control_points = np.array(control_
points)
# Define the x, y, and z values for the start and end positions
x_start, y_start, z_start = start_pos["X"], start_pos["Y"], start_pos["Z"]
```

[2] Reproduced with permission from Mr. Abdul Sami Urooj. Developed under guidance of Wasim Ahmed Khan.

x_end, y_end, z_end = end_pos["X"], end_pos["Y"], end_pos["Z"]
# Define the intermediate x, y, and z control points x_control = control_points[:,0]
y_control = control_points[:,1]
z_control = control_points[:,2]
# Calculate the toolpath using the parametric equation of a cubic Bezier curve
steps = len(control_points) * 10 #Increase the number of steps for smoother motion toolpath = []
for i in range(steps):
t = i / (steps - 1)
x = (1 − t)**3*x_start + 3*(1 − t)**2*t*x_control[0] + 3*(1 − t)*t**2*x_control[1] + t**3*x_end y = (1 − t)**3*y_start + 3*(1 − t)**2*t*y_control[0] + 3*(1 − t)*t**2*y_control[1] + t**3*y_end z = (1 − t)**3*z_start + 3*(1 − t)**2*t*z_control[0] + 3*(1 − t)*t**2*z_control[1] + t**3*z_end
toolpath.append({"X": x, "Y": y, "Z": z})
return toolpath

Bezier Curves—Code Explanation

1. import numpy as np: This line imports the numpy package and renames it as np.
2. def bezier_curve(start_pos, end_pos, control_points):: This line defines a function named bezier_curve that takes three arguments: start_pos, end_pos, and control_points.
3. control_points = np.array(control_points): This line converts the control_points argument into a numpy array.
4. x_start, y_start, z_start = start_pos["X"], start_pos["Y"], start_pos["Z"]: This line extracts the x, y, and z values for the starting position from the start_pos argument.
5. x_end, y_end, z_end = end_pos["X"], end_pos["Y"], end_pos["Z"]: This line extracts the x, y, and z values for the ending position from the end_pos argument.
6. x_control = control_points[:,0]: This line extracts the x values for the control points.
7. y_control = control_points[:,1]: This line extracts the y values for the control points.
8. z_control = control_points[:,2]: This line extracts the z values for the control points.
9. steps = len(control_points) * 10: This line calculates the number of steps to use in the Bezier curve. It sets the number of steps as ten times the number of control points.
10. toolpath = []: This line initializes an empty list to store the x, y, and z values of the toolpath.
11. for i in range(steps):: This line starts a loop that will run steps times.
12. t = i / (steps - 1): This line calculates the parameter value t for the current step.
13. x = (1 − t)**3*x_start + 3*(1 − t)**2*t*x_control[0] + 3*(1 − t)*t**2*x_control[1] + t**3*x_end: This line calculates the x value for the toolpath at the current step using the parametric equation of a cubic Bezier curve.

14. y = (1 − t)**3*y_start + 3*(1 − t)**2*t*y_control[0] + 3*(1 − t)*t**2*y_control[1] + t**3*y_end: This line calculates the y value for the toolpath at the current step using the parametric equation of a cubic Bezier curve.
15. z = (1 − t)**3*z_start + 3*(1 − t)**2*t*z_control[0] + 3*(1 − t)*t**2*z_control[1] + t**3*z_end: This line calculates the z value for the toolpath at the current step using the parametric equation of a cubic Bezier curve.
16. toolpath.append({"X": x, "Y": y, "Z": z}): This line adds a dictionary with the x, y, and z values for the current step to the toolpath list.
17. return toolpath: This line returns the toolpath list containing the x, y, and z values for the toolpath.

7.6 Learning Outcomes

The reader should have a considerable knowledge on the mathematics of geometric primitives and their manipulation through mathematical operators within the use of CAD/CAE/CAM. The basic terminology of mathematics in the CAD/CAE/CAM context should be learned. Implicit and parametric equations for geometric primitives should be grasped. With the help of the "Interpreter Element for Bezier Curve" case study embodying a 5-Axis CNC Milling Machine Interpreter Development, a real-life application should be experienced.

7.7 Review Questions

Q1. What are the basic types of CAD models? What are they composed of?
Q2. What are the different types of 2D and 3D geometric models?
Q3. List at least five different types of lines.
Q4. List at least five different types of curves.
Q5. What are geometric primitives? List at least five different types of solid geometric primitives.
Q6. What is surface modeling? List the advantages and disadvantages of surface modeling.
Q7. Differentiate between Analytic and Synthetic curves/surfaces?
Q8. What are the advantages of parametric representation of curves/surfaces?
Q9. What is the parametric equation of a line? How can it be programmed?
Q10. What is the parametric equation of a circle? How can it be programmed?
Q11. What is the parametric equation of parabola? How can it be programmed?
Q12. What is the parametric equation of the ellipse? How can it be programmed?
Q13. What are the surface manipulation techniques?
Q14. Construct the wireframe tetrahedron.
Q15. Describe the basic database hierarchical structure.
Q16. Elaborate on the basic types of logical circuits.

Q17. What are the reasons for finding a mathematical form for representing curved surfaces?

Q18. Describe the parametric representation of a line.

Q19. What is a Bezier curve?

Q20. What are the benefits of the Bezier curve over a cubic spline?

Bibliography

1. Preparata FP, Shamos MI (2012) Computational geometry: an introduction. Springer Science & Business Media
2. Zeid I (2005) Mastering CAD/CAM. McGraw-Hill Higher Education
3. O'Rourke J (1998) Computational geometry in C. Cambridge University Press
4. Zeid I (2009) Cad: theory and practice. Tata McGraw Hill Education Private Limited
5. de Berg M (2000) Computational geometry: algorithms and applications. Springer Science & Business Media
6. Minsky M, Papert S (1969) An introduction to computational geometry, vol 479, no 480, p 104. Cambridge Tiass., HIT
7. Rao PN (2010) Cad/Cam: Prin & Appl 3E. Tata McGraw-Hill Education
8. Devadoss SL, O'Rourke J (2011) Discrete and computational geometry. Princeton University Press

Chapter 8
Standards for CAD/CAE/CAM

Prelude

In the machine design and manufacturing sectors, the 1960s witnessed the change of transforming from conventional technical drawings on paper to electronic CAD drawings, embodying various advantages such as boosting productivity and improving flexibility in carrying out design amendments. CAD tools have also paved the way towards the advancements in CAM via generating drawing-oriented manufacturing commands for faster operations. Standardization in industrial sectors that utilize CAD/CAE and CAM has helped mitigate various issues associated with data exchange between computerized systems. These standardization efforts have taken place across the world so that computer-aided design and manufacturing can be carried out effectively on a global scale. This chapter is dedicated to the discussion of various information and geometry exchange standards in CAD/CAE and CAM.

8.1 Objectives

Standards are the basis of knowledge progression. By standardization, mankind moves further and does not have to reinvent the wheel. The same is the case with CAD/CAE/CAM. The basic standards involved here are the Information Interchange standards, Geometry Representation standards, Materials Standards, Metrology related standards and Environment related standards. The first part, i.e., Information Interchange, ASCII, and Unicode are explained. For the second part, i.e., Geometry Representation, DXF, IGES, SET, VDA-FS, PDES, and STEP are explained. The reader is then given the sources of information for other standards listed above. The reader

---

Supplementary Information The online version contains supplementary material available at https://doi.org/10.1007/978-3-031-77106-4_8.

W. A. Khan et al., *Computer Aided Engineering Design and Manufacturing*, Springer Series in Advanced Manufacturing, https://doi.org/10.1007/978-3-031-77106-4_8

is also introduced to the procedure of how Information Interchange and Geometry Representation Standards are used.

The reader must develop an instinct to identify and use pertinent standards for their application.

8.2 Standards for CAD/CAE/CAM

Standards are means for rapid design development and supplement to the task of sourcing and after sales service. As CAD/CAE and CAM for discrete products are encapsulated in some form of systems engineering, these standards cover billions of products being developed, marketed, and used by novice to a master user and change hands in the process. Standards from international, national, regional, and professional bodies for a very minor part (or a process) to an assembly (or a cycle composed of numerous processes) are available or being developed to formalize the knowledge acquired by the human being at any stage of modern living.

8.3 Information Interchange Standards

Information interchange standards basically allow humanly comprehensible information (e.g. alphabets, numerals, symbols, and visible and non-visible characters) representing human designed logic to computing machine readable binary format that in effect represents voltage levels (Low and High) allowing computing operations to be performed, today, at a phenomenal speed.

This is prominently done through American Standard Code for Information Interchange (ASCII) for English, and Unicode for different languages.

Speed and languages are explained in much detail in Part II, Chap. 5 and Part VII, Chap. 16.

8.4 Geometry Exchange Standards

With the widespread use of CAD/CAE and CAM, the technology resulted in many Software Application Packages having diverse features. The common element of all these packages is the use of CAD model geometry be it wireframe, surface, or solid model. To provide convenience to users, number geometry exchange standards are defined as geometry exchange processors generally providing an exact model to the target API and in some cases requiring marginal editing. These common geometry exchange standards are as follows.

8.4.1 DXF

Company Autodesk generated the Drawing Interchange Format (DXF) as part of its CAD software AutoCAD 1.0 in 1982. DXF has been accepted and used by many both in industry and academia since then to facilitate information and geometry exchange of 2D/3D CAD technical drawings. The file structure of DXF is fully detailed and established in ASCII data-encoding format.

Along with the other published exchange file formats of DWF and DWG; DXF is an open file format by Autodesk. These file formats are well known to incorporate developer options which cater for clear-cut and smooth means of handling the design information not just within other relevant software of Autodesk, but also in products of other CAD/CAE/CAM software companies.

8.4.2 IGES

Initial Graphics Exchange Specification (IGES) format is known to be developed by the United States Air Force (USAF) around the 1980s through their Integrated Computer Aided Manufacturing (ICAM) project. Other companies such as Autodesk have also contributed to developing IGES along with many other standards for electronic information and geometry exchange.

As in other accepted and widely used formats, the main idea of IGES is based on the exchange of product information amongst CAD systems. It is reported that the CAD/CAM integration information network (CIIN) of the company Boeing has helped construct the fundamentals of IGES format. IGES is described as a neutral data format, which attempts to construct technical drawings in various forms such as 3D wireframe, surface, and volume models, and in other symbolic depictions. The IGES standard is known to embody two file formats: (i) fixed-length ASCII, and (ii) compressed ASCII. Fixed-length ASCII deposits information in 80-character entries, which is utilized by the IGESIMPORT and IGESEXPORT commands.

8.4.3 SET

In 1984, French engineers introduced the SET format—known as SET 85—which is an interface CAD/CAE/CAM standard regarding exchange of all data. New versions of SET in the form SET F88, SET S88 and SET V88 have also been produced in the following years. SET is a comprehensive format transferring crucial geometry information as well as others such as surface and wire-frame models, scientific data, and technical drawings. Due to its structure, some information such as size and geometric tolerances, material attributes and properties, and regulative data cannot be incorporated into specific model files; however, they can be accommodated in the drawing file for transferring purposes. SET proves to be a versatile platform making data transfer possible not only among computer-aided systems, but also from individual computerized systems to pivotal databases. When compared to IGES, its data structure is claimed to be significantly more universal, keeping data in smaller sizes, and eventually making processing times relatively lower.

8.4.4 VDA-FS

VDA-FS, standing for *Verband der Automobilindustries Flachenschnittstelle*, literally translating as *Association of the Automotive Industry Surface Interface*, is an interface format cultivated by the Association of the German Automobile Industry. Several versions of this interface have been generated aimed at transferring 3D geometric entities such as surfaces and curves, for which other formats cannot propose adequate answers. VDA-FS interface possesses a fixed 80-character record length as in IGES. A VDA-FS interface file starts with a header, which contains essential information such as title of the project, date created, date to expiry, the source, and user identification. Header is followed by the data field, which is formed with the user interface APT, and contains geometric entities.

8.4.5 PDES

PDES is an interface platform that targets data transfer for the whole system or product containing its product development and manufacturing phases. PDES can be said to additionally contain technological and administrative information on top of the contents of IGES format that it embodies. An example can be PDES encompassing IGES v4.0; however, with dissimilarities among their logical and physical framework. To overcome this issue, conversion programs are made for converting IGES into PDES. Computer language EXPRESS is utilized to improve PDES for integrating product data.

8.4.6 STEP

Standard for the Exchange of Product (STEP) is an established exchange format that contains product data, and information on required mechanisms and rationales facilitating data transfer between various CAD media. STEP aims to implement an independent mechanism that can sustain information on the system or the product during its lifespan. STEP format covers various aspects of product lifecycle including but not limited to production, design, maintenance, and scrapping.

Along with other CAD/CAE/CAM companies, Autodesk states that it has taken part in the development of many information and geometry exchange standards including ISO 10303 (STEP). The standard ISO 10303-203 (AP203 of STEP) expresses that it bolsters the final geometry of the models; however, it does not uphold the exchange of properties, limitations, or history information. In other words, AP203 file only incorporates geometrical and topological information belonging to the latest geometry of the model, including no details of the history of the project.

Arrangement of the STEP components can be listed as:

- characterization methods,
- integrated assets,
- utilization customs,
- abstract test collections,
- application procedures, and
- compliance testing.

As in PDES, STEP also utilizes the systematic specification language EXPRESS to define the product information in a rational and precise fashion.

These standards vary in terms of being the most used standard (DXF), language specific standard (SET and VDA-FS) to international standards (IGES, PDES and STEP). These international standards also vary in terms of evolution of technology e.g. the case of going from IGES to STEP. The definition of the acronyms and detail format of these geometry exchange standards can be found from the specific standard document and some simplified detail can also be obtained from various sources on the web and in the literature.

8.5 Other Standards

Other common standards used in geometry exchange form instead of geometry to geometry transfer are e.g. ISO 4343-2000 (BS 5110-2: 1978) cutter location data (CLDATA) for defining position of tool traversing on a surface; ISO/IEC/IEEE 24765:2017(en) and ISO/IEC 13817-1:1996(en) for systems and software, and IT (STL) to convert pixel cloud data into an image readable by a CAD APIs.

Similarly numerous ISO standards exist with (equivalent national and professional bodies standards) that provide important design data e.g. ASHRAE Boiler Code, ASTM, and SAE Material data, ASM machining parameter values and ASME tolerance data. These are just a few examples. Several sources are available for other technical standards, benchmarking information and regulatory information.

This includes NIST data for metrology and ISO standards for the environment.

8.6 Case Study: Application of ISO 841:2001 and ISO 6983-1:2009—Scope of Interpreter and Generation of Pseudo Code

The following case study applies ISO 6983-1: 2009 standard to all machines defined under ISO 841:2001. As ISO 841:2001 defines axes orientation based on right hand rule, all machines that have movement in primary, secondary, rotational, and tertiary axes can use the following explanation. The machine tool configuration can be that of a strategic machine or a non-strategic machine including but not limited to subtractive manufacturing machinery, additive manufacturing machinery, testing machines, scanning machines, measurement machines and any machine that has axis movement orientation as per ISO 841:2001.

Scope of the Interpreter

1. *Selection of Commands*

G-Commands

G00	Point to point movement (Rapid traverse) (only in XY Plane)
G01	Linear interpolation (only in XY Plane)
G02	Circular interpolation (clockwise) (only in XY Plane)
G03	Circular interpolation (Counterclockwise) (only in XY Plane)
G04	Dwell
G06	Parabolic interpolation (only in XY Plane)
G70	Inch programming
G71	Metric programming
G90	Absolute coordinate system
G91	Incremental coordinate system

M-Commands

MM00	Program stop
M02	Stop/end of program
M03	Spindle ON (clockwise)
M04	Spindle ON (counterclockwise)
M05	Spindle OFF
M06	Tool change
M07	Coolant on (Mist)
M08	Coolant on (Flood)
M09	Coolant OFF

X	Movement in (+ve) x-direction
Y	Movement in (+ve) y-direction
Z	Movement in (+ve) z-direction
I	Polar coordinate in case of Circular Interpolation (X Component) Tangent point in Case of Parabolic Interpolation (X Component)
J	Polar coordinate in case of Circular Interpolation (Y Component) Tangent Point in Case of Parabolic Interpolation (Y Component)
F	Feed
S	Spindle speed
T	Tool

+	Plus
%	Begin
(	Control out
)	Control In
/	Optional block skip
-	Minus
Space	Command Separator
Carriage Return (CR)	End of Block
EOF	End of File (inserted by the ASCII Text Editor such as Notepad)

2. *Format Classification*

The control block for the machine tool controller is:

N3 G2 M2 X3.2 Y3.2 Z2.1 I2.1 J2.1 F2 S4 T2 CR

3. *Symbol Used*:

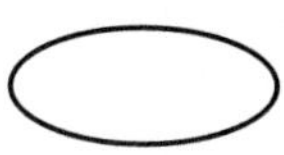	Oval indicates beginning or end of program
	Input or output from the program
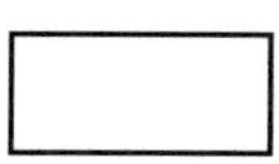	Assignment of value to variable or parameter

(continued)

(continued)

	Diamond indicates a point where decision is made
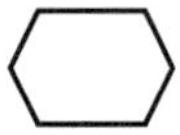	Hexagon indicates the beginning of a repetition
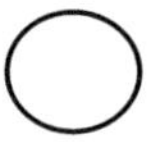	Combine flow line (Connector)
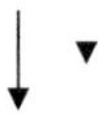	Direction in which program to be executed

The following are the main steps while reading the part program file for interpolation or Part Program into generating electronic signals to NEMA standard Stepper Motor:

(i) performs the command identification step.
(ii) performs the command formation step.
(iii) performs the command execution step.

The pseudo coding in the form of flow charts for the interpreter is performed in an open loop system for a light capacity knee and column type milling machine designed as per ISO 841:2001 and operating as per ISO 6983-1:2009. Both the standards have been reviewed by ISO in late 2020 and are valid standards. As it is a light machine, it is assumed that the milling machine is driven by stepper motors. Error trapping is not linked to the text editor. Preceding Zeros are not considered. The command in the Part Program must be in Control Block format.

1. **Command Identification:** Command identification identifies each letter in the part program.

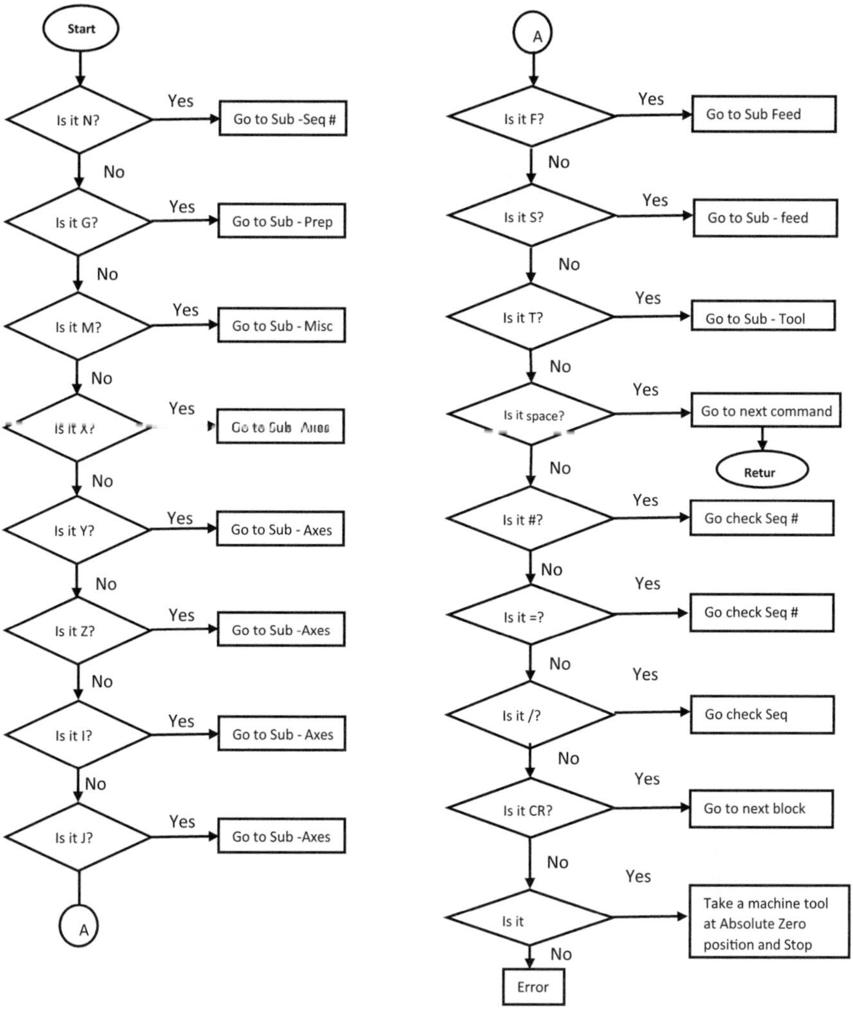

2. **Command Formation**

N3 (N Sequence):

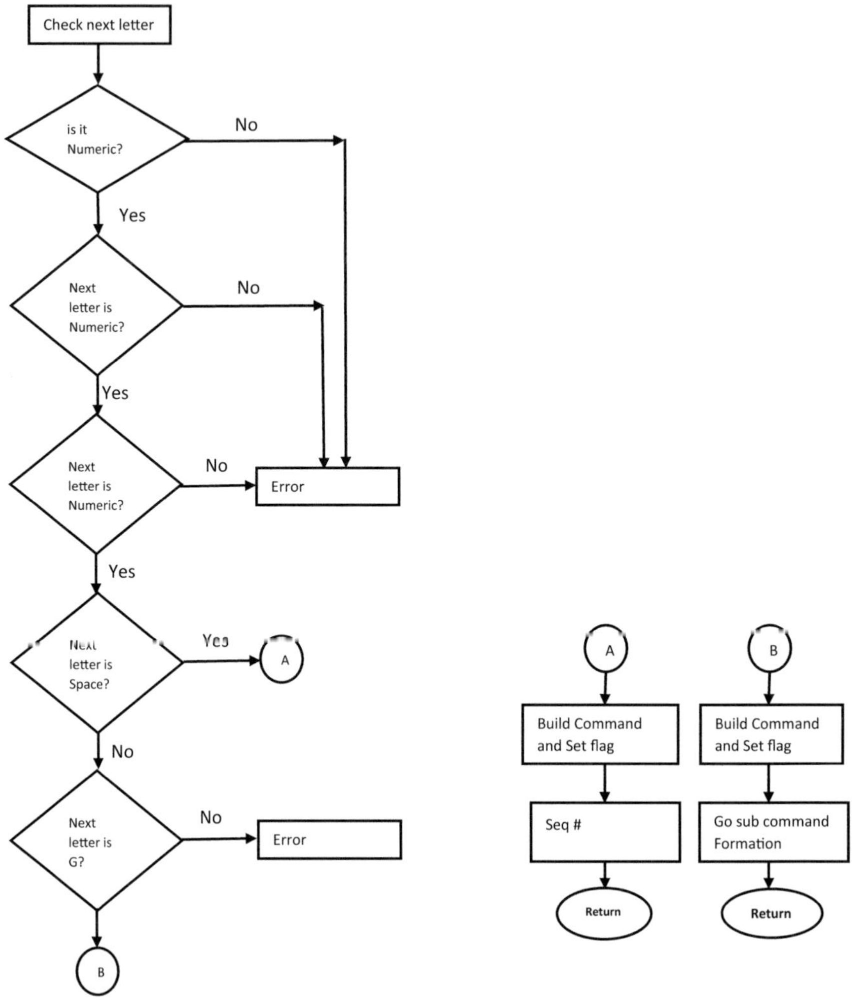

a. Sub Seq #
b. Sub Prep
c. Sub Misc

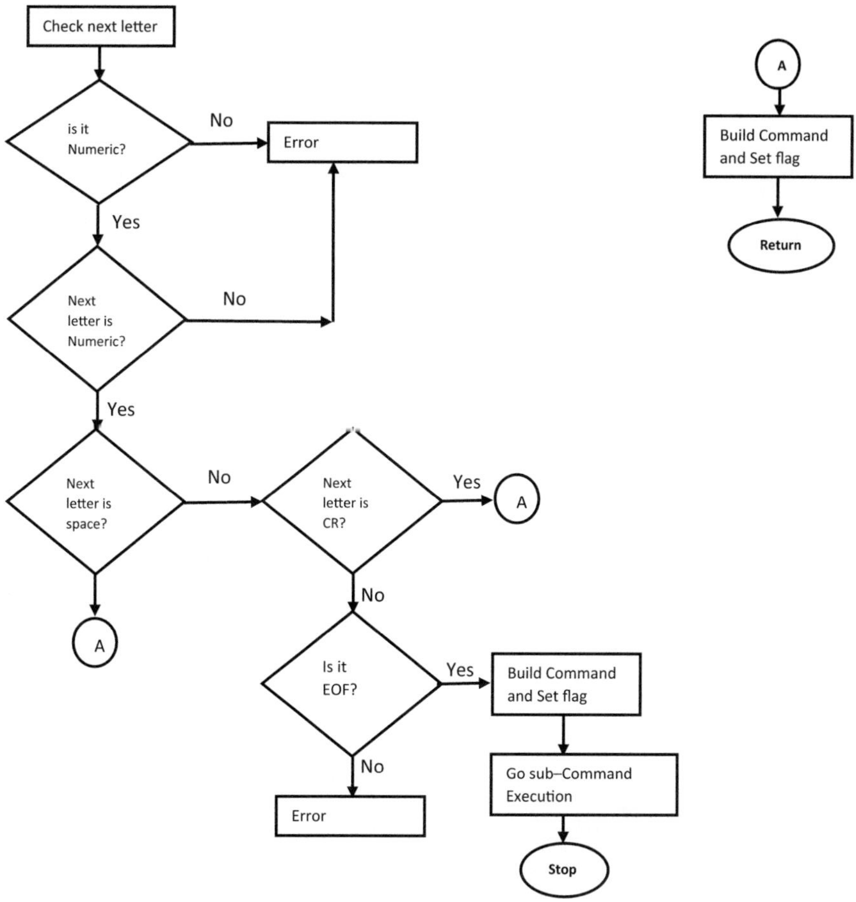

d. Sub Axes:

Axes and polar coordinate are all positive

For X, Y, Z, I and J (Important: Control Block Values for X, Y is same, and it is also same for Z, I and J)

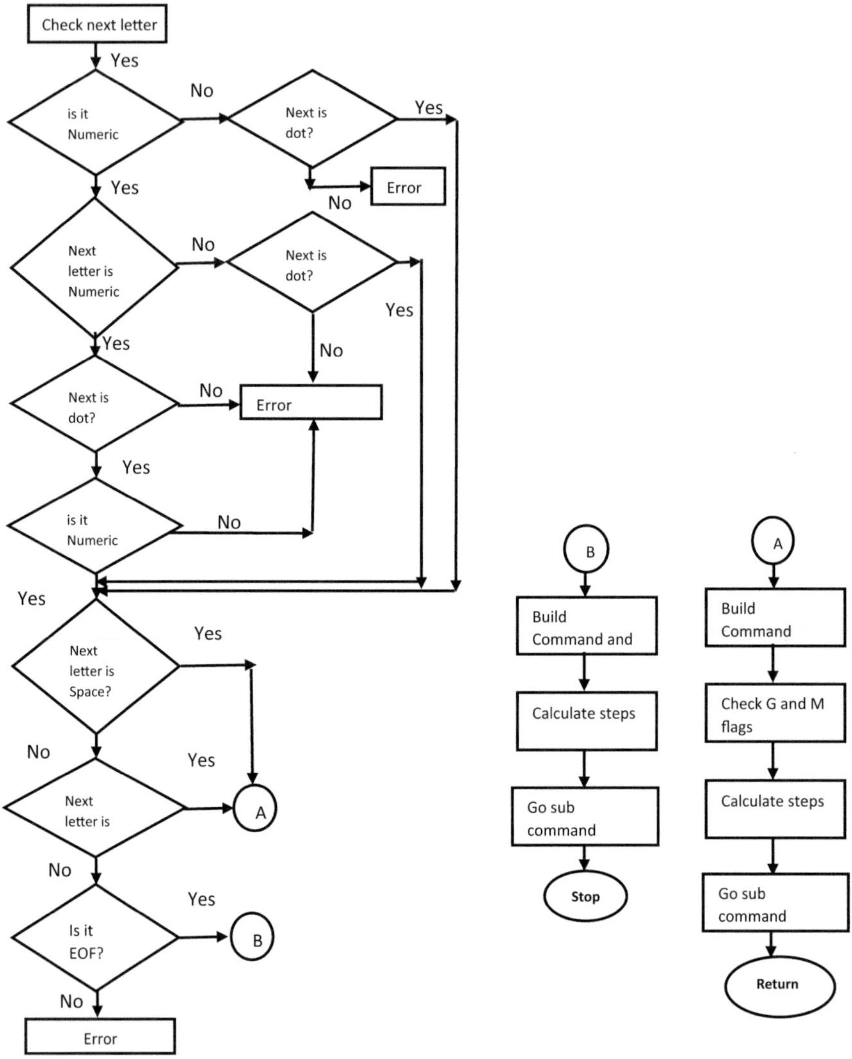

e. Sub Feed (for F)
f. Sub Tool (for T)

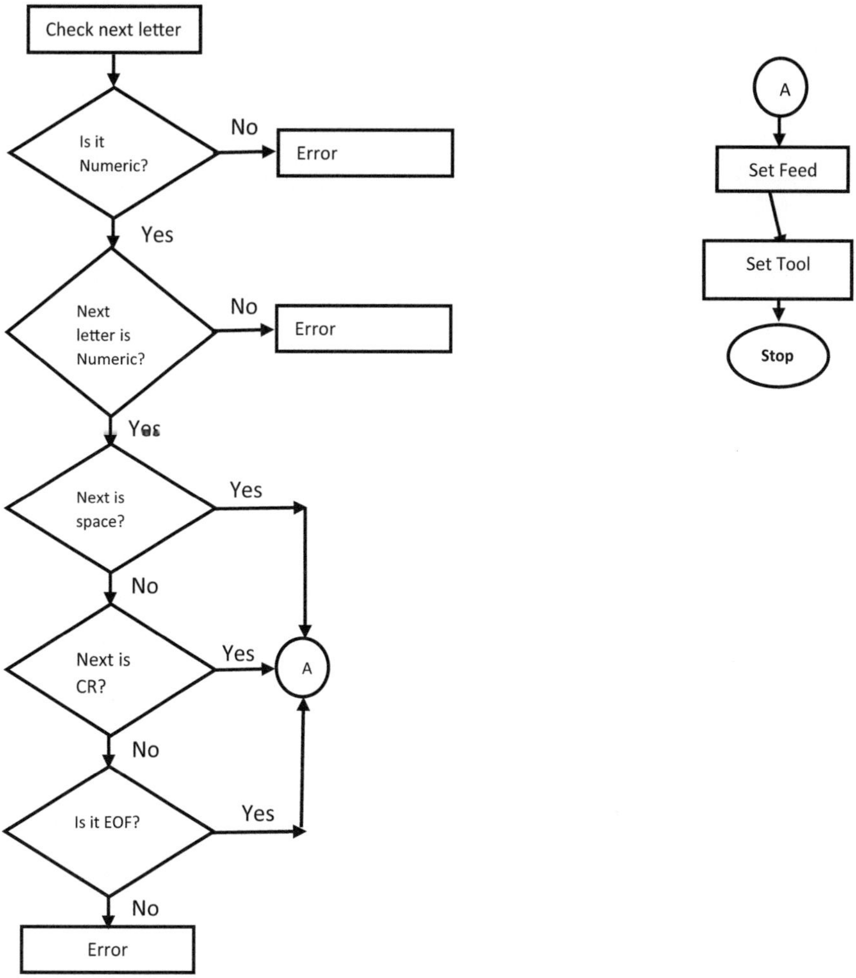

g. Sub Speed (For S)

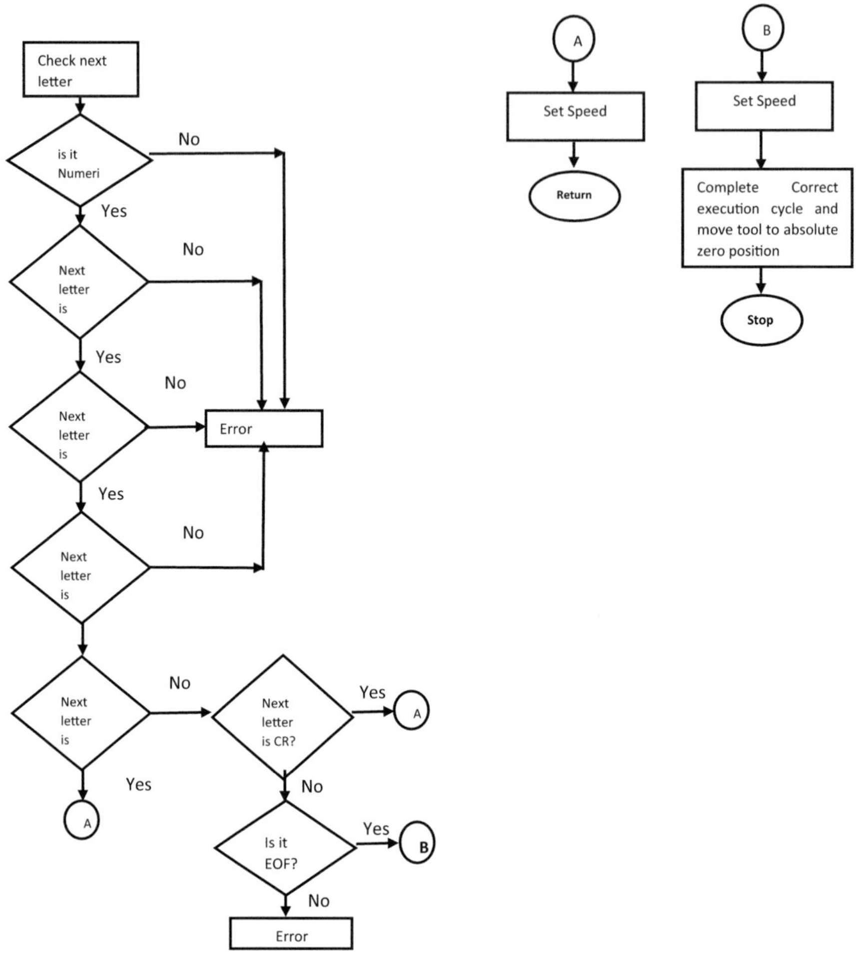
Check next letter
is it Numeri
No
Yes
Next letter is
No
Yes
Next letter is
No
Error
Yes
Next letter is
No
Next letter is
No
Next letter is CR?
Yes
A
Yes
A
No
Is it EOF?
Yes
B
No
Error
A
Set Speed
Return
B
Set Speed
Complete Correct execution cycle and move tool to absolute zero position
Stop

Rapid Traverse:

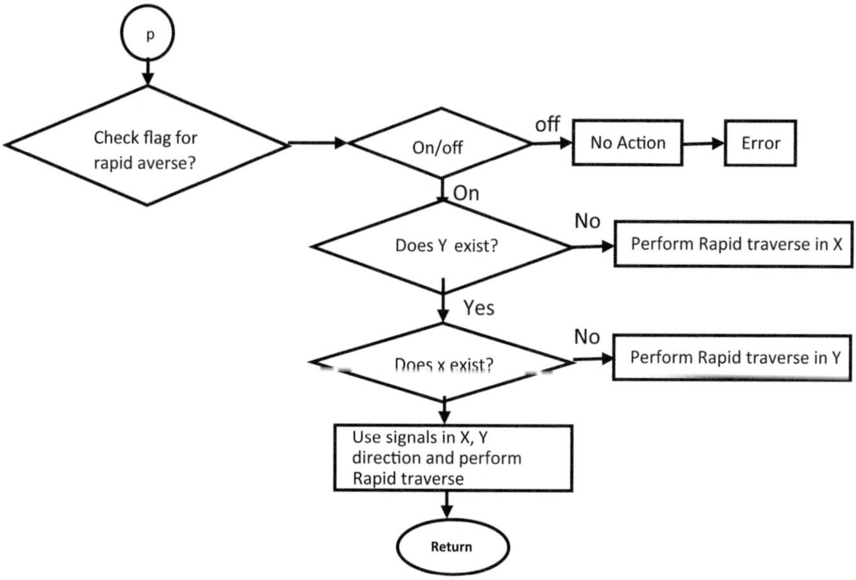

Calculate Steps

Assuming a 3 coil 1.8° step stepper motor as shown in the figure below.

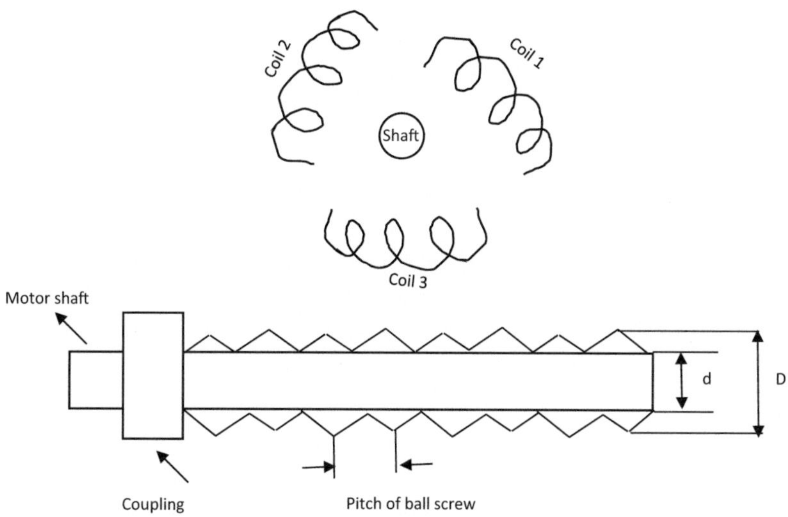

Energizing 1 coil gives 1.8° steps to shaft for 360° step to one complete rotation of shaft.

The number of steps will be $\frac{360^\circ}{1.8^\circ} = 200$ steps.

The linear distance traveled by ball screw for one complete rotation of shaft (crest to crest of ball screw) will be pitch of ball screw × 200 steps = πD.

h. Linear Interpolation:

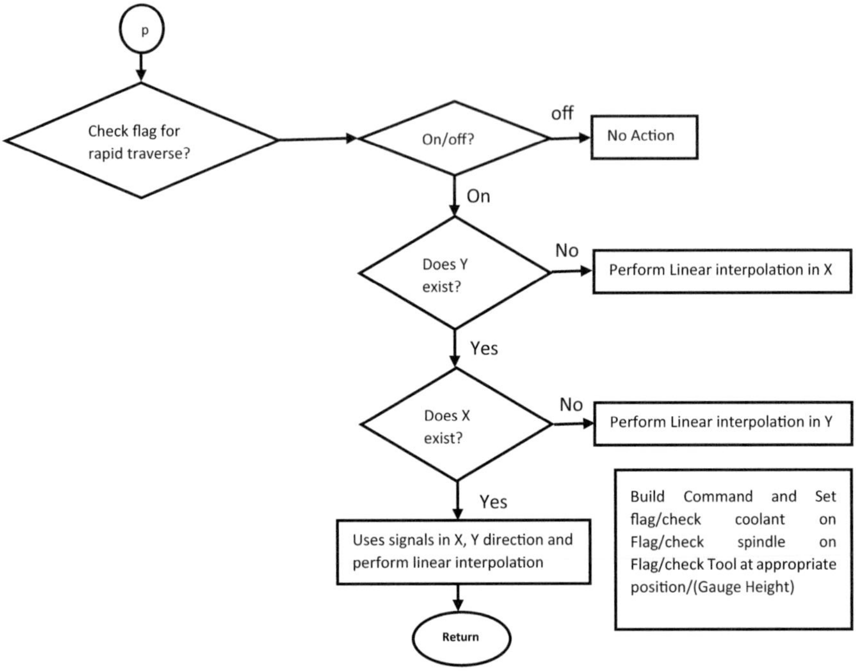

Linear Interpolation Python Code

```
# Calculate the tool path using parametric equation of straight line
dx = target_position.get("X", current_position["X"]) - current_position["X"]
dy = target_position.get("Y", current_position.get("Y", current_position["X"]))
- current_position["Y"]
dz = target_position.get("Z", current_position.get("Z", current_position["X"]))
- current_position["Z"]
steps = int(max(abs(dx), abs(dy), abs(dz)))
toolpath = []
for i in range(steps):
   x = current_position["X"] + (i / steps) * dx y = current_position["Y"] + (i /
steps) * dy z = current_position["Z"] + (i / steps) * dz toolpath.append({"X": x,
"Y": y, "Z": z})
toolpath.append(target_position)
```

- This part of the code calculates the tool path, which is the series of points that the tool (in this case, a stepper motor) will move through to follow the desired path.
- The toolpath is calculated using the parametric equation of a straight line, where the starting point is the current position, and the endpoint is the target position specified in the G01 command.

- The code first calculates the differences (dx, dy, dz) between the target position and the current position along the X, Y, and Z axes, respectively.
- It then calculates the number of steps required to move from the current position to the target position in the longest direction (i.e., the direction with the largest absolute difference). This is done using the maximum absolute value of dx, dy, and dz.
- Next, the code iterates through each step and calculates the position along the straight line using the parametric equation.
- The toolpath is stored as a list of dictionaries, where each dictionary represents a point on the toolpath with X, Y, and Z coordinates.
- Finally, the code appends the target position to the toolpath to ensure that the endpoint is included.

i. Clockwise Circular Interpolation XY plane:

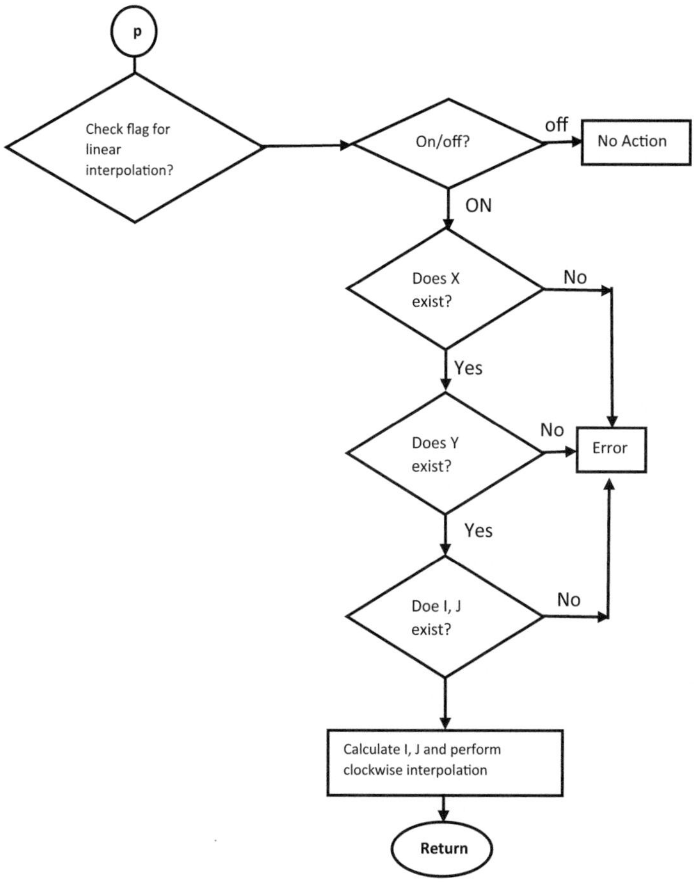

j. Counterclockwise Circular Interpolation:

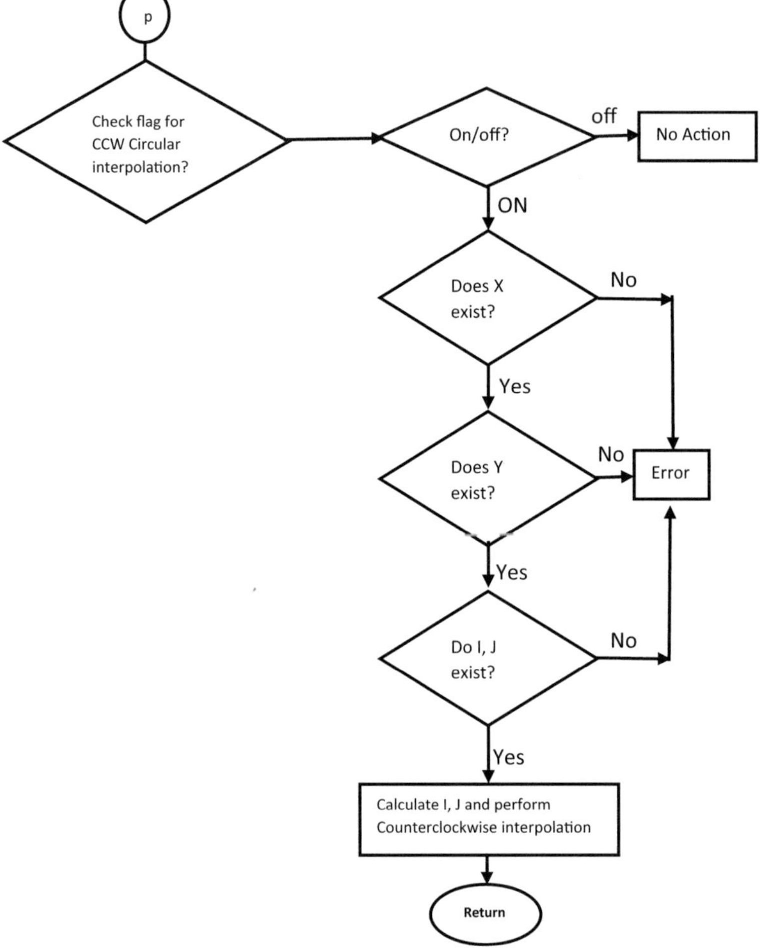

G02—Clockwise Arc Cutting

# Calculate toolpath using parametric equation of arc
center = {"X": arc["I"] + current_position["X"], "Y": arc["J"] + current_position["Y"], "Z":
current_position["Z"]}
radius = ((arc["I"] ** 2) + (arc["J"] ** 2)) ** 0.5
start_angle = math.atan2(current_position["Y"] - center["Y"], current_position["X"] - center["X"])

end_angle = math.atan2(arc_end[“Y”] - center[“Y”], arc_end[“X”] - center[“X”])
if arc[“Direction”] = = “CW”:
if end_angle > = start_angle:
end_angle - = 2 * math.pi else:
if end_angle < = start_angle:
end_angle + = 2 * math.pi angle_range = end_angle - start_angle
steps = int(abs(angle_range) * radius / 3.0) # 3 is a scaling factor, can be adjusted for smoother motion
toolpath = []
for i in range(steps):
angle = start_angle + i * angle_range / steps x = center[“X”] + radius * math.cos(angle)
y = center[“Y”] + radius * math.sin(angle)
z = current_position[“Z”]
a = current_position[“A”] - (angle - start_angle) * arc[“Radius”] * math.sin(arc[“Angle”]/2) * 2 c = current_position[“C”] - (angle - start_angle) * arc[“Radius”] * math.sin(arc[“Angle”]/2) * 2 toolpath.append({“X”: x, “Y”: y, “Z”: z, “A”: a, “C”: c})
toolpath.append(arc_end).

2: G03—CounterClockwise Arc Cutting

# Calculate tool path using parametric equation of arc
center = {“X”: arc[“I”] + current_position[“X”], “Y”: arc[“J”] + current_position[“Y”], “Z”:
current_position[“Z”]}
radius = ((arc[“I”] ** 2) + (arc[“J”] ** 2)) ** 0.5
start_angle = math.atan2(current_position[“Y”] - center[“Y”], current_position[“X”] - center[“X”])
end_angle = math.atan2(arc_end[“Y”] - center[“Y”], arc_end[“X”] - center[“X”])
if arc[“Direction”] = = “CCW”:
if end_angle < = start_angle:
end_angle + = 2 * math.pi else:
if end_angle > = start_angle:
end_angle - = 2 * math.pi angle_range = end_angle - start_angle
steps = int(abs(angle_range) * radius / 3.0) # 3 is a scaling factor, can be adjusted for smoother motion
toolpath = []

for i in range(steps):
angle = start_angle + i * angle_range / steps x = center[“X”] + radius * math.cos(angle).
y = center[“Y”] + radius * math.sin(angle)
z = current_position[“Z”].
a = current_position[“A”] + (angle - start_angle) * arc[“Radius”] * math.sin(arc[“Angle”]/2) * 2 c = current_position[“C”] + (angle - start_angle) * arc[“Radius”] * math.sin(arc[“Angle”]/2) * 2 toolpath.append({“X”: x, “Y”: y, “Z”: z, “A”: a, “C”: c})
toolpath.append(arc_end)

3: In-Depth Code Explanation (G02 AND G03)

- center = {“X”: arc[“I”] + current_position[“X”], “Y”: arc[“J”] + current_position[“Y”], “Z”: current_position[“Z”]}: This line calculates the center point of the arc using the current position, the arc’s I and J values, and sets the Z value to be the same as the current position’s Z value.
- radius = ((arc[“I”] ** 2) + (arc[“J”] ** 2)) ** 0.5: This line calculates the radius of the arc using the Pythagorean theorem.
- start_angle = math.atan2(current_position[“Y”] - center[“Y”], current_position[“X”] - center[“X”]): This line calculates the start angle of the arc using the current position and the center point.
- end_angle = math.atan2(arc_end[“Y”] - center[“Y”], arc_end[“X”] - center[“X”]): This line calculates the end angle of the arc using the arc endpoint and the center point.
- if arc[“Direction”] = = “CCW”: This line checks if the direction of the arc is counterclockwise.
- if end_angle < = start_angle: This line checks if the end angle is less than or equal to the start angle.
- end_angle + = 2 * math.pi: This line adjusts the end angle by adding 2 π to it if it is less than or equal to the start angle.
- Else if end_angle > = start_angle:: This line checks if the end angle is greater than or equal to the start angle.
- end_angle - = 2 * math.pi: Thi s li ne adj us ts the end angle by subtracting 2 π from it if it is greater than or equal to the start angle.
- angle_range = end_angle - start_angle: This line calculates the angle range of the arc.
- steps = int(abs(angle_range) * radius / 3.0): This line calculates the number of steps for the arc by multiplying the absolute value of the angle range, the radius, and dividing by 3.0. This value can be adjusted to change the smoothness of the motion.
- toolpath = [] #This line initializes an empty list/dictionary called toolpath.

for i in range(steps):

angle = start_angle + i * angle_range / steps x = center["X"] + radius * math.cos(angle)
y = center["Y"] + radius * math.sin(angle)
z = current_position["Z"]
a = current_position["A"] + (angle - start_angle) * arc["Radius"] * math.sin(arc["Angle"]/2) * 2 c = current_position["C"] + (angle - start_angle) * arc["Radius"] * math.sin(arc["Angle"]/2) * 2 toolpath.append({"X": x, "Y": y, "Z": z, "A": a, "C": c})
toolpath.append(arc_end)

- This portion of the code generates the toolpath for the given arc by creating a list of points that the tool should move to. It does this by dividing the angle range into a certain number of steps and calculating the corresponding x, y, and z coordinates for each step. It also calculates the corresponding A and C axis positions for each step using the arc radius and angle, as well as the starting and ending angles. Finally, it appends all of these points to the toolpath list, and then appends the final endpoint of the arc to the end of the toolpath list.

4: Difference Between G02 and G03 Codes

The difference between the two code snippets is that one of them generates a toolpath for clockwise cutting (G02) while the other generates a toolpath for counterclockwise cutting (G03). The difference can be observed in the condition-checking part of the code where the direction of the arc is determined. If the direction is counterclockwise, then the end angle is adjusted by adding 2 pi if it is less than the start angle to ensure that the toolpath stays within the arc. On the other hand, if the direction is clockwise, the end angle is adjusted by subtracting 2 pi if it is greater than the start angle. This is done to ensure that the toolpath follows the arc correctly in both cases.

Parabolic Interpolation[1]

Parabolic interpolation defines a parabolic span lying in any plane. It is recommended that the method of programming of the span should be by three points. The intermediate point and the end point shall be programmed in successive blocks. The example in Figure shows the geometric properties of the span and gives examples of meanings of the coordinate values to be programmed. The first block shall contain the G-function word (if not currently active):

(a) G06 Parabolic interpolation.
(b) the coordinates of the intermediate point.

[1] ISO 6983-1: 2009, Licensed to Wasim Ahmed Khan.

The following block shall contain the coordinates of the end point. The coordinates of all points shall be expressed in either absolute or incremental dimensions and addressed by any valid motion address, such as X, Y or Z.

Alternatively, the span may be defined in one block using interpolation parameters. The block shall contain a) the G-function word (if not currently active): G06 Parabolic interpolation; b) the coordinates of the end point, which shall be expressed in either absolute or incremental dimensions and addressed by any valid motion address, such as X, Y or Z; c) the interpolation parameters addressed by I, J, K. I, J, K should be the coordinates of the tangent intersection point. Key P0 start point Pt tangent intersection point P1 intermediate point (tangent parallel to P0, P2) P2 end point Using intermediate point (G06 Y X Z F) and end point (XYZ) Block Absolute dimensions Incremental dimensions:

First block

X = x1 Y = y1 Z = z1 X = x1 − x0 Y = y1 − y0 Z = z1 − z0

Second block

X = x1 Y = y1 Z = z1 X = x1 − x0 Y = y1 − y0 Z = z1 − z0

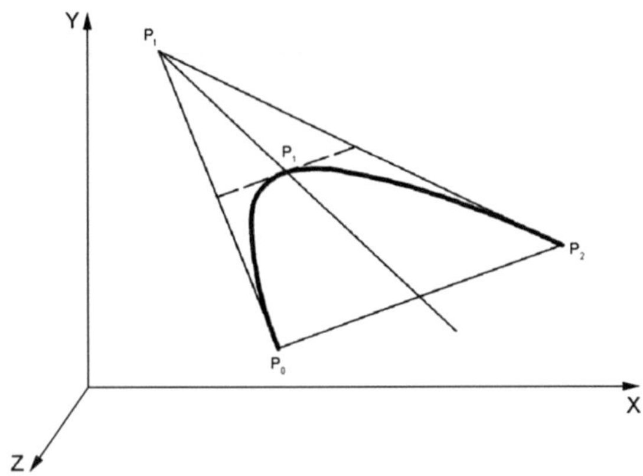

Example of parabolic interpolation in three axes by the method of programming in two successive blocks.

k.

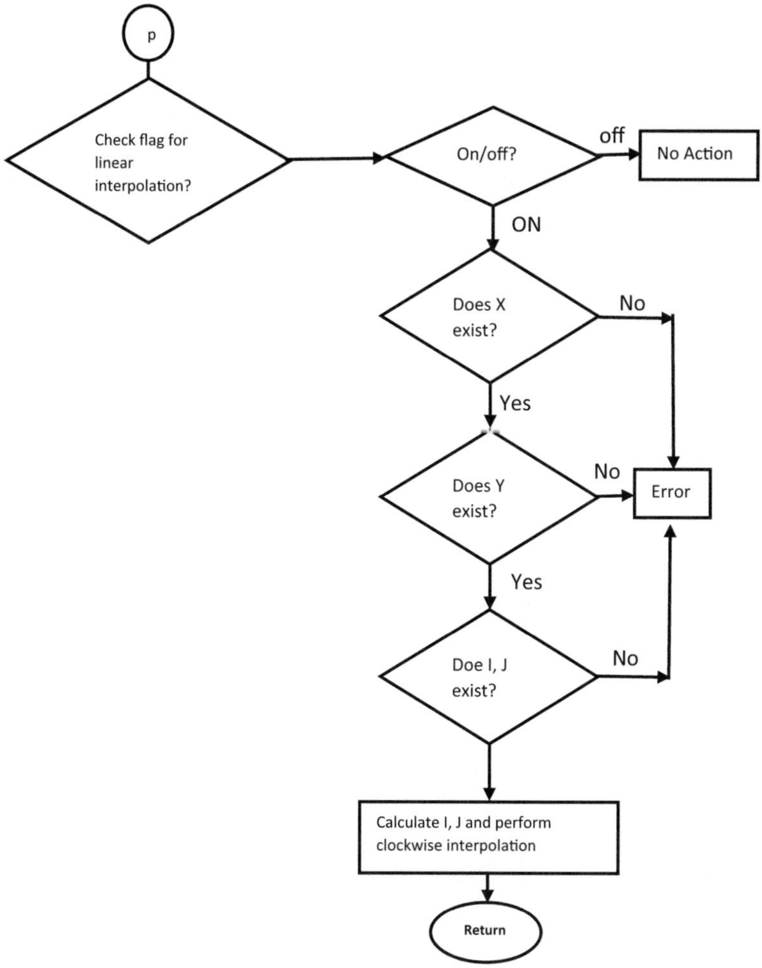

l. Counterclockwise Parabolic Interpolation in XY plane:

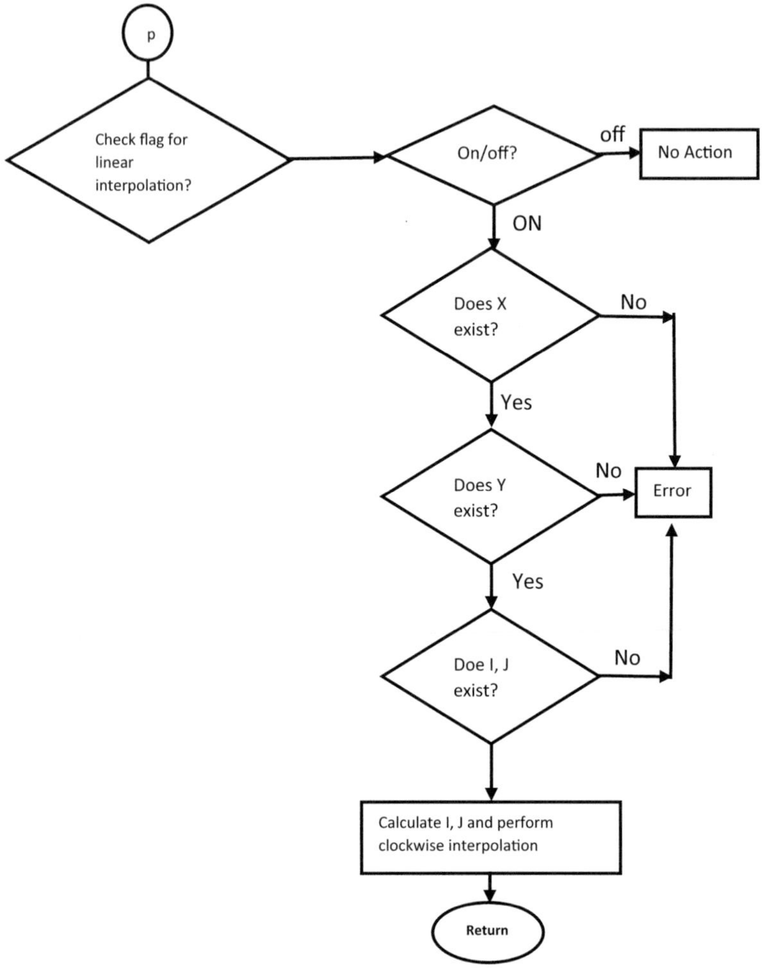

Parabolic Curve Interpolation

Def parabolic_interpolation(start_pos, end_pos, control_point, steps):
toolpath = []
for i in range(steps):
t = i / (steps - 1)
x = (1 - t) ** 2 * start_pos["X"] + 2 * (1 - t) * t * control_point["X"] + t ** 2 * end_pos["X"] y = (1 - t) ** 2 * start_pos["Y"] + 2 * (1 - t) * t * control_point["Y"] + t ** 2 * end_pos["Y"] z = (1 - t) ** 2 * start_pos["Z"] + 2 * (1 - t) * t * control_point["Z"] + t ** 2 * end_pos["Z"] a = start_pos["A"] + (1 - t) * (end_pos["A"] - start_pos["A"])

```
c = start_pos["C"] + (1 - t) * (end_pos["C"] - start_pos["C"])
toolpath.append({"X": x, "Y": y, "Z": z, "A": a, "C": c})
return toolpath
```

1 In-Depth Code Explanation

- The 'parabolic_interpolation' function takes four arguments: 'start_pos', 'end_pos', 'control_point', and 'steps'.
- 'start_pos' represents the starting position, 'end_pos' represents the ending position, and 'control_point' represents the control point for parabolic interpolation.
- 'steps' represents the number of steps to be taken during the interpolation process.
- The 'toolpath' list is initialized to store the interpolated positions.
- The 'for' loop iterates 'steps' times to calculate the interpolated positions.
- The parameter 't' is calculated as the current step divided by '(steps-1)'. It ranges from 0 to 1, representing the progress of the interpolation.
- The 'x', 'y', and 'z' positions are calculated using the parametric equation of a parabolic curve. The equation involves the start, control, and end positions, and it varies based on the parameter 't'.
- The 'a' value represents the rotation about the A-axis. It is calculated by linearly interpolating between the starting and ending A-axis positions based on the parameter 't'.
- The 'c' value represents the rotation of the C-axis. It is calculated by linearly interpolating between the starting and ending C-axis positions based on the parameter 't'.
- The calculated position, rotation about A-axis, and rotation about C-axis are added as a dictionary to the 'toolpath' list for each step.
- Once all the steps are completed, the 'toolpath' list containing the interpolated positions, A-axis rotations, and C-axis rotations is returned.

8.7 Learning Outcomes

The readers must understand that Industry 4.0 and Industry 5.0 standards are just the tip of the iceberg. There are thousands of legacy standards that are still being used and will remain so for the foreseeable future. Here in this book, only relevant standards are explained where there is a need to do so. The case study has briefly explained ISO 841: 2001 and ISO 6983-1: 2009. Both standards have been reviewed in 2017 and 2020 by ISO and are current versions of the standards.

Therefore, in the previous chapter standards related to the CNC were introduced, in this chapter Geometry Representation and Information Interchange standards are introduced while in the coming chapters standards related to Information Communication and other standards related to quantum computing will be introduced. At the same time, the authors admit that the different nations/corporate bodies may opt to adopt a different version (logic or language) of a standard that serves the purpose of the end user.

Just an indication for the reader regarding the development of a standard, the standard is a live document processed by a technical committee manned by the most experienced persons in the field. This committee keeps on looking at the standard over its life cycle where it is edited due to Technology Readiness Level and merged with other standards developed by international, national, and regional bodies, professional societies, and companies.

8.8 Review Questions

Q1. Specify the most common information interchange standards.
Q2. What are the most widely used, established geometry interchange standards?
Q3. Which company created DXF, DWG, and DXF, and when?
Q4. Who created IGES?
Q5. List down the forms of IGES for technical drawing construction.
Q6. What kinds of data does SET contain, and whatnot?
Q7. Where was VDA-FS generated?
Q8. What is the computer language used to improve PDES?
Q9. Name the components of the STEP format.

Bibliography

1. Sebastian-Coleman L (2012) Measuring data quality for ongoing improvement: a data quality assessment framework. Newnes
2. ISO 10303, Industrial automation systems and integration—product data representation and exchange—ISO 10303:1994. International Organization for Standardization (ISO), Geneva, Switzerland
3. "IGES 5.3 (ANSI-1996)", US Product Data Association, 1996-09-23, 650 p
4. Autodesk, Inc. (2011) Data exchange standards in the AEC industry, 4 p
5. Autodesk, Inc. (2023) About importing and exporting IGES files. Accessed in Nov 2023
6. Bharath VG, Rajashekar P (2015) Virtual manufacturing: a review. In: National conference emerging research areas mechanical engineering conference proceedings, pp 355–364
7. Padhy RP, Patra MR, Satapathy SC (2011) RDBMS to NoSQL: reviewing some next-generation non-relational database's. Int J Adv Eng Sci Technol 11(1):15–30
8. Lee YT, Luo Y (2004) A database design for the machine shop information model. US Department of Commerce, Technology Administration, National Institute of …

9. Koh K-H, de Souza R, Ho N-C (1995) Direct database-simulation of a job-shop. Int J Prod Econ 39(3):281–287
10. Chorafas DN (1993) Manufacturing databases and computer integrated systems. CRC Press
11. Kuehn W, Draschba C (2004) Simulation based job shop production analyser. In: Proceedings 18th European simulation multiconference

Chapter 9
Use of Databases for CAD/CAE/CAM/CIM

Prelude
The use of databases in Computer Integrated Manufacturing (correlating with Horizontal and Vertical Integration as a pillar of Industry 4.0) is essential to optimize the use of knowledge based on information gathered during several years and even more. The database types used in CIM are simple to complex in development and operation with their usage spanning over storing and sharing e.g., CAD models, final analysis in CAE, part programs in CAM.

At a SME level both relational and semantic databases are pertinent while at a more ambitious level bots may also operate along with semantic databases to align the businesses and minimize the effort that would otherwise be required. There needs to be a formal arrangement between SMEs or LSEs once the bots have identified a potential source of information.

9.1 Objectives

With the advent of modern computing technologies, databases in the form of a list during MS-DOS (Microsoft-Disc Operating System) time to today's semantic databases operating under Windows 365 and other operating systems with bots navigating the World Wide Web; the use of databases in Engineering Design and Manufacturing has taken a deep root. These databases are used to facilitate engineering design and manufacturing during systems approach to engineering design and systems approach to manufacturing and beyond.

Supplementary Information The online version contains supplementary material available at https://doi.org/10.1007/978-3-031-77106-4_9.

W. A. Khan et al., *Computer Aided Engineering Design and Manufacturing*, Springer Series in Advanced Manufacturing, https://doi.org/10.1007/978-3-031-77106-4_9

This chapter illustrates the basics of relational and semantic databases and defines the working of bots. It later indulges into some examples of technology databases helping designers and process planners to perform e.g., design using different materials with their mechanical and physical properties; and, e.g., selection of a machine element from machine elements manufacturer's database, etc.

The reader has the responsibility to update the case studies given in Chaps. 1 and 2 with the additional exercise of using relevant databases available at their place of learning or the company database which they can join free of cost.

As for the use of standards, the readers must develop an instinct that which database is relevant to their design and manufacturing application.

9.2 Relational and Semantic Databases

A simple definition of Relational Databases are entities that are stored in defined fields with appropriate number or character variable type. These fields correlate with each other and any change in the magnitude of one variable affects the other or the total outcome. The example of Relational Database is e.g. database of stored consumable and non-Consumable items at a Job Shop.

The Semantic databases in a specialized field such as discrete manufacturing have defined field names and variable type and size as well as a particular structure. It helps SMEs and LSEs to minimize duplicating the effort and share the knowledge base based on defined protocols for simplicity and ease.

9.3 Geometry Databases

'Geometry'—the central element of all discrete engineering products is the asset that has value and is maintained by SMEs and LSEs in some hierarchical manner that is accessible only to authenticated users. These users are of two types: one group belonging to the company's design office having complete access to this database and the second group are the personnel from the marketing department that can use the product data for marketing purposes and give its access to vendors as well for e.g. manufacturing of different parts of a car.

These hierarchical databases can be arranged using a coding a classification mechanism and are stored in a particular geometry format such as .dxf and .Iges or other pertinent geometry storage standards.

9.4 Materials Databases

Material databases are required to select physical and mechanical properties of all kinds of engineering materials for design calculation by the designers of discrete products. This information is mainly maintained to the full extent by professional institutions e.g. ASTM, SAE, ASME, ASHRAE and others.

The information is empirical and has appropriate units of measurement. The database is accessible electronically to members of the specific institution through paying a nominal fee.

9.5 Machine Elements Database

Machine elements such as gears, bearings, chains, sprockets, and others are commonly used in discrete products. Today's engineering design methodology is lean and does not embark upon designing the machine elements itself. It instead uses a design by selection procedure whereby the machine element manufacturer's database is used to acquire the major dimensions and related properties of the machine element and is amalgamated in the core design of the discrete product.

This was achieved through Constructive Solid Geometry (CSG) and the virtual reality pillar of Industry 4.0. This principle is described in Chap. 7. An example is provided there using SolidWorks 3D Experience.

These databases provide the geometry of the machine elements commonly in .dxf and .iges format. For 3D experience, currently (February 2024) the user either must use SolidWorks or have their own VR plugin.

9.6 Control Elements Database

Control elements are sensors, transducers, and electrical prime movers. The rest of the information is like machine elements. Currently, only software such as Proteus, Altium, orCAD allow Printed Circuit Board design simulation. Its integration with SolidWorks 3D experience type software is the future direction that will allow simulation of today's discrete products that are multidisciplinary in nature and are covered under the subject area, 'Mechatronics'.

9.7 Other Databases

Hydraulic element databases, Pneumatic element databases are currently available while other prime movers that use forms of input energy such as renewable energy forms like air, sea wave etc. are more futuristic and will attain more popularity as the world population becomes more conscious about sustainability.

9.8 Case Study: Technology Databases for Virtual Manufacturing System[1]

Discrete Manufacturing facilities arranged as Job Shop, generate a significant quantity of process data. This data is saved in databases and used as and when required for reporting to different management tiers. It is worth looking into the many sorts of databases and structures that are used to store discrete manufacturing-related data so that informed decisions can be made when choosing databases and schemas for reports. The main aim of this work is to construct a database consisting of physical and virtual machines at a job shop located at the Faculty of Mechanical Engineering, GIK Institute of Engineering Sciences and Technology, Pakistan. The process of creating a database starts from the collection of manuals from the workshop and extracting prime information related to different types of machine tools as well as for virtual machines referring to ISO 841-2001. A relational database management system is created having relational features. The relational database management system is designed for use in a discrete Virtual Manufacturing System (mERP).

9.8.1 Introduction

Manufacturing plants and large facilities can generate a lot of process data, which is saved in databases and used as data for reporting. It's worth looking into the many sorts of databases and structures that are used to store that data so that informed decisions can be made when choosing databases and schemas for reports. It is critical to consider the report's end user requirements while choosing a database type. Report generation speed, flexibility, and data availability must all be balanced in the reporting system. Small workshops, manufacturing, or assembly shops that plan and work on a batch of comparable projects, complete them, and then move on to a new set of jobs are referred to as job shops. The two jobs may not be identical and may demand distinct talents, as well as different operations planning.

[1] Developed by Hassan Ali, Hamza Abbas, Muhammad Umair under guidance from Wasim Ahmed Khan. Reproduced with Permission.

A production database contains the data which is used for production tasks such as creating, modifying, and updating features. Data in a production database can be utilized to build a digital or hard-copy map or chart, or a specific form of data, depending on the data model being employed. For example, databases for tools, jigs, fixtures, attachments, etc. in any production setup.

In a computer-generated environment that supports multidisciplinary collaborative product creation, virtual engineering integrates geometric models and related engineering tools like design, analysis, simulation, optimization, and decision-making tools. Virtual Manufacturing System (VMS) is the primary aspect of virtual engineering in manufacturing. VMS produces manufactured goods using computer-aided design models and manufacturing simulations that are integrated with real equipment. With the help of Augmented Reality/Virtual Reality, it is possible to "Manufacture in the Computer" and exchange models for use in simulation and control environments (mERP).

The routing and scheduling of jobs in job shop manufacturing are far more complicated than in other production systems. A job description outlines the needed sequence of the necessary production steps, which determines the workflow of each production task. The decision to schedule work and the resource to use for each production stage is based on the job requirements, real resource availability, and other factors.

A relational database is a type of data structure that enables you to link data from many "tables," or various kinds of data containers. A data bucket needs to include what is referred to as an index or key (that enables it to specifically identify any atomic component information in the bucket). Additional information buckets could make use of that key to build a connection between their data atoms and the atom that the key pointed out.

Manufacturing databases aid in the management of parts, components, and stocks for specific assemblies. It holds a vast amount of data that is required to offer the necessary information for manufacturing efficiency, automation, and speedy decision-making.

Intermittent production is a feature of the job-shop scheduling problem. A job shop is a collection of all equipment and other resources (such as tools, labor, and so on) that are linked with a specific set of jobs. The goal is to maximize some specified yet observable characteristics of the order by ordering the activities to be performed on each machine based on routing and shop restrictions. Because it must balance several objectives, job-shop scheduling is inherently complicated.

Overall data consistency is aided by the consistent adoption of data model standards. Data standards and naming conventions help to ensure that concepts are represented consistently. It is vital to analyze the condition of the data model as a starting point for understanding the quality of data in a database. It's crucial to understand what data representation standards exist and how well those have been enforced inside the data model itself.

9.8.2 *Methodology for Generating Database for Job Shop*

Job shops having machine shops, for example, consist of metal cutting, grinding, shaping, and polishing machines and work on a variety of jobs involving various geometries, shapes, dimensions, and materials. They are usually used as ancillaries or as part of the supply chain for large companies' maintenance. The following are characteristics of such a manufacturing system:

- Product engineering with a lot of flexibility.
- Exceptional growth potential (machines are easily added or substituted by adding/removing vendors or by having upgradation at the main production facility).
- High flexibility in production volume (owing to minimal increases in productive capacity)
- Low obsolescence (machines are often multifunctional) High resilience to machine breakdowns.

(a) **Database development**

A database model is created to correspond to the machine shop data model. A set of relational tables is provided in a tree-like layout in the database. A database's essential building components are tables. On a datasheet, a table is a collection of chosen data structured into fields (columns) and records (rows). A record maintains a set of values specified by fields, whereas a field indicates a data type for a group of values in a table. The guidelines that were utilized to create the database are described in this section. It also includes examples of mapping from an XML element to a relational table, a table design view, and table relationships.

(b) **Guidelines**

When creating the database structure, keep the following criteria in mind:

- In the database model elements are represented as "tables" in the information model. Exceptional growth potential (machines are easily added or substituted).
- In the following table, traits, and child elements are represented as "fields."
- In the accompanying table, parent elements are represented as "fields." These fields are necessary for system upkeep.
- If a child element exists, a cardinality connection is defined with the specification of each field.
- The INDEX field is a system-required field for each table.
- A field that is a referred table, a referring table, or a system-used item is in the upper case. All additional fields are displayed in the lower half of the page.

(c) **Data structure in a design view in Microsoft Access**

In Microsoft Access, each table's fields are created using Design View. In this database management software, Field, Data Type, and Description are the three columns that make up a design view. The "Field" column indicates the element's properties and distinct elements that the database system, or the referenced tables, employ. The "Data Type" column stipulates the field's data type format. The "Description" column is used to store evidence on the attribute's use, domain, definition (which establishes the internal association between tables), reference, and cardinality relationship, among other things.

(d) **Data collection and implementation**

A database is a logically organized collection of structured data that is usually saved electronically in a computer system or on the cloud. A Database Management System is usually in charge of a database (DBMS). The data, the DBMS, and the applications collectively are referred to as a database system, which is commonly abbreviated as just a database.

In this description of work performed, the task is to make a digital database of the existing machines (Physical and Virtual) in the job shop of Faculty of Mechanical Engineering, GIK Institute. The task was started by first taking the hard copy of the Machine Manuals, mostly provided with the machines at the time of purchase to ensure the smooth operation of the machine and to make it easy for the machine Operator to understand.

The basic parameters/features and working of the machine. Then the required data was extracted from the hard copies of the manuals such as the machine dimensions, number of tool posts, machine weight, operating axis, etc., and fed all that data in an excel form to organize it properly. The data related to the virtual machines were taken from the virtual machines database which is based on ISO 841: 2001 for defining axes orientation and machine tool configuration. The physical machines in the job shop were supplemented by virtual models of the machines as per axis orientation and machine tool configuration provided in ISO 841: 2001 standard. This essentially means that some physical manual machines also had virtual models and some virtual machines had integrating physical machine those were retrofitted in the considered job shop (Fig. 9.1).

Then the data was fed into Microsoft Access to make the required Database (Figs. 9.2, 9.3 and 9.4).

9.8.3 Conclusion

This research work focuses on the formation of databases for job shops in Microsoft Access 2007 (2016 file format) for the job shop of the Faculty of Mechanical Engineering, GIK Institute. The database for the job shop of FME was successfully created which includes different parameters for different machines like lathe machines, milling machines, sawing machines, drilling machines, boring machines, band saw, and manufacturing systems lab such as CNC milling, CNC turning, CNC router, general bench drilling machine, jig boring, radial drilling machine, shaper machine, surface grinding machine, additive manufacturing machines, and virtual

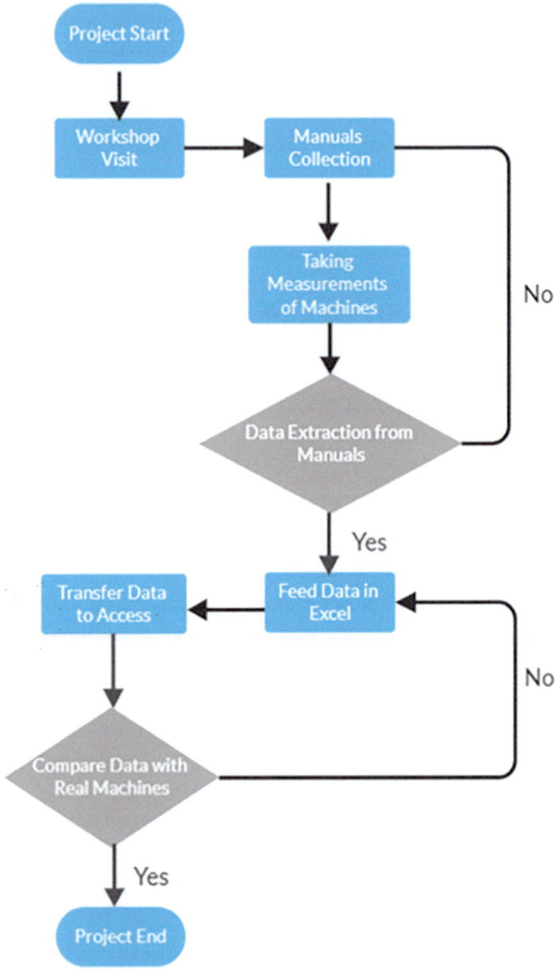

Fig. 9.1 Flow chart of the conducted work

machines, etc. In the future, it will be possible to add more data to the database and use it for the automation of the job shop as a part of the Virtual Manufacturing System being developed at the GIK Institute.

This database, by considering recognized field names and with ability to let bots (short programs those navigate the World Wide Web), can easily be converted in a semantic database.

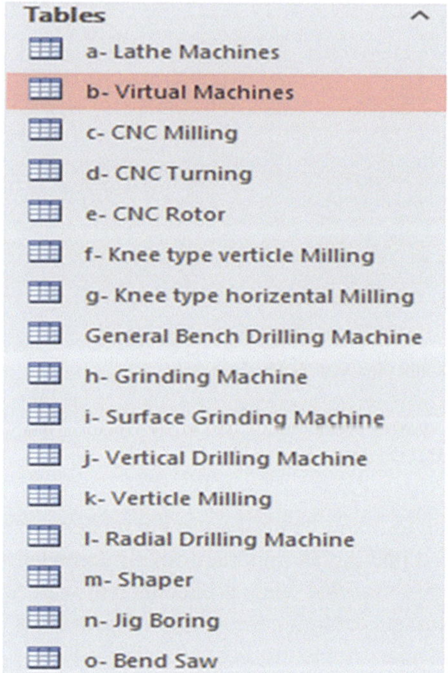

Fig. 9.2 Microsoft Access worksheet for different machines

a- Lathe Machines

Sr No	MC No	MC Type	Center to cer	Swing over b	Swing over sl	Swing over g	Axes Orienta	Tool cross	Machine dim	Tool changer	Width of bed	Cross feed (n	Longitudinal	Gap type
1	L-1630	Lathe	750	410	250	580	X,Z,A	Yes	1875*750*168 5	No	275	0.06-0.92	0.08-1.15	Swing in gap Width of gap
2	BL115	Lathe	610	226	136	371	X,Z,A	Yes	1265*155*138	No	155	0.023-0.67	0.05-1.38	Swing in gap Width of gap
3	CS6140 & CS6140B CS6240 &	Lathe	0	400	220	630	X,Z,A	Yes	2382*975*123 0	No	0	0.027-1.07	0.063-2.52	Swing in gap Width of gap
4	CS6150 &CS6150B CS6250	Lathe	0	500	300	710	X,Z,A	Yes	2382*975*127 0	No	0	0.027-1.07	0.063-2.52	Swing in gap Width of gap
5	CS6166 &CS6166B CS6266	Lathe	0	660	420	870	X,Z,A	Yes	2382*975*135 0	No	0	0.027-1.07	0.063-2.52	Swing in gap Width of gap
6	CY6140	Universal Lathe	750	400	220	0	X,Z,A	Yes	2375*1043*12 75	No	0	0.027-1.07	0.063-2.52	Swing in gap Width of gap
7	CY6240	Gap-bed Lathe	1500	400	220	630	X,Z,A	Yes	3125*1043*12 75	No	0	0.027-1.07	0.063-2.52	Swing in gap Width of gap
8	CY6150	Universal Lathe	750	500	300	0	X,Z,A	Yes	2375*1077*13 15	No	0	0.027-1.07	0.063-2.52	Swing in gap Width of gap
9	CY6250	Gap-bed Lathe	1500	500	300	110	X,Z,A	Yes	3125*1077*13 15	No	0	0.027-1.07	0.063-2.52	Swing in gap Width of gap

Fig. 9.3 Microsoft Access worksheet for Lathe Machine

a- Lathe Machines | b- Virtual Machines

Sr No	MC Type	Length	Width	Height	Axes Orient	Table Lengtl	Table Width
1	5 Axis Milling	851	600	800	X,Y,Z,A,C	229	100
2	Twin Turret Lathe (A3)	5063	1552	2247	X,Z,U,W,R,C	0	0
3	Vertical Turning & Boring Lathe (A4)	82	36	85	X,U,Z,C	0	0
4	3 Axis Milling	31	31	31	X,Y,Z	11	6
5	Portal Type Milling	12	24	26	X,Y,Z	12	14
6	Gantry Type Milling	37	28	36	X,Y,Z	24	28
7	Planar Type Horizental Boring Machine	142	78	123	X,Y,Z	101	78
8	Profile and Contouring Milling Machine	250	162	133	X,Y,Z	170	113
9	Profile and Contouring Milling Machine	142	132	123	X,Y,Z,A	101	78
10	Profile and Contouring Milling Machine	24	14	30	X,Y,Z,A	17	8
11	Profile and Contour Milling Machine wi	24	14	30	X,Y,Z,A	17	8
12	External Cylinderical Grinding Machine	22	4	12	X,Z,W,R,C,B	0	0
13	Tool and Cutter Grinding Machine	22	4	12	X,Z,W,R,C,B	0	0
14	Open side Planar	19	42	24	X,Y,Z	8	23

Fig. 9.4 Microsoft worksheet for Virtual Machines

9.9 Learning Outcomes

The reader is expected to orient themselves in the modern scenario where artificial intelligence has started playing an important role in knowledge delivery to human beings. It is important to note that while producing text with Apps like Chat GPT is an important step in human evolution the misuse of such tools will be monitored by modern versions of plagiarism monitoring software Turnitin.

There is still a time when the reader can produce Engineering Design and Production Methods automatically. However, the information provided here is not a stumbling block to success in this direction, instead it is a stepping stone in this direction.

9.10 Review Questions

Q1. Differentiate between relational and semantic databases.
Q2. What are geometry databases?
Q3. What are Machine Element and Control Element databases?
Q4. What are the different guidelines for creating a database?
Q5. Describe a general database development.
Q6. What is a semantic database?
Q7. What is a bot?
Q8. How does a bot operate?
Q9. Why are bots important and what is their productive use?
Q10. What are the applications of bots in the Computer Integrated Manufacturing (CIM) environment?

Bibliography

1. DQAF measurement types 6, p 99999
2. Argis Desktop (2001) Production database
3. Bharath VG, Rajashekar P (2015) Virtual manufacturing: a review. In: National conference emerging research areas mechanical engineering conference proceedings, pp 355–364
4. Padhy RP, Patra MR, Satapathy SC (2011) RDBMS to NoSQL: reviewing some next-generation non-relational database's. Int J Adv Eng Sci Technol 11(1):15–30
5. MIS Class Blog, Databases in manufacturing. http://misclassblog.com/database-application/databases-in-manufacturing/
6. Koh K-H, de Souza R, Ho N-C (1995) Direct database-simulation of a job-shop. Int J Prod Econ 39(3):281–287
7. Comeau S, Choosing the right database type, blog
8. Kuehn W, Draschba C (2004) Simulation based job shop production analyser. In: Proceedings 18th European simulation multiconference, Citeseer
9. Lee YT, Luo Y (2004) A database design for the machine shop information model. US Department of Commerce, Technology Administration, National Institute of …
10. Chorafas DN (1993) Manufacturing databases and computer integrated systems. CRC Press
11. Sebastian-Coleman L (2012) Measuring data quality for ongoing improvement: a data quality assessment framework. Newnes

Part IV
Process and Production Planning

Prelude

The bridge between engineering design and manufacturing is covered by the process and the production planning. Referring to the process flow diagram (PFD) of a small and medium enterprise (SME) detailed in Part I, the dynamics of multiple orders entering a SME requires mainly the following types of services:

1. Material
2. Energy, and
3. Information.

With a specified lead time, for the sustainable operation of the SME (which is quite akin to all types of manufacturing systems), the orders should keep flowing in creating technical, economic, and social impacts to all humans involved in a particular vicinity (Space).

Historically, manufacturing at SME by the low to middle class of a nation is the backbone of that nation. Large-scale enterprises (normally the project shop, flow line, and continuous manufacturing systems) scale up the technical and human resources required for production in terms of expertise, nature of equipment, and production rate. The impact of the operation of large-scale enterprises is much higher in terms of technical, economic, social, and environmental impacts.

Dynamics of Discrete and Continuous Manufacturing Facilities

From the order to delivery of a product to use of product to the end of product's life, all aspects are stochastic in nature, and it is the human interference that gives product a definite transition in each of the above time span.

Dynamics of discrete and continuous manufacturing facilities are best described by the methods of producing the product using certain methodology and its impact on nature during the period the product is used in its useful life. Producing a product while abiding by the factors such as complete system approach to engineering design and manufacturing for UI, UX, and; with no compromise on quality and being innovative, customized, multidisciplinary, and sustainable puts a toll on the cost which may be prohibitive for the user thus negating the basic ideology of design and manufacturing which is wealth creation. A balance is thus required to have optimized parameters while remaining within all benchmarks and legal frameworks. And consequently, it affects the process and the production planning.

Product Life Management

Product Life Management (PLM) is part of the product life cycle. It starts from the establishment of a need for a product and ends at after-sales service for the product. PLM is encapsulated in the product life cycle.

Manufacturing Execution Systems

Manufacturing Execution Systems (MES) is the process flow diagram of the manufacturing system including input requirement for the ordered product, traceability in the manufacturing system, material and energy management and work in progress (WIP), and other intangible services as they occur as well as the product dispatch to distributor/client. MES is encapsulated in PLM.

Chapter 10
Computer-Aided Process Planning

Prelude
Informally the process planning remained the part of interface between design and manufacturing either without recording it for future use or with recording it for the sake of use for similar products in philosophies such as group technologies having certain coding and classification scheme.

This is an evolution from the stone age to the computer age. Earlier only the most experienced worker was trusted with this function and today it is a process involving resources from different departments and databases of a different nature with Industry 4.0 technology such as Artificial Intelligence available conveniently.

10.1 Introduction

A process plan is a document in any suitable format that describes the steps required to convert raw materials into finished products. As such no process planning was done in the days of the conventional approach to engineering design and manufacturing, rather the most experienced person in these organizations had the de facto knowledge of these steps, and no proper documentation was made. In modern days both manual and computer-aided procedures exist to store this information.

In manual process planning different formats containing the manufacturing information derived for the manufacture of discrete products are presented. Some formats are presented in Sect. 10.2. The computer-aided process plans may be derived for the discrete products and can either be generative process plans or variant process plans. The generative process plan is a computerized version of the manual process plan.

Supplementary Information The online version contains supplementary material available at https://doi.org/10.1007/978-3-031-77106-4_10.

W. A. Khan et al., *Computer Aided Engineering Design and Manufacturing*, Springer Series in Advanced Manufacturing, https://doi.org/10.1007/978-3-031-77106-4_10

While the variant process plan is derived from the existing process plans, those are stored in a coding and classification-based manufacturing setup normally operating under group technology like production philosophy.

The reader is practically introduced to these concepts in the coming sections and through the case study on generative process planning.

10.2 Process Planning[1]

Manual Processes planning is performed on paper sheets. It may be a single sheet or multiple sheets describing details such as:

- The route to machines the part will take for transition from raw material to finished stage. The machine tools or other manufacturing facilities are numbered, and their sequence will determine the route the part will take to get to the finished stage. A typical Route Sheet is exemplified in Fig. 10.1.
- The procedure to hold the workpiece at each manufacturing facility is detailed in a procedure sheet. A typical Procedure Sheet is provided in Fig. 10.2.
- The manufacturing parameters at each manufacturing facility detailed in Route and Procedure sheets are listed. A typical process sheet is given in Fig. 10.3.

There may be multiple Route, Procedure and Process Sheets to describe all the manufacturing requirements to take the part, structure, assembly, mechanism to the stage of a machine (Product).

Process Planning applies equally to additive and subtractive manufacturing.

Project Code No:
Route Sheet No:
Procedure No:
Process Sheet No:

Route Sheet

Project		Planner		Material		Batch No. S.No. From To		Part Name Part No.	
Qty. to be Processed				Job Order No.				Page No.	
Op. No.	Order / Process	W.S. No.	SECT.	Qty. Load	Loading Date	Qty. Accp.	Qty. Rej.	Operator Sign	Read by Date

Fig. 10.1 A typical route sheet

[1] Process Planning Sheets used at Precision Engineering complex [an affiliated organization of Pakistan International Airline (PIA)] producing Boeing Parts. These Process Planning Sheets were used in the 1990s. Reproduced as an example.

Project Code No:
Process Sheet No:
Revision No:
Issued Ac:

Procedure Sheet

Project			Planner / Material	Batch No. S.No. From To			Part Name Part No.			
Qty. to be Manufactured			Job Order No.		STD		Actual		Sign	
Date	Op. No.	W.S. No.	Tools / Operations	Met Dtm	S/T Min	Rum Min	S/T Min	Rum Min	Opr	Sup Q/A

Fig. 10.2 A typical procedure sheet

Project Code No:
Process Sheet No:
Revision No:
Issued Ac:

Process Sheet

Project	Planner	Material	Batch No. S.No. From To		Part Name Part No.			
Qty. to be Processed		Job Order No.	STD		Actual		Sign	
Date	Pr. No.	Process Detail	S/T Min	Rum Min	S/T Min	Rum Min	Opr	Sup Q/A

Fig. 10.3 A typical process sheet

10.3 Computer-Aided Process Planning

Computer Aided Process Planning is done using two methods. These are:

10.3.1 Generative Process Planning

In Generative Process Planning, the methods used in Manual Process Planning are digitized without much change.

10.3.2 Variant Process Planning

In Variant Process Planning, Coding and Classification is used to categorize the part, structure, assembly, mechanism, and machine. Cross referencing is done to retrieve a similar process plan for a new part and is edited for the new part.

Computer Aided Process Planning is an established part of production engineering and commercial software is available to perform this activity.

10.4 Case Study: Variant Process Planning for a Job Shop

Process planning in discrete manufacturing using either manual method or performed using Computer Aided Process Planning (CAPP). CAPP is further divided into Generative Process Planning (GPP) and Variant Process Planning (VPP). The VPP approach uses group technology principles to classify parts into part families based on geometric and manufacturing properties. VPP helps the human operator to select and modify the process plan from a database of the machine. It is a much easier and more economical process of planning alternatives for manufacturing similar components for each family. In this research paper, we develop a variant process plan for the compressor piston using the VPP methodology. All manufacturing procedures and tools information are stored in a database. The VPP recognizes and retrieves the existing process plane of a similar part from the database and generates a processing plane of some alteration for the new part. Memory (database) is used to retrieve the manufacturing plan for that part. If the new part differs significantly from the previous ones, there will be no similar features. The plan can then be modified by a human operator to accommodate the new part. Developed VPP is used as a module of discrete manufacturing in a virtual manufacturing suite of software being developed at the GIK Institute.

10.4.1 Introduction

Production control is the application of scientific management concepts and ideas to ensure that industrial resources are used effectively. Production scheduling is a link between planning the steps of a process and sending out jobs. Process planning is what goes into production scheduling. The production schedule is the basis for task remitting. Process planning is one of the most important parts of a schedule, so there should be a lot of research on how the two work together. If they are looked at together instead of separately, better results will be seen, such as less inventory, no more order delays, less complicated scheduling, better use of resources, and so on.

Process planning turns the information in a part design into instructions for making the part. Planners of manual processes get better at their jobs over time. To automate process planning, experience and knowledge must be written down as procedures and methods that a computer can understand. These systems must know what processes are available and what they can do. Capabilities include the shapes and size restrictions that a process can produce, such as the upper and lower bounds of the machine tools and fixtures used in each process. The geometry and topology of the component must be analyzed by the automated process planning system to identify the high-level machining relevant shape information, such as slots, holes, and pockets. Since Chrysidoids, Chan, and Cobb's contribution to Integrated Process Planning and Scheduling (IPPS) [1], different practitioners have taken their ideas and put them into the following three categories:

- The flexible and nonlinear process planning approach used by Jain and Singh (2006), Kumar and Rajotia (2003), and Leung et al. (2010) ensure that all possible process plans are examined and considered [2–4]. The procedure with the highest priority could be implemented, and if the objective is achieved, it could be clarified. If not, the next highest priority will receive priority.
- Closed loop process planning is used by Wang (2009) and Zhang, Saravanan, and Fuh (Zhang, Saravanan, and Fuh) (2003) [5, 6]. Using this method, the process is developed based on the current state of the resources and feedback from the workshop.
- The distributed process plan is the most recent way to schedule and plan at the same time. Planning would be done in two stages: primary and secondary. In traditional production systems, the planning and scheduling of each process are done separately. Integration of these two decisions yields considerable advantages, like a more efficient use of manufacturing resources, lower total production costs, and shorter lead times. The most common way to plan a process is with CAPP.

There are some basic computer-aided processes planning variants.

The VPP uses group technology to make plans for similar parts by looking up information in a table. After that, the process planner adjusts as necessary to make a version that fits the needs of the part under consideration.

The generative method usually involves making a new plan for each part without looking back at the old ones. Its main benefits are how quickly and consistently plans can be made and how easy it is to add new processes, equipment, methods, and tools to plans.

According to group technology principles, the variant approach is used to find the part families based on their geometric and manufacturing properties. The standard process plans are made for each part family and saved in a database. The manufacturing of new part process planning is generated by retrieving a process plan for a similar part and modifying it as needed. In the present study, variant process planning will be used to optimize the process of the job shop of a compressor piston. The variant method based on the system knowledge, classification, and different machining operations will be investigated to improve the process plan for a component. The database is used to retrieve the variant process plan for a part of the new part that differs significantly from the previous ones.

10.4.2 Methodology

A process plan signifies that the production process of a component is planned based on operation by operation. A specific job can consist of various manufacturing operations. However, if alternate processing planning is accounted for, it may generate six or eight operation plans. As a result of using alternate process planning, to examine the impact of efficient algorithms.

A model is developed to factor in the fact of multi-objective and adaptability in production plans based on predecessor relationships, which results in MOFJSP-APP. Suppositions: The MOFJSP-APP on which the arithmetical problem in this paper is premised is as follows:

- Each machine could be working on one of the products.
- The production process of an operation on a machine persists, and the break is not approved.
- No predecessor correlations between operations from different jobs.
- Set-up times are independent of processing times and therefore are not included.
- Processing times would include time preparing operations.
- Machines are self-contained. The figure shows a problem with three jobs, each of which has four, two, and three operations. There is a set of machinery that can be used to process every operation.

There are a lot of optimization techniques with discrete and non-linear attribute activities present in variant process planning. The developed optimization method is based on discrete-event simulation and combinatorial optimization. The optimization algorithm depicted in Figs. 10.4, 10.5, 10.6, 10.7 and 10.8 represents the initial configuration structure of variant process planning. The source data could be obtained from an existing supply mechanism or estimated process. In the algorithms, each processing node describes the distinct entities. This allows us to set the input values of the controller parameters for each reconditioning node separately, and we made a clear supposition. The number of servers for each servicing node is determined by balancing the rates of said incoming workflow process with the catering node's overall economy; the input limit can also be measured.

(a) **Establishing Priorities with Criterion**

The efficient approach is utilized to recognize the next step of the process plan, which helps to manage the sequence of variant process planning.

(b) **Simulation of the Manufacturing Process**

The objective of this event is to collect data on the simulated output service's actions. Modeling testing should always be borne out to collect the data. The warm-up period, propagation multitude, and propagation size will still be determined; methods for determining these specifications are well established.

(c) **Recognizing the Bottleneck**

A bottleneck is described as a base of the product output node.

(d) **Minimization of Local Slowdown**

The goal of this track is to transform the bottleneck processing cluster hierarchy. According to each iteration of the optimization techniques, a decision about sustaining the optimization process should be made. It will be resolved regardless of refinement could raise the price of the attribute process to slow or stop the technique.

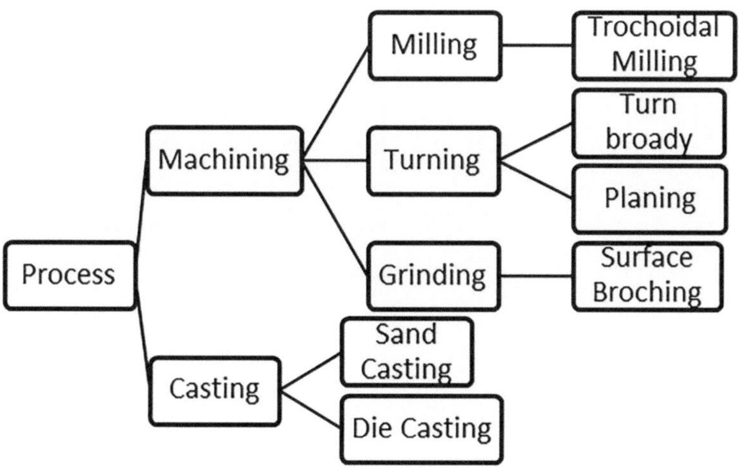

Fig. 10.4 Process planning of compressor piston

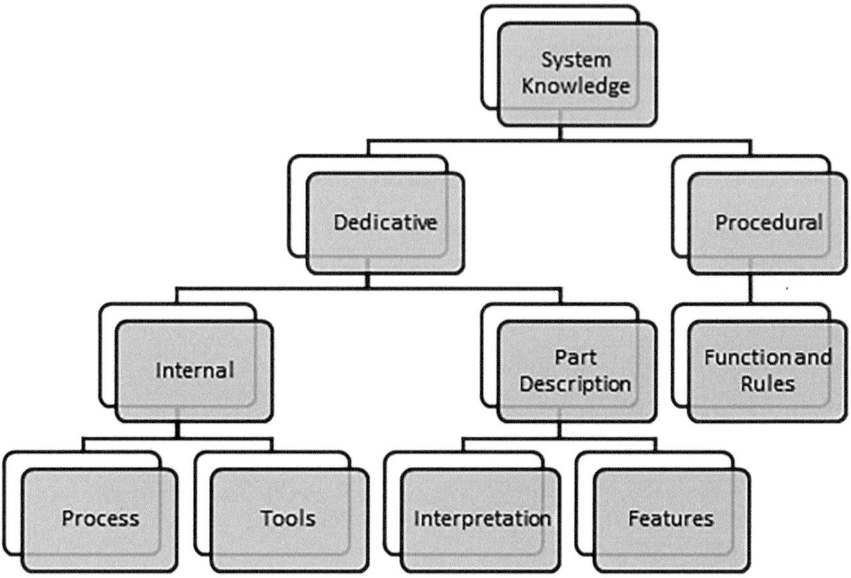

Fig. 10.5 Variant process planning

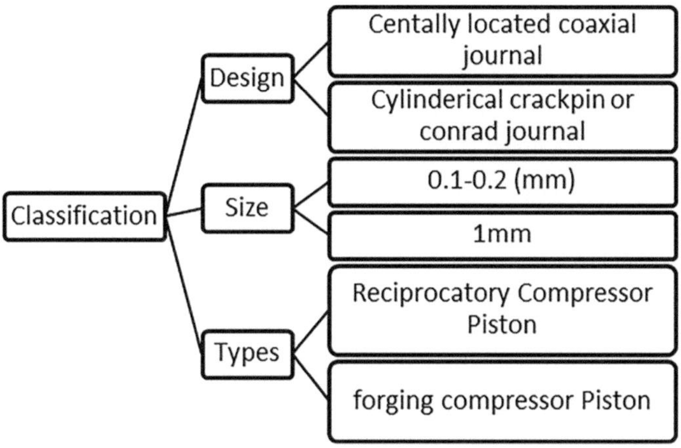

Fig. 10.6 Classification of VPP for compressor piston

10.4.3 Process Planning

Frames that directly correspond to objects in the object-oriented sense are used to depict the knowledge process. In modeling and representing process knowledge, hierarchical abstraction is efficiently used. It is a member of the kernel process hierarchy for process planning. The processes are separated into two abstract types at the

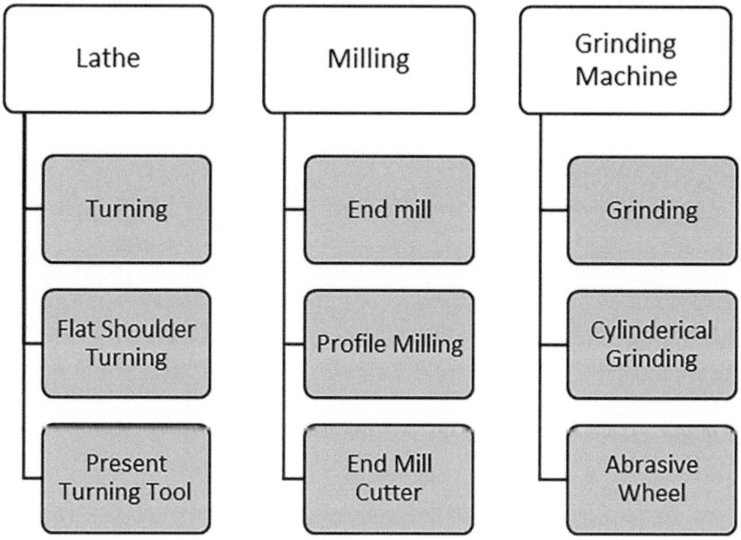

Fig. 10.7 Operation performed during manufacturing of compressor piston

highest level of abstraction: surface-generating processes and surface-finishing activities. Each of these classes is further broken down into milling processes, planning, shaping, and broaching procedures, among other operations (Fig. 10.4).

(a) **Variant Process Planning**

The geometric capabilities and engineering limitations of any process must be captured, represented, and used in the process knowledge. The shape, proportions, and features that a process is capable of producing are included in its geometric capabilities. The geometry tolerances, surface quality, and material hardness for which a process is suitable are all examples of engineering restrictions. Additionally, we have used "the permit for a procedure" in addition to these attributes (Fig. 10.5).

(b) **Classification of VPP**

See Fig. 10.6.

(c) **Machining Operations**

See Figs. 10.7 and 10.8.

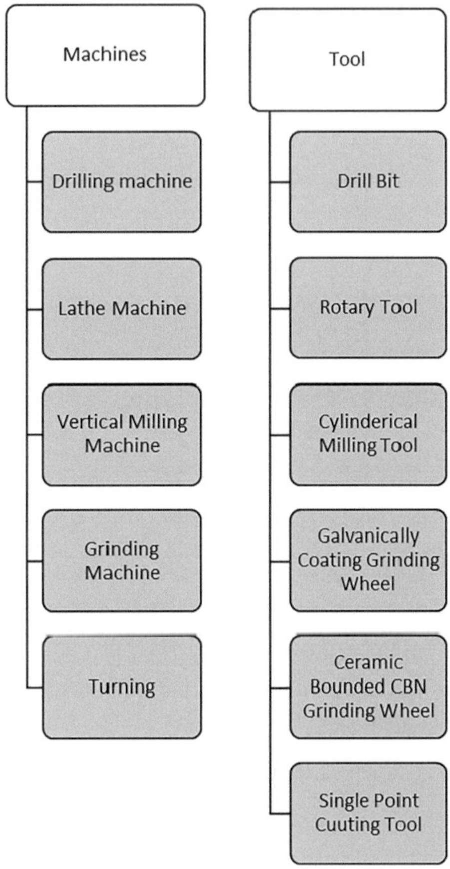

Fig. 10.8 Machines and tools used in VPP of compressor piston

10.5 Conclusion

In the present case study, in absence of a process plan database, the exercise performed by another group of students developing a process plan for the compressor piston using Generative Process Planning was used to develop the variant process plan. It was concluded that the VPP can recognize and retrieve already used process plans for similar parts from databases while also generating modified process plans for new parts. There will not be any equivalent features if the new component differs significantly from the older ones. The variant process plan can also be modified by a human operator to accommodate the new part.

10.6 Learning Outcomes

The reader is expected to gain knowledge of this primary part of the interface between engineering design and manufacturing. This aspect of the production cycle is presented in a more vocational manner to give the reader a practical understanding of the topic.

Generative process planning and variant process planning use databases in different ways as explained in relevant sections and the reader must be able to account for this knowledge.

10.7 Review Questions

Q1. What type of services are required by the dynamics of multiple orders entering a SME?
Q2. Describe the dynamics of Discrete and Continuous Manufacturing Facilities.
Q3. What is Product Lifecycle Management (PLM)?
Q4. What are Manufacturing Execution System (MES)?
Q5. How is Manual Process Planning performed?
Q6. What is Computer Aided Process Planning (capp)?
Q7. What is the difference between Generative and Variant Process Planning?
Q8. What turns the information in a part design into instructions for making the part?
Q9. Describe the three categories of ideas by different practitioners for identifying the high-level machining relevant shape information.
Q10. According to group technology principles, which approach is used to find the part families based on their geometric and manufacturing properties?
Q11. Describe the Process Planning of the Camshaft.
Q12. What is the route sheet? Explain with an example.
Q13. What is the procedure sheet? Explain with an example.
Q14. What is the process sheet? Explain with an example.
Q15. Describe the Variant Process Planning.

Bibliography

1. Jain A, Jain PK, Singh IP (2006) An integrated scheme for process planning and scheduling in FMS. Int J Adv Manuf Technol 30:1111–1118
2. Leung CW, Wong TN, Mak K-L, Fung RYK (2010) Integrated process planning and scheduling by an agent-based ant colony optimization. Comput Ind Eng 59(1):166–180
3. Shi G, Yang Z, Xu Y, Quan Y (2022) Solving the integrated process planning and scheduling problem using an enhanced constraint programming-based approach. Int J Prod Res 60(18):5505–5522

4. Wang L, Holm M, Adamson G (2010) Embedding a process plan in function blocks for adaptive machining. CIRP Ann 59(1):433–436
5. Kumar M, Rajotia S (2003) Integration of scheduling with computer aided process planning. J Mater Process Technol 138(1–3):297–300
6. Zhang YF, Saravanan AN, Fuh JYH (2003) Integration of process planning and scheduling by exploring the flexibility of process planning. Int J Prod Res 41(3):611–628

Chapter 11
Production Planning and Control

Prelude
Production Planning and Control (CAPP) is an interface between Computer Aided Design (CAD) and Computer Aided Manufacturing (CAM). It provides intangible services like Computer Aided Process Plans (CApp), Part Programs, Manufacturing Instructions, (Material/Tool/Tool Bits/Jigs and Fixtures/Measurement Gadgets/Coolants/Lubricants etc.) Requisition Forms, Master Schedule, current Job/s Schedule/s, Master Maintenance Schedule, Emergency Maintenance Schedule, Record of all Inventories including Material/Tool/Tool Bits/Jigs and Fixtures/Measurement Gadgets, In-Process Inventory, Finished Product Inventory, Lead Times for in-Process Jobs, Data for new Bids, Requisition for any new requirement of Material/Tool/Tool Bits/Jigs and Fixtures/Measurement Gadgets/coolants/Lubricants etc. It is also referred to as Computer Aided Production Management.

11.1 Objectives

Production planning and control is the intangible part that makes manufacturing production. It is the provision of services provided at the manufacturing setup that includes a part program or manufacturing instruction, process plan, raw materials/tool/jigs and fixtures/coolant/lubricant requisition forms in subtractive manufacturing (will change in additive manufacturing), short term and long term (master) schedule, maintenance schedule and management control procedures for orders and their completion in required lead time, an account of different types of inventories at production setup, and; assembly and packaging and postage procedures that make

---

Supplementary Information The online version contains supplementary material available at https://doi.org/10.1007/978-3-031-77106-4_11.

W. A. Khan et al., *Computer Aided Engineering Design and Manufacturing*, Springer Series in Advanced Manufacturing, https://doi.org/10.1007/978-3-031-77106-4_11

the dynamics of a production setup. Today, most of it and its interaction with other functions of the organization is covered under the Manufacturing Execution Services (MES) covering the value chain and becoming part of the supply chain. For discrete products, Product Life Management (PLM) starts from need analysis for design to supply of product to distributor to after sales service within warranty period taking care of finished product inventory to optimize tied-up capital.

11.2 Production Planning and Control[1]

Production Planning and Control can be described as direction and coordination of firms' resources towards attaining the prefixed goals. "PPC helps in efficient utilizations of 6M's (Man, Machine, Material, Method, Money and Management) through; planning, coordinating and controlling production activities that transform the raw material into finished products".

IMS-SOP-8/4
Issue 01, Rev.03

INFINITY Engineering	**Production Planning & Control**		**ISO Norms:** IATF 8.5.1.7	**Department:** PPC	
	Code No: IMS-SOP-8/4	**Issue No:** 01	**Date:** 09-09-2017	**Revision No:** 03	**Date:** 15-12-2022
	Prepared by (AM IMS): ______	Reviewed by(MPPC):________		Approved by(MIMS):________	

1. Objective & Scope:
This procedure describes; the production scheduling, production planning, production optimization, productivity & process improvement and tooling management under the scope of production and service provision.

2. Review History

Sr. No.	Issue No.	Revision No.	Review Date
1	01	00	09-09-2017
2	01	01	01-08-2018
3	01	01	09-12-2019
04	01	02	30-10-2020
05	01	03	15-12-2022

[1] Production Planning and Control Procedure as practiced at Infinity Engineering (Pvt) Limited, Lahore, Pakistan. Reproduced with permission.

Definitions and Abbreviations

The following definitions in addition to formal definitions in IATF16949 are applicable to this procedure:

CE	Chief Executive
GMT	General Manager Technical
MP	Manager Production
MSCM	Manager Supply Chain Management
PPC	Productivity Planning and control
MPPC	Manager Production Planning and control
Dy. MT	Deputy Manager Tooling
POs	Process Owners
MQA	Manager Quality Assurance
MRD	Manager Research and Development
MIMS	Manager Integrated Management System
AM P&O	Assistant Manager Purchase and Outsource
SMED	Single-Minute Exchange of Dies (Figs. 11.1 and 11.2).

Production Scheduling

a. Production scheduling is carried out based on customer order considering inventory levels which is chalked out by MPPC and shared to MP after approval of CEO.

Capacity Analysis:

a. Capacity analysis is carried out by MPPC based on production schedule, which is analyzed in meeting conducted among the HODs of the following departments:
    - Operations,
    - Production,
    - R&D
    - PPC.
b. Capacity analysis activity can also be initiated as per need identified by the GMT and/or CEO.
c. Following steps are carried out for production capacity analysis:
    i. Production scheduling
    ii. Review and update Process Flow and Process Layout
    iii. Review and update Cycle Time Studies (SOP-F-8.3-PPC/03)
    iv. Machines availability:
        - Total uptime,
        - Overall Equipment Effectiveness.
    v. Raw material availability
    vi. Available Manpower.

Production Planning

a. MPPC makes production planning to achieve cost effective production efficiency against the monthly production schedule, considering the following information at minimum:

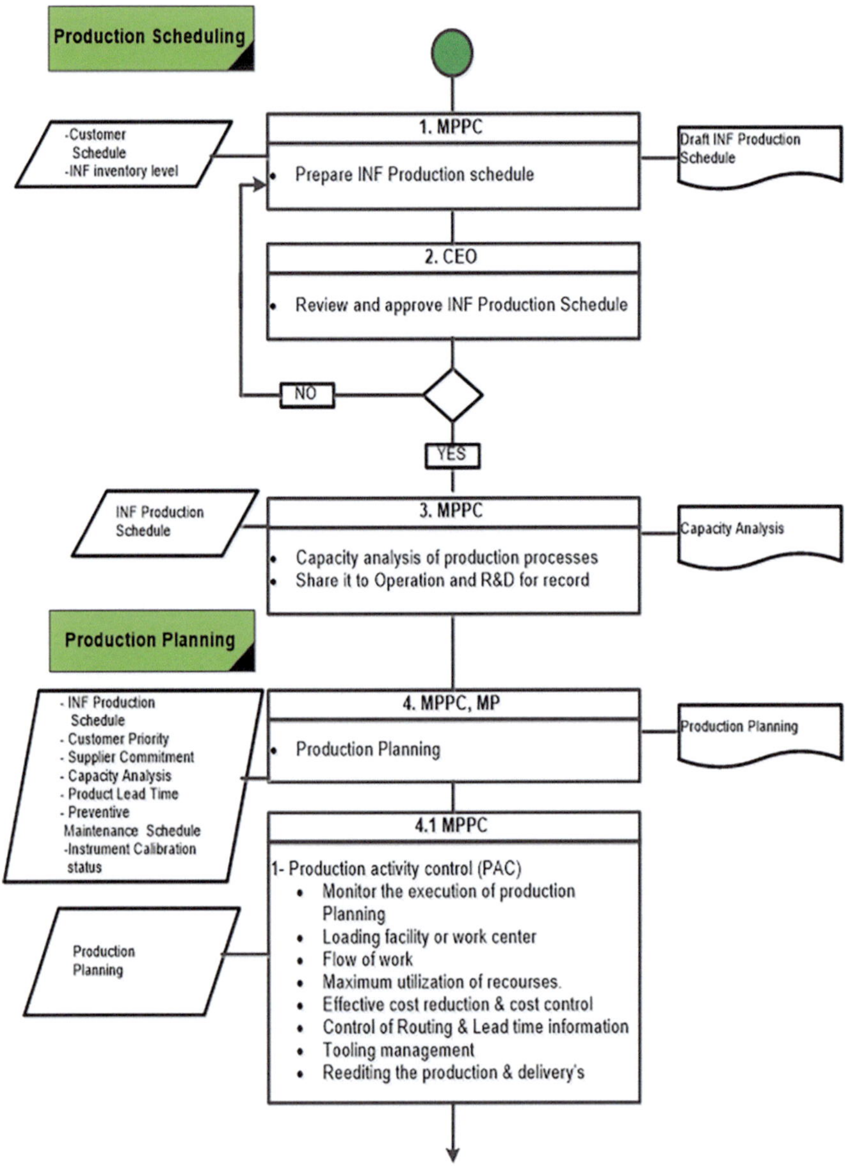

Fig. 11.1 Production planning

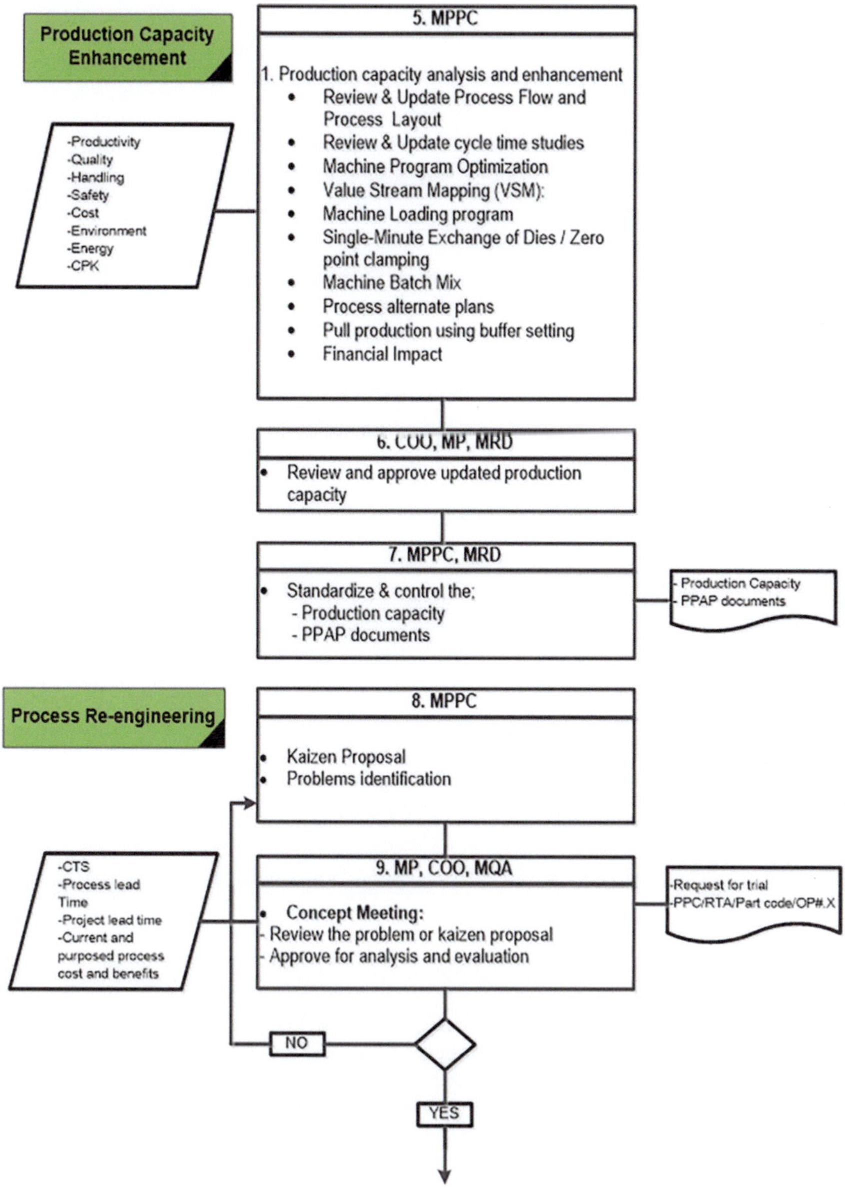

Fig. 11.1 (continued)

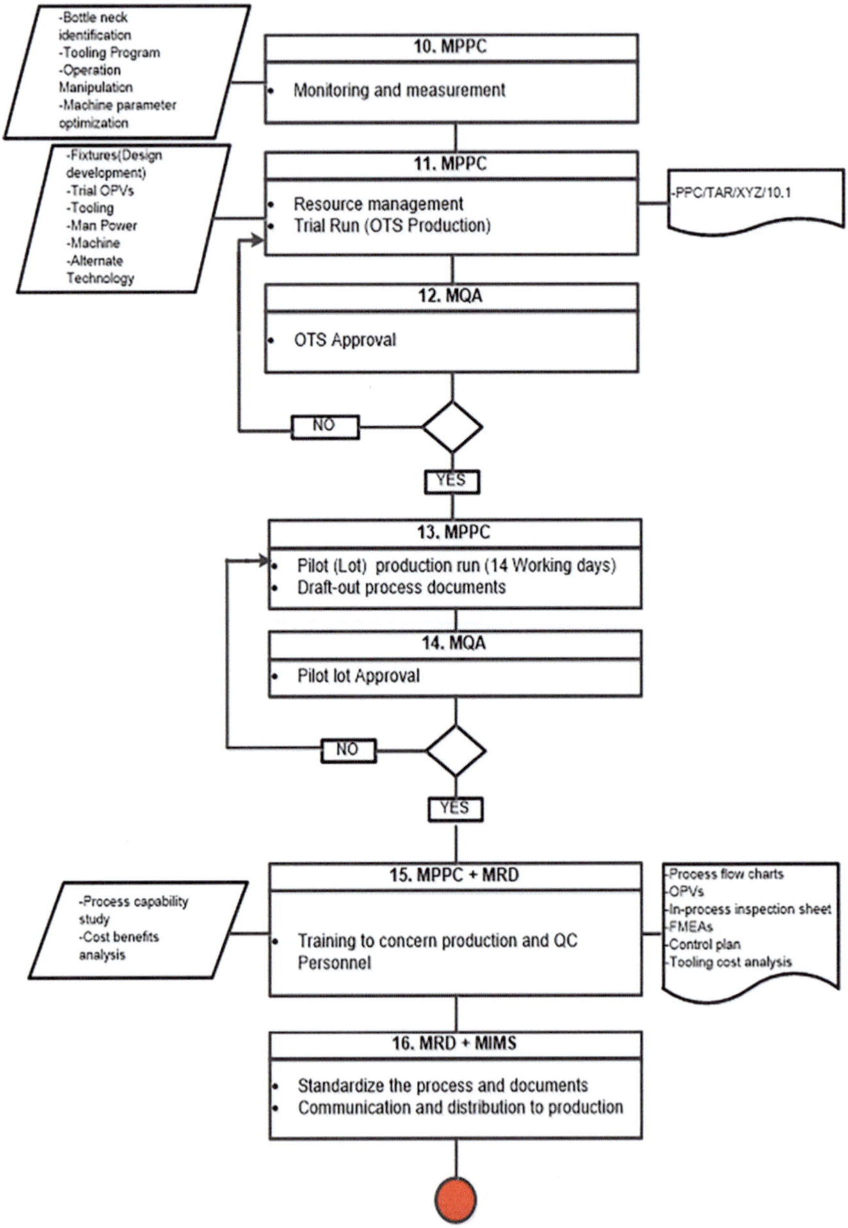

Fig. 11.1 (continued)

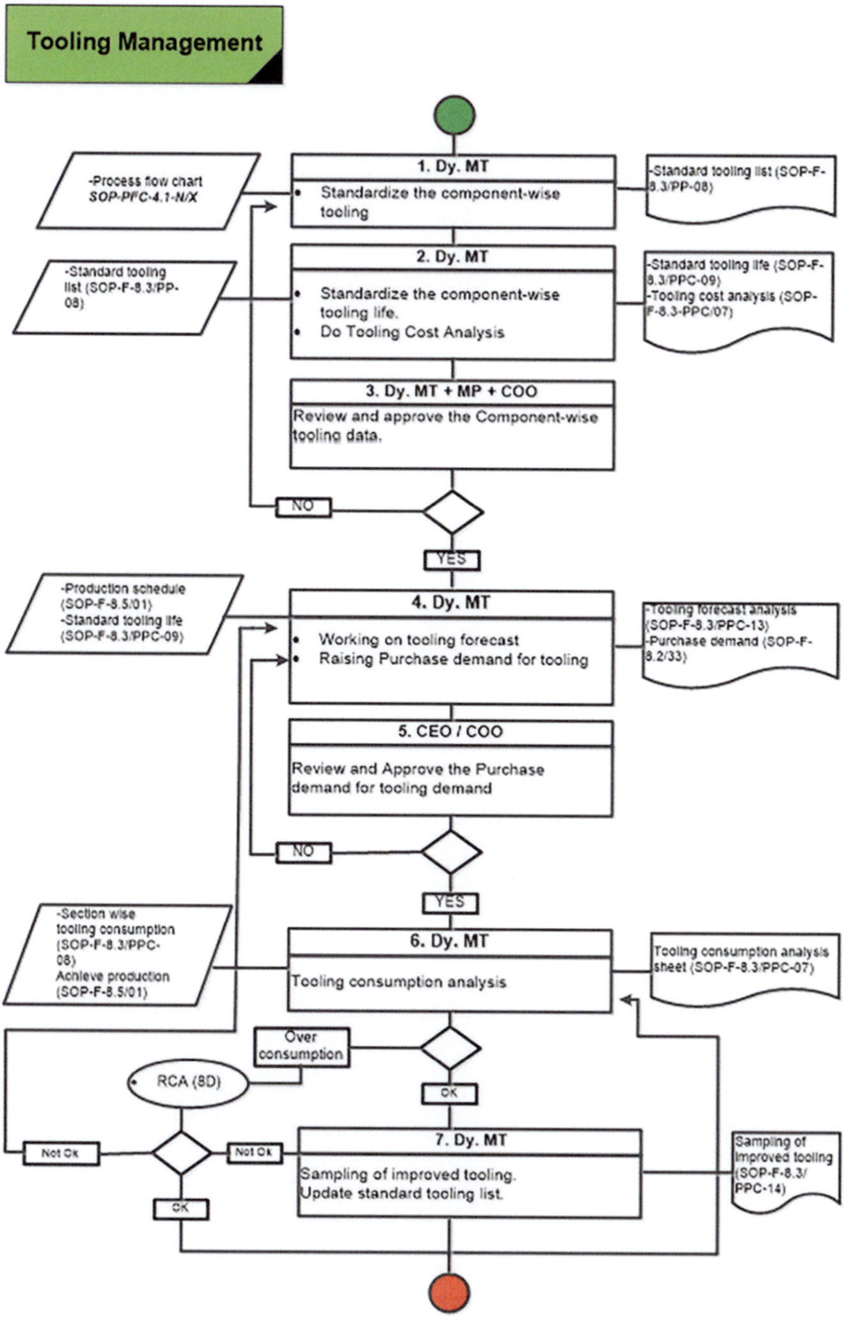

Fig. 11.2 Tooling management

- Customer demand and requirements,
- Supplier on time delivery performance,
- Capacity Analysis,
- Shared loading (multi-part station),
- Product lead time,
- Preventive maintenance status,
- Measuring Instruments calibration status.

Production Activity Control (PAC):

a. MPPC execute the production planning and control by adopting followings steps:

   i Raw Material:
      a. Raw material planning is done considering the following information:
         - Inventory status from MS
         - Order delivery date
         - Quality rate of manufacturing process.
      b. It also includes material standardization i.e., quality, quantity, variety reduction, etc.
   ii. Machines and equipment:
      a. MPPC ensures availability of machines and equipment with minimum down time. It includes analysis of past records of facility as follows:
         - Breakdown
         - OEE
         - Total up time.
   iii. Method:
      a. This function is concerned with;
         - Analysis of alternatives,
         - Selection of the best method with due consideration to constraints imposed,
         - Developing specifications for operations.
   iv. Routing:
      a. The selection of efficient and effective path/route which the raw material follows to get transformed into finished product is ensured through following:
         - Improving layout of operations,
         - Determination of sequence of operations,
         - Breaking down of operations to define each operation in detail,
         - Deciding the setup time for each operation.
   v. Estimating:
      a. After extensive analysis of; operations, methods and routing, a standard time for operations is established using work measurement techniques.

vi. Loading:
   a. It is decided which job is assigned to which work center along with starting and completion times of each operation.
   b. Machines are loaded according to their capacity, and capability of performing the given task.

vii. Dispatching:
   a. It authorizes the start of production activities by:
      - Issuing required materials, components, tooling, etc.
      - Issuing operating instructions, OPVs and inspection sheets to the operator.
      - Release necessary work orders, time tickets, etc., to authorize timely start of operations.
      - Recording start and finish time of each job on each machine or by each man.
      - Ensuring inspection, repair, maintenance, and management of tools, dies, machining fixtures, etc.
   b. MP ensures maximum utilization of all resources using money and management techniques:
      - Manpower,
      - Equipment,
      - Material,
      - Method/procedure.

viii. Expediting:
   a. Expediting that the work is carried out as per plan and delivery schedules are met, by:
      - Identifying bottlenecks, delays and interruptions because of which the production schedule may be disrupted.
      - Devising action plans (remedies) for correcting the errors.

ix. Inspection:
   a. Following data is collected and analyzed for the purpose of progress reporting:
      - Rejection and rectification,
      - Process variations,
      - Equipment downtime,
      - Operator efficiency,
      - Operator absenteeism,
      - Tool life.

x. Evaluation:
   a. The production performance is evaluated for effectiveness.
   b. Based on analysis of process data, corrective actions are taken for improvement, which may include the following at minimum:

- Expediting the work,
- Schedule modifications,
- Capacity modifications,
- Changing maintenance policy, procedure, and schedule,
- Selection and determination of the alternative method and sequence of operations,
- Developing specifications for processes,
- Improve layout,
- Improving set up time and process time for each operation.
- Standardizing material quality,
- Make or buy decisions.

Productivity Improvement/Capacity Enhancement

a. INF defined the process flow chart for Production Planning and Control (SOP-PFC-8.3A1) and direction to improvement (SOP-PFC-8.3A2).
b. Following steps are carried out by MPPC for productivity improvement or production system capacity enhancement:

1. Review and Update Process Flow and Process Layout
2. Review and Update cycle time studies
3. Capacity Analysis
4. Production Scheduling/Machines Loading Plan
5. Machine Program Optimization
6. Value Stream Mapping (VSM):

- Define process boundaries: from raw material to customer
- Collect Process Data:

  – Inventory
  – Cycle time
  – Change over time
  – Up-time
  – Number of operators
  – Shifts worked
  – Net available working time
  – Scrap rate
  – Pack size/pallet sizes
  – Batch Size.

- Interpreting the Current Value Stream Map
- Create Future and Idea Value Stream
- Line Balancing
- Throughput/shift
- WIP Reduction
- Material flow
- Information flow.

7. Job Setup Analysis; SMED/Zero-point clamping
8. Production and Measuring tooling management
9. Ergonomics
10. MUDA elimination
11. Machine Batch Mix
12. Process alternate plans
13. Pull production using buffer setting
14. Financial Impact of improvement activity.

c. During PPC activities, quality standards and customer specifications are maintained by the PPC team.

Process Re-engineering

a. If there are substantial problems with a manufacturing process it may need to be completely redesigned.
b. The PPC team develops a new process to most effectively use the inputs to meet the demands of the process output.
c. Any trail for study requires approval on Request for Trail Activity RTA (SOP-F-8.3-PPC/01) from Sr MO and MP.
d. During activity MPPC fills TAR (Trial Activity Report "SOP-F-8.3-PPC/02) for maintaining the record of trail.
e. On successful trials MPPC runs the pilot production of the part according to the PSS (Production Standardization sheet) for two weeks.
f. After a successful pilot run MPPC seeks pilot lot approval by MRD, MQA, and respective MP.
g. All the process documentation until standardized is controlled by MPPC using a TRAIL stamp.
h. MPPC updates and standardizes the technical documents of the respective part from R&D.
i. The PPAP documents are reviewed and updated at least once a year for continuing suitability.
j. MPPC is responsible for any change in operation or sequence related to machines or parts during the Trial activity. Furthermore, after implementation, MPPC ensures production of that part for only two weeks.
k. MPPC writes down the procedure of a job set up for specific operation and its monitoring and verification.
l. MPPC trains the CNC programmers to change tools for perishable tools.
m. Hourly Production Targets (SOP-F-8.3-PPC/05) are set for each process and machine.

Tooling Management

a. Tooling requirement is fulfilled by preparing a component-wise standard tooling list (SOP-F-8.3-PPC/06) following the process flow chart SOP-PFC-4.1-N/X.
b. Regular tooling demand is processed through the store while new/special tooling is processed through the store after approval of CE.

c. Tooling Cost Analysis Sheet (SOP-F-8.3-PPC/07) is prepared for selection of cost-efficient tools.
d. Selection and timely arrangement of consumable tooling for regular and new production keeping in view the lead time and production schedules.
e. Inspection, Repair, Maintenance, and management of tools, dies and machining fixtures is ensured on Monitoring of Tool consumption (SOP-F-8.3-PPC/08).
f. Tooling is arranged as per the purchase demand from the store and in advance as per tentative production schedule issued by MPPC.
g. Purchase demand (SOP-F-8.2/33) is thoroughly studied, analyzed, and revised if necessary. Then concerned suppliers are requested for Proforma Invoice/Quotation.
h. Proforma Invoice is finalized after discussion with CE and approved Proforma invoice is handed over to the finance department for payment.
i. Inspection and handling of procured tools is ensured 3.
j. Monthly tools stock report is prepared and submitted by the store.
k. Tool issuance and return is managed as per daily production plan.
l. In-house tool or fixture development/improvement/changes are initiated through job order.
m. Dy. Manager Tooling ensures proper monitoring of any outsourced activities regarding production tooling.

Formats/Records

List of documents (IMS-SOP-7.5/01)

while the product life cycle starts from design and ends at reuse directly or through remanufacturing, recycle and/or dispose of completely or in part. The supply chain is a production philosophy that starts with ore extraction to the end of the life of the product through reuse, recycling or disposing of waste. These concepts have been discussed in part in earlier sections and are being placed like solving a jigsaw puzzle.

The reader will be introduced to all these concepts slowly and gradually making sure that by example all the concepts and their flow are understood.

INFINITY Engineering	Department: PPC	Description Part No.	Code No: SOP-F-8.3PPC-01
	Section: PPC	Issue No: 01	Date: 17/03/2018
	Page No: 01	Revision No: 00	Date: 17/03/2018
	Prepared by:__________	Reviewed by: _________	Approved by: ____________
	TRIAL ACTIVITY REPORT		

Machine Utilization Time & Production loss durning Study						**Date:**	
Sr No.	Machine to be Engaged	Start Time	End Time	Total Hours	Prod/Hr	Prod Loss	Total Loss (RS)

Sr No.	Tooling Requirement	Quantity	No. of parts Used for study			No. of parts damaged during study		
			Parts Serial No.			Parts Serial No.		

Sr No.	Personnal to be Engaged During Trial	Department	Designation	Incharge Approval

Tools & Accessories damage cost					
Sr No.	Insert Name/Code	No. of Insert Damage	Cost (Rs)	Other accessories damage	Cost (Rs)
	Total Cost (RS)			**Total Cost (RS)**	

Summary	**Remarks**
Result of Trial	
Total cost of production loss during study	
Total cost of inserts break during study	
Total cost of other accessories damage during study	
Total cost (RS)	

SOP-1

INFINITY Engineering	Department : PPC	Description Part No. N/A	Code No:SOP-F-8.3PPC-02
	Section: PPC	Issue No: 01	Date: 17-03-2018
	Page No: 01	Revision No: 00	Date: 17-03-2018
	Prepared by: ___________	Reviewed by: ___________	Approved by: ___________
	REQUEST For Trial ACTIVITY		

Part Description :	Operation :
Trail Description	

Sr.#	Machines To Be Engaged:	Start Time	Expected End Time	Actual End Time	Expected Total Hours	Actual Total Hours	Remarks

Sr.#	Tooling Requirement	Quantity	Availabile / Not Available

Personal to be Engaged during Trial	Department	Designation

Request Initiated By : ___________	Approved :	Trial Date:	
		Trial Time:	
Request Received By : ___________		Trial Duration :	

SOP-2

Dept. Production	Code No: SOP-F-8.3/PPC/19	
Section: PPC	Issue No: 1	Date: 10-03-2016
Page No: 01 of 01	Revision No: 02	Date: 10-02-2017
Prepared by:	Reviewed by:	Approved by:
	Cycle Time Study of Hub FR	

Operation Name	Machine	Chipping Time	M/c Movement time	Logistic Time	Total Cycle Time	Pr/Hr	Remarks
1st side turning/facing /boring	YCM - 120803	229	24	29	282	12.1	
2st side turning/facing /boring	Hitachiseiki - 130101	137	16	37	190	18.0	
Broaching	Broaching M/c	40	45	17	102	33.5	
Final Turning/facing/boring	YCM - 120801	187	24	69	280	12.2	
5 hole Drilling / Chamfering	Tapping Centre - 110201	96	21	55	172	19.9	
Chamfering	Gang Drill	35			35	97.7	
Deburring + Polishing	Manual	45			45	76.0	
Crack Detection	Crack Detector	72			72	47.5	
Paint	Manual	80			80	42.8	
Stud Fitting	Hydraulic press - 381401	85			85	40.2	
Final Inspection	Manual	125			125	27.4	
Washing	Manual	55			55	62.2	
Rust prevention	Manual	65			65	52.6	
Packing	Manual	90			90	38.0	

SOP-3

Sr. No	Machine	M/C Code	Component	Process	Schedule	Prod./hr	Setup Time hr	Prod / Day	Days Required	Machine Utilization Days	Spare Time (Days)	Remarks
1	Double Head MW20	51/20	Flywheel	Both sides Turning/Facing/Boring	4000	15.0	12.0	322.5	12.9	24.1	1.9	
			Hub FR	Rough Turning facing first side	3000	13.0	12.0	279.5	11.2			
			Hub FR	Second Side Rough Facing Boring								
2	KIA#1	51/15	Hub FR	Rough Turning facing first side	1000	13.0	0.0	279.5	3.6	13.34	12.7	
			SB-308	2nd side Turning/Facing/Boring	500	13.0	12.0	279.5	2.3			
			Idler Hub	Grinding Both Sides	900	6.0	12.0	129.0	7.5			
3	KIA#2	51/16	Innentail JB	Facing/Drilling/ Boring/ Countering	0	12.0	6.0	258.0	0.3	0.25	25.8	
4	YCM (HTC)	51/13	HUB FR	Final Both sides Turning/Facing/Boring	4000	10.0	12.0	215.0	19.1	20.8	5.2	
			SB-308	1st side Turning/Facing/Boring	300	12.0	12.0	258.0	1.7			
5	YCM (HTC)	51/14	Idler Hub	Grinding Both Sides	2100	6.0	12.0	129.0	16.8	25.5	0.5	
			SB-308	1st side Turning/Facing/Boring	2000	12.0	12.0	258.0	8.3			
			Hub Idler	Rough turning first and second side	0	6.0	12.0	129.0	0.5			
6	YCM (HTC)	51/21	SB-308	1st side Turning/Facing/Boring	5200	12.0	12.0	258.0	20.7	25.0	1.0	
			HUB RH-413	Rough Turning facing first side	1000	12.0	12.0	258.0	4.4			
7	Hitachisiki Turning	51/8	SB-308	2nd side Turning/Facing/Boring	7000	13.0	12.0	279.5	25.5	25.5	0.5	
8	Hitachisiki Turning	51/7	Hub FR	Second Side Rough Facing Boring	1000	18.0	12.0	387.0	3.1	7.2	18.8	
			HUB RH-413	2nd side Turning/Facing/Boring	1000	13.0	12.0	279.5	4.1			

SOP-4

Hourly Production Targets - Unit 2				
Section	**Sr.**	**Part Name**	**Pr / Hr (Previous)**	**Pr / Hr (Current)**
CNC Turning	1	HUB RH-413	12	12
	2	HUB SB-308	12	15
	3	Fly Wheel	13	13
	4	HUB Idler	6	6
	5	Cap Water	35	42
	6	Pipe Water	22	22
	7	Case Distributor	10	12
	8	HUB FR	8.5	12
B&B Section	1	Intake Manifold ST	7	8
	2	Intake Manifold SB	3.5	3.5
	3	Brake MF 375(GT1 & GT2)	3.5 set	4 set
	4	Brake MF-375(GS1 & GS2)	3.5 set	4 set
	5	Brake MF-260	6 set	6 set
	6	Brake FIAT-480	5 set	5 set
	7	Bracket Caliper	4 set	4 set
	8	Bracket 1.3	12	12
Connecting Rod	1	Connecting Rod MF-385	18	25.5
	2	Connecting Rod FIAT-480	10	14
	3	Steering Body	1.6	2.3
	4	Steering Cover	4	5

SOP-5

Dept.Production			Code No: **SOP-F-8.3PPC-06**
Section: PPC	Issue No: 01		Date:05-11-2015
Page No: 01 of 03	Revision No: 02		Date:10-10-2017
Prepared by:	Reviewed By :		Approved by:

Standard Tooling List of Hub FR

Process	**Cutting Operation**	**Make**	**Manufacturer Tool Code**	**Spares**	**Tool Type (detail)**	**Size (mm)**	**Control Plan Tool Code**
Op#10 1st Side Turning	Rough Turning	ECHAIN	EWLNR-2525M08	Roughing Tool Holder		25	TLB01
				SMW422	Shim		
				WN08-4	Wedge Clamp		
				CS80	Screw		
				LW4	Seat Pin		
				T15	Hex Key		
		Taegutech	WNMG 080408 TT8125	LW4	Insert For Roughing		
	Boring	INFINITY	S16R SCLCR-09	Boring Bar for finishing		16	TLB 02
				CS35	Insert Screw		
				T15	Key		
		Taegutech	CCMT 09T308 TT8125	Insert For Boring			
	Boring	INFINITY	S20R SCLCR-09	Boring Bar for finishing		20	TLB 03
				CS35	Insert Screw		
				T15	Key		
		Taegutech	CCMT 09T308 TT8125	Insert For Boring			
				CS80			

SOP-6

INFINITY Engineering	Dept.Production	Code No: SOP-F-8.3PPC-07	
	Section: PPC	Issue No: 01	Date:05-11-2015
	Page No: 01 of03	Revision No: 00	Date: 10-10-2017
	Prepared by:	Reviewed By:	Approved by:
	Tooling Cost Analysis of Hub FR		

Process	Cutting Operation	Make	Manufacturer Tool Code	Spares	Tool Type (detail)	Tool Life	Tool Cost	Cost per Piece	Size (mm)
Op#10 1st Side Turning	Rough Turning	ECHAIN	E WLNR-2525M08	Roughing Tool Holder		10000.00	3300.00	0.33	25
				SMW422	Shim	1500.00	400.00	0.27	
				WN08-4	Wedge Clamp	2000.00	400.00	0.20	
				CS80	Screw	2000.00	180.00	0.09	
				LW4	Seat Pin	2000.00	420.00	0.21	
				T15	Hex Key	2500.00	100.00	0.04	
		Taegutech	WNMG 080408 TT8125	Insert For Roughing		300	410	1.37	
	Boring	INFINITY	S16R SCLCR-09	Boring Bar for finishing		10000.00	2000.00	0.20	16
				CS35	Insert Screw	2200.00	180.00	0.08	
				T15	Key	2500.00	100.00	0.04	
		Taegutech	CCMT 09T308 TT8125	Insert For Boring		300	300	1.00	
	Boring	INFINITY	S20R SCLCR-09	Boring Bar for finishing		10000.00	2000.00	0.20	20
				CS35	Insert Screw	2200.00	180.00	0.08	
				T15	Key	2500.00	100.00	0.04	
		Taegutech	CCMT 09T308 TT8125	Insert For Boring		300	300	1.00	
OP#20 2nd Side Turning	Rough Turning	ECHAIN	E WLNR-2525M08	Roughing Tool Holder		10000.00	3300.00	0.33	25
				SMW422	Shim	1500.00	400.00	0.27	
				WN08-4	Wedge Clamp	2000.00	400.00	0.20	
				CS80	Screw	2000.00	180.00	0.09	
				LW4	Seat Pin	2000.00	420.00	0.21	
				T15	Hex Key	2500.00	100.00	0.04	
		Taegutech	WNMG 080408 TT8125	Insert For Roughing		300	410	1.37	
	Boring	INFINITY	S20R SCLCR-09	Boring Bar for finishing		10000.00	2000.00	0.20	20
				CS35	Insert Screw	2200.00	180.00	0.08	
				T15	Key	2500.00	100.00	0.04	
		Taegutech	CCMT 09T308 TT8125	Insert For Boring		300	300	1.00	

SOP-7

Dept.Production		Code No: **SOP-F-8.3PPC-08**	
Section: PPC		Issue No: 01	Date:05-11-2015
Page No: 01 of 03		Revision No: 02	Date:10-10-2017
Prepared by:		Reviewed	Approved by:

Tooling Consumption Sheet of Hub FR

Process	Cutting Operation	Make	Manufacturer	Spares	Tool Type (detail)	Tool Life	Size (mm)	Control Plan Tool Code
Op#10 1st Side Turning	Rough Turning	ECHAIN		Roughing Tool Holder		10000		
				SMW422	Shim	1500		
				WN08-4	Wedge Clamp	2000		
				CS80	Screw	2000		
				LW4	Seat Pin	2000		
			EWLNR-2525MC	T15	Hex Key	2500		
		Taegutech	WNMG 080408 T	Insert For Roughing		300	25	TLB01
	Boring			Boring Bar for finishing		10000	16	
		INFINITY	S16R SCLCR-09	CS35	Insert Screw	2200		
				T15	Key	2500		
		Taegutech	CCMT 09T308 TT	Insert For Boring		300		TLB 02
	Boring			Boring Bar for finishing		10000	20	TLB 03
				CS35	Insert Screw	2200		
		INFINITY	S20R SCLCR-09	T15	Key	2500		
		Taegutech	CCMT 09T308 TT	Insert For Boring		300		
				CS80				
OP#20 2nd Side Turning				Roughing Tool Holder		10000		TLB04
				SMW422	Shim	1500		
				WN08-4	Wedge Clamp	2000		
				CS80	Screw	2000		
				LW4	Seat Pin	2000		
		ECHAIN	EWLNR-2525MC	T15	Hex Key	2500		
	Rough Turning	Taegutech	WNMG 080408 T	Insert For Roughing		300	25	
	Boring			Boring Bar for finishing		10000	20	
		INFINITY		CS35	Insert Screw	2200		
			S20R SCLCR-09	T15	Key	2500		
		Taegutech	CCMT 09T308 TT	Insert For Boring		300		TLB 05
OP#30	Broaching	Aceeds	Broach for Hub	Broach		20000	22.4	TLB 06
OP#40 Final Turning				Final Turning Tool Holder		50000		
				SMT-322	Shim	3000		
				LT-3	Screw of Shim	5000		
				TN16-3	Clamp	2000		
				CS80	Screw	3000		
		ECHAIN	ETQNR-2525M1	T15	Key	5000		
	Final Turning	Taegutech	TNMG 160408 T	Insert For Final Turning		180	25	TLB 07
				Final Turning Tool Holder		50000		
				SMT-322	Shim	3000		
				LT-3	Screw of Shim	5000		
				TN16-3	Clamp	2000		
				CS80	Screw	3000		
		ECHAIN	ETQNL-2525M1	T15	Key	5000		
	Final Turning	Taegutech	TNMG 160408 T	Insert For Final Turning		240	25	TLB 08
	Boring	INFINITY		Boring Bar for finishing		10000		
				CS35	Insert Screw	2200		
			S20R SCLCR-09	T15	Key	2500		
		Taegutech	CCMT 09T308 TT	Insert For Boring		300	16	TLB 09
OP#50 Drilling	5-Hole Drilling	Yestool	YSD13.95	Solid Carbide Drill		800		
		Anway	ER32 (13~14)	Collet		8000		
		Anway	BT30 ER32 60	Adoption		10000	13.95	TLB 10
OP#60 Countering	5-Hole Countering	Market	Dia 20mm	Counter		3000	20	
		Anway	ER32 (12~13)	Collet		8000		
		Anway	BT30 ER32 60	Adoption		10000		TLB 11

SOP-8

11.3 Work in Progress

All orders currently being catered to by the SMEs or LSEs constitute Work in Progress. It is a transitional state as m jobs are completing their lead time, n new jobs are entering the SME or LSE while o jobs are under process when m jobs were completed and will leave the system according to their lead time and the cycle will repeat in a catalytic economy.

11.4 Scheduling

All jobs entering the SMEs Or LSEs after successful award of job needs scheduling depending on the lead time agreed with the client and that was worked out by marketing department in consultation with production planning and control must be scheduled with appropriate time buffers to meet the deadlines without any delays. A classic job shop scheduling problem is probably easiest to handle. On the other hand, Scheduling in other types of manufacturing systems and at assembly areas may prove mathematically difficult to compute and therefore human intervention is necessary.

11.4.1 Master Scheduling

Master scheduling is the stochastic scheduling to handle scheduling conditions defined in Sect. 11.3. It helps to have a probability-based schedule in hand that is monetized and confirmed after short spans of time to have a pragmatic level of certainty.

11.4.2 Maintenance Scheduling

All machines at the SMEs and LSEs are subject to a continuous cycle of maintenance be it cursory or detailed. The maintenance schedules are drawn to have this continuous check while not to interfere with any production schedule or if it had to disrupt the production schedule then it had to be completed within the time buffers available thus not disturbing the production lead times in any way. In case of emergency maintenance requirements, the maintenance jobs should finish while the time buffers available permits it and no compromise can be made on this maintenance requirement. Modern Industry 4.0 Pillar: Augmented Reality/Virtual Reality is very useful in such scenarios.

11.5 Inventory Management

Inventory of any object and of any type is nothing but tied up capital. It therefore needs to be optimized continuously to keep the account books properly balanced. Any compromise on inventory management will lead to reduced revenue and will also affect the profits.

11.6 Value Chain

Value chain is the process of adding value through manufacturing methods to raw material thus increasing its value to a significant level especially if it is today's innovative, customary, and multidisciplinary product.

11.7 Supply Chain

Supply Chain as a production philosophy is the complete life of the product starting from cradle (Raw Material to Value Chain) to distributor, to client, to use of the product, to possible change of hands, to end of useful life, to remanufacturing or reuse, to recycling, to disposal at a safe site. Industry 4.0 and Industry 5.0 supports supply chain in many respects, almost using all its pillars, especially the Internet of Things (IoT).

11.8 Case Study: Short Term Scheduling of Job Shops in Industry 4.0 Paradigm[2]

Job shop scheduling is an important procedure to optimize resources. The aim of this study is to perform short-term scheduling of job shops that cover the planning and scheduling for new orders that need to be processed soon. The linear programming approach is used to solve the problem that involves multiple constraints to the objective function. The number of machines to be operating inside the job shop is 35 and the number of products is 3 and there are various constraints. Using the shortest processing time (SPT) rule, the linear program is solved through lingo software. The results show the time of various products on different machines. The variable X_{11} (Job1 on Machine1) has a value of 5, which means product 1 will take 5 min time on

[2] Developed by Ishaq Alam, Owais Durani, Azfar ud Din, Yasir Khan, Hassan Ali under guidance of Wasim Ahmed Khan. Reproduced with permission.

machine 1. Similarly, the processing time on other required machines, considering the available constraints, is also worked out. This methodology is aimed at utilization in the Virtual Manufacturing System being developed at the GIK Institute.

11.8.1 Introduction

For the operation of a production system to be successful, it depends on economic planning and the consumption of resources to be used in the production of various goods. This study deals with the long- and short-term scheduling of manufacturing at an operational level. It is very important to understand that plant-wide decisions are about the order and timing in which jobs must visit the various specified machines and rescheduling when new information becomes available [1]. The study includes deciding how and when to reschedule short-term and long-term plans as new jobs become available. The problem faced in job shop scheduling is that every project in a set of sequences needs handling on a distinct group of resources available, which is an essential operation research subject that covers a wide variety of other types of issues [2].

A job shop is a type of manufacturing system where machines with the same or similar material handling capacity are grouped. Lathes constitute a turning shop, milling machines constitute a milling shop, etc. In Job shops, structures are generally for general use, and can contain a wide variety of job categories or types of parts. In this system, the part or batch of parts moves through the system by visiting the various work centers based on the part's process plan. To accommodate different types of parts, material handling needs to be very flexible, so it's usually done with manually controlled tools. Various techniques are used for job shop scheduling.

To take full advantage of the flexibility of a job shop, complex decisions must be made and implemented in real-time. Job Shop has several advantages such as Ease of operation, as they are assigned to a single machine, balancing of machine workloads, and ease of maintenance and repair in case of interruptions between production. Proper work planning, which depends on the nature of the product, its batch size, the transference of energy and material, and the execution of intangible jobs such as planning of production and process, the inventory control, and logistics, is crucial for efficient use of manufacturing facilities. In most industries, task planning is done by skilled workers manually by using pen and paper and different graphical approaches and some industry databases. The planning dilemma in a Job shop is the most common and difficult planning problem in the Job shop. There are no restrictions on the requirements associated with each job, so a task may require processing on any subset of processors, in any order. Most research assumes that all jobs are non-preventive and that the goal is to reduce the maximum flow time. Dealing with a large criteria function containing a minimally weighted delay criterion is an exception. All optimization techniques used to solve this problem appear to be branch-and-bound. Although much attention is paid to this problem, the main problems that have been solved are fewer than ten jobs.

When a machine becomes available after completing an operation, the question arises as to which of the many possible operations it should perform next. Additionally, a lack of coordination between work centers can hinder the smooth flow of parts from one work center to another. This results in a longer dwell time of the parts in the Job shop and high inventories. For most scheduling problems, that consider the intermediate maintenance costs, the solution of the challenge is not completely defined by the arrangement of operations on the machines: the start time is compulsory to be specified for each operation on each machine, as the inactive time of each work between the processing steps determine the holding costs [2]. The challenge of planning tasks in a job shop is twofold, involving long-term planning and short-term planning. Long-term planning implies planning for a period that can vary from a week to a month and its inputs are the number of products needed for planning, the availability of personnel, and the status of the sectors involved before and after the assembly line effective planning. Short term planning, on the other hand, deals with planning in the various areas of an assembly line and its input parameters are the specifics of long-term planning and the availability of staff and machines [4].

11.8.2 Literature Review

The most well-known scheduling approaches in the manufacturing business are flow shop, and job shop scheduling. For the greatest results, all jobs are handled on a single machine and scheduled in ascending order [4]. In a flow shop, each machine must process every work, and the turnaround time for each task varies per machine. Orders are combined into numerous orders in a flow shop, and the best order is chosen based on the quickest batch completion time. Any task may be processed on any machine in a job shop, and the time it takes to process each job varies per machine [5].

It's one of the toughest problems to tackle, and calculating optimal planning is challenging. Many strategies for optimal workshop planning have been developed over time. These days, workshop planning is a hot issue [6]. The bulk of literature studies concentrate on shortening job completion times and boosting use. Today, most research focuses on the use of heuristic and meta-heuristic methodologies due to the magnitude of the obstacles and limits.

Job shops are utilized in shops and factories that, on a practical level, produce many customized orders in a process layout. Each job only visits the resources on its preset path once in this arrangement [8]. The key operational challenge a dispatcher confronts is determining a processing order for each of the resources, given the routes and processing needs of the jobs [9].

In contrast to job shop scheduling, the order of work processes in flow shop scheduling must be the same. Flow shop performs the same order in the same order every time. While each machine must process a maximum of one work, each job must be handled by only one machine under this scheduling paradigm [10]. Flow shop issues are common in mechanical manufacturing and industrial production environments. The flow shop's stability and iteration make it simple to adapt this technology to robots. However, this strategy is not adaptable. As a result, changing the continuous manufacturing line is difficult. To properly utilize a contractor's flexibility, complex choices must be taken and implemented in real time. For effective utilization of production facilities, proper labor planning is required, which is dependent on the kind of product, batch size, energy and material transfer, and the execution of intangible activities such as production and process choices, inventory management, and logistics. Job planning is done by trained individuals utilizing pen and paper, different graphical techniques, and certain industry databases in most sectors.

It's crucial to create a feasible mathematical model for a scheduling problem, particularly if it's a new one. This is because mathematical models may help with the mining of the scheduling problem's intrinsic knowledge and can solve relatively minor concerns to optimality. Furthermore, the best responses can be used to build approximation algorithms as a benchmark. Linear programming paradigms come in a variety of shapes and sizes. Consider the difficulties of determining a company's most profitable product mix as an example of one common kind. Let n represent the maximum number of things that can be produced. For each product j (j = 1, 2, …, n), a decision variable X_j is introduced to reflect the decision on its production rate. Let Z be the total profit rate from the product mix choice, C_j represents the profit per unit of product j produced. This choice is limited due to the limited capacity of the production facilities available for these goods. Let me stand for the number of different types of facilities that will be needed.

11.8.3 Methodology

In this study, an optimization technique has been used to find out the optimal solution for the job shop problem. A linear programming approach was used to solve the problem that involved multiple constraints to the objective function. Linear programming is the mathematical tool for maximizing or minimizing a quantity (usually profit or quantity of production), subject to certain constraints. A manufacturing plant must know the given number of items of a certain product each month for the next n months. They can be produced either in regular time, subject to a maximum each month, or in overtime. We have assumed the number of machines to be operating inside the job shop to be 35 and the number of jobs were there. The data was collected through a previous project of databases on a job shop. Table 11.1 shows the number of machines (in) and the maximum lead time that it needs to complete a job. Whereas Table 11.2 shows the number of jobs (J_n) and the maximum lead time that it needs to be manufactured. Table 11.2 shows the assumed data of the maximum lead time taken by each machine that would require a specific product. A linear programming approach was used to find the optimal solution. In linear programming, there is an objective function, based on which the whole optimization problem is based. The objective function of this problem (Eq. 11.1) was developed with the help of Table 11.2.

Table 11.1 Machines specifications

Machine (j)	Max operating time	Machine (j)	Max operating time
1	20	19	32
2	25	20	34
3	17	21	26
4	18	22	35
5	32	23	34
6	21	24	30
7	29	25	31
8	21	26	19
9	15	27	15
10	29	28	27
11	23	29	34
12	15	30	15
13	28	31	23
14	20	32	22
15	17	33	29
16	27	34	20
17	32	35	18
18	30		

t_{ij} = time for ‘i’ on machine ‘j’.

Objective function:

$$\mathrm{Min} \sum_{ij}^{t} t_{ij} \tag{11.1}$$

Subjected to constraints (Figs. 11.3 and 11.4):

$$\sum_{i} t_{ij}\, T_j \tag{11.2}$$

$$\sum_{i} t_{ij}\, T_i \tag{11.3}$$

$$T_{ij} < t_{i+1,j}.$$

Table 11.2 Job processing time of machines

Machines	Jobs			Machines	Jobs		
	i1	i2	i3		i1	i2	i3
j1	9	5	6	j18	–	7	–
j2	6	1	–	j19	9	2	7
j3	3	–	–	j20	–	–	3
j4	–	1	3	j21	6	5	4
j5	3	–	–	j22	–	8	–
j6	5	8	–	j23	2	8	8
j7	–	4	4	j24	7	–	6
j8	7	–	9	j25	–	8	2
j9	–	–	–	j26	2	2	–
j10	2	8	3	j27	–	1	5
j11	–	9	5	j28	8	–	8
j12	8	–	–	j29	–	8	–
j13	9	7	6	j30	5	1	5
j14	–	9	6	j31	–	4	1
j15	4	6	–	j32	1	9	–
j16	5	–	1	j33	–	–	6
j17	–	8	7	j34	8	8	7
				j35	7	8	–

Fig. 11.3 Constraints of jobs on machines

Lingo Model - Lingo1

```
x129+x229+x329<=15;
x130+x230+x330<=23;
x131+x231+x331<=22;
x132+x232+x332<=29;
x133+x233+x333<=20;
x134+x234+x334<=18;
x135+x235+x335<=20;
x11<x21;
x21<x31;
x12<x22;
x22<x32;
x13<x23;
x23<x33;
x14<x24;
x24<x34;
x15<x25;
x25<x35;
x16<x26;
x26<x36;
x17<x27;
x27<x37;
x18<x28;
x28<x38;
x19<x29;
```

Fig. 11.4 Lead time of jobs on machines

Variable	Value
X11	5.000000
X12	4.000000
X13	1.000000
X15	3.000000
X16	0.550000
X18	4.000000
X110	2.000000
X112	1.300000
X116	2.110000
X119	4.000000
X121	3.660000
X123	4.100000
X124	3.910000
X130	4.010000
X132	0.010000
X134	2.560000
X135	4.320000
X21	1.110000
X22	3.210000
X24	5.000000
X26	3.000000
X210	2.110000
X213	4.210000
X217	2.000000
X218	1.010000
X221	2.650000
X223	6.100000
X226	0.550000
X229	7.120000
X230	0.050000
X231	1.320000
X232	0.550000
X234	1.320000
X235	2.010000

11.8.4 Results and Discussion

Using the shortest processing time (SPT) rule, the linear program is solved through lingo software. The results show the time of various products on different machines. For our selected job shop, the total machines are 35 in number and the number of products is 3. This model can be adapted for utilization in the Virtual Manufacturing System being developed at the GIK Institute.

11.9 Learning Outcomes

This chapter is particularly important as it speaks about the fusion of major concepts and lays down the foundation for estimating computing requirements in Turing Machine and Quantum Machine.

The reader is required to concentrate on multiple case studies provided in this chapter and get a solid understanding of these concepts while experimenting with actual facilities at their place of learning if possible.

11.10 Review Questions

Q1 Describe production planning and control.
Q2 Define WIP (Work in Progress).
Q3 What is Master Scheduling?
Q4 Differentiate between value chain and supply chain.
Q5 What is a Job Shop?
Q6 Why is job order random in nature?
Q7 How is job order affected by the general economic condition?
Q8 Why do the five basic parameters (type, size, accessories, parameter and control) contribute to randomness in a job shop?
Q9 What is the most basic random number generator that reflects the operation of a job shop?

Bibliography

1. Adams J, Balas E, Zawack D (1988) The shifting bottleneck procedure for job shop scheduling. Manage Sci 34(3):391–401
2. Gomes MC, Barbosa-Povoa AP, Novais AQ (2005) Optimal scheduling for flexible job shop operation. Int J Product Res 43(11):2323–2353
3. Sharma P, Jain A (2016) A review on job shop scheduling with setup times. Proc Inst Mech Eng B J Eng Manufact 230(3):517–533
4. Palacios-Gomez F, Lasdon L, Engquist M (1982) Nonlinear optimization by successive linear programming. Manage Sci 28(10):1106–1120
5. Murota K (2019) Linear programming. In: Computer vision: a reference guide. Springer International Publishing, Cham, pp 1–7
6. Bülbül K, Kaminsky P (2013) A linear programming-based method for job shop scheduling. J Sched 16:161–183
7. Burggräf P, Dannapfel M, Adlon T, Schukat E, Kahmann H, Holtwiesche L (2020) Modeling and evaluating agile assembly systems using mixed-integer linear programming. Proc CIRP 93:1073–1078
8. Jones A, Rabelo LC, Sharawi AT (1998) Survey of job shop scheduling techniques. NISTIR, National Institute of Standards and Technology, Gaithersburg, p 12
9. Azemi F, Tokody D, Maloku B (2019) An optimization approach and a model for job shop scheduling problem with linear programming
10. Abbas M, Abbas A, Khan WA (2016) Scheduling job shop—a case study. In: IOP conference series: materials science and engineering. IOP Publishing, p 012052
11. Liu A, Luh PB, Sun K, Bragin MA, Yan B (2023) Integrating machine learning and mathematical optimization for job shop scheduling. IEEE Trans Autom Sci Eng
12. Meng L, Zhang C, Ren Y, Zhang B, Lv C (2020) Mixed-integer linear programming and constraint programming formulations for solving distributed flexible job shop scheduling problem. Comput Ind Eng 142:106347
13. Applegate D, Cook W (1991) A computational study of the job-shop scheduling problem. ORSA J Comput 3(2):149–156
14. Zhang YF, Saravanan AN, Fuh JYH (2003) Integration of process planning and scheduling by exploring the flexibility of process planning. Int J Prod Res 41(3):611–628

15. Cebi C, Atac E, Sahingoz OK (2020) Job shop scheduling problem and solution algorithms: a review. In 2020 11th international conference on computing, communication and networking technologies (ICCCNT). IEEE, pp 1–7
16. Rosen II, Lane RG, Morrill SM, Belli JA (1991) Treatment plan optimization using linear programming. Med Phys 18(2):141–152
17. Graves SC (1981) A review of production scheduling. Oper Res 29(4):646–675
18. Gass SI, Fu MC (2013) Encyclopedia of operations research and management science, 3rd edn. Springer US, New York
19. Cavalieri S, Crisafulli F, Mirabella O (1999) A genetic algorithm for job-shop scheduling in a semiconductor manufacturing system. In: IECON'99. Conference proceedings. 25th annual conference of the IEEE industrial electronics society (Cat. No. 99CH37029). IEEE, pp 957–961
20. Karnati SR (2012) An approach to solve job shop scheduling problem
21. Ziaee M (2018) A new mixed integer linear programming model for flexible job shop scheduling problem. In: Proceedings of the international conference on industrial engineering and operations management (SEP), pp 2079–2087

Part V
Future Manufacturing Techniques

Prelude

Fourth industrial revolution provides the opportunity to integrate digital, physical, and biological technology, in the evolving fields of engineering and manufacturing. This part, *Future Manufacturing Techniques*, invites us to investigate recent developments where new materials, robotics, nanotechnology, artificial intelligence, and additive manufacturing come together to transform the manufacturing sector. In addition to presenting technological developments, this part challenges us to think about their larger significance. We need to tackle ethical, societal, and environmental issues when we consider the seemingly limitless potential of 3D printing and nanofabrication, as well as the collaborative nature of robotics and AI's cognitive powers. More than just a glimpse into the future, the *Future Manufacturing Techniques* part offers a front-row seat to the merging of humans.

Additive Manufacturing Applications

Following are some of the major additive manufacturing applications:

1. Customized medical implants for precise patient needs.
2. Smart fabrics and wearables with adaptable properties.
3. Microscale electronics for wearable devices and sensors.
4. Sustainable and intricate architectural designs.
5. Personalized consumer goods through mass customization.
6. Lightweight and efficient spacecraft components.
7. Bioprinted tissues and organs for transplantation.
8. Energy-efficient devices like flexible solar panels.
9. Lightweight components for hybrid vehicles and aircraft.
10. Innovative food alternatives, including 3D-printed foods.
11. Environmental solutions such as customized water filters.
12. Enhanced virtual and augmented reality devices.

Chapter 12
Future Manufacturing Techniques

Prelude
The manufacturing industry is undergoing a revolutionary transformation, powered by cutting-edge technologies and innovative approaches. From 3D printing to AI-driven automation, the future of manufacturing is bursting with possibilities that can change how we design, produce, and consume goods. Imagine a world where you can print your car parts at home, customize your sneakers on demand, or get your medications delivered by a fleet of autonomous drones. This is not science fiction, but a glimpse of the future for manufacturing. In this chapter, we will explore the most exciting and promising techniques shaping the future of manufacturing and paving the way for a more efficient, sustainable, and consumer-driven manufacturing industry.

12.1 Objectives

Objectives of this chapter are not limited to providing an overview of the emerging manufacturing techniques but also explaining the fundamentals of modern techniques that are transforming the industry, including their advantages and limitations. Additive manufacturing/3D printing is envisaged as the future of manufacturing for discrete and customized applications in aerospace, automotive, and biomedical applications.

Supplementary Information The online version contains supplementary material available at https://doi.org/10.1007/978-3-031-77106-4_12.

W. A. Khan et al., *Computer Aided Engineering Design and Manufacturing*, Springer Series in Advanced Manufacturing, https://doi.org/10.1007/978-3-031-77106-4_12

12.2 Subtractive Manufacturing

Subtractive manufacturing, frequently referred to as machining or traditional manufacturing, is an approach to producing a desired shape or product by removing material from a solid block or workpiece. Cutting, drilling, milling, or grinding processes are used in this method to gradually remove surplus material and reveal the final form. Large and intricate objects can be precisely created by metal removal processes. In sectors like automotive, aerospace, and medical where it's essential to create distinct components with close tolerances and precise geometries, subtractive manufacturing is commonly used. Even with the progress of additive manufacturing, subtractive manufacturing is still necessary to produce high-quality, functional parts that adhere to rigorous specifications.

12.3 Additive Manufacturing

Additive manufacturing commonly known as 3D printing is an innovative way of producing goods layer by layer by depositing material rather than removing it. Using this method, intricate three-dimensional structures can be directly produced from a computer design file termed as STL file [1, 2]. A virtual model that has been divided into tiny layers is the starting point for the process of additive manufacturing. A 3D printer then deposits material—such as plastic, metal, or resin—layer by layer until the desired object is created. Because intricate interior frameworks and shapes can be made without ease, this approach allows for greater freedom of design. Reduced material waste, faster prototyping, customization capabilities, and the ability to manufacture highly complex and lightweight parts are just a few of the benefits of additive manufacturing. It has found use in a variety of industries, including aerospace and biomedical. Additive manufacturing starts with a 3D CAD model developed by any modeling software. The model is further converted to STL file format which gives a thorough geometrical description of CAD model. STL file is supplied to 3D printing with appropriate slicing to define processing parameters e.g. layer thickness, constraints, deposition time etc. Finally, the printing process started within the chamber and continuous material deposition started with material consolidation directed by a predefined path [1, 2].

The first patent associated with AM was filed in 1920 by Baker. However, Modern 3D Printing started in the mid-1980s. Then, this technology started to grow in the late 1980s and start of 1990s. The demand for layered manufacturing machines increased exponentially in 1994. Now additive manufacturing is believed to be a new industrial revolution. This manufacturing technique is obtaining extensive popularity in industry and recent research, as it accomplishes manufacturers' aim in minimizing lead times and supply chains and it also helped to improve automation level

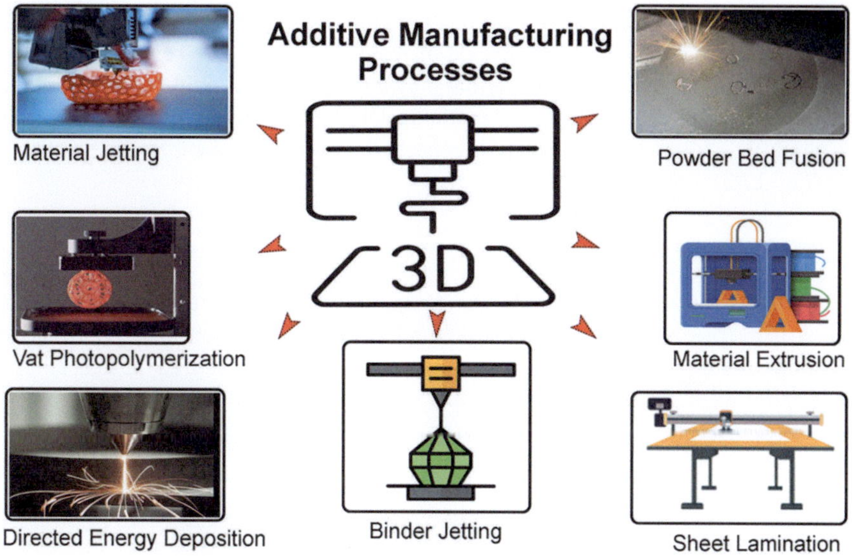

Fig. 12.1 Classification of additive manufacturing technologies as per ISO/ASTM 52900:2015 [3]

in modern manufacturing industry and production rate as it replaces many manufacturing processes by just one step process. ISO/ASTM 52900:2015 has already identified seven types of additive manufacturing as shown in Fig. 12.1 [3].

- **Directed Energy Deposition**

Directed Energy Deposition (DED) additive manufacturing is a cutting-edge technology that uses a directed energy source, such as a laser or an electron beam, to deposit material layer by layer to fabricate complex three-dimensional structures. Laser metal deposition (LMD) or laser-engineered net shaping (LENS) are alternative acronyms for DED. This method has distinct advantages and is employed in a variety of industries, including aerospace, automotive, and defense [2].

Contrary to traditional additive manufacturing techniques, DED allows for the deposition of large amounts of material at high deposition rates. DED begins with a computer-aided design (CAD) model that is sliced into layers. The DED system then precisely controls the deposition process using a robotic arm or a multi-axis machine. The energy source, typically a laser or an electron beam, is directed onto the substrate and simultaneously, material in the form of powder or wire is fed into the melt pool to consolidate in desired geometry.

Since DED has a rapid deposition rate, it is appropriate for large-scale production and enables the fabrication of complex fragments in relatively short periods of time. DED can work with a wide variety of materials, including metals, alloys, and composites. Components with variable mechanical, thermal, and chemical characteristics can be manufactured easily due to process flexibility. DED is frequently used for repair and modification applications. It can restore functionality and extend the life of worn-out or damaged parts by adding material to them. DED enables the fabrication of intricate geometry and overhanging structures that would be impossible to produce with typical manufacturing methods. DED reduces waste and material costs by carefully adding material where it is needed, as opposed to subtractive methods [4] (Fig. 12.2).

- **Powder Bed Fusion**

PBF process is an advanced production technology that can be utilized to fabricate actual parts from the desired 3D solid CAD model data based on layer-by-layer addition concept. This process uses either a laser or electron beam as the heat energy sources for irradiation, fusion and melting of powder particles. The most used heat sources in PBF are various types of lasers or electron beams [5] (Fig. 12.3).

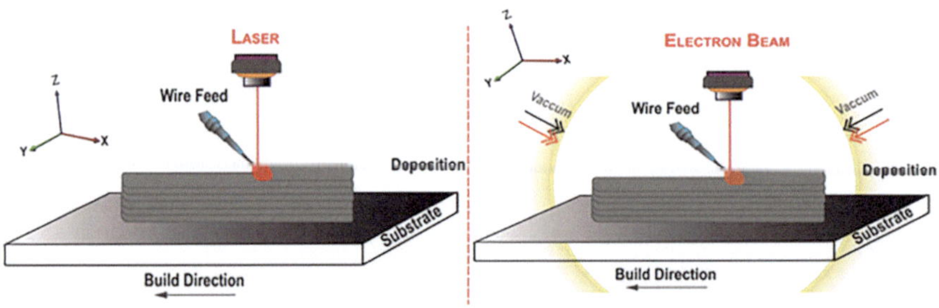

Fig. 12.2 Directed energy deposition wire based additive manufacturing process schematic

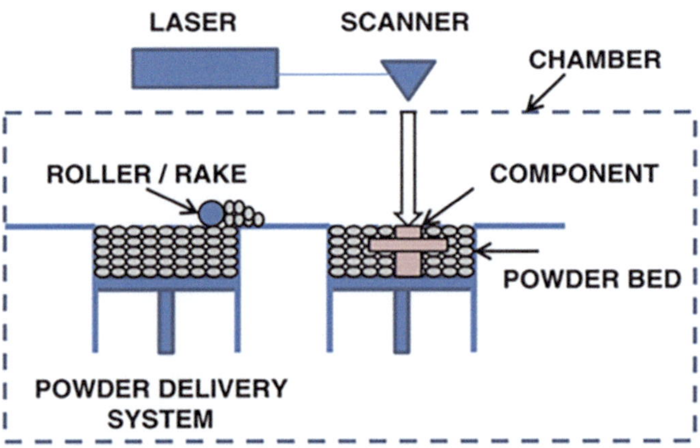

Fig. 12.3 Powder bed fusion process mechanism

The types of PBF processes are

- **Selective Laser Sintering (SLS)**

Selective Laser Sintering (SLS) is a PBF technology that utilizes a high-density laser to selectively fuse powdered materials, typically polymers, into a three-dimensional object layer by layer. Typical process involves spreading a deposition of a thin powder layer, commonly nylon or polyamide, across a build platform. A high-powered laser selectively scans and sinters (heats without completely melting) the powdered material based on the cross-sectional pattern of the 3D model being produced. The laser fuses the powder particles, forming a solid layer. The build platform is then lowered, and a new layer of powder is spread over the previous one. This new layer is scanned by the laser, and repeated until the whole item is created layer by layer. Each layer that has been sintered cools and solidifies, forging a solid bond with the layers which came preceding it. Excess powder is typically left around the object after printing is finished, aiding in maintaining the structure. The extra powder is removed after printing, and the finished product might go through other post-processing processes like surface finishing or cleaning. Because the surrounding powder provides support, Selective Laser Sintering is well renowned for its ability to produce parts with complex geometries and advanced interior systems without the requirement for support structures. It is frequently utilized for quick prototyping and producing working parts in various industries. Furthermore, a variety of materials can be used with SLS, such as polymers and, in certain cases, metal powders by Selective Laser Melting (SLM). Among the various PBF technologies, selective laser sintering is speeding up its maturity (Fig. 12.4).

- **Selective Laser Melting (SLM)**

SLM is a widely adopted PBF technology that uses a high-power laser to fuse and melt the powder particles. SLM can process a variety of alloys, such as aluminum, titanium, steel and nickel. The laser beam melts and consolidates metal powder particles together during the printing process. The laser beam selectively joins or fuses the particles as it hits a thin layer of the material. The printer integrates a fresh layer of powdered material on top of the previous one after finishing a print cycle. The object is then lowered to a particular depth equal to one layer's thickness. Someone will manually remove the unused powder from the product after the print process is finished. The main difference between SLM and SLS is that SLM uses fully melted powder, whereas SLS uses partially melted or sintered powder. SLM final products typically have fewer or no voids, leading to them being stronger generally. SLM is used by the aerospace sector in some of its cutting-edge efforts. Usually, these are the ones that concentrate on accurate, strong, and light parts. These days, SLM is widely used in the biomedical, automotive and aerospace fields.

Fig. 12.4 Technological development in powder bed fusion processes

- **Electron Beam Melting (EBM)**

Electron Beam Melting is a powder bed fusion process that uses a high-density electron beam as heat source to melt the metallic powders together, creating complex and customized products. EBM is used for complex and customized parts and uses conductive powder materials. Layer-by-layer fabrication of the parts is in a vacuum-built chamber. Almost all PBF methods e.g. SLS, SLM and EBM involved the spreading of powders over previous layers. This can be done by a roller or a blade mechanism. PBF techniques start from the spreading of powders and the beam scans the layer as per the dimensions provided by a digital file. The powders that fuses the powders and solidify it as end components. These methods manufacture accurate and good surface finish parts for small scale production because of smaller beam size (Fig. 12.5).

- **Sheet Lamination**

Sheet lamination (SL) or Laminated Object Manufacturing (LOM) is one of the first commercialized AM techniques and it stands as a method employing metallic sheets as raw materials [6]. It harnesses a localized energy source, typically ultrasonic or laser, to join a stack of precisely shaped metal sheets, creating a three-dimensional structure. A widely adopted variant is Ultrasonic Additive Manufacturing (UAM) or Ultrasonic Consolidation (UC). This involves applying ultrasonic waves and mechanical pressure to sheet metal stacks at room temperature, leading to the bonding of adjacent sheets through diffusion, avoiding the need for melting. Layer by layer,

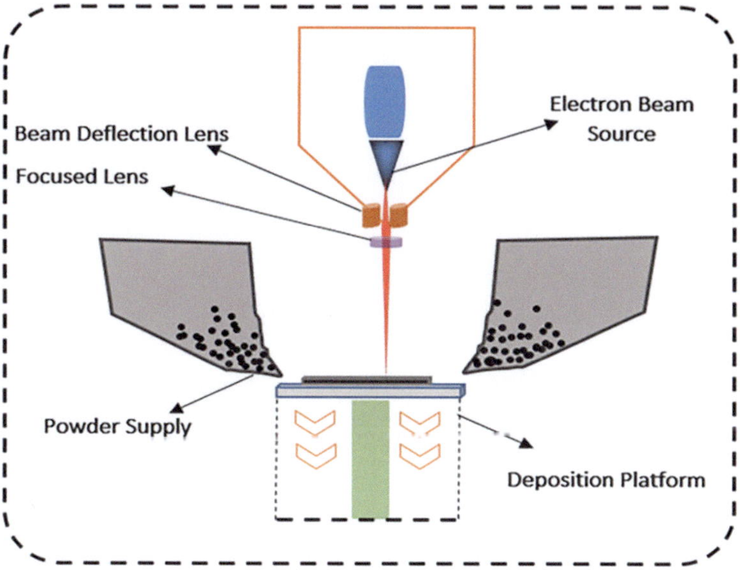

Fig. 12.5 Electron beam melting PBF additive manufacturing

these stacked sheets coalesce into a 3D object without the reliance on frictional heat. Prior to UC bonding, the metallic sheets are often cut according to the desired design. Traditional polishing can be optionally integrated into the process, either during or after consolidation, to achieve meticulous finishing [6] (Fig. 12.6).

Thin materials, such as paper or aluminum foil, are coated with glue and bonded to one another in layers. Before joining the layers together, each section is carefully sliced with a mechanical cutter or laser. The form-then-bond procedure allows to produce internal features and successful thermal bonding of metallic and ceramic materials by removing excess material before joining. After cutting, the residual material might be recycled or used for support. This coating contains the bonding chemical required for the succeeding layers to attach to one another. The coated layers are cemented together in stages, forming a three-dimensional object layer by layer. The successive bonding procedure enables the creation of complicated geometry and interior structures.

The remaining material after cutting can be repurposed or utilized as support after being removed. This coating provides the bonding agent needed for the successive layers to adhere to one another. The coated layers are successively glued together, creating a three-dimensional object layer by layer. This sequential bonding process allows for the construction of complex geometries and internal structures. A mechanical cutter or laser is utilized to precisely cut each layer before bonding. This step ensures accuracy in the final shape of the object and allows for the removal of excess material prior to the bonding process. The form-then-bond method distinguishes sheet lamination by enabling the removal of excess material before bonding. This

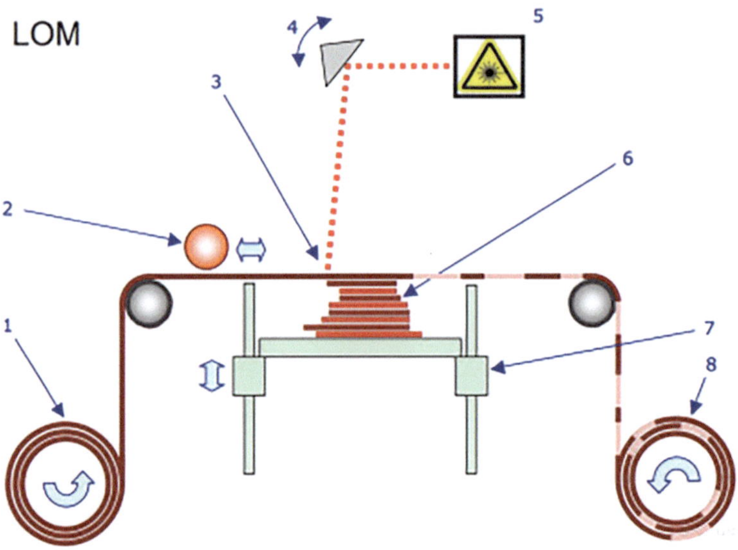

Fig. 12.6 Sheet lamination process mechanism

approach facilitates effective thermal bonding, particularly in the case of ceramic and metallic materials, and allows for the creation of intricate internal features. Excess material remaining after cutting can serve as a temporary support during the bonding process. Subsequently, these support structures can be removed, or in an eco-friendly watch, recycled for future use.

- **Material Jetting**

Material Jetting Additive Manufacturing, often referred to as Material Jetting, is an advanced additive manufacturing technique that involves the precise deposition of liquid or molten materials onto a build platform. Precisely, the droplets of 3DP material are selectively deposited in material jetting. In this process, material is jetted through tiny nozzles, similar to inkjet printing, and is solidified or cured to create layers that ultimately form a three-dimensional object. Material jetting (MJ) is a key enabling and competitive manufacturing technique for a broad range of industrial and scientific applications due to its cost-effectiveness, high-throughput and scalability of production, along with its ability to extend functionality via multi-material printing [7]. This technology is a full-color 3DP process, which directly deposes resins in CMYK-W (cyan, magenta, yellow, black, and white) when required. Furthermore, MJT technology reproduces color gradients and variations unavailable with other 3DP technologies.

Key features of material jetting additive manufacturing include high resolution, fine detailing, and the ability to produce intricate geometries. This technique is especially valuable when creating prototypes, models, and parts that require high surface quality and accuracy. The versatility of material jetting allows for the use of a variety of materials, including plastics, photopolymers, ceramics, and even metals.

- **Binder Jetting**

It involves selectively depositing a liquid binder onto a powder material in order to create a three-dimensional object layer by layer. With this process, intricate and complex parts can be rapidly produced with a variety of materials, making it suitable for prototypes and final products. A key advantage of binder jetting is its speed, versatility, and ability to produce large parts. It is particularly useful for creating functional prototypes, investment casting patterns, and even components for industries such as automotive, aerospace, and consumer goods. This technique works with a wide range of materials, from metals to ceramics to polymers. This technique works with a wide range of materials, from metals to ceramics to polymers.

Applications for binder jetting include casting molds and patterns, making complex jewelry pieces, and fabricating metal parts for the automotive and aerospace industries. Although binder jetting provides high-speed production and material flexibility, it is crucial to take into account the unique properties of the material as well as the need for post-processing steps in the manufacturing workflow as a whole [8].

- **Material Extrusion**

Material extrusion, sometimes referred to as fused filament fabrication (FFF) or fused deposition modeling (FDM), is a popular additive manufacturing process that uses the layer-by-layer deposition of thermoplastic materials to produce three-dimensional objects. This technique is accepted for both functional part production and rapid prototyping due to its simplicity, adaptability, and ease of use [9].

The process is quite straight forward as shown in Fig. 12.7 and the fundamental characteristics of material extrusion include its relatively low cost, wider flexibility of materials, and the capability to create both prototypes and end-use parts. Material extrusion is widely used across various industries for rapid prototyping, functional testing, and creating concept models. It is frequently used in low-cost prototypes, patternmaking, aerospace, consumer goods, education, and more [10].

Several parameters are involved in material extrusion where, Layer thickness, build orientation, infill density, air, temperature of the printing bed or substrate, number of shells, top and bottom thickness, infill patterns, raster angle, and raster pattern are just a few of the customizable printing parameters. The method primarily relies on customizing printing parameters for a specific kind of material. These process parameters have a significant impact on both the mechanical properties and quality of the printed-items.

- **Vat-Photopolymerization**

Vat Photopolymerization is a 3D printing technique that employs photopolymerization to create three-dimensional objects. It operates by selectively solidifying liquid

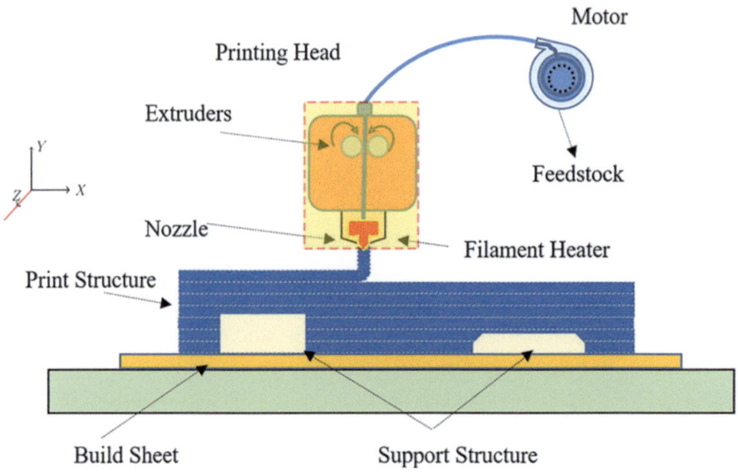

Fig. 12.7 Fused filament fabrication schematic [9]

photopolymer resin using light, typically UV light, layer by layer. This process is recognized for its ability to produce high-resolution parts with fine detail and smooth surface finishes [11].

Key characteristics of VAT photopolymerization include its precision, suitability for intricate designs, and its applications in industries requiring precise prototypes and small-scale production. Vat Photopolymerization is employed in two main subcategories: Stereolithography (SLA) and Digital Light Processing (DLP). In SLA, a laser or UV light source is precisely controlled to solidify the resin layer by layer. In DLP, a digital light projector exposes an entire layer at once, speeding up the process [12].

Vat-photopolymerization is widely used in industries such as jewelry, dentistryl, healthcare, and aerospace for creating intricate and detailed parts that require high accuracy and surface quality. However, considerations like resin selection, post-processing, and the need for support structures should be considered for successful VAT photopolymerization-based additive manufacturing.

12.4 Virtual Manufacturing

Virtual manufacturing is another important paradigm in future industrialization schemes and is supposed to be implemented in the manufacturing of all kinds of discrete products. It relates to increased customization of products for the customer, flexibility in the production of multidisciplinary products, and customer access to production status after due authentication [13].

12.5 Case Study: Analysis of Distortions in Selective Laser Melted TiAl6V4 Part[1]

12.5.1 Abstract

The metal additive manufacturing process known as selective laser melting (SLM) frequently produces components with significant geometric distortions due to its noticeable temperature gradient. These deformations play a vital role in ensuring the structural integrity of the part; however, they can adversely affect the component's performance. This study uses a combination of experimental analysis and numerical simulation to examine the deformations in a TiAl6V4 sample produced using the SLM process. The investigation starts with a component-level finite element analysis (FEA) simulation, with an emphasis on using the inherent strain method (ISM). This is followed by an experimental process. The utilization of ISM is highlighted due to its ability to notably reduce the calculation time compared to traditional transient thermo-mechanical analysis methods. The results reveal the presence of heightened stresses concentrated along the central region of opposing edges in the build orientation. Remarkably, no significant visible deformations are observed in the printed titanium samples due to the SLM process. In conclusion, this study highlights the significance of numerical simulation in the fields of product development and technological advancement by demonstrating a strong correlation between experimental observations and simulations.

Keywords: Selective laser melting; distortions; inherent strain method; contour approach.

12.5.2 Introduction

Selective laser melting (SLM) has gained increasing popularity as a manufacturing process for producing intricate, moderately sized, and high-value metal components. The mechanical properties of SLM-produced parts are on par with those of bulk materials [1, 2]. Over the recent decades, SLM has found extensive use in industries such as aviation, automotive, biomedical, and power generation [3–6] due to its ability to efficiently manufacture high-quality metal components with complex geometries and inherent characteristics. However, it is evident that the SLM process introduces residual stresses throughout the manufacturing process [7–9]. These inherent residual stresses, an unavoidable byproduct of the SLM process, can potentially lead to manufacturing deformations or even build failures [10].

[1] Developed by Hassan Ali, reproduced with permission.

While Selective Laser Melting (SLM) technology has found widespread adoption in practice over the past few decades, it is essential to recognize that the distortions arising during the SLM build phase are unavoidable consequences of the significant temperature gradient involved. These distortions have a detrimental impact on the dimensional accuracy and functional performance of the final components [11, 12]. It is important to note that distortions resulting from residual stress are inherent to thermal manufacturing methods, with components produced through SLM being particularly susceptible to construction errors and structural weaknesses due to these process-induced stresses during manufacturing [10, 13]. Consequently, the elevated internal residual stresses increase the risk of component failure both during and after the SLM process [14]. Various simulations have been conducted to assess the residual stress and deformation of SLM-manufactured components at the micro or mesoscale, examining either single or multiple paths [15, 16].

Presently, there are numerous methods available for assessing the distortions in SLM parts caused by residual stresses. These include both experimental techniques [21–24] and numerical modeling approaches [25, 26] to evaluate residual stresses in SLM-manufactured components. For instance, X-ray diffraction (XRD) has been previously utilized to assess residual stresses in metal SLM, specifically in AISI 316L [24]. Residual stresses obtained through neutron diffraction (ND) and contour methods (CM) have shown strong correlations in the case of laser-direct metal-printed Waspaloy [23]. Optical techniques have also been employed to analyze stresses in SLM-produced parts [27]. Researchers have explored various approaches to quantify residual stresses in SLM components. Salmi et al. [28]. Liu et al. [29] examined stress distribution using a thermal gradient method, while Childs et al. [30] delved into the influence of several factors on the volume of molten thin layers in SLM, noting an increase in melting density with higher scanning rates. Read et al. [31] and Kempen et al. [32] have focused on optimizing SLM process variables for the fabrication of nearly solid AlSi10Mg components.

While in-situ evaluation of residual stress during the manufacturing process shows promise, it faces significant technological and practical challenges [34]. Consequently, as a viable alternative, numerical modeling of the complete additive manufacturing (AM) cycle has become a critical topic of discussion among AM researchers, with the goal of accurately representing the evolution of residual stress and distortions throughout the manufacturing process.

This article primarily focuses on using the Inherent Strain Method (ISM) and the contour method to predict residual stresses and deformations in SLM. ISM is a mechanical technique that avoids transient thermal analysis, substantially reducing computation time, making it suitable for larger parts. For validation, the contour method was chosen to analyze stress distribution along the build orientation, offering a comprehensive map of residual stress distribution, less influenced by material microstructure compared to XRD and ND methods. The contour method is also more widely available. TiAl6V4 powder was used in a Farsoon SLM FS421M machine, and Simufact Additive served as the FEM modeling simulation tool for the entire SLM process. Post-manufacturing residual stresses and distortions were evaluated to validate the numerical predictions.

12.5.3 Materials and Methods

A practical examination was carried out to assess the distortions and residual stresses present in parts produced through the Selective Laser Melting (SLM) process. Cube specimens, with dimensions measuring 15 × 15 × 15 mm^3 and made from TiAl6V4 using SLM, were produced for the purpose of investigating the distortions caused by residual stresses. To introduce irregularity into the geometry, notches were intentionally incorporated into the upper portion of the cube samples. The size and specifications of the titanium cube, as well as the manufactured specimen, are illustrated in Fig. 12.8.

The experimental process was conducted with the utilization of a Farsoon FS421M SLM machine featuring a single laser, a laser power of 300 W, and a maximum laser fill speed of 1000 mm/s. The produced samples were suitable for the examination of stresses through the contour approach and distortions through the 3D scanning method. The simulation of the SLM process was executed by employing the inherent strain method to analyze residual stresses and distortions, utilizing Simufact Additive 2022.

12.5.3.1 SLM Process Parameters

Spherical TiAl6V4 powders, each with a size of 45 μm, and possessing the following material properties: Poisson's ratio of 0.28, density of 4.31 g/cm^3, Young's modulus of 112 GPa, tensile strength of 1288 MPa, and yield strength of 1138 MPa, were employed in the SLM process for part fabrication. The inherent strains of TiAl6V4, determined from experimental data and calibrated for use in Simufact Additive in this study, are as follows: εxx = – 0.0054, εyy = – 0.0023, εzz = – 0.03. The SLM

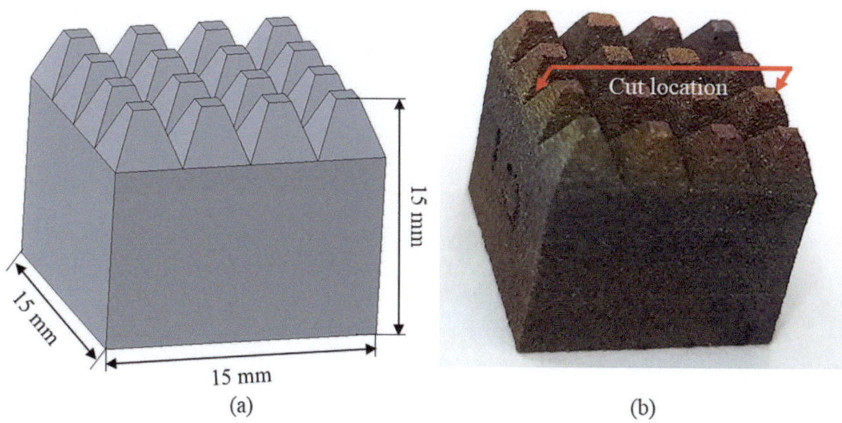

Fig. 12.8 SLM manufactured sample **a** CAD with dimensions **b** cut location

Table 12.1 Process parameters for SLM

Parameters	Values
Laser power (W)	300
Layer thickness (mm)	0.04
Scan speed (mm/s)	1000
Scan width (mm)	10
Hatch strategy	Stripe wise
Material type	TiAl6V4

procedure was executed using the Farsoon FS421M metal laser powder bed fusion (LPBF) machine, with a strip-wise scanning strategy adopted for part manufacturing. The specific SLM printing parameters applied in this study can be found in Table 12.1.

12.5.3.2 Simulation Setup for ISM

In this paper, the Finite Element Method (FEM) was employed to conduct a comprehensive mechanical simulation. Simu-fact Additive utilizes a purely mechanical analysis approach, which requires inputting inherent strain rates for the numerical simulation. The specific values for these inherent strains are influenced by the material's properties and process parameters. Figure 12.9 demonstrates that in mechanical simulations, only the magnitude of inherent strain is essential for computing stresses and distortions. Consequently, factors such as phase transformation and thermal strains were omitted, resulting in a reduction in computational time.

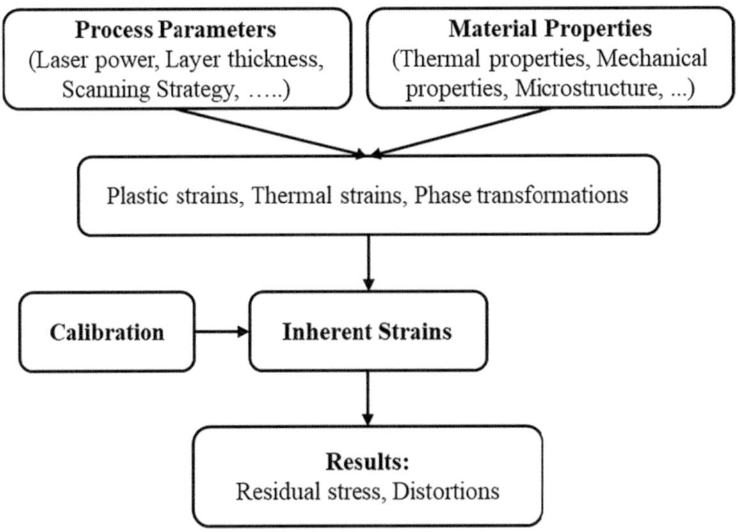

Fig. 12.9 ISM using finite element simulation

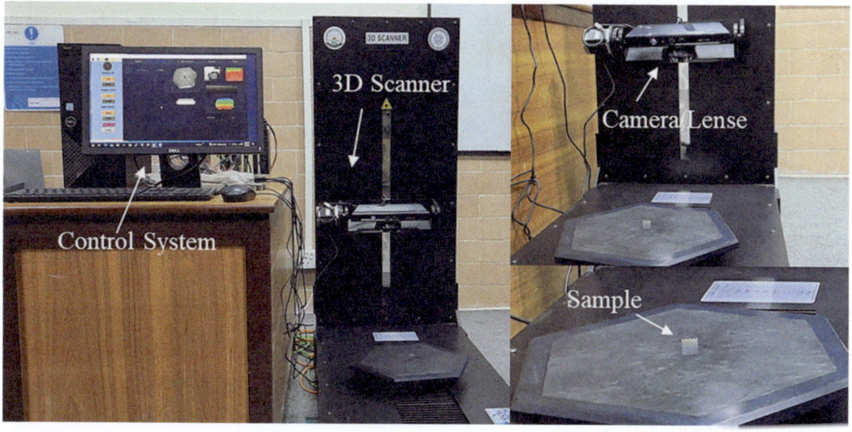

Fig. 12.10 3D scanning for distortions evaluation

12.5.3.3 Experimental Setup

The main objective of this research was to assess and examine residual stress and distortions occurring after the manufacturing of parts. Consequently, the specific influence of process parameters was not considered. For the analysis of process-induced stresses and deformations, TiAl6V4 cubic samples were produced using an SLM machine. In this investigation, the contour approach was utilized to assess residual stresses, and the resulting deformations were measured with a 3D scanner, as depicted in Fig. 12.10.

The contour approach is a practical technique for quantifying residual stresses. Figure 12.11 illustrates the wire electro-discharge machining (EDM) operation applied to the SLM-manufactured sample. During this operation, the specimen of interest is split in two by the wire EDM process.

The residual stress induced in the sample results in distortion of the sliced surface. Figure 12.12 depicts the fundamental process of the contour approach. The analyzed stress portion is oriented perpendicular to the sliced surface. By employing a 3D scanner, the profile of the relaxed surface for both sliced halves are obtained. The data on surface deformation, once collected, undergo smoothing, averaging, and cleaning. The inverse of the observed pattern is then used as displacement boundary conditions in a 3D finite element model of the sliced surface. Subsequently, a Finite Element Analysis (FEA) is conducted to examine the stress distribution within the sample [35].

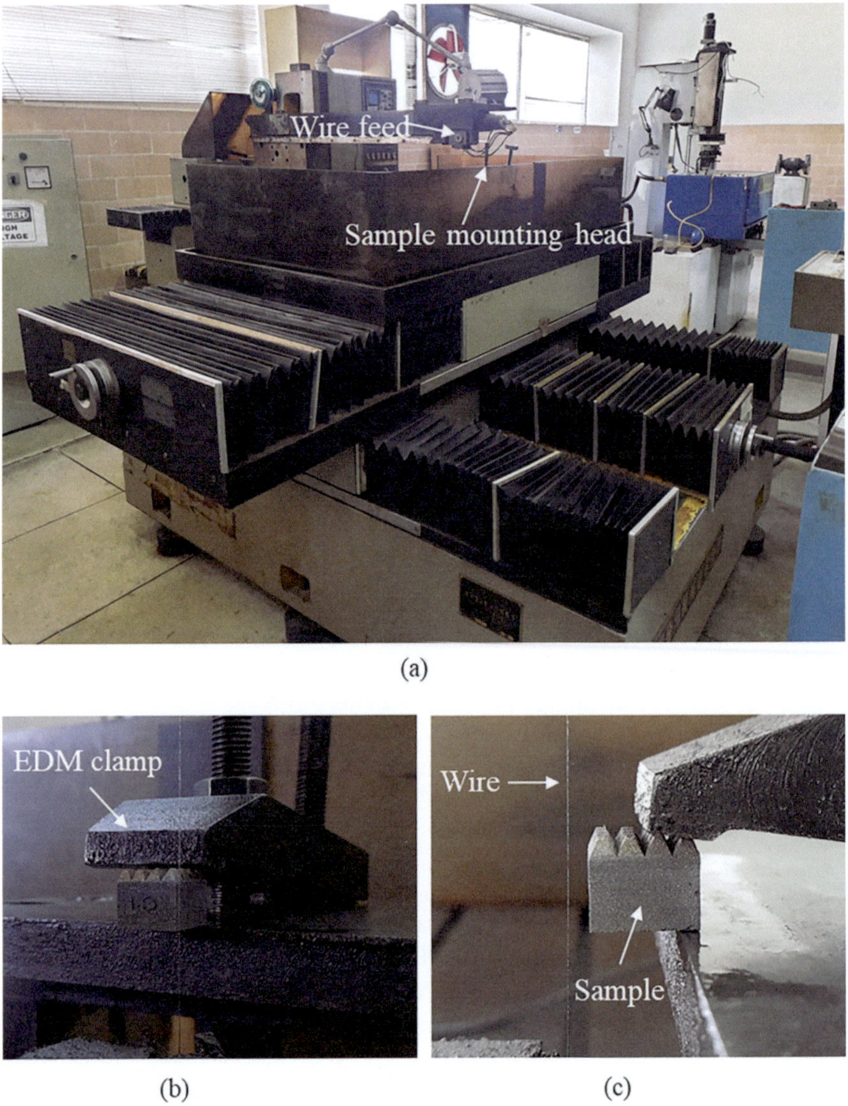

Fig. 12.11 **a** Wire EDM operation **b** sample clamping in EDM, and **c** EDM operation

12.5.4 Results and Discussion

In this study, the contour approach was selected to assess the overall stress distribution along the build direction of the SLM-manufactured sample and to validate the simulation outcomes. Employing the contour technique, residual stresses were assessed along the build orientation in cubic TiAl6V4 samples produced using SLM

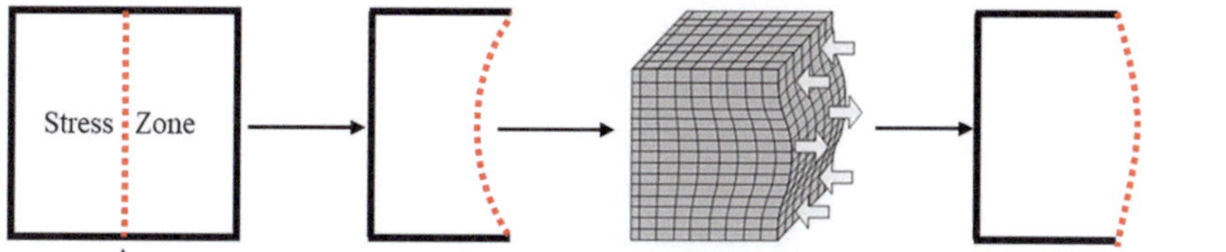

Fig. 12.12 Contour approach principle for stress calculation

additive manufacturing technology. There were no significant visible distortions in the titanium samples produced through the manufacturing process. Figure 12.13 displays the surface deviation curve for the observed stress specimen in the sliced TiAl6V4 cube samples.

The data shown in Fig. 12.13 represents the average of data obtained from the sliced pieces, resulting in a reasonably smooth and adaptable deviation profile. Residual stress is observed along the edges of the sliced specimen, particularly in the part's build orientation and at its mid-height. Line 1, as illustrated in Fig. 12.16, exhibits higher tensile residual stresses compared to other regions within the sample. Figure 12.14 provides a visualization of the total surface deformations.

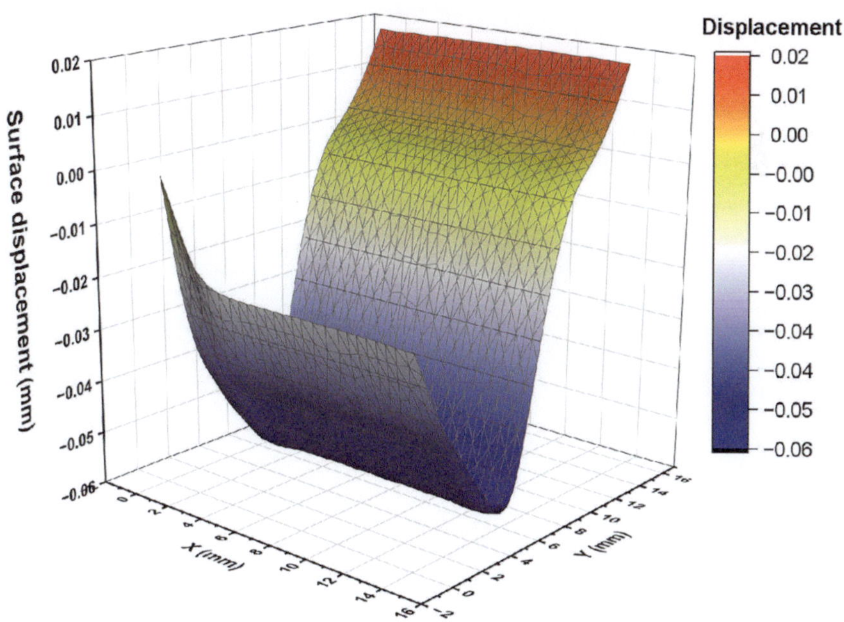

Fig. 12.13 Surface deviation profile

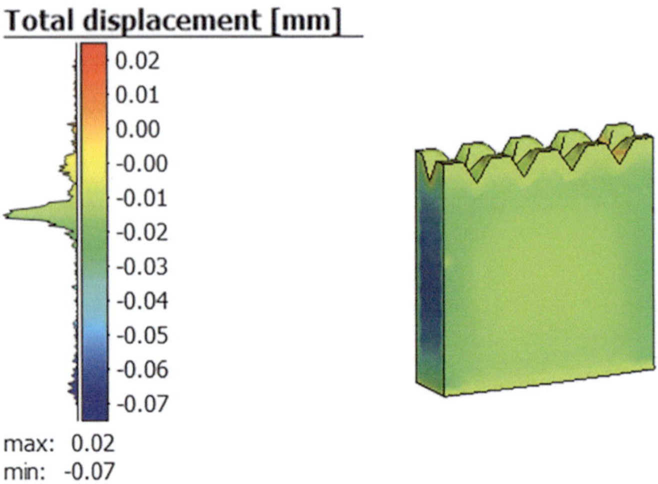

Fig. 12.14 Total surface displacements

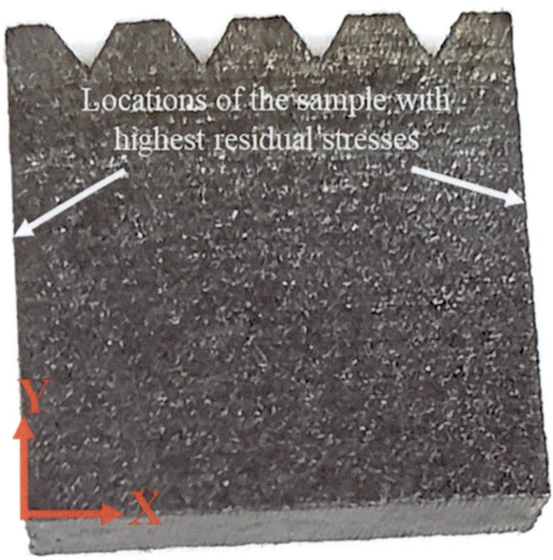

Fig. 12.15 Residual stresses location

Figure 12.15 displays the locations of the sliced specimen with greater residual stress, which are near the midpoints of lines 1 and 2 (as per Fig. 12.16). As shown in Fig. 12.16, a set of areas are indicated for the generation of line maps to provide a more thorough representation of the stress concentration in the sample.

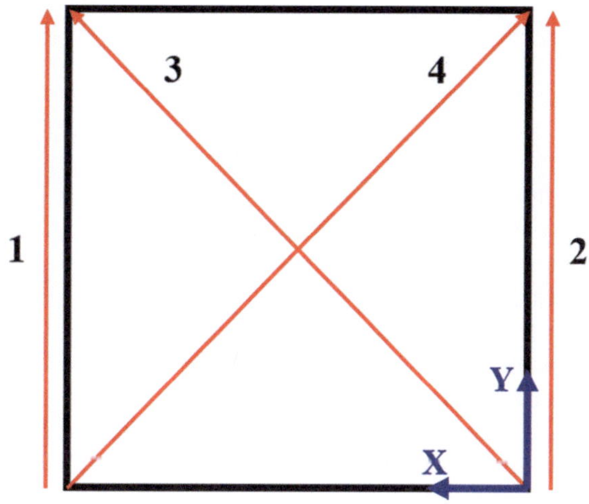

Fig. 12.16 Locations of the stress line profiles

Figure 12.17 displays the stress distribution map generated by the numerical simulation for the TiAl6V4 stress component of interest, which shows a maximum residual stress of 1101.4 MPa. Compared to the contour method, the numerically simulated results show a nearly similar type of residual stress distribution map, except for slightly more variation in stresses at the upper location. This might be due to the SLM manufacturing phase (i.e., the end of printing).

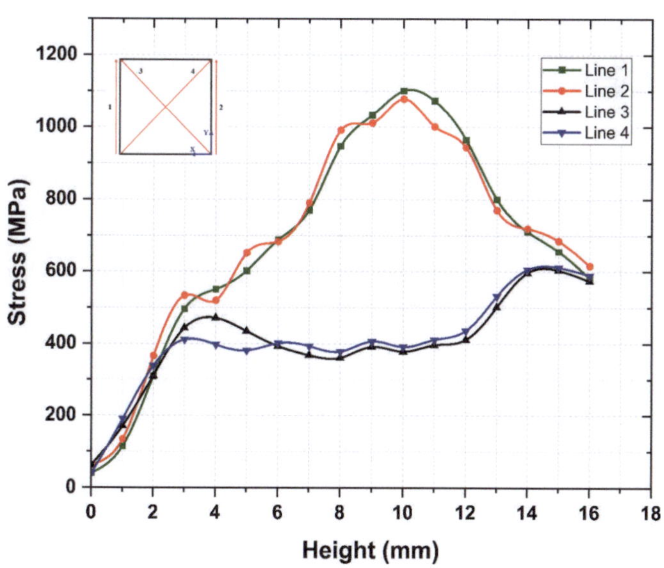

Fig. 12.17 Stress line profiles by simulation method

Apart from the differences in the additive manufacturing (AM) processing parameters, the slight variations in stress distributions within the sliced samples, particularly at the edges, could be attributed to imperfections stemming from data smoothing, wire EDM cutting, and 3D scanner measurements [36]. The alignment between the data obtained through the contour approach and the ISM simulation is quite robust, particularly concerning stress distribution, except for the upper portion where the numerical stresses are notably higher compared to the contour method.

Figure 12.17 presents a comparison of stress distribution profiles between the simulation study and the contour approach for the sliced TiAL6V4 specimen along the edges (specifically, lines 1 and 2, as indicated in Fig. 12.16). The line numbers correspond to those in Fig. 12.16. The agreement between the results obtained through the contour approach and the ISM simulation is quite strong in terms of stress distribution, except for the upper region where the numerical stresses are notably higher than those obtained through the contour method.

Regarding the specific stress element of interest, there is a good correlation between the stress distribution profiles assessed using the contour approach and those analyzed with the ISM simulation model. However, due to errors introduced by wire EDM, 3D scanner measurements, and data smoothing, it can be challenging to obtain an exact distribution map of residual stresses near the distorted surface using the contour approach. Considering the cost-effectiveness and shorter computational time of numerical simulations based on the ISM, the alignment between the ISM and the contour approach holds significant industrial significance.

12.6 Conclusion

In the realm of additive manufacturing, Selective Laser Melting (SLM) has risen as one of the most prevalent techniques, widely employed in industrial applications. However, the SLM process gives rise to notable levels of geometric distortions in the fabricated parts, potentially resulting in manufacturing defects and printing failures. In this investigation, we analyzed the distribution distortions in TiAl6V4 samples created through SLM, employing ISM numerical modeling and experimental measurements via the contour approach. In summary, the simulated results and the contour approach exhibited a strong alignment. The key findings can be summarized as follows:

- The inherent strain approach provides reasonably accurate predictions for process-induced distortions in selectively laser-melted parts.
- Both the numerical simulation and experimental measurements indicated a higher concentration of process-induced deformations along the central region of the opposing edges in the build direction.
- Overall, there was a notable consistency in results between the ISM simulation model and the contour approach.

12.7 Learning Outcomes

The reader will be able to understand the fundamentals of subtractive and additive manufacturing technologies. More focus remained on additive manufacturing, its classifications based on the material and heat source along with potential applications. Several additive manufacturing technologies are thoroughly reviewed from indigenous designs and fabrications of end products. It will enhance the understanding and impart the skills to incorporate subtractive, additive, and virtual manufacturing in the current industrial stature.

12.8 Review Questions

Q1 What is additive manufacturing and how does it differ from traditional manufacturing?
Q2 Enlist and describe the common additive manufacturing technologies?
Q3 Discuss the various materials used in metallic additive manufacturing.
Q4 Discuss the effects of process parameters on the SLM produced part manufacturing time.
Q5 Describe the key parameters that influence the additive manufacturing process and how these parameters affect the quality.
Q6 Describe the experimental techniques that can be used for analyzing the residual stresses in SLM produced parts?
Q7 Describe the experimental techniques that can be used for analyzing the geometric distortions in SLM produced parts?
Q8 Describe the numerical techniques that can be used for analyzing the residual stresses in SLM produced parts?
Q9 Describe the numerical techniques that can be used for analyzing the geometric distortions in SLM produced parts?
Q10 Identify the current challenges associated with the metal AM, identify the solutions.

Bibliography

1. Rashid A (2019) Additive manufacturing technologies. CIRP Encycl Prod Eng 1034:10–1007
2. ISO/ASTM52900-15, Standard terminology for additive manufacturing—general principles—terminology. American Society for Testing and Materials, West Conshohocken
3. Pagac M et al (2021) A review of vat photopolymerization technology: materials, applications, challenges, and future trends of 3D printing. Polymers 13(4):598
4. Chua CK, Wong CH, Yeong WY (2017) Standards, quality control, and measurement sciences in 3D printing and additive manufacturing. Academic Press

5. Srinivas M, Babu BS (2017) A critical review on recent research methodologies in additive manufacturing. Mater Today Proc 4(8):9049–9059
6. DebRoy T et al (2018) Additive manufacturing of metallic components–process, structure and properties. Prog Mater Sci 92:112–224
7. Frazier WE (2014) Metal additive manufacturing: a review. J Mater Eng Perform 23:1917–1928
8. Gibson I et al (2021) Additive manufacturing technologies, vol 17. Springer
9. Gülcan O, Günaydın K, Tamer A (2021) The state of the art of material jetting—a critical review. Polymers 13(16):2829
10. Mustafa MS, Muneer MA, Zafar MQ, Arif M, Hussain G, Siddiqui FA (2022) Process parameter optimization for fused filament fabrication additive manufacturing of PLA/PHA biodegradable polymer blend. Int Polym Proc 37(1):1–14
11. Soori M, Arezoo B, Dastres R (2024) Virtual manufacturing in industry 4.0: a review. Data Sci Manag 7(1):47–63
12. Carneiro OS, Silva AF, Gomes R (2015) Fused deposition modeling with polypropylene. Mater Des 83:768–776
13. Zhang F et al (2021) The recent development of vat photopolymerization: a review. Addit Manuf 48:102423

Part VI
Communication

Prelude

Humans, since 14,000 BC, have been working on speech forms. In the age of the fourth industrial revolution, it is covered by two pillars, cloud and cyber security, in the presence of legacy communication technologies developed since the first industrial revolution and much before that once it comes to encoding a stream of pulses.

Communication of signals as meager as between 0 V and 0.8 V with respect to the ground terminal is a transistor logic 'low' and a transistor logic input signal is defined as "high" when between 2 V and 5 V is driving the whole digital world.

Network topologies transmitting signals from one node to another either through a fiber optics cable or through networks of satellite is reality today.

Communication from Earth to Mars takes anywhere between 3 min and just over 22 min, depending on where the two planets are in their orbits.

This short part summarizes the communication in a nutshell as it is required for engineering design and manufacturing today.

Chapter 13
Communication Protocols—Legacy and Industry 4.0

Prelude
Today, transmitting low-voltage digital and analog signals is standard practice. Following traditional and Industry 4.0 protocols, electronic devices use these signals to control prime movers, sensors, and transducers. This is done through peripheral devices, like machine tools, operated by microcontrollers of varying capacities.

Physical and Cellular Communication

In physical and cellular communication, signals are commonly transmitted between nodes, which can be very close or millions of miles apart. This book focuses specifically on communication between Turing machines over networks and their communication with peripheral devices.

13.1 Objectives

The objectives include understanding data communication, which can occur serially or in parallel using physical or cellular mediums. Cellular communication is prevalent in the current cloud-based era, though physical Wide Area Networks (WANs) are also vital. Data transmission employs both traditional and Industry 4.0 standards. However, data remains secure from potential threats only when cybersecurity algorithms exist.

This chapter aims to educate the reader on the modes and mediums of data transmission, the types of networks involved, the security risks present, and the strategies to mitigate these risks in data transmission.

Supplementary Information The online version contains supplementary material available at https://doi.org/10.1007/978-3-031-77106-4_13.

W. A. Khan et al., *Computer Aided Engineering Design and Manufacturing*, Springer Series in Advanced Manufacturing, https://doi.org/10.1007/978-3-031-77106-4_13

13.2 OSI—Open System Interconnection Standard

The OSI standard, formulated by the International Organization for Standardization (ISO) in the late 1970s, acts as a framework, setting forth protocols and guidelines for designing and implementing network communication systems. This section delves into the OSI standard, examining its seven distinct layers, their respective roles, and the overall influence of the OSI model on network communication. The OSI model comprises seven layers, each with specific network communication responsibilities are shown in Fig. 13.1. These layers include [1]:

i. **Physical Layer**: This layer is responsible for direct data transmission across a physical medium; it might include fiber optics, copper cables, or wireless networks. It addresses different factors like voltage levels, encoding of data, and the timing of signals. It also outlines such transmissions required electrical, mechanical, and procedural attributes.
ii. **Data Link Layer**: This layer handles the consistent transmission of data frames between network nodes nearby. It also ensures error-free data transfer by rectifying errors originating from the physical layer. This layer contributes to data framing, data flow management, error detection, and access control.
iii. **Network Layer**: It is tasked with forwarding data packets through various networks. This layer helps to identify the optimal route for data transmission while considering the traffic management, network structure, and addressing systems. It also sets out logical addressing and routing protocols to ensure efficient and dependable data delivery.
iv. **Transport Layer**: This layer controls reliable data transmission from one host to another. It offers solutions for error correction, congestion control, and data flow management by separating data into smaller segments. It is critical in maintaining service quality, sequencing data, and establishing connections.
v. **Session Layer**: This layer is focused on maintaining, initiating, and concluding sessions between applications across different network devices. It enables the coordination of various communication streams by synchronizing data. It also oversees authorization, authentication, and the recovery of sessions.
vi. **Presentation Layer**: This layer is focused on encryption, compression, and data translation. It helps to prepare the data in a specific format to convert it into readable form for the application layer. That is why this layer efficiently does the format conversion of the data.
vii. **Application Layer**: This layer handles end-user applications by allowing users to access network services. It directly interacts with specific protocols like SMTP for email and HTTP for web browsing. It contributes to multiple activities like file transfers, user interfaces, and network services.

TCP/IP, prevalently used in current networks, adopts a layering approach akin to the OSI model. However, TCP/IP simplifies the model by merging several OSI layers into fewer ones, making it more practical. The TCP/IP model consists of four layers and, despite differences in layers and protocols, shares the OSI model's aim of enhancing network communication and interoperability.

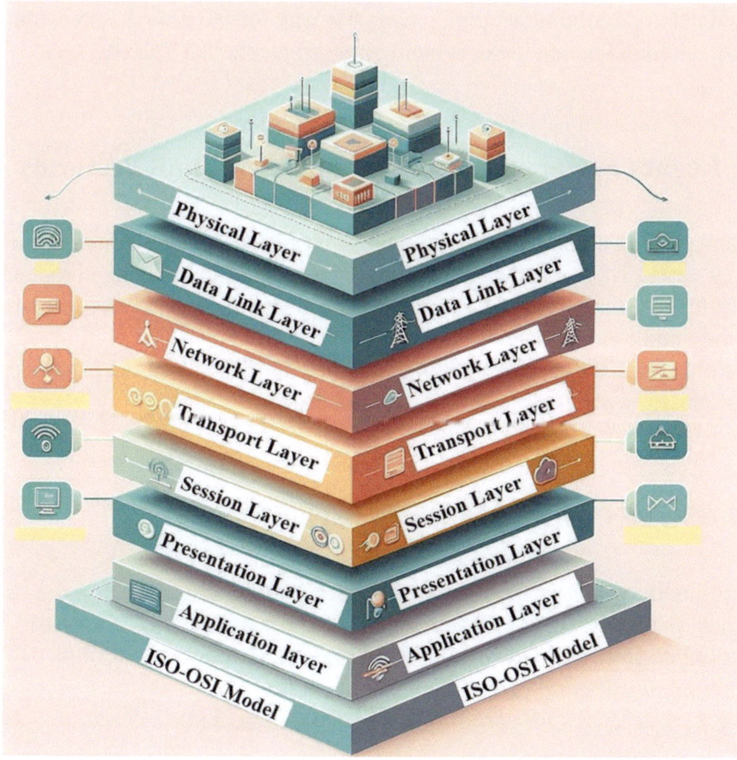

Fig. 13.1 ISO-OSI layers

The OSI model has influenced the development of real-world networking protocols. Although OSI protocols are not commonly used, their conceptual framework has been crucial in developing protocols like TCP/IP, the backbone of the modern internet.

The OSI model adopts a layered approach, segregating various functions for modular design. Each layer has distinct tasks, and their interfaces dictate their interaction. This design offers flexibility, scalability, and interoperability, permitting independent implementation and updates of layers without impacting others.

A major benefit of the OSI model is its facilitation of interoperability among different networking technologies from various vendors. It sets a standard for protocols and interfaces, enabling smooth communication between devices of different makes. This interoperability has significantly contributed to the expansion 2 and evolution of computer networks, fostering networks with diverse components.

The OSI standard outlines a systematic approach to network architecture, defining protocols and guidelines for network communication system design and implementation. It consists of seven layers, each with specific network communication functions. While not widely used in practice, the OSI model's theoretical framework has greatly impacted network communication, paving the way for the growth and evolution of networks, and influencing the development of protocols like TCP/IP.

13.3 Legacy Serial and Parallel Communication Protocols

Serial and parallel communication protocols have been crucial in data transmission for many years. These methods have been instrumental in allowing computers and other devices to share information. This has been key in the early stages of network technology and moving data between various components. This section will delve into the old serial and parallel communication protocols, discussing their features, benefits, and drawbacks.

13.3.1 Serial Communication Protocols

Serial communication protocols send data one bit at a time, in sequence, using a single data pathway. This approach is common in various uses, from basic device connections to intricate network infrastructures. The following examples of older serial communication protocols [2] are shown in Fig. 13.2.

Table 13.1 describes the serial communication protocols.

13.3.2 Parallel Communication Protocols

Parallel communication protocols send multiple bits simultaneously across several data lines. This approach, prevalent in early computer systems and external devices, enabled fast data transmission. Below are some examples of older parallel communication protocols [3], as shown in Fig. 13.3.

Table 13.2 describes the parallel communication protocols.

13.4 Advantages and Limitations

Serial and parallel communication protocols offered distinct advantages and faced specific limitations [4, 5].

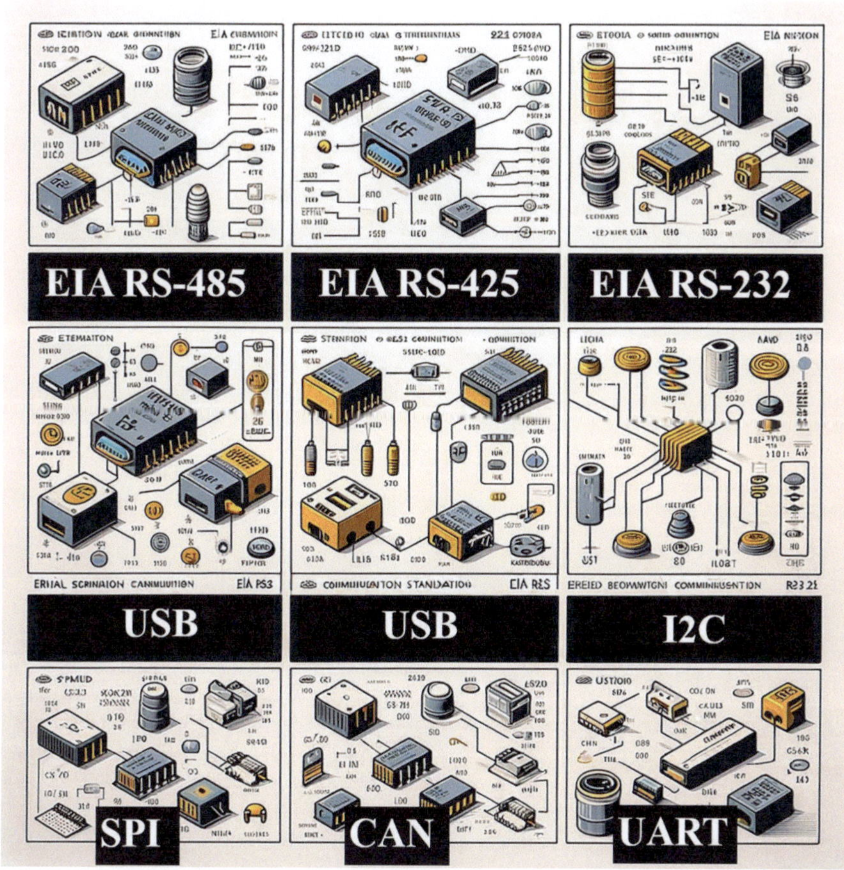

Fig. 13.2 Serial communication protocols

13.4.1 Advantages of Serial Communication

Table 13.3 describes the advantages of serial communication protocols.

13.4.2 Limitations of Serial Communication

Table 13.4 describes the limitations of serial communication protocols.

Table 13.1 Serial communication protocols with their description

Protocol	Description
EIA RS-485 (Electronic Industries Alliance, Recommended Standard 485)	• Used for longer distances and multi-node communication • Extensively used in industrial automation and process control systems • Differential signaling for improved noise immunity and longer cable runs • Supports multi-drop communication • Master–slave architecture • Higher data rates (100 kbps–10 Mbps) depending on cable length
EIA RS-425	• Outlines electrical properties for balanced voltage digital communication • Used for serial data communication between DCE and DTE • Prefer to use over short distances
EIA RS-232	• Often referred to as the serial port or COM port • Primarily used to connect devices such as modems, printers, and early personal computers • Voltage level signaling system (positive and negative voltages for binary data) • Full-duplex communication • Max data rate of 20 kbps over short distances • Performance decreases over longer distances • Uses multiple signal lines including data, control, and ground lines
USB	• Used for data and power transmission • Supports high-speed data transfer, • Supports plug-and-play and hot-swapping • Used for asynchronous and isochronous data transfer
I2C	• Designed for short-range, on-board communication using two lines: data (SDA) and clock (SCL) • Operates in a master–slave architecture • Used to connect peripheral integrated circuits to CPUs and microcontrollers
SPI	• Used for short-range communication in embedded systems • Operates in full-duplex mode • Requires more wires and generally faster than I2C

(continued)

Table 13.1 (continued)

Protocol	Description
CAN	• Vehicle bus protocol used extensively in automotive applications and industrial automation • Enables microcontrollers and devices to communicate without a central computer
UART	• Hardware component that converts data between parallel and serial formats • Used for short-distance, low-speed, and low-cost data exchange • Commonly found in microcontrollers

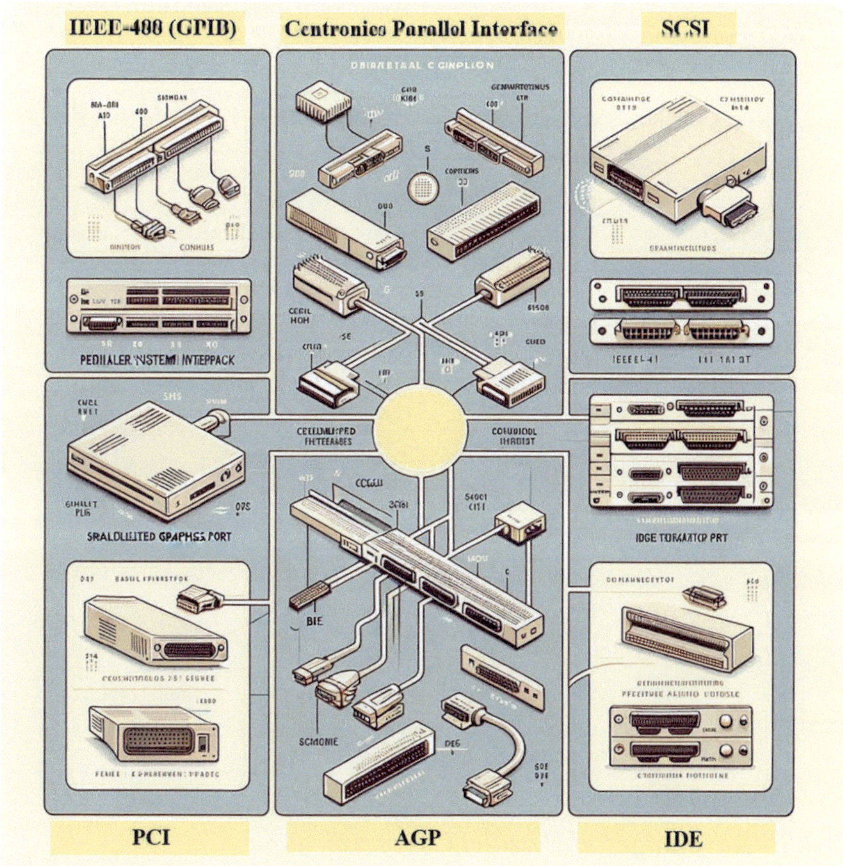

Fig. 13.3 Parallel communication protocols

Table 13.2 Parallel communication protocols with their description

Protocol	Description
IEEE-488 (GPIB) (Institute of Electrical and Electronics Engineers) (General Purpose Interface Bus)	• Initially for programmable instruments • Parallel multi-master communication • Up to 16 bidirectional data lines • Used in labs and industrial automation
Centronics (Parallel Printer Port)	• Industrial standard for connecting printers to PCs • Supports parallel data transmission up to 8 or 16 lines • Operated synchronously with handshake mechanism • Faster data transfer than serial protocols • Data transfer speed limited to megabits per second
SCSI (Small Computer System Interface)	• Connects computers to external devices • Supports parallel bus architecture • Used for hard drives, scanners, optical drives • Higher data transfer rates • Offers a wide range of device control and data transfer commands
PCI (Peripheral Component Interconnect)	• High-speed parallel interface for internal hardware • Used inside computers
AGP (Accelerated Graphics Port)	• Exclusively for graphics cards • High-speed parallel connection
IDE (Integrated Drive Electronics)	• Parallel interface for storage devices • Connects to computer's motherboard

Table 13.3 Advantages of serial communication protocols

Aspect	Merits
Simplicity	• Requires fewer data lines • Easier to implement • Simplifies connectors and cables
Longer distances	• Suitable for longer cable runs • Less prone to signal degradation compared to parallel protocols
Multi-drop capability	• Allows multiple devices to share a single data line (e.g., RS-485) • Reduces the number of connections needed

13.4.3 Advantages of Parallel Communication

Table 13.5 describes the advantages of parallel communication protocols.

Table 13.4 Limitations of serial communication protocols

Aspect	Limitations
Lower data rates	• Generally slower data transfer rates compared to parallel protocols • Not ideal for high-speed data transmission requirements
Limited bidirectional communication	• Often supports half-duplex communication, allowing one-way data transmission at a time • Can cause delays in bidirectional communication
Complexity in multi-node communication	• Requires additional control mechanisms for multi-node communication • This can increase complexity and potentially reduce performance

Table 13.5 Advantages of parallel communication protocols

Aspect	Merits
Higher data rates	• Enabled faster data transfer by transmitting multiple bits simultaneously • Ideal for high-speed data transmission applications
Simultaneous data transfer	• Allowed for simultaneous data transfer between devices • Potentially reduced latency • Improved overall system performance

13.4.4 Limitations of Parallel Communication

Table 13.6 describes the limitations of serial communication protocols.

In the early stages of data transmission, traditional serial and parallel communication protocols were key. Serial protocols like RS-232 and RS-485 were effective for short and long-distance communication. Meanwhile, parallel protocols like Centronics and SCSI enabled quick data transfer between devices. Each protocol type had unique benefits and drawbacks, and their application was based on specific needs. Nonetheless, with technological progress, newer protocols have mostly supplanted

Table 13.6 Limitations of parallel communication protocols

Aspect	Limitations
Increased complexity	• Required a larger number of data lines • More complex connectors and cables
Signal integrity challenges	• Difficult to maintain signal integrity with higher data rates and longer cables • Potential for signal degradation and data errors
Cost	• Additional hardware complexity increased costs • Less suitable for cost-sensitive applications

Table 13.7 Modern serial communication protocols with their descriptions

Protocol	Description
High-definition multimedia interface (HDMI)	• Transmits high-definition audio and video • Supports high-resolution video
Thunderbolt	• High-speed interface • Combines data, video, audio, and power • Known for high data transfer rates
Serial peripheral interface (SPI)	• Synchronous serial communication • Used in short-distance communication • Simple and faster • Used in embedded systems
Ethernet	• Uses serial technology in modern implementations • Includes Gigabit Ethernet for high-speed network communications

these older serial and parallel methods, providing enhanced efficiency, expandability, and user-friendliness.

13.5 Modern Serial and Parallel Communication Protocols

In the ever-changing landscape of technology, serial and parallel communication protocols remain crucial for data transmission. However, contemporary protocols have been developed to overcome the constraints of older systems and fulfill the growing need for fast, dependable, and efficient communication. This section will explore current serial and parallel communication protocols, examining their features, benefits, and uses. The modern serial and communication protocols [6] are described in the following sections.

13.5.1 Modern Serial Communication Protocols

Table 13.7 describes the modern serial communication protocols.

13.5.2 Modern Parallel Communication Protocols

Table 13.8 provides the modern parallel communication protocols.

Table 13.8 Modern parallel communication protocols with their descriptions

Protocol	Description
PCI express (PCIe)	• Evolved from PCI • High-speed interface • Uses a serial, point-to-point connection but noted for parallel data processing capabilities • Used in motherboards for graphics cards, SSDs, etc.
InfiniBand	• High throughput and low latency • Uses a switched fabric topology • Supports point-to-point and parallel processing • Used in supercomputing
HyperTransport	• Developed by AMD • High-speed, low latency link • Reduces the number of buses within a computer, blending serial and parallel characteristics • Used in advanced computing
Front side bus (FSB)	• Connects CPU to main memory and other components • Key in earlier computer architectures but now largely replaced • Used in CPU to memory communication systems

13.6 Advantages and Applications

Modern serial and parallel communication protocols offer several advantages over their legacy counterparts and are used in different sectors.

13.6.1 Advantages of Modern Serial Protocols

Table 13.9 tabulates the advantages of modern serial communication protocols.

Table 13.9 Advantages of modern serial communication protocols

Key feature	Advantages
Higher data transfer rates	• Significantly faster data rates • Efficient transmission of large data volumes in shorter time frames
Simplicity and ease of use	• Simpler cabling requirements • Supports plug-and-play functionality • User-friendly and convenient deployment
Enhanced compatibility	• Designed for backward compatibility • Works with legacy devices while offering improved capabilities

Table 13.10 Advantages of modern parallel communication protocols

Key feature	Advantages
Higher bandwidth	• Increased bandwidth • Faster data transfer rates • Suitable for high-performance applications dealing with large data streams
Scalability	• Allows for scalable configurations • Accommodates a wide range of devices • Expands system capabilities as needed
Low latency	• Minimizes latency due to simultaneous transfer of multiple bits • Suitable for real-time and high-performance applications

13.6.2 Advantages of Modern Parallel Protocols

Table 13.10 exhibits the advantages of modern parallel communication protocols.

13.6.3 Application of Modern Serial and Parallel Protocols

Table 13.11 contains the application of modern serial and parallel protocols.

To summarize, contemporary serial and parallel communication protocols have developed to cater to the growing need for fast, dependable, and effective data transfer. Serial protocols such as USB and Thunderbolt deliver flexible connection choices for numerous devices, whereas parallel protocols like PCIe and SAS offer substantial

Table 13.11 Application of modern serial and parallel communication protocols

Key feature	Application
High-speed storage systems	• Used in storage systems like Thunderbolt, PCIe, SAS • Provide fast and reliable data transfer rates between storage devices and host systems
Multimedia and video production	• Serial protocols like Thunderbolt and USB • Connect high-resolution displays, cameras, audio interfaces, etc. • Allow for efficient data transfer and real-time streaming
Networking and communication	• Serial protocols like USB and Ethernet • Utilized in networking equipment, routers, switches, network adapters • Facilitate reliable and high-speed communication between devices
High-performance computing	• Parallel protocols like PCIe • Applied in high-performance computing systems, gaming PCs, workstations • Essential for fast data transfer rates and low latency in processing-intensive tasks

bandwidth and minimal latency for high-demand uses. These protocols have revolutionized data transfer, facilitating smooth communication and aiding progress in storage systems, multimedia, networking, and high-performance computing.

13.7 Communication Networks

Communication networks facilitate data transfer and connectivity among various devices and systems. They can be configured in several topologies, each with unique characteristics and applications. The most common network topologies include circle (ring), square (mesh), star, bus, tree, and hybrid [7], as shown in Fig. 13.4. The table describes each topology:

Fig. 13.4 Communication networks topologies

Table 13.12 explains each topology.

These topologies can be selected based on the specific requirements, scale, cost considerations, and desired robustness of the network. Each has its advantages and disadvantages, making them suitable for different scenarios.

Table 13.12 Communication network topologies with their descriptions

Topology type	Description	Configuration
Circle (ring)	• Each node connects to exactly two other nodes, forming a ring for signal transmission • Data travels in one direction	• Requires each device to have two connections: one for input, one for output • Network disruption can occur if a node fails, unless a 'dual ring' is used for redundancy
Square (mesh)	• Mesh topology involves each node connecting to every other node, creating multiple pathways for data, visualized as a square	• Full mesh involves all devices interconnecting, which can be complex and costly in large networks • Partial mesh, with some nodes fully interconnected and others not, is more practical
Star	• All nodes are individually connected to a central node or hub, which acts as a repeater for data flow • Common in home and business networks	• Each device connects to a central hub • Network fails if the central hub goes down, but individual node failures don't impact others
Bus	• All devices connect to a single central cable or bus, with signals traveling in both directions to all nodes	• Easy installation and less cabling needed than in star or mesh topologies • A failure in the main cable can bring down the entire network
Tree	• Combines elements of star and bus topologies, with star-configured networks connected to a linear bus backbone	• Suitable for large, wide-area networks • Allows network expansion and independent branch management
Hybrid	• Combines two or more topology types, like mesh at the core and star in departments of a large network	• Versatile and customizable for specific needs and scalability • Integrates various topology advantages but is complex to design and manage

13.8 Case Study: Cyber Security of Communication Networks [1]

13.8.1 Serial Communication Standards and Their Cybersecurity Measures

Table 13.13 describes the security measures of serial communication standards [8–23].

Table 13.13 Cybersecurity measures of serial communication standards

Standard	Security considerations
EIA RS-485	• It lacks native encryption or security features • Security is typically implemented at a higher software layer with encryption and decryption of data. Physical security measures like secured cabling or fiber optics can also be used
EIA RS-425	• Isolation techniques like opto-isolators or transformers are used to protect against electrical interference and potential attacks
EIA RS-232	• It does not include encryption; thus, sensitive data should be encrypted at higher levels of the communication stack
USB	• It lacks a security layer for data transfer but supports code signing and device authentication • Secure USB drives with hardware-based encryption are available for more secure data storage and transfer
I2C	• It lacks built-in security features and is usually considered secure enough for short-range, internal device communication • Additional measures like data encryption at the software layer might be necessary for security-sensitive applications
SPI	• It lacks built-in security features • In secure applications, data encryption prior to transmission and secure initialization processes for devices are advisable
CAN	• It does not provide strong security features, making it susceptible to various attacks • Newer CAN variants aim to improve security, but typically additional security measures are implemented at higher layers
UART	• It does not have inherent security features • Additional encryption methods or protocols are advisable for secure data communication

[1] Developed by Muhammad Hammad, reproduced with permission.

Table 13.14 Cybersecurity measures of parallel communication standards

Standard	Security Considerations
IEEE-488 (GPIB)	• No built-in encryption/cybersecurity • Security via software/network safeguards • Requires additional measures like firewalls, access control
Centronics	• No native security features • Security added through secure print management systems
SCSI	• Basic hardware-based authentication • Requires additional security like software-based encryption
PCI	• Evolved to include data integrity checks • Security added via hardware/software firewalls
AGP	• No built-in security features • Relies on general computer security measures
IDE	• Lacks built-in security features • Legacy standard, replaced by SATA • Security through disk encryption software/hardware

13.8.2 Parallel Communication Standards and Their Cybersecurity Measures

Table 13.14 describes the security measures of parallel communication standards [24–35].

13.8.3 Internet Protocols with Cybersecurity Measures

Table 13.15 describes the security measures of different internet protocols [36–44].

Table 13.15 Cybersecurity measures of Internet protocols

Protocol category	Description	Security considerations
TCP/IP (Transmission Control Protocol/Internet Protocol)	• Fundamental suite of protocols that powers the Internet • Handles reliable data transfer; IP is responsible for routing data packets • Ubiquitous in industrial Ethernet solutions • Divides data into packets for reliable transmission	• Firewalls, intrusion detection systems • Encryption standards like SSL/TLS
Industrial ethernet	• Range of Ethernet protocols for industrial settings (e.g., PROFINET, EtherCAT, Ethernet/IP) • Offers real-time control and data collection • Used in manufacturing, process control, robotics, etc.	• VLANs, VPN, firewalls • Specialized industrial security appliances • Network segmentation, access controls, intrusion detection, encryption
Sercos IV (serial real-time communication system IV)	• Protocols for quick serial data exchange in real-time control systems • Designed for motion control and automation • Provides deterministic communication and synchronization • Used in high precision applications (e.g., CNC machines, robotics)	• Limited native security features • Data stream encryption, VPN • Message authentication, data integrity checks, secure configuration
Cloud protocols (HTTP/HTTPS, MQTT, CoAP)	• Category including HTTP/HTTPS, MQTT, CoAP • For web communication, lightweight messaging, and constrained devices. Enables on-demand cloud resource access • Used for data analytics, backup, monitoring	• SSL/TLS for encryption. Additional username/password authentication for MQTT • Data encryption, access controls, identity management, regular security audits by cloud providers
Modbus, DNP3, CANopen	• Protocols for specific industrial use-cases (e.g., Modbus for PLC, DNP3 for utilities)	• Additional layers of security due to lack of built-in features • One-way communication devices • Specialized IDS systems for industrial protocols

13.9 Review Questions

Q1 Describe the ISO-OSI layers.
Q2 Describe the function of each ISO-OSI layer briefly.
Q3 What are different serial communication protocols?
Q4 Describe the limitations of serial communication protocols.
Q5 What are modern serial communication protocols?
Q6 What are the modern parallel communication protocols? How do these function?
Q7 What Cyber security measures are necessary for serial communication protocols?
Q8 What Cyber security measures are necessary for parallel communication protocols?
Q9 What are communication network topologies?
Q10 How best to secure different communication networks from cyber threats?

Bibliography

1. Ab Ghani MR et al (2013) A review of communication protocols for intelligent remote terminal unit development. TELKOMNIKA (Telecomm Comput Electron Control) 11(4):819–826
2. Alam KT et al (2013) Low cost ethernet based data acquisition system in Linux platform. Int J Eng Adv Technol (IJEAT) 2(5):2249–8958
3. Alba E, Tomassini M (2002) Parallelism and evolutionary algorithms. IEEE Trans Evol Comput 6(5):443–462
4. Aldea CL, Bocu R, Vasilescu A (2022) Relevant cybersecurity aspects of IoT microservices architectures deployed over next-generation mobile networks. Sensors 23(1):189
5. Alin ANTON (2015) Bandwidth enhancement between graphics processing units on the peripheral component interconnect bus. J Elect Electron Eng 8(2):5–10
6. Atluri VR (1988) A multiprocessor node for communication and control. Ph.D. dissertation, University of Maryland
7. Axelson J (1996) Parallel port complete: programming, interfacing & using the PC's parallel printer port, vol 1. Lakeview Research LLC
8. Bharati S et al (2020) Threats and countermeasures of cyber security in direct and remote vehicle communication systems. arXiv preprint arXiv:2006.08723
9. Carranza A et al, Wireless sensor network security for smart home IoT systems (unpublished)
10. Chen W et al (2019) Network protocol analysis base on WeChat PC version. In: Artificial intelligence and security: 5th international conference, ICAIS 2019, New York, NY, USA, July 26–28, 2019, Proceedings, Part IV, 2019, pp 1–9
11. Cruz T et al (2015) Improving network security monitoring for industrial control systems. In: 2015 IFIP/IEEE international symposium on integrated network management (IM), Ottawa, ON, Canada, pp 878–881
12. Czechowski R (2016) Dedicated communication protocols and network interfaces in time synchronization cyber security in complex distribution smart grid. In: 2016 electric power networks (EPNet), Sydney, NSW, Australia, pp 45–50

13. Dawoud DS, Dawoud P (2020) 5 RS-232 variants RS-422/RS-423, RS-449, RS-485, and EIA-530. Serial communication protocols and standards, CRC Press, pp 169–190
14. Dawoud DS, Dawoud P (2020) 6 serial peripheral interface (SPI). Serial communication protocols and standards, CRC Press, pp 191–244
15. Dawoud DS, Dawoud P (2022) Serial communication protocols and standards. CRC Press
16. Englander I, Wong W (2021) The architecture of computer hardware, systems software, and networking: an information technology approach, 6th edn. Wiley, Hoboken
17. Ghirardello K et al (2018) Cyber security of smart homes: development of a reference architecture for attack surface analysis. In: Living in the Internet of things: cybersecurity of the IoT-2018, London, UK, pp 1–7
18. Graham J, Hieb J, Naber J (2016) Improving cybersecurity for industrial control systems. In: 2016 IEEE 25th international symposium on industrial electronics (ISIE), Santa Clara, CA, USA, pp 870–875
19. Graube M et al (2017) Information models in OPC UA and their advantages and disadvantages. In: 2017 22nd IEEE international conference on emerging technologies & factory automation (ETFA), Limassol, Cyprus, pp 1–8
20. He Q-F, Zeng Q-J, Tang X-M (2011) Research and implement on industry control networks based on embedded SERCOS-III protocol. In: 2011 international conference on electronics, communications and control (ICECC), Ningbo, China, pp 2381–2384
21. Heydarishahreza N et al (2020) Wireless sensor networks fundamentals: a review. In: 2020 11th IEEE annual information technology, electronics and mobile communication conference (IEMCON), Vancouver, BC, Canada, pp 0331–0337
22. Jeon Y-S, Lee S-W, Kim K-Y (2006) An efficient design of high speed network security platform using network processor. In: Proceedings of the ESA
23. Kay JA, Entzminger RA, Mazur DC (2014) Industrial ethernet-overview and best practices. In: Conference record of 2014 annual pulp and paper industry technical conference, Charleston, SC, USA, pp 1–8
24. Kumar H, Alvarez OA, Kumar S (2023) Experimental evaluation of smart electric meters resilience under cyber security attacks. IEEE Access 11:82578–82588
25. Kumar N, Jamwal P (2021) Analysis of modern communication protocols for IoT applications. Karbala Int J Mod Sci 7(4):14–22
26. Lawrenz W (1997) CAN system engineering: from theory to practical applications, 1st edn. Springer, New York
27. Leszczyna R (2018) A review of standards with cybersecurity requirements for smart grid. Comput Secur 77:262–276
28. Luft M, Cioć R, Pietruszczak D (2012) Integrated measurement system based on the IEEE-488 bus. In: 2012 ELEKTRO, Rajecké Teplice, Slovakia, pp 351–354
29. Mahmoodi Khaniabadi S et al (2023) An intelligent sustainable efficient transmission internet protocol to switch between user datagram protocol and transmission control protocol in IoT computing. Expert Syst 40(5), e13129
30. Mankar J et al (2014) Review of I2C protocol. Int J Res Advent Technol 2(1):1–5
31. Marais H (2008) RS-485/RS-422 circuit implementation guide. In: AN-960 application note, analog devices
32. Mathkar S, Vernekar N, Jotkar G, Comparative study of IoT protocols and classification of cloud platforms (unpublished)
33. Melkonian AO (1996) Small computer system interface (SCSI) universal services for the turbonet parallel computer. Ph.D. dissertation. California State University
34. Miao C, Du H (2018) Computer electromagnetic leakage emission automatic test system—based on CVI platform research and design. In: 2018 IEEE 4th information technology and mechatronics engineering conference (ITOEC), Chongqing, China, pp 1–5
35. Peña E, Legaspi MG (2020) UART: A hardware communication protocol understanding universal asynchronous receiver/transmitter. Visit Analog 54(4):22–25

36. Pulugurta U et al (2007) Security in IP storage networks
37. Rahman ML et al (2016) Computer fundamentals and ICT, 1st edn. Springer, New York
38. Ramaswamy R (1990) Data confidentiality service on top of transmission control protocol/internet protocol. In: COMPEURO'90: proceedings of the 1990 IEEE international conference on computer systems and software engineering—systems engineering aspects of complex computerized systems, Tel Aviv, Israel, pp 298–302
39. Ramirez R, Chang C-K, Liang S-H (2022) PLC cyber-security challenges in industrial networks. In: 2022 18th IEEE/ASME international conference on mechatronic and embedded systems and applications (MESA), Pavia, Italy, pp 1–6
40. Sanders DA et al (2002) Redundant arrays of IDE drives. IEEE Trans Nucl Sci 49(4):1834–1840
41. Su W, Antoniou A, Eagle C (2017) Cyber security of industrial communication protocols. In: 2017 22nd IEEE international conference on emerging technologies and factory automation (ETFA), Limassol, Cyprus, pp 1–6
42. Verma A, Dahiya PK (2017) PCIe BUS: a state-of-the-art-review. IOSR J VLSI Signal Process 7(2):24–28
43. Verstoep K et al (2004) Cluster communication protocols for parallel-programming systems. ACM Trans Comput Syst 22(3):281–325
44. Wu C-HJ, Irwin JD (2016) Introduction to computer networks and cybersecurity. CRC Press, Boca Raton

Part VII
Future of CAD/CAM

Prelude

The conventional and modern CAD/CAM cycle with the amalgamation of Industry 4.0 and Industry 5.0 with processing at turing machines is the reality today. Extension of this process with embedded processes within these landmark achievements of mankind, information communication on physical and cellular mediums, and information storage and frequent access from relational and semantic databases is requiring an urgent need for methods and procedures used in quantum computing.

The quantum computing principles and applications have been tested at a laboratory scale and commercial systems have the same Technology Readiness Level (TRL) as that of 4–8-bit silicon-based central processing units showing a promise that the technology shall overwhelm the CAD/CAE/CAM/CIM technology arena soon. It must be remembered that the edge devices in computing (router, switches, etc.) are still operating at a significant speed that will be fortified for quantum computing. In contrast, the edge devices in manufacturing (e.g., machine tools) have a significantly low data transfer rate due to the speed of mechanical devices. However, quantum information systems will surely provide tremendous benefits to the shop floor-waste disposal scenario for the product.

Enormous data and its analysis (big data and data analytics in Industry 4.0) will be generated like tracking billions of discrete products during their life cycle where these products have cradle (raw material extraction to processing into raw material for downstream industries, going through value addition in Manufacturing Engineering Services (MES), a core activity of Product Life Management (PLM) getting to the client directly or through a distributor/retailer) to beginning life after birth where the product is covered by after-sales service within or after the warranty period, used by the owner where it can also change hand, transferred to another city, province or country in the state it is in or after renovation, completing its life cycle assessed during the design phase or depending on how the product was used by the customer and transferring it to remanufacturing cycle or disassembly into components level for reuse, recycle or disposal at a safe site. This needs to be done for billions of discrete products circulating in the world and requiring attention for the sake of sustainability in life on earth. This requires new technology, training for workers using the technology, and new technology and financial streams serving the basic purpose of creating wealth, the very basic purpose of design and manufacturing to have a sustainable comfortable life. This is only possible with the speed with which the quantum computers operate and their access speed to the cloud storage through WIFI or fiber optics cable.

Please refer to the extended Product Life Cycle Code (ePLCC) case study in Part II Chap. 5.

Several other applications of STEAM require the use of quantum computing.

Quantum Computing

Quantum computing is an amalgamation of quantum mechanics, mathematics, and computing. In contrast to digital computing, quantum computing can operate on multiple states. Quantum computing uses the qubit as the basic unit of information rather than the conventional bit. The main characteristic of this alternative system is that it permits the coherent superposition of ones and zeros, the digits of the binary system around which all computing revolves.

Quantum superposition is a fundamental principle of quantum mechanics. In classical mechanics, things like position or momentum are always well-defined. It may not be known what they are at any given time, but that is an issue of understanding and not an issue of the physical system. A quantum system interacts in ways that the superposition of different discrete states can be explained. Measurements of quantum systems give a statistical result corresponding to any one of the possible states appearing at random.

Quantum computing is bringing in a new revolution in handling problems faced by the complex areas of all disciplines of STEAM.

World Economic Forum

The World Economic Forum (WEF) is an international non-governmental organization for public—private sector collaboration based in Cologne, Canton of Geneva, Switzerland. It was founded on 24 January 1971 by German engineer Klaus Schwab. The foundation, which is mostly funded by its 1000 member companies—typically global enterprises with more than US$5 billion in turnover—as well as public subsidies, views its own mission as "improving the state of the world by engaging business, political, academic, and other leaders of society to shape global, regional, and industry agendas".

Klaus Schawab authored the book *The Fourth Industrial Revolution* in 2016.

Chapter 14
Pillars of Industry 4.0 and CAD/CAM

Prelude

Paradigm shifts and dependence on new engineering and engineering management in the advancement of Britain, North America, and most Western European countries characterize the first to fourth Industrial Revolutions. The first to fourth industrial revolutions are shown in the figure.

Germans Defined Industry 4.0 with its nine pillars based on Cyber-Physical Systems (CPS). The current state of the art is the inclusion of pillars of Industry 5.0 that add a human factor to the 4th Industrial Revolution. In North America, this activity is known as 'Smart Manufacturing'.

14.1 Objectives

Industry 4.0 is the set of emerging technologies based on cyber, physical, and biological spaces and their combinations, as projected by Claus Schwab in his 2017 book 'The Fourth Industrial Revolution'. Sections 14.2 and parts of this book numbering 14.3.3 lament details of the current Technology Readiness Level of these pillars of Industry 4.0.

It further relates applications of relevant pillars to the study of computer-aided design (CAD) and computer-aided manufacturing (CAM) and attempts to provide an example of this nexus here. A case study based on Augmented Reality for Computer Aided Manufacturing is included.

It must be noted that terminologies such as Industry 4.0 are used in Europe, while Smart Manufacturing is the name given to the practice in the Americas.

Supplementary Information The online version contains supplementary material available at https://doi.org/10.1007/978-3-031-77106-4_14.

W. A. Khan et al., *Computer Aided Engineering Design and Manufacturing*, Springer Series in Advanced Manufacturing, https://doi.org/10.1007/978-3-031-77106-4_14

14.2 Pillars of Industry 4.0

Industry 4.0 centers on the integration of cutting-edge intelligent technologies within the manufacturing landscape. The fundamental foundations of Industry 4.0 encompass the following:

- **Autonomous Robots**

Endowed with the ability to learn and collaborate with humans, these robots play a pivotal role in reshaping manufacturing procedures and activities. They excel in tackling tasks that are either hazardous or strenuous for human workers. As a cornerstone of Industry 4.0, autonomous robots usher in advantages that profoundly reshape the manufacturing sector. These advantages encompass elevated levels of automation and productivity, unparalleled flexibility and adaptability, seamless human–robot collaboration, data-driven decision-making, heightened quality assurance, lowered labor expenses, enhanced safety measures, and the facilitation of tailored and personalized approaches [1].

- **Simulation**

Simulation uses advanced computer-based models to simulate various processes, systems, and operations in a manufacturing environment. This entails the generation of digital counterparts mirroring real-world systems, processes, and operations, allowing for the modeling, examination, and simulation of diverse scenarios. It helps industries analyze complex systems in a virtual environment before implementation in the real world, ultimately saving costs, reducing risks, and lessening the time to market [2].

- **Augmented Reality (AR)**

Augmented Reality (AR) is emerging as a revolutionary technology that elevates the interaction between humans and machines. It can aid in intricate assembly tasks, facilitate maintenance operations, and deliver real-time data and analytics to floor operators. AR seamlessly merges virtual elements with tangible surroundings, amplifying how individuals engage with machinery, data, and physical objects. This allows for more efficient, accurate, and engaging processes, from design and prototyping to manufacturing, maintenance, and training [3].

- **Virtual Reality (VR)**

VR is dramatically shifting how industrial operations are conceived, planned, and executed. It focuses on the integration of digital technologies into manufacturing processes. Unlike AR, which superimposes digital information over the real world, VR creates an entirely immersive, computer-generated environment where users can interact seemingly realistically. The following features show how VR fits into the Industry 4.0 landscape: design and prototyping, training and skill development, virtual manufacturing and simulation, human–machine interface, and data visualization [4].

- **Integration**

Integration is also an important pillar of Industry 4.0. It contains two parts: horizontal system integration, and vertical system integration.

- **Horizontal System Integration**

Horizontal System Integration primarily concerns the seamless interoperability and data exchange across various departments and functions within the organization and with external partners like suppliers, customers, and service providers. It focuses on connecting different processes, functions, and components within a single production level. The goal is to create a value chain where data flows freely from one end to the other without barriers, allowing for real-time decision-making and optimization. The key aspects of horizontal system integration are as follows: interconnected processes, collaborative manufacturing, efficient workflows, agile manufacturing, and end-to-end visibility.

- **Vertical System Integration**

It focuses on connecting systems and data flows within a single organization, from the shop floor to the top floor. It aims to integrate all the hierarchical levels within a manufacturing facility. This integration enables real-time data exchange, decision-making, and coordination between hierarchical layers. The key aspects of vertical system integration include shop floor-to-enterprise connectivity, data transparency and traceability, optimized resource allocation, real-time decision-making, and predictive maintenance [5].

- **Internet of Things (IoT)**

IoT is a network of tangible devices linked to the Internet, enabling them to gather and exchange data. These devices are equipped with sensors, software, and related technologies to collect and share data via the Internet. The spectrum of IoT devices encompasses a wide range, from intelligent sensors and actuators to interconnected machinery and wearable devices designed for workforce integration. These devices can intercommunicate with each other and central systems, facilitating the sharing and utilization of the accumulated data [6].

- **Industrial Internet of Things (IIoT)**

IIoT constitutes a subset of IoT, specifically designed for industrial contexts, focusing on amalgamating cutting-edge technologies into manufacturing workflows. IIoT technologies connect machines and systems and allow them to be remotely monitored, controlled, and optimized. Its objective lies in enhancing efficiency, productivity, and safety across diverse industrial domains such as manufacturing, energy, transportation, and more [7].

- **Cloud Computing**

Cloud computing is a critical enabler for digital transformation, providing the technological infrastructure and computational power necessary to manage the vast amounts of data generated in modern industrial environments. It involves delivering various computing services over the internet, including storage, processing power, networking, and software. It empowers industries to harness the power of data, analytics, and collaborative technologies in a streamlined and cost-effective manner. The key aspects of cloud computing in the context of Industry 4.0 are data management and analytics, real-time access and remote monitoring, scalability and flexibility, integration and collaboration, cost efficiency, and more [8].

- **Cybersecurity**

It is an integral part of the design and operation of industrial systems. As traditional manufacturing and industrial processes become increasingly interconnected through the IoT, cloud computing, and other digital technologies, the vulnerability to cyber threats and attacks also becomes more pronounced. It requires a multilayered approach involving technology, processes, and people to mitigate risks effectively. Cybersecurity aims to protect these interconnected systems from unauthorized access, data breaches, malware infections, and other malicious activities that can disrupt operations, compromise sensitive information, and lead to financial losses [9].

- **Additive Manufacturing**

Additive manufacturing, known as 3D printing, is considered a transformative technology that aligns seamlessly with modern industrial ecosystems' digital, interconnected nature. It represents a fundamental shift in manufacturing processes by enabling the creation of complex and customized objects layer by layer directly from digital designs. Additive manufacturing enhances production processes' agility, customization, and efficiency, making it a cornerstone technology for the Industry 4.0 paradigm [10].

- **Big Data**

It is pivotal in enabling smarter, more efficient, and more flexible manufacturing processes. It is a critical enabler of data-driven decision-making, optimization, and innovation across various industrial processes. Following are the key ways Big Data is influencing Industry 4.0: data-driven decision-making, real-time monitoring and predictive maintenance, personalization and customer-centric models, quality assurance, supply chain and logistics, resource optimization, and more [11].

- **Artificial Intelligence (AI)**

AI is crucial in Industry 4.0, fundamentally transforming manufacturing and other industrial sectors by driving automation, optimization, innovation, and data utilization. By leveraging AI technologies, manufacturers can achieve higher efficiency, flexibility, and competitiveness, transforming traditional manufacturing into smart and adaptive production environments [12].

- **Block-Chain**

It offers a decentralized ledger system that ensures transparent, immutable, and tamper-proof records of transactions. Industry 4.0 represents the integration of digital technologies and data-driven processes into manufacturing, and block-chain plays a significant role in enhancing trust, security, and efficiency across this landscape. Its potential to revolutionize supply chains, enhance security, and transform various industrial processes positions block-chain as a critical enabler of the smart manufacturing revolution [13].

- **Molecular Biology**

It focuses on understanding biological molecules' structure, function, and interactions, particularly at the molecular level. The intersection of Industry 4.0 and molecular biology is becoming increasingly relevant, particularly in biotechnology, pharmaceuticals, and personalized medicine. It is a cornerstone of Industry 4.0, driving innovations across various sectors by providing insights into the molecular mechanisms of life. Integrating molecular biology with emerging technologies enhances precision, efficiency, and customization in biotechnology, healthcare, agriculture, and environmental applications [14].

14.3 Relation of Industry 4.0 Pillars with CAD/CAM; A Case Study[1]

14.3.1 Integration of Industry 4.0 with CAD/CAM

The framework of the product life cycle (P), Industry 4.0 (I), and Technology Readiness Levels (TRLs) is shown in Fig. 14.1. Today's eleven pillars of Industry 4.0 directly relate to Computer Aided Design and Computer Aided Manufacturing, as shown in the following framework. The pillars of Industry 4.0 are shown in figure.

[1] Developed by Muhammad Hammad, reproduced with permission.

Fig. 14.1 Industrial revolutions: industry 1.0 to 4.0

The following section describes how the eleven pillars of Industry 4.0 relate to CAD/CAM (Table 14.1; Fig. 14.2).

14.3.2 Integration of CAD/CAM with Product Life Cycle

This section describes how CAD/CAM incorporates in each stage of the product life cycle. The pictorial view of the product life cycle is shown in Fig. 14.3.

i. **Establishment of Need**: In this initial phase, On the basis of the defined requirements, conceptual designs and models are created using CAD. It facilitates idea visualization and refinement prior to detailed design [25].
ii. **Design**: CAD is applied to detailed design. It enables product modelling, simulation, and testing within a simulated environment. In this stage certain operations can be performed including stress analysis, optimize design, and the creation of 3D models [26].
iii. **Manufacturing Processes**: CAD models need to translate into detailed instructions for manufacturing equipment, CAM took the responsibility to translate it. It may include the robotic arms, CNC machines and other automated tools to produce parts as per design requirements [27].
iv. **Production Planning and Control**: CAD/CAM systems collectively enable to optimize production processes. To ensure efficient resource utilization and workflow, CAD/CAM help in planning the sequence of material requirements, operations, tooling, and machinery needed [28].

Table 14.1 Relation of industry 4.0 with CAD/CAM

Industry 4.0 pillar/[References]	Relation with CAD/CAM
Internet of things (IoT) [15]	• Communication and data sharing are facilitated by IoT among devices • Within CAD/CAM, real-time data exchange is possible between design software and manufacturing equipment through IoT • Enhances efficiency in the design and manufacturing processes • Enables quicker and more responsive design changes
Artificial intelligence (AI) [16]	• CAD/CAM is improved through automation of intricate design tasks while using AI • By integrating AI with CAD/CAM, optimization of manufacturing workflows can be achieved • With the help of AI, maintenance requirements can be predicted that are essential for manufacturing equipment • Design strategies can be enhanced through AI algorithms on the basis of data
Additive manufacturing [17]	• CAD/CAM shows a tremendous role in additive manufacturing by utilizing 3D printing techniques • Direct utilization of CAD designs in the generation of print instructions for 3D printers characterizes this integration • Parts having complex geometries can be produced by the integration that looks a tough task for conventional methods
Augmented reality/virtual reality (AR/VR) [18]	• In CAD/CAM, AR/VR finds application in providing immersive design visualization • Designers can interact with 3D models within a real-world context using this technology • It assists in recognizing design issues and enhancing ergonomics, all before the creation of physical prototypes
Autonomous robotics [19]	• Design interpretation can be done through autonomous robotics while using CAD/CAM systems • After adjusting and aligning their data with CAD, they performed certain precise manufacturing tasks • It leads to more accurate and efficient production processes
Cloud-computing [20]	• Cloud computing enables offsite storage and processing of large CAD files • It helps to provide access to advanced computing resources remotely for the complex CAM simulations

(continued)

Table 14.1 (continued)

Industry 4.0 pillar/[References]	Relation with CAD/CAM
Cyber-security [21]	• With the growing connectivity of CAD/CAM systems to networks, cybersecurity is crucial • Ensuring protection against cyber threats is vital to safeguard intellectual property and sensitive data
Integration [22]	• Focusing on seamless integration, this pillar ensures the smooth flow of design data into manufacturing processes in CAD/CAM • In CAD/CAM, the emphasis on integration ensures the efficient transfer of design data into manufacturing, ultimately reducing errors and enhancing efficiency
Block-chain [15]	• Block-chain enhances the secure sharing of CAD files, guaranteeing authenticity and safeguarding intellectual property • It offers a transparent and tamper-proof record of design and manufacturing processes by tracking changes
Simulation [23]	• To predict product behavior and validate designs in different conditions before production, simulation is extensively used in CAD/CAM • Simulations lessen the need of physical prototypes by anticipating the performance of the product
Big data analytics [24]	• Certain possible outcomes can be achieved like predicting maintenance requirements, enhancing manufacturing procedures, and optimizing designs by leveraging the big data analytics in CAD/CAM • Leveraging big data analytics on the extensive data generated in CAD/CAM causes several advantages

v. **Production**: CAD/CAM facilitates automation in production that lessen errors and speeds up the manufacturing process. To ensure precision and adherence to CAD designs, CAM-driven machinery produces components [29].

vi. **Assembly Systems**: CAD data proves to be really helpful while guiding automated assembly lines and design assembly systems. Robots can be programmed to perform certain assembly tasks like welding, bolting, and aligning components [30].

vii. **Testing**: To meet design specifications, safety standards, and functional requirements, products undergo multiple testing processes. This is done after the design and production phases. CAD software is utilized for virtual testing that leads to various simulations to predict performance under different conditions. It may contain thermal behavior, stress analysis, fluid dynamics and

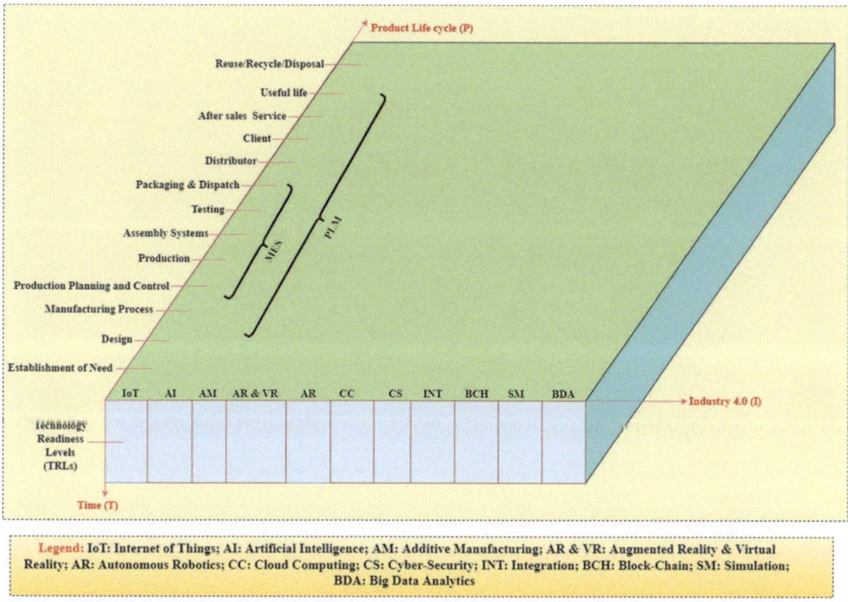

Fig. 14.2 Convergence of product life cycle, industry 4.0, and technology readiness levels

more. CAM plays a significant role in swiftly creating prototypes on the basis of CAD designs. Later on, these prototypes represent physical versions of the product used for hands-on-testing [31].

viii. **Packaging and Dispatch**: While CAD/CAM is less directly involved in packaging, it can be used for designing packaging systems and tools, ensuring that the product is well-protected and efficiently packed for distribution [32].

ix. **Distributor**: CAD models and documentation can be shared with distributors to provide a clear understanding of the product, its features, and its assembly/disassembly procedures [33].

x. **Client**: Clients can review CAD models and simulations to understand the product better. This can aid in marketing and pre-sales activities [34].

xi. **After Sales and Services**: CAD data helps in creating maintenance manuals and guides. Service technicians can use 3D models to understand the product structure and perform maintenance or repairs more effectively [27].

xii. **Useful Life**: CAD/CAM can aid in planning the disassembly and decommissioning of products. This includes designing products for ease of disassembly or recycling at the end of their lifecycle [35].

xiii. **Reuse/Recycle/Disposal**: CAD designs can incorporate features that facilitate recycling or safe disposal. CAM processes may be involved in automating recycling processes, such as separating materials for reuse [36].

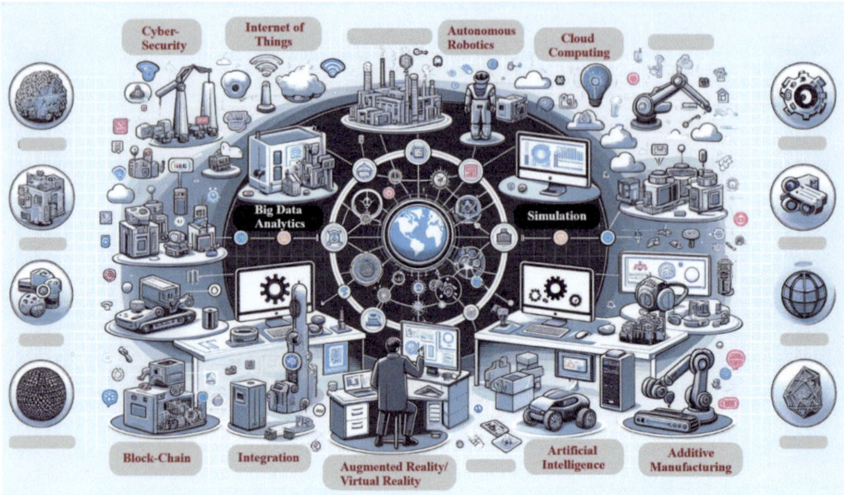

Fig. 14.3 Pillars of industry 4.0

14.3.3 Integration of Technology Readiness Levels (TRLs) with CAD/CAM and Industry 4.0

TRLs serve as a means of assessing the technological maturity at the acquisition stage of a program as shown in Fig. 14.4. In the context of CAD/CAM and Industry 4.0 pillars, TRLs play a crucial role to enable these technologies. The description of TRLs are as follows [37–39]:

- **Principles Observed (TRL 1)**: This is the most basic level of technology readiness. In the context of CAD/CAM and Industry 4.0, this might involve foundational research in areas like new algorithms for design, initial concepts of IoT integration, or early-stage AI for design automation.
- **Concept Formulation (TRL 2)**: At this level, the basic concepts of integrating Industry 4.0 technologies with CAD/CAM are formulated. For example, conceptualizing how AR/VR can be used for CAD visualization or how block-chain can secure CAD data.
- **Proof-of-Concept (TRL 3)**: In this level, proof-of-concept models are developed. For CAD/CAM, this could involve initial prototypes of AI-driven design tools, or early demonstrations of how additive manufacturing can be integrated with CAD systems.
- **Lab Validation (TRL 4)**: Technologies at this level are tested in a lab. In CAD/CAM, this might mean testing IoT connectivity in manufacturing equipment or validating AR/VR tools for design collaboration in a controlled environment.

- **Relevant Environment Validation (TRL 5)**: Moving beyond the lab, this level involves testing technology in a relevant environment. For example, using cloud-based CAD/CAM tools in a real-world design process, or integrating cybersecurity measures in a manufacturing setting.
- **Prototype Demonstration (TRL 6)**: At this level, a prototype system is demonstrated in an operational environment. This could include deploying autonomous robotics in a manufacturing process driven by CAD/CAM systems or running simulations for CAD designs under real-world constraints.
- **Operational Prototype (TRL 7)**: In this level, the technology is near completion. For CAD/CAM, it involves integrating multiple Industry 4.0 pillars like AI, IoT, and big data analytics into a fully operational manufacturing and design process.
- **System Qualification (TRL 8)**: The technology is finished and has been thoroughly tested in this level. In CAD/CAM, this would mean a fully integrated, Industry 4.0-enhanced design and manufacturing process that is proven to work reliably.
- **Verification Through Successful Implementation (TRL 9)**: The highest level of readiness, where the technology is not only complete but has been successfully used in real-world operations. For CAD/CAM, this means the technologies are fully integrated and used in everyday manufacturing and design, showing tangible benefits in efficiency, accuracy, and innovation (Fig. 14.5).

Fig. 14.4 Three dimensional view of product life cycle

Fig. 14.5 Technology readiness levels

14.4 Learning Outcomes

As Industry 4.0 is emerging quickly in terms of its use in modern nations and advancement in these countries, it is important to learn these methods as they play an important role in the gig economy and the modern economy. It is also important to note that major investments are being made in these areas even during the current recession.

14.5 Review Questions

Q1 Enlist the pillars of Industry 4.0.
Q2 What is the relation of Industry 4.0 with CAD/CAM?

Bibliography

1. Goel R, Gupta P (2020) Robotics and industry 4.0. In: A roadmap to industry 4.0: smart production, sharp business and sustainable development, pp 157–169
2. Gunal MM (2019) Simulation for industry 4.0: past, present, and future. Springer
3. De Pace F, Manuri F, Sanna A (2018) Augmented reality in industry 4.0. Am J Comput Sci Inf Technol 6(1):17
4. Kovar J et al (2016) Virtual reality in context of industry 4.0 proposed projects at Brno University of Technology. In: 2016 17th international conference on mechatronics-mechatronika (ME), Prague, Czech Republic, pp 1–6

5. Pérez-Lara M et al (2020) Vertical and horizontal integration systems in industry 4.0. Wireless Netw 26:4767–4775
6. Lampropoulos G, Siakas K, Anastasiadis T (2019) Internet of things in the context of industry 4.0: an overview (unpublished)
7. Malik PK et al (2021) Industrial internet of things and its applications in industry 4.0: state of the art. Comput Commun 166:125–139
8. Atobishi T, Gábor SZ, Podruzsik S (2018) Cloud computing and big data in the context of industry 4.0: opportunities and challenges (unpublished)
9. Lezzi M, Lazoi M, Corallo A (2018) Cybersecurity for industry 4.0 in the current literature: a reference framework. Comput Ind 103:97–110
10. Haleem A, Javaid M (2019) Additive manufacturing applications in industry 4.0: a review. J Ind Integr Manag 4(4):1930001
11. Xu LD, Duan L (2019) Big data for cyber physical systems in industry 4.0: a survey. Enterpr Inf Syst 13(2):148–169
12. Awasthi S et al (eds) (2021) Artificial intelligence for a sustainable industry 4.0. Springer International Publishing
13. Bodkhe U et al (2020) Blockchain for industry 4.0: a comprehensive review. IEEE Access 8:79764–79800
14. Javaid M, Haleem A (2019) Industry 4.0 applications in medical field: a brief review. Curr Med Res Pract 9(3):102–109
15. Ikubanni PP et al (2022) Present and future impacts of computer-aided design/computer-aided manufacturing (CAD/CAM). J Eur Syst Autom 55(3)
16. Rabemanantsoa M, Pierre S (1996) An artificial intelligence approach for generating assembly sequences in CAD/CAM. Artif Intell Eng 10(2):97–107
17. Barone S et al (2016) Interactive design of dental implant placements through CAD-CAM technologies: from 3D imaging to additive manufacturing. Int J Interact Des Manuf 10:105–117
18. Novak-Marcincin J, Barna J (2011) Augmented virtual reality applications. Proc Manuf Syst 6(2)
19. Dombre E et al (1986) Trends in CAD/CAM systems for robotics. In: Proceedings of the 1986 IEEE international conference on robotics and automation, San Francisco, CA, USA, vol 3, pp 1–7
20. Andreadis G, Drossos A (2013) A cloud based framework for automated CAD design. Int J Mod Manuf Technol 5(2):15–20
21. Sipos J, Apostol A (2008) CAD/CAM, CNC technology applied in the field of engineering, security technology and mechanical engineer training I. Hadmérnök 3(4):43–52
22. Gao J, Zheng D, Gindy N (2004) Mathematical representation of feature conversion for CAD/CAM system integration. Robot Comput Integr Manuf 20(5):457–467
23. Li ZZ et al (2003) A solid model-based milling process simulation and optimization system integrated with CAD/CAM. J Mater Process Technol 138(1–3):513–517
24. Ramnath S et al (2019) Automatically generating 60,000 CAD variants for big data applications. In: Proceedings of the international design engineering technical conferences & computers and information in engineering conference, Anaheim, CA, USA, vol 59179, pp 1–10
25. Myers W (1982) CAD/CAM: the need for a broader focus. Computer 15(1):105–117
26. Mikhalev ON, Yanyushkin AS (2020) CAD/CAM-system module for the design of automatic production. Industry 4.0 5(2):59–62
27. Theodorou P, Florou G (2008) Manufacturing strategies and financial performance—the effect of advanced information technology: CAD/CAM systems. Omega 36(1):107–121
28. Culler DE, Burd W (2007) A framework for extending computer aided process planning to include business activities and computer aided design and manufacturing (CAD/CAM) data retrieval. Robot Comput Integr Manuf 23(3):339–350
29. Corney J et al (2005) The CAD/CAM interface: a 25-year retrospective (unpublished)
30. Cicek A, Gülesin M (2007) A part recognition-based computer aided assembly system. Comput Ind 58(8–9):733–746

31. Wendler M et al (2017) Chairside CAD/CAM materials. Part 2: flexural strength testing. Dent Mater 33(1):99–109
32. Chang JW, Luh YP, Chiou SS (1997) Integrated application in CAD/CAM, scheduling and control. Integr Manuf Syst 8(6):378–387
33. Kao Y-C, Lin GC (1998) Development of a collaborative CAD/CAM system. Robot Comput Integr Manuf 14(1):55–68
34. Wang H-F, Zhang Y-L (2002) CAD/CAM integrated system in collaborative development environment. Robot Comput Integr Manuf 18(2):135–145
35. Wendler M et al (2018) Chairside CAD/CAM materials. Part 3: cyclic fatigue parameters and lifetime predictions. Dent Mater 34(6):910–921
36. Jha N, Litkouhi B (2001) Optimal life cycle analysis and design of thermal systems. In: Proceedings of the 2001 Annual Conference, Albuquerque, NM, USA, pp 1–10
37. Olechowski A, Eppinger SD, Joglekar N (2015) Technology readiness levels at 40: a study of state-of-the-art use, challenges, and opportunities. In: 2015 Portland international conference on management of engineering and technology (PICMET), Portland, OR, USA, pp 2084–2094
38. Piraintorn P (2017) Knowledge and technology transfer: levels and key factors in implementation of CAD/CAM/CAE systems in context of manufacturing in Thailand. Ph.D. dissertation. Asian Institute of Technology
39. Saad SM, Bahadori R, Jafarnejad H (2021) The smart SME technology readiness assessment methodology in the context of industry 4.0. J Manuf Technol Manag 32(5):1037–1065

Chapter 15
Industry 5.0

Prelude
Industry 5.0 is an important addition that was a natural progression of Industry 4.0 in the 4th Industrial revolution.

15.1 Objectives

Industry 5.0 is being proposed within academia and industry by defining independent intangible pillars in basic Industry 4.0 pillar definitions that are human-centric.

15.2 Industry 5.0 Pillars

15.2.1 Human-Centric Approach

This pillar of Industry 5.0 is focused on Ergonomics, RAMS of both processes and products, Customization and Learning required by the human resources. Considering the horizontal and Vertical Integration pillars of Industry 4.0, this pillar of Industry 5.0 has prime importance in digitally transformed design and manufacturing and remanufacturing scenarios [1, 2].

Supplementary Information The online version contains supplementary material available at https://doi.org/10.1007/978-3-031-77106-4_15.

W. A. Khan et al., *Computer Aided Engineering Design and Manufacturing*, Springer Series in Advanced Manufacturing, https://doi.org/10.1007/978-3-031-77106-4_15

- **Ergonomics**: Human factor engineering, while especially dependent on anthropometry, is the basic component of ergonomics. It opens further when the human is exposed to different spaces requiring safe exposure to different mediums where humans are operating for any type of work.
- **RAMS**: Reliability, Availability, Maintainability and Safety are the basic features that today's product must possess while having developed using System Engineering Design in Computer Aided environments. These products are innovative, multidisciplinary and may have been customized.
- **Customization**: Customization while considering ergonomics and RAMS for mass production in business-to-business environments and in business-to-client environments is becoming a common practice today. The product is normally designed in developed countries and production is done where the cheap labor is available or trained manpower or right manufacturing resources are not available.
- **Learning**: The learning factor is important in the knowledge economy as human resources must operate in gig economy mode thus after having basic skill the human resource has to have the ability to shift from one skill to another in a particular domain.

15.2.2 Resilience

This pillar of Industry 5.0 takes care of the resilience of the product and the processes used to manufacture the product. It investigates Product, Processes, Digital Twins, Configuration and Reconfiguration of product and processes and attempts to find a solution using Virtual Manufacturing.

- **Product**: Today's innovative, multidisciplinary, and customized product should have been designed and manufactured using System Engineering design with Re-Manufacturing principles incorporated in the exercise. The remanufacturing aspect incorporated design of product gives a global type of resilience to the product. The cradle-to-grave product life cycle varies even for a product from the same manufacturing batch leaving an unpredictable life cycle behavior for each product. It has been proposed[1] that each product should have a distinct identification code while remembering that there are billions of different types of products circulating in the world. The Product Life Cycle Code is defined by exactly keeping this factor in mind. With the current pace in the development of quantum computers it will become possible to implement Product Life Cycle Codes thus helping in a sustainable future by minimal use of all kinds of resources.
- **Processes**: The development of manufacturing processes into manufacturing equipment makes these a product as well that goes through the same cycle as described for the Product. Additionally, the processes require some form of material, energy, and information to produce products that can be used by an ordinary customer. The process's resilience is to keep operating as per RAMS practices

[1] Refer to the Case Study on extended Product Life Cycle (ePLCC).

and other conditions such as heavy smog, fog, very low temperature for human's exposure, very high temperature for a human to sustain and many other types of exposures. Traffic on the road to reach the factory is another factor. Major pandemics such as COVID-19 and others are also forced to try to control the factory remotely through Digital Twins to incorporate resilience in manufacturing processes, cells, and Systems.

- **Digital Twin**: Digital Twin is an exact replica of an equipment based on a process that was designed using systems engineering. Digital Twin mimics full operation of the equipment that includes the basic process designed using an open loop control or feedback control, provides a User Interface (UI) and User eXperience (UX) and can be operated remotely. Remotely theoretically means, distances as distance between Earth and Moon or distance between Earth and Mars.
- **Configuration and Reconfiguration**: A new discrete product, upon entering a manufacturing system follows a route defined in activities performed during the production planning. Production planning completely dictates provision of intangibles such as process plan, part program, requisitions for raw material, tool/s, jigs and fixtures, coolants and lubricants if required, and but not limited to standard manufacturing time determined using time motion study. This will form an operational configuration for the product and all its components for the first of its components to assembly of the product after complete quality inspection, packaging and dispatch to distributor or client. This operational configuration, be it operating in a conventional manufacturing system or digitally transformed manufacturing system, has many uncertainties e.g. unavailability of required tools on workshop floor and so on so forth disturbing the operational configuration defined by the production planning. The manufacturing processes and systems should have resilience to cope with this configuration and reconfiguration. Using digitally transformed methods the manufacturing processes and manufacturing systems can easily cope with this. The associated mechanical problems, such as setup change, can also be handled through automation. For conventional manufacturing processes and manufacturing systems based on conventional manufacturing processes such tasks can only be performed manually leaving inherent inflexibility affecting cost of production and thus profits.
- **Virtual Manufacturing**: Discrete and continuous manufacturing systems are common in today's industry constituting small, medium, and large-scale enterprises. The SMEs and LSEs are the backbone of the economy of any nation. With the advent of the 4th Industrial Revolution, the AR/VR, one of the pillars of Industry 4.0, is being utilized for many purposes. The traditional operation of SME or LSE is marred by several problems, especially operational problems, and pandemics. The promise of new colonies on the Moon and Mars and industrial operation, there is a case of controlling industry from a distance. The new operational pattern of operating the office from home has already set trends in developed nations and taking roots in Pakistan as well. The problem of operating the SME and LSE from home needs concepts like Virtual Manufacturing based on AR that has already led to in depth research in Digital Twins.

15.2.3 *Sustainability*

Sustainability is a major concern in today's world. Without respect for mother nature, humans have taken more from it than giving it back something. The weather patterns have changed, excessive excavation has led to erosion thus disturbing the balance of e.g., natural homogeneity of earth surface affecting irrigation and water flow patterns. Also, deforestation and emission of uncontrolled industrial gases and draining industrial effluents in stream as well as domestic solid waste and domestic liquid waste is affecting supply of potable water and life in rivers, seas, and oceans. The earth temperature is continuously rising thus leading to undesired melting of snow peaks thus increasing sea water level diminishing and threatening human settlements and causing havoc on its way from snow peaks to seas with floods of untold intensity [3].

United Nation Development Program (UNDP) Sustainable Development Goals (SDGs) if fully implanted by all the signatory countries (the whole world) can lead to a comparatively better outlook for the world.

World Economic Forum is doing work that is admirable. Other activities such as COP28 Moot is significantly supporting this cause of sustainable development.

- **LCA**: Life Cycle Assessment (LCA) is a technique used to evaluate the environmental impacts linked to each phase of a product's, process's, or service's life cycle. This approach is relatively new in engineering design and focuses on issues such as carbon footprint. The carbon footprint is a calculated metric that allows for the comparison of the total greenhouse gas emissions produced by an activity, product, company, or country.
- **Green/Lean Manufacturing**: Green manufacturing is produced using methods that are eco-friendly for the product and the processes that produce the product. Lean Manufacturing produces optimal configurations and any reconfiguration that may be required.
- **ePLCC**: **extended** Product Life Cycle Code (ePLCC) is a futuristic concept that traces the cradle to grave cycle of any product that is in billions in the world. To have sustainability all products must be tracked. This is possible through record keeping using Quantum Computing.
- **Reuse, Recycle, Waste Disposal**: The logical end of product whether designed using ordinary system engineering methods or whether consideration to remanufacturing has been implemented along with basic system engineering design and manufacturing is the end of product life cycle where reuse, recycle and disposal are the options. To avoid post disposal effects suitable disposal sites are necessary (Fig. 15.1).

Fig. 15.1 The three-dimensional visualization of Industry 5.0

15.3 Linkage of Industry 4.0 with Industry 5.0

Industry 5.0, while building upon the advancements of Industry 4.0, introduces a more human-centric approach. It emphasizes the collaboration between humans and machines, where intelligent technologies augment human capabilities rather than replace them. This paradigm shift aims to create more sustainable and resilient industrial systems by focusing on three core principles: sustainability, resilience, and human-centricity [4, 5].

15.3.1 Key Linkages Between Industry 4.0 and Industry 5.0

- **Technological Foundation**: Industry 5.0 leverages the technological advancements established by Industry 4.0, such as IoT, AI, and big data analytics. These technologies serve as the backbone for developing more sophisticated systems that enable human–machine collaboration.
- **Enhanced Automation**: While Industry 4.0 aims for extensive automation, Industry 5.0 takes it a step further by integrating human creativity and problem-solving skills into automated processes. This synergy enhances decision-making and innovation, allowing for more adaptive and flexible production systems.
- **Sustainability Focus**: Both Industry 4.0 and Industry 5.0 recognize the importance of sustainability. However, Industry 5.0 places a stronger emphasis on achieving environmental goals and ensuring that industrial activities are sustainable and environmentally friendly. This involves using smart technologies to optimize resource use and minimize waste.
- **Resilience and Adaptability**: Industry 4.0 introduced systems that are highly efficient and adaptable to changes in production demands. Industry 5.0 builds on this by incorporating resilience, ensuring that industrial operations can withstand and quickly recover from disruptions. This is achieved through the integration of advanced analytics and predictive maintenance technologies.
- **Human-Centric Approach**: The most significant shift from Industry 4.0 to Industry 5.0 is the focus on human involvement. Industry 5.0 envisions a future where human skills and creativity are complemented by intelligent systems, leading to more meaningful and engaging work environments. This approach not only enhances productivity but also improves job satisfaction and overall well-being.

15.4 Advance Topics

15.4.1 Ubiquitous Manufacturing

Ubiquitous manufacturing refers to the pervasive and seamless integration of manufacturing processes and systems across various locations, leveraging advanced connectivity, IoT, and cloud computing technologies. This approach ensures that manufacturing activities are not confined to specific geographical locations but can occur anywhere, at any time, with real-time data sharing and coordination [6].

Ubiquitous manufacturing helps to achieve the following goals:

- Human-Centric Collaboration
- Sustainability
- Resilience
- Real-Time Data and Connectivity.

15.4.1.1 Examples of Ubiquitous Manufacturing

i. **Distributed Production**: Companies can set up smaller, flexible production units closer to their customers, allowing for faster response times and customized production. This approach is particularly useful in industries where demand fluctuations are common.
ii. **Remote Monitoring and Control**: Engineers and operators can monitor and control manufacturing processes from remote locations using advanced IoT and cloud-based platforms. This capability enhances operational efficiency and allows for expert intervention without the need for physical presence.
iii. **Additive Manufacturing (3D Printing)**: Additive manufacturing technologies enable localized and on-demand production of components, reducing the need for extensive inventory and long supply chains. This technology is a key enabler of ubiquitous manufacturing.

The three-dimensional view capturing the essence of ubiquitous manufacturing is shown in Fig. 15.2.

15.4.2 Symbiotic Manufacturing

Symbiotic manufacturing is an innovative approach that emphasizes the collaboration and mutual benefit between humans and machines. This concept extends beyond mere automation to create a harmonious relationship where human intelligence and creativity are augmented by advanced technologies such as AI, robotics, and IoT.

Fig. 15.2 A vibrant, dynamic 3D view of ubiquitous manufacturing

Symbiotic manufacturing aims to leverage the strengths of both humans and machines to optimize production processes, enhance innovation, and improve overall efficiency [7].

Symbiotic manufacturing aligns perfectly with Industry 5.0's principles of sustainability, resilience, and human-centricity, offering several key benefits:

- Enhanced Human–Machine Collaboration
- Sustainability
- Resilience
- Human-Centric Approach
- Continuous Learning and Adaptation.

15.4.2.1 Examples of Symbiotic Manufacturing

i. **Collaborative Robots (Cobots)**: Cobots work alongside human operators, assisting with tasks that require precision, strength, or repetitive actions. These robots are designed to be safe and easy to work with, enhancing human capabilities and productivity.
ii. **AI-Driven Decision Support**: AI systems analyze vast amounts of data to provide decision support for human operators. For instance, AI can predict maintenance needs, optimize production schedules, and suggest process improvements, while humans make the final decisions based on their expertise and intuition.
iii. **Augmented Reality (AR) and Virtual Reality (VR)**: AR and VR technologies provide immersive training and support for workers, allowing them to visualize complex processes and interact with virtual models. This enhances learning and enables workers to perform tasks with greater accuracy and confidence.

Figure 15.3 shows the futuristic illustration of symbiotic manufacturing, having cobots working seamlessly with human operators.

15.4.3 Bionic Manufacturing

Bionic manufacturing refers to the integration of biological principles with advanced manufacturing technologies to create systems and processes that mimic or enhance natural biological functions. This approach often involves the use of biomimicry, where natural designs and processes are emulated to solve complex industrial problems, and bio-enhanced technologies, where biological elements are integrated with machinery to improve performance and functionality [8].

Fig. 15.3 A futuristic illustration of symbiotic manufacturing

Industry 5.0 emphasizes a sustainable, human-centric, and resilient approach to manufacturing, and bionic manufacturing aligns well with these principles:

- Human-Centric Enhancement
- Sustainability
- Resilience
- Innovation through Biomimicry
- Integration of Bio-enhanced Technologies.

15.4.3.1 Examples of Bionic Manufacturing in Practice

i. **Exoskeletons**: Wearable exoskeletons enhance human strength and endurance, enabling workers to perform heavy lifting and repetitive tasks with reduced physical strain. These devices improve worker safety and productivity in manufacturing environments.
ii. **Self-healing Materials**: Materials inspired by biological systems that can repair themselves after damage, reducing the need for frequent maintenance and extending the lifespan of products. These materials are particularly useful in harsh manufacturing environments where wear and tear are common.
iii. **Biohybrid Robots**: Robots that integrate biological tissues with synthetic components, offering enhanced capabilities such as soft and adaptive movements.

Figure 15.4 shows the innovative description of bionic manufacturing.

Fig. 15.4 An innovation depiction of bionic manufacturing

15.4.4 Self-healing Manufacturing Systems

Self-healing manufacturing systems refer to advanced manufacturing processes and materials that have the capability to detect and repair damage automatically, without human intervention. These systems leverage advanced technologies such as AI, machine learning, and smart materials to maintain optimal functionality and minimize downtime. Self-healing systems are designed to enhance the reliability, efficiency, and longevity of manufacturing operations [9, 10].

Self-healing manufacturing systems align well with these principles and offer several key benefits:

- Human-Centric Enhancement
- Sustainability
- Resilience
- Cost Efficiency.

15.4.4.1 Examples of Self-healing Manufacturing Systems

i. **Self-healing Polymers**: Polymers that can autonomously repair cracks and damages. These materials are used in various manufacturing applications, including coatings, adhesives, and structural components, to enhance durability and longevity.
ii. **Smart Sensors and AI-Driven Diagnostics**: Integration of smart sensors and AI algorithms to continuously monitor the health of manufacturing systems.

Fig. 15.5 A detailed 3D visualization of self-healing manufacturing systems

These technologies can detect early signs of wear or failure and trigger self-healing processes, such as activating repair mechanisms or rerouting workflows to prevent disruptions.

iii. **Bio-inspired Self-healing Materials**: Materials inspired by biological systems that can self-repair through mechanisms similar to wound healing in living organisms. These materials are used in manufacturing to create components that can recover from damage autonomously.

A detailed three-dimensional visualization of self-healing manufacturing systems is shown in Fig. 15.5.

15.5 Merits of Industry 5.0 for CAD/CAM

Pillars of Industry 5.0 are equally applicable to the complete conventional or modern CAD/CAM cycles. It covers [11, 12]:

- Need Analysis
- Concept Design
- Embodiment Design
- Detail Design
- Design Documentation
- Prototyping
- Pilot Production
- Production Documentation
- Production.

15.6 Case Study: Virtual Factory and Industry 5.0[2]

The Pillars of Industry 5.0 i.e., Human Centric Approach, Resilience and sustainability have a better control over practices required under these pillars through the AR/VR based on system engineering principles, Digital Twins based on AR/VR and associated sensors and transducers to acquire feedback control, and communication channels such as Industrial ethernet, SERCOSE IV and other Industry 4.0 and legacy communications standards.

Following URLs show the basic Virtual factory and then it is configured in to two different instances:

Virtual Factory

https://drive.google.com/file/d/1JO3ePKl1qpQcUcDmU0T_284e_MEOe5OV/view?usp=sharing.

Configuration into Factory A

https://drive.google.com/file/d/1CUYII2mpaVy2BzTsAzmP4gQGK8fX9ozX/view?usp=drivesdk.

Configuration into Factory B

https://drive.google.com/file/d/1uS7TlhNFmd8rUVozfYH-GTdokZZy5G7X/view?usp=drivesdk.

Virtual Factory (a product of System Engineering Design and Manufacturing, System Engineering Control, AR/VR and other pillars of Industry 4.0) highly supports Industry 5.0 pillars due to digital transformation.

Various aspects of Industry 5.0 pillars are described above.

15.7 Learning Outcomes

The concepts presented in this chapter are completely novel. It is also a tutorial-like chapter to enhance the understanding of the young reader. Please refer to the Review Questions.

[2] Logic designed by Wasim A. Khan, programmed by Aqeel Ahmad. protected by Pakistan Patent application number 37/2022 acceptance number 144217. US Patent application number 17864442 (Pending).

15.8 Review Question

Q1 What is the Human-centric approach?
Q2 What are RAMS?
Q3 Define Customization in context of ergonomics and RAMS.
Q4 Differentiate between product and processes.
Q5 What is a Digital Twin?
Q6 What is meant by configuration and reconfiguration?
Q7 What is Life Cycle Assessment (LCA)?
Q8 What are the merits of Industry 5.0 for CAD/CAM?

Bibliography

1. Alves J, Lima TM, Gaspar PD (2023) Is industry 5.0 a human-centred approach? A systematic review. Processes **11**(1), 193
2. Leng J et al (2022) Industry 5.0: prospect and retrospect. J Manuf Syst 65:279–295
3. Narkhede G et al (2023) Industry 5.0 and the future of sustainable manufacturing: a systematic literature review. Bus Strategy Dev 6(4):704–723
4. Nahavandi S (2019) Industry 5.0—a human-centric solution. Sustainability 11(16), 4371
5. Xu LD, Xu EL, Li L (2018) Industry 4.0: state of the art and future trends. Int J Prod Res 56(8):2941–2962
6. Lee J, Bagheri B, Kao H-A (2015) A cyber-physical systems architecture for industry 4.0-based manufacturing systems. Manuf Lett 3:18–23
7. Wang L, Törngren M, Onori M (2015) Current status and advancement of cyber-physical systems in manufacturing. J Manuf Syst 37:517–527
8. Gronau N (2013) Bionic manufacturing. Ind Manag 6:12–16
9. Ahner J et al (2016) Self-healing functional polymeric materials. In: Self-healing materials, pp 247–283
10. White SR et al (2001) Autonomic healing of polymer composites. Nature 409(6822):794–797
11. Ikram A, Pušavec F (2021) Control, monitoring, modeling, and optimization of manufacturing and machining technologies. In: Soft computing in smart manufacturing: solutions toward industry 5.0, vol 8, pp 1–20
12. Popović N, Mladenović I (2001) Some industry 4.0 technologies supporting the advancements in dentistry

Chapter 16
Quantum Computing

16.1 Objectives

Quantum computing today is comparable to the early silicon-based digital revolution. These are the early days of Quantum Computing where new designs and applications are being experimented with. The cost, capacity, and speed matter as was the case in the 1980s for digital computing.

As history dictates, the future lies with quantum computing enabling humans to capture new horizons in information technology especially simulation, big data analytics, Artificial Intelligence and Machine Learning.

This chapter is more of a description to explain to the readers the very basics of quantum computing and projection and what it can achieve that digital technology is unable to achieve. In this pursuit, the authors indulge in the description of basics of Quantum Computing and architecture of Quantum computing machines.

The users are led to websites of the major companies involved in the development of Quantum Machines. Major startups and conglomerates working in this area are Xanadu, ColdQuanta, and QC Ware, along with well-established corporations such as Alphabet Inc. (NASDAQ: GOOGL), Microsoft Corporation (NASDAQ: MSFT), Amazon.com, Inc. (NASDAQ: AMZN), International Business Machines Corporation (NYSE: IBM), D-Wave Systems (NYSE: QBTS), and IonQ, Inc. (NYSE: IONQ).

Supplementary Information The online version contains supplementary material available at https://doi.org/10.1007/978-3-031-77106-4_16.

W. A. Khan et al., *Computer Aided Engineering Design and Manufacturing*, Springer Series in Advanced Manufacturing, https://doi.org/10.1007/978-3-031-77106-4_16

16.2 Basics of Quantum Computing

Quantum computing is an advancing technological area that utilizes the intricate and captivating concepts of quantum mechanics for information processing. In contrast to traditional computing that relies on bits for information storage and processing, quantum computing utilizes quantum bits, known as 'qubits'. The basics of quantum computing encompass the following [1–3]:

- **Qubits**: Quantum bits, commonly referred to as qubits, serve as the basic elements of quantum computing, like how bits function in traditional computing. Unlike classical bits that are confined to a state of 0 or 1, qubits can occupy several states at once, due to the principle of superposition. This ability allows qubits to represent and process a vast amount of information far more efficiently than traditional bits. When a qubit is measured, it 'collapses' into one of the classical states (0 or 1), but until then, it can exist in a superposition of both states.
- **Superposition**: Superposition is a principle at the heart of quantum computing. It refers to the capability of quantum systems, like qubits, to exist in multiple states (say, 0 and 1) at the same time. This contrasts sharply with classical computing, where a bit must be either 0 or 1. Superposition allows a quantum computer with n qubits to store and process information in $2n$ possible states simultaneously, offering an exponential increase in computational capacity compared to classical computers.
- **Entanglement**: Quantum entanglement is another cornerstone of quantum computing. This phenomenon occurs when the states of two or more qubits become interconnected or correlated, leading to a situation where the state of one qubit immediately and directly affects the state of another, irrespective of the physical distance separating them. Such interdependence allows quantum computers to execute intricate calculations more efficiently and rapidly. Entangled qubits represent a key resource for quantum information processing, enabling phenomena like quantum teleportation and playing a crucial role in quantum algorithms and error correction techniques.
- **Quantum Gates and Circuits**: Quantum gates and circuits are the mechanisms through which operations are performed on qubits. Like the logical gates used in classical computing, quantum gates manipulate the state of qubits. However, unlike their classical counterparts, quantum gates operate on the principles of superposition and entanglement, enabling more complex and nuanced operations. Quantum circuits are formed by sequencing these gates and qubits. This arrangement allows the execution of algorithms in a quantum computing environment, enabling complex computations that are difficult or impossible for classical computers to perform efficiently.

Fig. 16.1 An informative infographic on the basics of quantum computing

An infographic of Quantum computing is shown in Fig. 16.1. Quantum computing's reliance on these fundamental concepts opens new horizons in computing power, making it vastly different and potentially more powerful than classical computing. Its potential applications range from solving complex mathematical problems to advancing materials science and drug discovery.

16.3 Free Resources for Quantum Computing (January 2024)

Following resources are available (December 2024) for learning and experimentation with quantum machines and their assembler:

- IBM Quantum Experience (Platform)
- Microsoft Quantum Development Kit (Framework)
- Edx (Course)
- Quantum Katas by Microsoft (Tutorial)
- Qiskit Textbook by IBM (Textbook)
- Quantum Computing Playground (Playground)
- Quantum Computing for the Very Curious (Articles)
- Quantum Open-Source Foundation (Support)
- The Quantum Game (Game)
- Introduction to Quantum Information Science (Course).

16.4 Future of Commercial Quantum Computing and Quantum Machines

Quantum computing has witnessed significant advancements in recent years. One of the most notable achievements was Google's announcement of achieving 'quantum supremacy' in 2019. Google's quantum computer, named Sycamore, achieved a notable milestone by completing a certain computational task in just 200 s, an accomplishment that would require about 10,000 years for a top-tier supercomputer to replicate. This marked a major milestone in demonstrating the practical potential of quantum computing. IBM has also made substantial contributions, particularly in the realm of quantum hardware. They unveiled their IBM Quantum Hummingbird processor in 2020, featuring 65 qubits, and later the Eagle processor with 127 qubits in 2021. These advancements are critical steps towards scalable quantum computing, aiming to eventually build a quantum processor with 1000 qubits or more. Another area of significant progress is in quantum error correction and coherence time. Researchers have been developing more sophisticated quantum error correction codes and techniques to maintain the stability of quantum information over longer periods, which is essential for practical quantum computing applications.

As for the future of quantum computing, experts speculate that the field will continue to grow rapidly, with potential breakthroughs on the horizon. One major area of ongoing research is in developing quantum algorithms optimized for specific applications, such as materials science, drug discovery, and complex optimization problems. These algorithms, once fully realized, could drastically change the landscape in these fields. There is also a strong focus on making quantum computing more accessible. Efforts are underway to develop quantum computing platforms that can be accessed through the cloud, democratizing access to quantum resources and enabling broader collaboration and innovation in the field. Quantum computing could lead to the development of entirely new materials, aid in solving complex global challenges like climate change, and even transform our understanding of quantum physics itself. However, these outcomes are contingent on overcoming current technological hurdles, particularly in improving qubit coherence, reducing error rates, and developing efficient quantum algorithms.

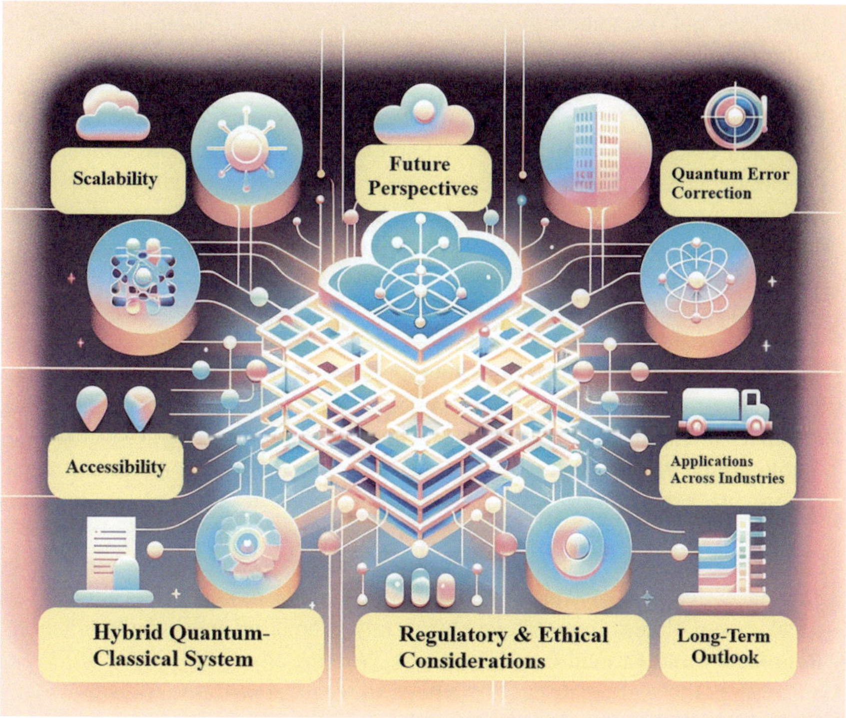

Fig. 16.2 A futuristic visualization of quantum computing

The future of commercial quantum computing machines is a subject of great interest and intense research [4–6]. In the coming years, we are likely to see significant developments that will shape the landscape of quantum computing and its applications across various industries. The futuristic visualization of quantum computing is shown in Fig. 16.2 and the description is as follow:

- **Scalability and Accessibility**: A primary focus in the evolution of commercial quantum computing is scalability. The capability of present-day quantum computers is constrained by their ability to effectively handle and sustain a coherent state across a certain number of qubits. Companies like IBM and Google are working towards increasing the qubit count, aiming to achieve 'quantum advantage', where quantum computers can solve problems beyond the reach of classical supercomputers. The goal is to develop large-scale quantum computers with thousands of qubits, which will significantly enhance their computational capabilities. In parallel, there is a push to make quantum computing more accessible. Cloud-based quantum computing services, such as IBM's Quantum Experience and Amazon's Braket, are already providing access to quantum processors via the cloud. This trend is expected to continue and expand, allowing more businesses and researchers to experiment with and develop quantum algorithms without the need for their own quantum hardware.

- **Quantum Error Correction and Coherence**: Improving quantum error correction and extending the coherence time of qubits are crucial challenges that need to be addressed for practical and reliable quantum computing. Researchers are developing new error correction codes and techniques to maintain the stability of quantum information. Extending coherence times will allow for more complex and longer computations, which is essential for commercial applications.
- **Hybrid Quantum-Classical Systems**: The near-term future will likely see the rise of hybrid systems that combine classical and quantum computing. These systems will leverage the strengths of both technologies, using quantum computing for specific tasks that benefit from its unique properties, while relying on classical computing for other tasks. Hybrid systems could be more practical and cost-effective for commercial applications in the short term.
- **Applications Across Industries**: As quantum technology matures, its applications are expected to expand into various industries. In finance, quantum computing could optimize asset management and risk assessment. In logistics, it could solve complex optimization problems for supply chain management. In the pharmaceutical industry, quantum computing holds the promise of speeding up the development of new drugs by more efficiently simulating molecular interactions.
- **Regulatory and Ethical Considerations**: With the advancement of commercial quantum computing, there will be a need for regulatory frameworks to address data security, privacy, and ethical concerns. Quantum computing has the potential to render existing encryption techniques obsolete, thereby requiring the creation of cryptography that is secure against quantum computing threats.
- **Long-Term Outlook**: In the long term, quantum computing could become a transformative technology, changing how we approach problem-solving in various fields. The development of fault-tolerant, scalable quantum computers would mark a new era in computing, leading to breakthroughs in science, engineering, and beyond.

16.5 Standards for Quantum Computing

The establishment of standards for quantum computing is an essential step towards the maturation and widespread adoption of this technology. As quantum computing evolves, the need for comprehensive standards becomes increasingly important to ensure compatibility, reliability, security, and ethical use of quantum computing technologies [7–9].

- **Interoperability and Compatibility**: One of the primary goals of standardization in quantum computing is ensuring interoperability and compatibility among different quantum computing systems. This includes standardizing quantum programming languages, interfaces, and communication protocols. Efforts are being made to develop universal quantum programming languages, similar to classical programming languages, that can work across various quantum computing platforms.

- **OpenQASM (Open Quantum Assembly Language)**: Developed by IBM, it's an intermediate representation for quantum instructions. OpenQASM is designed to express quantum circuits and support high-level constructs for procedure calls and qubit allocation.
- **Qiskit**: An open-source quantum computing framework by IBM that supports OpenQASM and is designed to operate with quantum computers at the level of pulses, circuits, and algorithms.
- **Performance Metrics**: Defining performance metrics is crucial for evaluating and comparing the capabilities of quantum computers. Standard metrics include qubit count, coherence time, gate fidelity, and quantum volume. Quantum volume, for instance, measures the overall capability and error rates of a quantum computer, considering both the number of qubits and the quality of quantum operations.

  – **Quantum Volume (QV)**: Developed by IBM, this is a hardware-agnostic metric that evaluates the performance of quantum computers, considering both the number of physical qubits, the depth of quantum circuits, gate and measurement errors, device cross talk, and connectivity.
  – **NISQ (Noisy Intermediate-Scale Quantum) Technology**: While not a standard metric, the term NISQ, coined by John Preskill, is often used to describe the current state of quantum computing technology, emphasizing the imperfections and limitations of existing quantum devices.

- **Quantum Security Standards**: With the potential of quantum computing to break current cryptographic protocols, developing quantum-resistant algorithms is a priority. Standards for post-quantum cryptography are being established by organizations like the National Institute of Standards and Technology (NIST). The three-dimensional visualization of quantum computing standards is shown in Fig. 16.3. These standards aim to develop cryptographic systems that are secure against both classical and quantum computing threats.
- **NIST Post-quantum Cryptography Standardization**: A project initiated by the NIST to standardize post-quantum cryptographic algorithms. This project is in the process of evaluating and standardizing quantum-resistant cryptographic algorithms for public key encryption, digital signatures, and key establishment.
- **Error Correction and Fault Tolerance**: Standards for quantum error correction and fault tolerance are essential for the practical use of quantum computing. These standards would guide the development of error correction codes and fault-tolerant quantum computing protocols, which are crucial for maintaining the integrity of quantum computations. There are currently no universally accepted standards for quantum error correction and fault tolerance. However, various error correction codes and techniques are in development, including:

  – **Surface Codes**: Widely studied for their potential to offer fault-tolerant quantum computing.

Fig. 16.3 3D visualization of quantum computing standards

  - **Toric Code and Color Codes**: Other forms of topological codes used in quantum error correction.

- **Regulatory and Ethical Standards**: Regulatory and ethical standards are needed to address the broader implications of quantum computing. This includes regulations on the use of quantum computing in sensitive applications, data privacy, and security. Ethical guidelines are also necessary to guide research and usage decisions in quantum computing.
  - **IEEE P1913—Standard for Quantum Computing Definitions**: Provides a set of definitions and concepts necessary for a common understanding of quantum computing.
  - **ISO/IEC JTC 1/SC 40 IT Service Management and IT Governance**: While not specific to quantum computing, this standard could potentially be expanded to include quantum computing services in terms of governance and management frameworks.

- **Industry and Research Collaboration**: Collaboration between industry, academia, and regulatory bodies is key in developing these standards. Organizations like the IEEE Quantum Computing Standards Working Group are actively working to create a set of standards for quantum computing. Such collaborative efforts ensure that standards are comprehensive, practical, and facilitate the growth of the quantum computing field.
  - **IEEE Quantum Computing Standards Working Group**: Part of IEEE P7130, this group works on creating a set of standards for quantum computing.

– **Quantum Economic Development Consortium (QED-C)**: Facilitates collaboration among industry, government, and academia to identify and address gaps in technology, standards, and workforce readiness relevant to quantum computing.

16.6 Merits of Quantum Computing for CAD/CAE/CAPP/CAM/CIM

CAD/CAM in a CIM environment at SMEs and LSEs, in all types of manufacturing system configurations, will benefit enormously with quantum computing once the quantum computing becomes within the price proportion to that of silicon technology in its stable industrial application inception era. It ranges from all transparent interfaces in CAD/CAE/CAPP and CAM, repetitive optimizations step in numerical analysis, communication across CIM, inter and intra Enterprise Resource Planning (ERP) for services and manufacturing industry; and tracking products during their lifecycle span for a sustainable life on earth.

The Technology Readiness Level steps for quantum technology will, like in silicon technological based digital transformation, going to consume a lot of capital unless the lessons learned from development of digital technology are applied while remembering that the digital transformation using silicon-based technology has not yet unfolded to its complete potential. Also, the silicon technology developed in the modern world is slowly being relegated to BRICS and Next Eleven Countries for fulfilling this gap that will be left by the modern world, following the centuries old principle of wealth creation while keeping a lid on the new set of knowledge for quite a few years to come unless Next Eleven and BRICS nations try to catch the quantum revolution now along with Industry 4.0 and Industry 5.0 (January 2024).

Quantum computing offers a range of potential advantages for Computer-Aided Design (CAD) and Computer-Aided Manufacturing (CAM), areas that are critical in engineering, design, and manufacturing industries. The illustration of quantum computing merits is shown in Fig. 16.4. The unique capabilities of quantum computing can address several challenges and enhance the efficiency and effectiveness of CAD/CAM applications [10–12].

- **Complex Problem Solving**: CAD/CAM often involves solving complex optimization and simulation problems, such as design optimization, material properties simulation, and manufacturing process optimization. Quantum computing, with its ability to handle vast and complex datasets through quantum parallelism, can significantly reduce the time required for these computations. This capability is especially valuable in handling multi-variable optimizations common in CAD/CAM, where traditional computers might struggle or take longer to produce results.
- **Improved Design and Simulation**: In the design phase, quantum computing can simulate and analyze material properties and behaviors at the molecular or

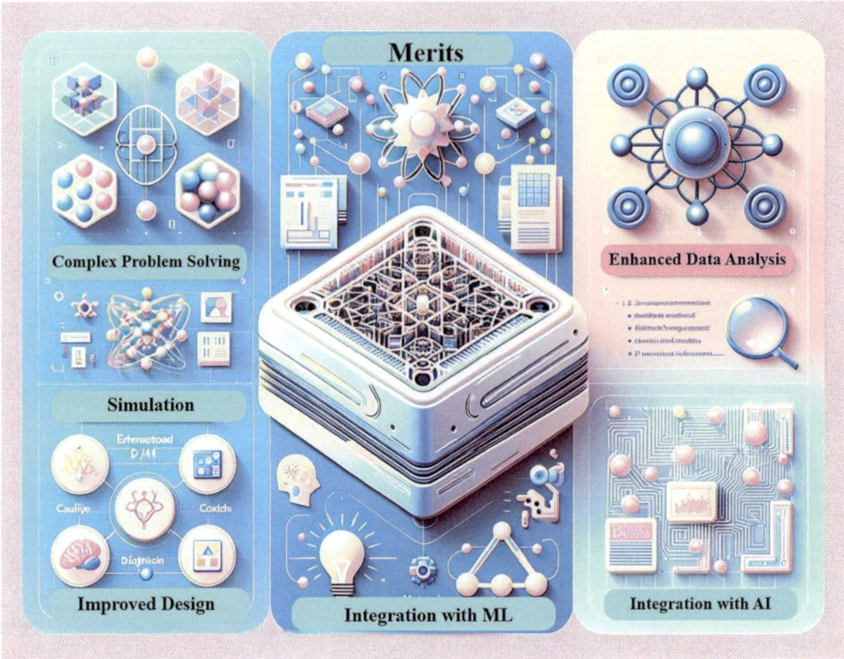

Fig. 16.4 Illustration of quantum computing benefits

atomic level, offering insights that are not feasible with classical computing. This precision is particularly beneficial for advanced materials engineering, enabling more efficient and innovative designs. In CAM, quantum computing can optimize manufacturing processes by simulating various scenarios and outcomes, leading to improved production efficiency and reduced costs.

- **Enhanced Data Analysis**: CAD/CAM systems generate large amounts of data. Quantum computing's ability to process and analyze this data rapidly can lead to more informed decision-making, better design choices, and optimized manufacturing processes. This rapid data processing capability is crucial in scenarios where real-time data analysis can significantly impact the design and manufacturing outcomes.
- **Integration with AI and Machine Learning**: Integrating quantum computing with AI and machine learning can further enhance CAD/CAM applications. Quantum algorithms can potentially improve the performance of machine learning models, leading to better predictive analytics, pattern recognition, and decision-making processes in design and manufacturing.

While the potential benefits are significant, the application of quantum computing in CAD/CAM is still in its early stages, with many theoretical aspects yet to be realized in practice. Advances in quantum hardware, error correction, and the development of specific quantum algorithms for CAD/CAM applications are necessary for these benefits to be fully realized.

16.7 Case Study: A General Purpose Quantum Information System[1]

A pertinent case study in the realm of quantum information systems is the development and application of IBM's quantum computer and its cloud-based quantum computing service, IBM Quantum Experience. This case study exemplifies the practical implementation of quantum computing technology and provides insight into its potential impacts and current limitations.

- **IBM Quantum Experience; Overview and Development**: IBM Quantum Experience represents a significant step in making quantum computing accessible to a broader audience. Launched in May 2016, it was the first initiative to offer public access to a set of quantum computers via the cloud, enabling users, researchers, and developers to run algorithms and experiments on IBM's quantum processors. This platform includes a simulator and a set of five quantum computers with different numbers of qubits.
- **Applications and Research**: The IBM Quantum Experience has facilitated a wide range of research and development in quantum computing. It has been used for educational purposes, helping to train the next generation of quantum engineers and scientists. Researchers have utilized the platform for experiments in quantum algorithms, cryptography, and materials science. One notable example is the use of quantum computers to simulate small molecules, a task that could revolutionize drug discovery and materials science.
- **Challenges and Learnings**: The IBM Quantum Experience has also highlighted several challenges in quantum computing. The qubits in IBM's quantum computers, like others, are susceptible to errors and have limited coherence times. These factors restrict the complexity of the algorithms that can be effectively run on these machines. The platform has underscored the need for advancements in quantum error correction, algorithm development, and hardware improvements [13–15].

[1] Produced by Muhammad Hammad, reproduced with permission.

16.8 Learning Outcomes

The Reader is expected to Learn the basics of quantum computing in the context of available quantum machines. It is explained in the same manner as that of digital computing in its early days.

Quantum computing is in its transition stage and the reader needs to understand that this transition is forming a new technological revolution that will further change the ways things are done today while the world has not fully capitalized the potentials of Industry 4.0 and Industry 5.0.

16.9 Review Questions

Q1. What is the term Qubits used for?
Q2. What is Hybrid Quantum-Classical Systems?
Q3. What are the different types of Error Correction and Fault Tolerance codes used in quantum computing?
Q4. Discuss Industry and Research Collaboration in the context of quantum computing.
Q5. What are the merits of Quantum Computing for CAD/CAM/CIM?
Q6. What are the Regulatory and Ethical Standards used in quantum computing?

Bibliography

1. Pérez-Delgado CA, Kok P (2011) Quantum computers: definition and implementations. Phys Rev A 83(1):012303
2. Hirvensalo M (2013) Quantum computing. Springer Science & Business Media
3. Vedral V, Plenio MB (1998) Basics of quantum computation. Prog Quant Electron 22(1):1–39
4. Bravyi S et al (2022) The future of quantum computing with superconducting qubits. J Appl Phys 132(16)
5. Gill SS et al (2022) Quantum computing: a taxonomy, systematic review and future directions. Softw Pract Exp 52(1):66–114
6. Shor PW (1994) Algorithms for quantum computation: discrete logarithms and factoring. In: Proceedings 35th annual symposium on foundations of computer science, Santa Fe, NM, pp 124–134
7. van Deventer O et al (2022) Towards European standards for quantum technologies. EPJ Quant Technol 9(1):33
8. Kan K, Une M (2021) Recent trends on research and development of quantum computers and standardization of post-quantum cryptography. Unpublished
9. Rietsche R et al (2022) Quantum computing. Electron Mark 32(4):2525–2536
10. Daley AJ et al (2022) Practical quantum advantage in quantum simulation. Nature 607(7920):667–676
11. Mohammadi M, Eshghi M (2009) On figures of merit in reversible and quantum logic designs. Quant Inf Process 8:297–318

12. Elfving VE et al (2020) How will quantum computers provide an industrially relevant computational advantage in quantum chemistry? arXiv preprint arXiv:2009.12472
13. Barabasi S et al (2020) Student user experience with the IBM Qiskit quantum computing interface. In: Advances in information and communication: proceedings of the 2019 future of information and communication conference (FICC), vol 2. Springer International Publishing, pp 891–905
14. Steffen M et al (2011) Quantum computing: an IBM perspective. IBM J Res Dev 55(5):13:1–13:12
15. Silva V (2018) Practical quantum computing for developers: programming quantum rigs in the cloud using Python, quantum assembly language and IBM QExperience. Apress

Chapter 17
Nexus of CIM, Industry 4.0, Industry 5.0, and Quantum Computing

Prelude

The fourth Industrial revolution is witnessing an era of exceptional technical advancement. Not a single field of STEAM will remain unchanged. This will however, while putting enormous pressure on normal human beings all over the world, will tear the social fabric, and people will vie for control from a micro domain to global domain resulting in wars: mini to mega—spanning over communities to nations.

To be content with what one has and something that weaves the socio-economic fabric in a legal framework is the only way out. The second part is industrialization that makes the whole argument a circular argument.

Nature is random. Structuralism is done to exercise control. However, too much structuralism leads to robotic life with work sleep work cycle and no room for self. There is a soft war with self to gain more and more capital that presumably will solve all the problems faced by self.

Joy, whatever form it may be in, should be an integral part of life, be it from strolling on the streets, walking/jogging on a track, reading a book, cooking for self and friends or any other that pleases the self.

Thus, structuralism is necessary but nature with its randomness is also important. So, it is just that: do it easy.

Supplementary Information The online version contains supplementary material available at https://doi.org/10.1007/978-3-031-77106-4_17.

W. A. Khan et al., *Computer Aided Engineering Design and Manufacturing*, Springer Series in Advanced Manufacturing, https://doi.org/10.1007/978-3-031-77106-4_17

17.1 Objectives

Integration of CAD/CAM/CAE/CAPP in a CIM environment has been attempted (ESPRIT UK-1980s) resulting in islands of automation. The problem was essentially a lack of standards. The condition improved in the era of the 4th industrial revolution and several manufacturing systems can be found that are fully integrated locally and globally in terms of manufacturing and production with the major production philosophy being a dynamic Supply Chain. All other production philosophies revolve around the supply chain.

The problem is addressed from the perspective of the level of complexity and need of computing power that can only be facilitated by quantum computing.

The reader is exposed to these complexities with more theoretical concepts that remain the dream for mankind.

17.2 CAD/CAE/CAPP/CAM Complexity Level

From an ensuing competition for better algorithm to better display in CAD to enough computing power for complex numerical analysis methods for design analysis to services requirement of production planning and control to competition to empower shop floor with maximum information communication between operator to designer in design office the basic Turing machine to advance quantum computing is the need of SMEs and LSEs.

The level of complexity in above mentioned problems may reach a level when Turing machine is completely intractable (the problem cannot be solved i.e. Polynomially Not Complete (NP-Complete) having no solution in finite time on Turing machine). It happens in all the problems listed in the previous paragraph.

Humans, by their very instinct, observe the problem, experiment with solutions, and eventually perfect the solution. The solution is still on a continuum and will be applicable only under certain conditions. If conditions change the solution algorithm needs to be changed as well.

The NP-Completeness may be in the form of NP, NP-Hard and NP-Combinatorial. The situation becomes more severe if other problems are also associated with the current situation at hand.

Mathematical formulation and algorithm development is on continuum and is accommodated by human intervention. The Technology Readiness Level stages in different parts of the world also play an important role.

When applied to real practical problems at SMEs and LSEs the solution required is not academic, but it should be one that should not in any way hinder the production cycle and should lead to profits.

For all pillars of Industry 4.0 having a suitable interface with CAD/CAE/CAPP and CAM the software interface should have an optimal UI and provide a pragmatic UX using Turing machine and in case of larger problem size requires an optimal cost quantum machine.

17.3 Amalgamation of Industry 5.0 Methodologies for Addressing CAD/CAE/CAPP/CAM Complexities

Industry 5.0 poses a more holistic and realistic approach to the total production cycle and because most of its intangible content requires a metric in different conditions its quantification requires intense computing in case of a hierarchical CIM environment leading to necessity of some types of quantum machine.

The current basic quantum machines are quite feasible for solving problems related to this situation.

17.4 Top 10 Emerging Technologies of World (2024)

As per World Economic Forum's Strategic Intelligence, following are emerging technologies of the World in 2024:

1. AI for scientific discovery
2. Privacy-enhancing technologies
3. Reconfigurable intelligent surfaces
4. High altitude platform stations
5. Integrated sensing and communication
6. Immersive technology for the built world
7. Elastocalorics
8. Carbon-capturing microbes
9. Alternative livestock feeds
10. Genomics for transplants.

17.5 Case Study: eERP—Quantum Information System for Discrete Manufacturing[1]

In Design, manufacturing, and production scenarios the information system refers to Enterprise Resource Planning software while accounting for System Engineering, CAD/CAE/CAPP/CAM, Industry 4.0 and Industry 5.0 Pillars.

With Quantum Computing, the Quantum Information System (QIS) in the form of an Enterprise Resource Planning software for manufacturing is proposed. Following configurable components of a complete suite of software systems are currently computable in the Digital Computing environment. This quantum information systems covers all the applications described in different chapters of this book.

17.5.1 VMS Administrator

The VMS Administrator covers all the administrative functions of a manufacturing concern including but not limited to:

- Human Resources
- Accounts
- Finance
- Security
- Administration.

https://drive.google.com/file/d/14I8ufDcyfc8wKYinP_xwB_exoV0Lls3a/view?usp=sharing.

17.5.2 VMS Design

The activities performed in the Design office are covered in this component of eMRP. It includes all components of system engineering design and their interactions in manufacturing organization and beyond covering product life cycle.

https://drive.google.com/file/d/1U7tsfqLVq8z1MeZd0gbTnoVRs14ytkqP/view?usp=sharing.

[1] Logic Designed by Wasim A. Khan and programmed by Amir Hussain, Protected by Pakistan patent number 37/2022 Acceptance Number 144217 and US patent applications number 1786442 (Pending); reproduced with permission.

17.5.3 VMS Planning

The activities performed in VMS planning are the intangible services required to produce product bids to actual production. It covers Process Planning, Production Planning and Control including product manufacturing Schedule within Lead Time, Master Production Schedule, Maintenance schedule and inventory status of consumable items, non-consumable items; raw material inventory, In-Process inventory, and Finished Product Inventory.

https://drive.google.com/file/d/1GlD5RpwW9zXZPSaeB9K-p9atCeK9nfIj/view?usp=sharing.

17.5.4 VMS Vendors

The registered vendors for sourcing and for production of components, structures, assembly, or mechanism as per agreed benchmarking are covered in VMS Vendor. These are in significant numbers to generate positive competition [Software Under Development].

17.5.5 VMS Monitor

VMS Monitor is the digitally twined operation of the manufacturing machine, manufacturing cell or Manufacturing System giving real time status of the manufacturing concern. VMS Monitor like all other components of eMRP is interconnected with all other components of eMRP in the manufacturing concern forming interconnectivity of microdomains and it is also connected to external microdomains where the manufacturing concern interacts for its optimal operation like Vendors, Banks, and legal entities etc.

https://drive.google.com/file/d/1shkoz8Gbv6-z2vbZJUDON2DLFf2DmOPV/view?usp=sharing.

17.5.6 VMS Quality

Depending on the nature of the product, either quality control or quality assurance is exercised. In today's manufacturing concern no compromise on quality can be made therefore VMS Quality is the most important of all eMRP components [Software Under Development].

17.5.7 VMS Fault Diagnostics

Predictive maintenance operation is the key to continuous operation of the manufacturing concern and part of maintaining the reputation of the manufacturing concern for delivery of quality products on time [Software Under Development].

17.6 Learning Outcomes

These Learning Outcomes are a short description of what authors expect from the reader after studying this volume. Several concepts have been introduced in this book and the authors expect that the reader is able to relate to these concepts and have a knowledge level to reflect in the future to come.

17.7 Review Questions

Q1. Briefly discuss the Complexity Level of CAD/CAE/CAPP/CAM.
Q2. What is a VMS Administrator?
Q3. Discuss the role of ERP in design and manufacturing.
Q4. How can QIS handle problems such as ePLCC?
Q5. What is the current status of development in Quantum Computing and how long it will take to use Quantum Computers in applications such as QIS?
Q6. Experiment with available quantum computing resources at IBM to model a small QIS for manufacturing.

Bibliography

1. Finance A (2015) Industry 4.0: challenges and solutions for the digital transformation and use of exponential technologies. Finance, Audit Tax Consulting Corporate, Zurich, Swiss, pp 1–12
2. Ford M (2015) Rise of the robots: technology and the threat of a jobless future. Basic Books
3. Frank AG, Dalenogare LS, Ayala NF (2019) Industry 4.0 technologies: implementation patterns in manufacturing companies. Int J Prod Econ 210:15–26
4. Geissbauer R, Vedso J, Schrauf S (2016) Industry 4.0: building the digital enterprise
5. Hammad M, Islam MS, Salam MA, Jafry AT, Ali I, Khan WA (2023) Framework for the implementation of smart manufacturing systems: a case in point. Processes 11(5):1436
6. Ito Y, Matsumiya R, Endo T (2017) ooc_cuDNN: accommodating convolutional neural networks over GPU memory capacity. In: 2017 IEEE international conference on big data (big data). IEEE, pp 183–192
7. Lasi H, Fettke P, Kemper H-G, Feld T, Hoffmann M (2014) Industry 4.0. Bus Inf Syst Eng 6:239–242
8. Lee J, Kao H-A, Yang S (2014) Service innovation and smart analytics for industry 4.0 and big data environment. Procedia CIRP 16:3–8

9. Pittschellis R (2015) Multimedia support for learning factories. Procedia CIRP 32:36–40
10. Porter ME, Heppelmann JE (2016) How smart, connected products are transforming companies. Harv Bus Rev 94(1–2):24
11. Schwab K (2017) The fourth industrial revolution. Crown Currency
12. Weyer S, Schmitt M, Ohmer M, Gorecky D (2015) Towards industry 4.0—standardization as the crucial challenge for highly modular, multi-vendor production systems. IFAC-PapersOnLine 48(3):579–584

Index

W. A. Khan et al., *Computer Aided Engineering Design and Manufacturing*, Springer Series in Advanced Manufacturing, https://doi.org/10.1007/978-3-031-77106-4